PHYSIOLOGICAL DISORDERS OF FRUITS AND VEGETABLES

Causes and Remedies

Postharvest Biology and Technology Series

PHYSIOLOGICAL DISORDERS OF FRUITS AND VEGETABLES

Causes and Remedies

Ghulam Khaliq, PhD
Sajid Ali, PhD
Mohammed Wasim Siddiqui, PhD
Editors

AAP | APPLE ACADEMIC PRESS

First edition published 2026

Apple Academic Press Inc.
1265 Goldenrod Circle, NE,
Palm Bay, FL 32905 USA

760 Laurentian Drive, Unit 19,
Burlington, ON L7N 0A4, Canada

CRC Press
2385 NW Executive Center Drive,
Suite 320, Boca Raton FL 33431

4 Park Square, Milton Park,
Abingdon, Oxon, OX14 4RN, UK

Apple Academic Press exclusively co-publishes with CRC Press, an imprint of Taylor & Francis Group, LLC

Library and Archives Canada Cataloguing in Publication

CIP data on file with Canada Library and Archives

Library of Congress Cataloging-in-Publication Data

CIP data on file with US Library of Congress

ISBN: 978-1-99851-132-7 (hbk)
ISBN: 978-1-99851-133-4 (ebk)
DOI: 10.1201/9781998511334

About the Book Series: Postharvest Biology and Technology

Editor-in-Chief

Mohammed Wasim Siddiqui, PhD
Scientist-cum-Assistant Professor, Department of Food Science and Technology, Bihar Agricultural University, Sabour, Bhagalpur, Bihar, India
AAP Acquisitions Editor | Horticultural Science
Founding/Managing Editor | *Journal of Postharvest Technology*

As we know, preserving the quality of fresh produce has long been a challenging task. In the past, several approaches were in use for the postharvest management of fresh produce, but due to continuous advancement in technology, the increased health consciousness of consumers, and environmental concerns, these approaches have been modified and enhanced to address these issues and concerns.

The Postharvest Biology and Technology series features edited books that address several aspects related to postharvest technology of fresh produce. The series presents existing and novel management systems that are in use today or that have great potential to maintain the postharvest quality of fresh produce in terms of microbiological safety, nutrition, and sensory quality.

The books are aimed at professionals, postharvest scientists, academicians researching postharvest problems, and graduate-level students. This series is intended to be a comprehensive venture that provides up-to-date scientific and technical information focusing on postharvest management for fresh produce.

Books in the series address the following themes:

- Nutritional composition and antioxidant properties of fresh produce
- Postharvest physiology and biochemistry
- Biotic and abiotic factors affecting maturity and quality
- Preharvest treatments affecting postharvest quality
- Maturity and harvesting issues
- Nondestructive quality assessment
- Physiological and biochemical changes during ripening
- Postharvest treatments and their effects on shelf life and quality

- Postharvest operations such as sorting, grading, ripening, de-greening, curing etc.
- Storage and shelf-life studies
- Packaging, transportation, and marketing
- Vase life improvement of flowers and foliage
- Postharvest management of spice, medicinal, and plantation crops
- Fruit and vegetable processing waste/byproducts: management and utilization
- Postharvest diseases and physiological disorders
- Minimal processing of fruits and vegetables
- Quarantine and phytosanitory treatments for fresh produce
- Conventional and modern breeding approaches to improve the postharvest quality
- Biotechnological approaches to improve postharvest quality of horticultural crops

BOOKS IN THE SERIES

Postharvest Biology and Technology of Horticultural Crops: Principles and Practices for Quality Maintenance
Editor: Mohammed Wasim Siddiqui, PhD

Insect Pests of Stored Grain: Biology, Behavior, and Management Strategies
Ranjeet Kumar, PhD

Postharvest Management of Horticultural Crops: Practices for Quality Preservation
Editors: Mohammed Wasim Siddiqui, PhD, and Asgar Ali, PhD

Innovative Packaging of Fruits and Vegetables: Strategies for Safety and Quality Maintenance
Editors: Mohammed Wasim Siddiqui, PhD,
Mohammad Shafiur Rahman, PhD, and Ali Abas Wani, PhD

Advances in Postharvest Technologies of Vegetable Crops
Editors: Bijendra Singh, PhD, Sudhir Singh, PhD, and
Tanmay K. Koley, PhD

Plant Food By-Products: Industrial Relevance for Food Additives and Nutraceuticals
Editors: J. Fernando Ayala-Zavala, PhD, Gustavo González-Aguilar, PhD, and Mohammed Wasim Siddiqui, PhD

Emerging Technologies for Shelf-Life Enhancement of Fruits
Editors: Basharat Nabi Dar, PhD, and Shabir Ahmad Mir, PhD

Sensor-Based Quality Assessment Systems for Fruits and Vegetables
Editors: Bambang Kuswandi, PhD, and Mohammed Wasim Siddiqui, PhD

Emerging Postharvest Treatment of Fruits and Vegetables
Editors: Kalyan Barman, PhD, Swati Sharma, PhD, and Mohammed Wasim Siddiqui, PhD

Packaging and Storage of Fruits and Vegetables: Emerging Trends
Editor: Tanweer Alam, PhD

Postharvest Physiological Disorders of Fruits
Editors: Ghulam Khaliq, PhD, and Sajid Ali, PhD

Bioengineered Fruit and Vegetables: Trends in Postharvest Quality Improvement
Editors: Mohammad Irfan, PhD, and Pankaj Kumar, PhD

Novel Packaging Systems for Fruits and Vegetables
Editors: Kirtiraj K. Gaikwad, PhD, and Suman Singh, PhD

Ozone Applications in the Food Industry: Trends and Future Prospects
Editors: Ravi Pandiselvam, PhD, Anjineyulu Kothakota, PhD, and Kaavya Rathnakumar, PhD

Nonthermal Food Processing Technologies: Impact on Color Profile
Editors: Kaavya Rathnakumar, PhD, and R. Pandiselvam, PhD

Ready-to-Eat Snacks: Emerging Technologies for Production and Safety
Editors: R. Pandiselvam, PhD and Gülşah Çalışkan Koç, PhD

Handbook of Research on Mango Fruit: Postharvest Science, Production, Nutrition, and Processing Technology
Editors: Sajjad Hussain, PhD, Kamrun Nahar, PhD, Ishtiaq A. Rajwana, PhD, Sezai Ercisli, PhD, and Shakeel Ahmad, PhD

Sustainable Approaches in Postharvest Handling and Management
Editor: Pankaj B. Pathare, PhD

Physiological Disorders of Fruits and Vegetables: Causes and Remedies
Editors: Ghulam Khaliq, PhD, Sajid Ali, PhD, and Mohammed Wasim Siddiqui, PhD

ABOUT THE SERIES EDITOR

Mohammed Wasim Siddiqui, PhD
Scientist-cum-Assistant Professor, Department of Food Science and Technology, Bihar Agricultural University, Sabour, Bhagalpur, Bihar, India

Mohammed Wasim Siddiqui, PhD, is an Assistant Professor and Scientist in the Department of Food Science and Post-Harvest Technology, Bihar Agricultural University, Sabour, India, and author or co-author of peer-reviewed research articles, book chapters, two manuals, and conference papers. He has over 20 edited and an authored books to his credit, published by Elsevier, CRC Press, Springer, and Apple Academic Press. Dr. Siddiqui has established the international peer-reviewed *Journal of Postharvest Technology.* He is Editor-in-Chief of two book series *(Postharvest Biology and Technology and Innovations in Horticultural Science)*, published by Apple Academic Press. Dr. Siddiqui is also a Senior Acquisitions Editor for Apple Academic Press, USA, for Horticultural Science. He has been serving as an editorial board member and active reviewer of several international journals.

Dr. Siddiqui has received several grants and respected awards for his research work by a number of organizations. He is an active member of the organizing committees of several national and international seminars, conferences, and summits. He is one of key members in establishing the World Food Preservation Center (WFPC), LLC, USA, and is currently an active associate and supporter.

Dr. Siddiqui acquired a BSc (Agriculture) degree from Jawaharlal Nehru Krishi Vishwa Vidyalaya, Jabalpur, India. He received his MSc (Horticulture) and PhD(Horticulture) degrees from Bidhan Chandra Krishi Viswavidyalaya, Mohanpur, Nadia, India, with specialization in Postharvest Technology. He is a member of Core Research Group at the Bihar Agricultural University (BAU), which provides appropriate direction and assistance in prioritizing research.

About the Editors

Dr. Ghulam Khaliq obtained his BSc and MSc (Hons) degrees in Horticulture from Gomal University, Dera Ismail Khan, Pakistan. He obtained his PhD degree in Horticulture from Universiti Putra, Malaysia. He focused on postharvest physiology and technology in a wide range of fruits, vegetables, and ornamentals during his BSc and MSc (Hons) degrees. As a PhD student, he concentrated on understanding the mechanisms regulating fruit susceptibility to low-temperature stress during storage. He also worked on the approaches to identify key genes modulating fruit resistance mechanisms against biotic and abiotic stresses.

Currently, Dr. Khaliq is working as an Associate Professor and Chairman of the Department of Horticulture, Faculty of Agriculture, Lasbela University of Agriculture, Water and Marine Sciences, Uthal, Pakistan. Before working at Lasbela University, he was a research associate in the Agriculture and Cooperatives Department of the Government of Pakistan. He has experience in food science, biochemistry, and plant molecular biology. His research is focused on the postharvest physiology of different fruits, vegetables, and ornamental plants. He is working on different postharvest technologies such as edible coatings, elicitors, biopreservative packaging, and the development of novel approaches to maintain quality and extend the shelf life of highly perishable products. All studies are focused on developing strategies toward a better understanding of the preharvest and postharvest factors that affect the quality, storage life of fresh produce, and also to develop innovative technologies to overcome problems of industry.

He is involved in teaching undergraduate and postgraduate students with a special emphasis on postharvest physiology, plant growth regulators, mineral nutrition of plants, and plant biochemistry. He has supervised several postgraduate students. He has executed research projects supported by the Higher Education Commission of Pakistan. He also collaborates with many other research institutions around the world. He published research articles, book chapters, conference papers, and reports in many national and international high-impact factor journals. He received several research awards for the development of new technologies and for meritorious performance in the field of science. He was awarded an overseas scholarship from the Ministry of Science and Technology for higher study. He is a lifetime member of several scientific societies like Pakistan Society for Horticultural Sciences, Pakistan Botanical Society, and Pakistan Society of Plant Protection. He is working as an editor, associate editor, and editorial board member of several international journals. He is a reviewer of more than 70 international peer-reviewed journals. He is a member of several scientific committees related to conferences, seminars, curriculum development, quality enhancement, and project evaluation. He has worked closely with extension departments, regional fruit specialists, storage operators, and industry professionals to disseminate research findings through newsletters and to organize awareness sessions. Dr. Khaliq has also conducted numerous training seminars and workshops on fruit processing and preservation, integrated nutrient management, and climate-smart agriculture. These initiatives have helped uplift the socioeconomic conditions of farming communities by introducing innovative agricultural techniques for improved livelihoods.

Sajid Ali, PhD, is an Assistant Professor in the Department of Horticulture, Faculty of Agricultural Sciences and Technology at Bahauddin Zakariya University, Multan, Pakistan. Dr. Ali teaches undergraduate and postgraduate classes with a major focus on fruit production and postharvest physiology. His current focus is on improving the postharvest quality of fresh horticultural produce with minimum loss. He worked as a research officer and as a project development officer on two research projects at the University of Agriculture, Faisalabad, Pakistan. He has won several research projects. Dr. Ali has authored and coauthored more than 100 peer-reviewed international journal publications and around 25 book chapters. He is an academic editor of *Plos One* and editor of *BMC Plant Biology.* Dr. Ali acquired a BSc (Honors) in Agriculture, majoring in Horticulture, from Bahauddin Zakariya University, Multan, Pakistan. He received his MSc in Horticulture and PhD in Horticulture (Postharvest Physiology) from the University of Agriculture, Faisalabad, Pakistan. He also received advanced training in postharvest physiology of horticultural crops at the US Horticultural Research Laboratory, US Department of Agriculture-Agricultural Research Service, Fort Pierce, Florida, USA.

Mohammed Wasim Siddiqui, PhD, is an accomplished postharvest researcher, academic, and editor with a strong international presence. Currently, he holds the position of Assistant Professor and scientist at Bihar Agricultural University, Sabour, India. In addition, he serves as the Executive Director of World Food Preservation Center LLC, USA, and an adjunct associate professor at Benue State University, Nigeria. Dr. Siddiqui has been recognized for his expertise and listed as one of the top 2% of researchers in the world by Stanford University in both 2021 and 2022. He has been an ambassador of the World Food Preservation Center, USA, since 2017, and previously contributed as a member of the board of directors for the Postharvest Education Foundation, USA, in 2020.

With an entrepreneurial spirit, Dr. Siddiqui is the founding editor of two esteemed book series: *Postharvest Biology and Technology and Innovations in Horticultural Science,* published by Apple Academic Press. In addition, he holds the distinguished position of editor-in-chief for the *Journal of Postharvest Technology.* Dr. Siddiqui has received numerous fellowships, awards, and accolades for his exceptional research and teaching contributions from renowned national and international organizations throughout his career. He has authored or coauthored over 75 scientific articles and has published more than 25 books in collaboration with international presses including Elsevier, Springer, and CRC Press. Dr. Siddiqui's contributions to the field are highly regarded, evident from his impressive metrics, including over 4000 citations, an h-index of 32, and an i10 index of 71. His dedication to advancing postharvest knowledge and his commitment to teaching and research make him an esteemed professional in his field.

Contents

Contributors ... *xv*

Abbreviations ... *xix*

Acknowledgments ... *xxiii*

Preface ... *xxv*

Introduction ... *xxvii*

1. **Postharvest Physiology of Horticultural Crops ... 1**
   Oscar W. Mitalo, Patrick O. Juma, and Ghulam Khaliq

2. **Postharvest Losses in Fruits and Vegetables ... 37**
   Phebe Ding and Ghulam Khaliq

3. **Postharvest Physiological Disorders in Fruits and Vegetables ... 67**
   Nirmal Kumar Meena, K. Choudhary, Sajeel Ahmad, Menaka M., and Vivek Kumar Saurabh

4. **Preharvest Factors Affecting Postharvest Physiological Disorders ... 103**
   Ghulam Khaliq, Sajid Ali, Shaghef Ejaz, Oscar W. Mitalo, and Phebe Ding

5. **Postharvest Factors Affecting Physiological Disorders ... 133**
   Poe Nandar Kyaw, Ei Hay Man, Wai Wai Lwin, Khin Thida Myint, and Zora Singh

6. **Temperature-Related Physiological Disorders of Fruits and Vegetables ... 153**
   Jahan Zaib Ashraf, Hafiz Noor ul Ain, Maira Anam, and Aysha Saleem

7. **Low-Temperature-Related Physiological Disorders ... 177**
   Yoshihiro Imahori

8. **Calcium-Related Physiological Disorders ... 223**
   Babak Madani, Amin Mirshekari, Ahmad Shakerardekani, Farzad Azadshahraki, Abolfazl Golshan Tafti, and Shokrollah Hajivand

9. **Water Loss and Humidity Physiological Disorders ... 239**
   Yasaman Esmaeili and Hilma Sabet

10. Storage Atmosphere-Related Physiological Disorders 259
Sajid Ali, Aamir Nawaz, Safina Naz, Shaghef Ejaz, Ghulam Khaliq, and Muhammad Azam

11. Physiological and Biochemical Aspects of Physiological Disorders..... 291
Harjinder Kaur, Ab Waheed Wani, Atul B. Khalangre, Khan Jabroot J., Pallvi Verma, Safura Nabi, Meraj Ahmad, Anis A. Merza, Glomeraz Abdi, Jatinder Singh, and Rahul R. Rodge

12. Causes of Postharvest Physiological Disorders 327
Shaghef Ejaz, Muhammad Azam, and Beenish Chaudhary

13. Mechanism and Management of Postharvest Physiological Disorders 347
Rachid Ezzouggari, Najwa Seddiqi Kallali, Ilham Dehbi, Khadija Benamar, and Rachid Lahlali

14. Future Aspects of Postharvest Physiological Disorders....................... 391
Yasir Abbas Shah, Muhammad Afzaal, Farhan Saeed, Noor Akram, and Zargham Faisal

Index 413

Contributors

Glomeraz Abdi
Department of Biotechnology, Persian Gulf Research Institute, Persian Gulf University, Bushehr, Iran

Muhammad Afzaal
Department of Food Science, Government College University, Faisalabad, Pakistan

Meraj Ahmad
Department of Soil Water and Engineering Lovely Professional University, Jalandhar, Punjab, India

Sajeel Ahmad
Division of Food Science and Postharvest Technology, ICAR-Indian Agricultural Research Institute, New Delhi, India

Hafiz Noor ul Ain
Department of Food Science and Technology, University of Central Punjab, Lahore, Pakistan

Noor Akram
Department of Food and Nutrition, Government College University, Faisalabad, Pakistan

Sajid Ali
Department of Horticulture, Faculty of Agricultural Sciences and Technology, Bahauddin Zakariya University, Multan, Punjab, Pakistan

Maira Anam
Department for Innovation in Biological, Agrofood and Forest Systems, University of Tuscia, Viterbo, Italy

Ahmad Shakerardekani
Yazd Agricultural and Natural Resources Research and Education Center, AREEO, Yazd, Iran

Jahan Zaib Ashraf
Department of Agriculture, Food, Natural Resources and Engineering (DAFNE), University of Foggia, Italy

Farzad Azadshahraki
Agricultural Engineering Research Institute, Agricultural Research Education and Extension Organization, Karaj, Iran

Muhammad Azam
Institute of Horticultural Sciences, Faculty of Agriculture, University of Agriculture, Faisalabad, Pakistan

Khadija Benamar
Phytopathology Unit, Department of Plant Protection, National School of Agriculture, Meknes, Morocco; Department of Biology, Laboratory of Microbial Biotechnology and Bioactive Molecules, Sciences and Technologies Faculty, Sidi Mohamed Ben Abdellah University, Fez, Morocco

Beenish Chaudhary
Institute of Horticultural Sciences, Faculty of Agriculture, University of Agriculture, Faisalabad, Pakistan

K. Choudhary
Agriculture University, Jodhpur, India

Ilham Dehbi
Phytopathology Unit, Department of Plant Protection, National School of Agriculture, Meknes, Morocco; Laboratory of Plant Biotechnology and Molecular Biology, Faculty of Sciences, Moulay Ismail University, Meknes, Morocco

Phebe Ding
Department of Crop Science, Faculty of Agriculture, Universiti Putra Malaysia, Selangor, Malaysia

Shaghef Ejaz
Department of Horticulture, Faculty of Agricultural Sciences and Technology, Bahauddin Zakariya University, Multan, Pakistan

Yasaman Esmaeili
Department of Food Science and Technology, Isfahan (Khorasgan) Branch, Islamic Azad University, Isfahan, Iran

Rachid Ezzouggari
Phytopathology Unit, Department of Plant Protection, National School of Agriculture, Meknes, Morocco; Laboratory of Biotechnology, Conservation and Valorization of Natural Resources (LBCVNR), Faculty of Sciences Dhar El Mehraz, Sidi Mohamed Ben Abdallah University, Fez, Morocco

Zargham Faisal
Institute of Food Science and Nutrition, Bahauddin Zakariya University Multan, Pakistan

Shokrollah Hajivand
Department of Genetics and Breeding, Temperate Fruits Research Center, Horticultural Research Center, Agricultural Research Education and Extension Organization, Karaj, Iran

Yoshihiro Imahori
Graduate School of Life and Environmental Sciences, Osaka Prefecture University, Gakuen-cho, Nakaku, Sakai, Osaka, Japan

Khan Jabroot
Department of Horticulture, Lovely Professional University, Jalandhar, Punjab, India

Patrick O. Juma
Department of Horticulture and Food Security (HFS), Jomo Kenyatta University of Agriculture and Technology, Nairobi, Kenya

Najwa Seddiqi Kallali
Phytopathology Unit, Department of Plant Protection, National School of Agriculture, Meknes, Morocco; Laboratory of Biotechnology and Valorization of Phyto-Resources, Faculty of Sciences, Moulay Ismail University of Meknes, Meknes, Morocco

Harjinder Kaur
Department of Horticulture, Lovely Professional University, Jalandhar, Punjab, India

Atul B. Khalangre
Department of Horticulture, Lovely Professional University, Jalandhar, Punjab, India

Ghulam Khaliq
Department of Horticulture, Faculty of Agriculture, Lasbela University of Agriculture, Water and Marine Sciences, Uthal, Pakistan

Poe Nandar Kyaw
Department of Horticulture, Yezin Agricultural University, YAU, Myanmar

Rachid Lahlali
Phytopathology Unit, Department of Plant Protection, National School of Agriculture, Meknes, Morocco

Wai Wai Lwin
Department of Horticulture, Yezin Agricultural University, YAU, Myanmar

Menaka M
Division of Food Science and Postharvest Technology, ICAR-Indian Agricultural Research Institute, New Delhi, India

Babak Madani
Horticultural Crops Research Department, Natural Resources Research and Education Center of Chaharmahalobakhtiari, Agricultural Research Education and Extension Organization, Shahrekord, Iran

Ei Hay Man
Department of Horticulture, Yezin Agricultural University, YAU, Myanmar

Nirmal Kumar Meena
Department of Fruit Science, College of Horticulture and Forestry, Jhalawar Agriculture University, Kota, India

Anis A Merza
Department of Horticulture, Lovely Professional University, Jalandhar, Punjab, India

Khin Thida Myint
Department of Horticulture, Yezin Agricultural University, Myanmar

Amin Mirshekari
Department of Agronomy and Plant Breeding, Faculty of Agriculture, University of Yasouj, Yasouj, Iran

Oscar W. Mitalo
Tropical Agriculture Research Front, Japan International Research Center for Agricultural Sciences, Ishigaki, Okinawa, Japan

Safura Nabi
Department of Fruit Science, SKUAST-Kashmir, Jammu and Kashmir, India

Aamir Nawaz
Department of Horticulture, Faculty of Agricultural Sciences and Technology, Bahauddin Zakariya University, Multan, Punjab, Pakistan

Safina Naz
Department of Horticulture, Faculty of Agricultural Sciences and Technology, Bahauddin Zakariya University, Multan, Punjab, Pakistan

Rahul R Rodge
Department of Horticulture, Lovely Professional University, Jalandhar, Punjab, India

Hilma Sabet
Department of Horticulture, Faculty of Agriculture, Isfahan (Khorasgan) Branch, Islamic Azad University, Iran

Farhan Saeed
Department of Food Science, Government College University, Faisalabad, Pakistan

Aysha Saleem
Department of Agriculture, Food, Natural Resources and Engineering (DAFNE), University of Foggia, Italy

Vivek Kumar Saurabh
Division of Food Science and Postharvest Technology, ICAR-Indian Agricultural Research Institute, New Delhi, India

Yasir Abbas Shah
Department of Food Science, Government College University Faisalabad, Pakistan

Jatinder Singh
Department of Horticulture, Lovely Professional University, Jalandhar, Punjab, India

Zora Singh
Foundation Professor, Horticulture, School of Science, Edith Cowan University, Joondalup, Western Australia

Abolfazl Golshan Tafti
Agricultural Engineering Research Institute, Agricultural Research Education and Extension Organization, Karaj, Iran

Pallvi Verma
Department of Horticulture, Lovely Professional University, Jalandhar, Punjab, India

Ab Waheed Wani
Department of Horticulture, Lovely Professional University, Jalandhar, Punjab, India

Abbreviations

1-MCP	1-methylcyclopropene
AATs	alcohol acyl transferases
ABA	abscisic acid
ACC	1-aminocyclopropane-1-carboxylic acid
ADC	arginine decarboxylase
ADP	adenosine diphosphate
AOX	alternative oxidase
APX	ascorbate peroxidase
AsA	ascorbic acid
AsA-GSH	ascorbate-glutathione
ATP	adenosine triphosphate
BAP	6-benzylaminopurine
BER	blossom-end rot
BP	bitter pit
CA	controlled atmosphere
CAP	controlled atmosphere packaging
CAT	catalase
CBF	C-repeat binding factor
CI	chilling injury
DCA-CF	dynamic controlled atmosphere-chlorophyll fluorescence
DPA	diphenylamine
EL	electrolyte leakage
ERFs	ethylene response factors
FAO	food and agriculture organization
FLI	food loss index
GA	gibberellic acid
GRAS	generally recognized as safe
HSPs	heat shock proteins
HWT	hot water treatment
IB	internal browning

LD	lenticel discoloration
LOX	lipoxygenase
LTB	low temperature breakdown
MA	modified atmosphere
MAP	modified atmosphere packaging
MD	moderate damage
MDA	malondialdehyde
MeJA	methyl jasmonate
ND	no damage
NDFSD	number of fruits showing signs of disturbance
NFT	nutrient film technique
NILs	near-isogenic lines
NO	nitric oxide
NOS	nitric oxide synthase
OAT	ornithine aminotransferase
PAL	phenylalanine ammonia-lyase
PAO	polyamine oxidase
PAR	photosynthetically active radiation
PAs	polyamines
PCA	pollination constant astringent
PCNA	pollination constant non-astringent
PG	polygalacturonase
PGA	polygalacturonic acid
PHL	postharvest losses
PLD	phospholipase D
PME	pectin methylesterase
POD	peroxidase
PPO	polyphenol oxidase
PVA	pollination variant astringent
PVNA	pollination variant nonastringent
QTL	quantitative trait locus
RDI	regulated deficit irrigation
RH	relative humidity
RLOS	repeated low oxygen stress
RNA	ribonucleic acid
ROS	reactive oxygen species

RS	russeting
SAD	sugar accumulation disorder
SAM	S-adenosyl-methionine
SB	sunburn browning
SBD	storage breakdown disorder
SFA	saturated fatty acids
SN	sunburn necrosis
SNP	sodium nitroprusside
SOD	superoxide dismutase
SSC	soluble solids concentration
TA	titratable acidity
TBZ	thiabendazole
TEM	transmission electron microscope
TF	transcription factors
TMTs	tonoplast monosaccharide transporters
TPC	total phenolic compound
TTI	time-temperature indicator
UNSFA	unsaturated fatty acids
USAID	United States Agency for International Development
USB	under-skin browning
UV	ultraviolet
VHT	vapor heat treatment
VLD	very little damage
VSD	very severe damage
VTB	vein tract browning
WAA	weeks after anthesis
WCI	white core inclusions

Acknowledgments

We express our warm gratitude to all those who have contributed to the development and completion of this book, *Physiological Disorders of Fruits and Vegetables: Causes and Remedies*. This endeavor would not have been possible without the collective efforts, insights, and expertise of a diverse group of researchers, academicians, scientists, and professionals dedicated to the advancement of horticultural science.

We are particularly indebted to the chapter contributors, who have generously shared their knowledge and research findings. Their scholarly contributions have provided a strong foundation for the scientific rigor and comprehensive scope of this book. Each chapter reflects the depth of their expertise and the years of meticulous research conducted in their respective domains. Their commitment to excellence and willingness to collaborate have been instrumental in shaping this book into a valuable resource for students, researchers, extension specialists, and industry professionals alike.

We also extend our sincere thanks to the many researchers and colleagues whose work and perspectives have enriched the contents of this volume. Their input through collaborative discussions, critical reviews, and constructive feedback has added value and clarity to the understanding of physiological disorders affecting fruits and vegetables. We are grateful for their willingness to share unpublished data, offer expert opinions, and suggest improvements throughout the development process.

Special appreciation goes to the professionals in the horticulture and postharvest industries who offered practical insights and field-level experiences that helped bridge the gap between theory and real-world application. Their perspectives have been invaluable in ensuring that the book remains relevant to the current challenges faced in postharvest handling of horticultural crops.

We acknowledge the support of our respective institutions, which have encouraged and facilitated this scholarly project. We are thankful for the research facilities, access to academic resources, and administrative support provided throughout the journey. We are thankful to the organizations, institutions, and researchers who granted permission to reproduce their illustrations and research materials in this book. Their generosity has significantly enhanced the content and visual impact of the work.

We are deeply appreciative of all the peer reviewers who took the time to evaluate the manuscript and provide detailed suggestions that greatly enhanced the overall quality of this work. for their help in editing, proofreading, and design of the present book. We are also grateful to the publishing team for their invaluable support in editing, proofreading, and designing this book. Their professionalism, attention to detail, and dedication have greatly contributed to the clarity, coherence, and overall presentation of the final manuscript.

To all who contributed their time, knowledge, and enthusiasm thank you. Your collaboration has been vital to the realization of this book, and we hope it serves as a useful and enduring reference in the field of postharvest physiology and fresh produce management.

Preface

Fruits and vegetables play an essential role in human nutrition and health, serving as vital sources of vitamins, minerals, fiber, and antioxidants. However, postharvest losses in fruits and vegetables represent one of the most pressing challenges facing global food systems today. Despite remarkable advancements in agricultural production, a significant portion of harvested produce never reaches consumers in optimal condition. These losses are particularly critical in perishable commodities such as fruits and vegetables, which are biologically active and highly susceptible to rapid deterioration. The situation existed in both developing and developed countries. In an era of increasing concern over food security, postharvest losses, and sustainable agriculture, understanding and addressing post-harvest physiological disorders is more important than ever. Among the many factors contributing to postharvest losses, physiological disorders stand out as a major concern. These disorders have profound effects on the visual appeal, texture, flavor, quality and nutritional value of produce. Unlike infectious diseases caused by pathogens, physiological disorders are triggered by internal physiological imbalances or external abiotic stresses encountered during growth, harvesting, handling, storage, and transportation. Their manifestation not only reduces marketability and consumer acceptance but also leads to substantial economic losses and food waste.

This book, *Physiological Disorders of Fruits and Vegetables: Causes and Remedies,* focused on addressing this critical issue by exploring the complex interactions that lead to physiological disorders and by providing practical strategies for their mitigation. The primary aim of this volume is to foster a comprehensive understanding of the causes behind these disorders and to share effective, science-based management techniques that can be adopted at various stages of the supply chain. Each chapter presents detailed descriptions of specific disorders affecting various fruits and vegetables, supported by illustrations, case studies, and evidence-based recommendations. The content is designed to serve as a reliable reference for researchers, students, extension agents, supply chain professionals, and policymakers working to improve postharvest handling and storage systems.

This book is a timely and comprehensive resource that bridges this critical knowledge gap. It offers in-depth insight into the occurrence, diagnosis, and management of major postharvest physiological disorders affecting a wide range of fruit and vegetable crops. The chapters collectively draw on the latest scientific advances, field observations, and applied technologies to provide a holistic understanding of these disorders and strategies for their prevention and mitigation. The content is organized to serve a diverse readership. Researchers and graduate students will find updated literature, mechanistic explanations, and research directions, while extension personnel, storage operators, and growers will benefit from practical management guidelines and case studies. We have also included high-quality illustrations, photographs, and tables to enhance the clarity and visual understanding of complex disorders.

Each chapter has been contributed by experts who have extensively worked in the field of postharvest physiology, crop science, and horticultural management. Their collective expertise brings scientific depth and practical relevance to the topics discussed. The book also draws upon real-world experiences from research stations, commercial orchards, and storage facilities, ensuring that the information presented is both academically sound and practically applicable. We have endeavored to ensure it is useful not only for specialists but also for students, extension workers, and policymakers involved in horticultural development.

This book offers a unique blend of innovative preventive and diagnostic approaches, integrating the latest scientific research with practical applications for managing postharvest physiological disorders. Richly illustrated with high-quality images, flowcharts, and tables, it enables easy identification and understanding of disorders across various fruits and vegetables. Emphasizing sustainable and environmentally friendly solutions, the book also explores emerging technologies such as bio stimulants, nanotechnology, and digital tools. With a user-friendly layout, real-world case studies, and multidisciplinary insights, it serves as a comprehensive and accessible resource for students, researchers, educators, and professionals in horticulture and postharvest science.

We hope that this book will not only serve as an informative guide for understanding and managing physiological disorders in fruits and vegetables but also act as a catalyst for further research and innovation in the field of postharvest physiology. By highlighting both the underlying causes and practical remedies of these complex disorders, we aim to encourage interdisciplinary collaboration and the development of new technologies that will enhance fresh produce quality, reduce losses, and support sustainable horticultural practices.

—Ghulam Khaliq, PhD
Sajid Ali, PhD
Mohammed Wasim Siddiqui, PhD

Introduction

Fruits and vegetables are vital components of a balanced diet, offering a rich source of essential nutrients including vitamins, minerals, dietary fiber, and a wide array of phytochemicals and antioxidants. These compounds contribute to numerous health benefits, such as reducing the risk of chronic diseases including cardiovascular diseases, certain cancers, obesity, type 2 diabetes, and digestive disorders. Regular consumption of fruits and vegetables has also been associated with enhanced immune function, improved health, and better mental well-being. Beyond their nutritional and health-promoting qualities, fruits and vegetables play a critical role in the global economy. They are high-value crops that generate significant income for millions of smallholder farmers, contribute to rural employment, support food security, and stimulate agri-businesses such as processing, packaging, and export industries. The growing demand for fresh and minimally processed produce further emphasizes their economic importance and the need for sustainable production and postharvest practices to reduce losses and maintain quality from farm to consumer.

Fruits and vegetables play a vital role in feeding the world's growing population and addressing the pressing challenge of global food security. As the global population continues to reach nearly 10 billion by 2050. Ensuring access to sufficient, safe, and nutritious food becomes increasingly critical. At the same time, the degradation of natural resources such as soil, water, and biodiversity, along with the impacts of climate change, are threatening agricultural productivity and sustainability. In this context, fruits and vegetables offer an efficient and sustainable means of improving food availability, enhancing dietary diversity, and reducing hunger and malnutrition. Moreover, fruits and vegetables can be produced on small plots of land, have relatively short growth cycles, and can be cultivated using sustainable practices, including urban farming and integrated crop systems, making them accessible even in resource-limited settings. By increasing the availability and consumption of fruits and vegetables, countries can not only improve the nutritional status of their populations but also build more resilient food systems capable of supporting both present and future generations.

On the other hand, postharvest losses in fruits and vegetables are a major global concern due to their perishable nature and sensitivity to handling, storage, and transportation conditions. These losses can occur at various stages of the supply chain from harvest, handling, and storage to transportation, retail, and consumption. According to the Food and Agriculture Organization (FAO) of the United Nations, it is estimated that up to 45–55% of fruits and vegetables produced globally are lost or wasted before reaching the consumer. These postharvest losses are particularly significant in developing countries where inadequate infrastructure, lack of cold chain facilities, poor handling practices, and limited market access contribute to higher spoilage rates. Such losses not only compromise food availability and access but also pose a serious threat to global food security, especially in regions facing population growth, climate change, and limited agricultural land.

The implications of postharvest losses are multi-faceted, economic, nutritional, and environmental. Economically, these losses result in reduced incomes for growers, traders, and retailers, particularly smallholder farmers who rely heavily on horticultural crops for their livelihoods. In many cases, market rejection of produce due to minor quality defects leads to significant income disparities. Nutritionally, the loss of fruits and vegetables rich in essential vitamins, minerals, and phytochemicals translates into reduced availability of vital nutrients needed to combat malnutrition and support public health. From an environmental perspective, postharvest losses represent a waste of valuable resources used in production, such as water, fertilizers, land, labor, and energy. Furthermore, decayed produce contributes to greenhouse gas emissions, particularly methane, when decomposed in landfills, thus exacerbating climate change.

One of the critical factors contributing to postharvest losses in fresh produce is the occurrence of physiological disorders. These are non-pathogenic, internal or external abnormalities that develop due to preharvest factors, genetic predispositions, or postharvest environmental conditions such as temperature extremes, humidity fluctuations, or mechanical injuries. Common physiological disorders include chilling injury, internal browning, cracking, pitting, and shriveling, which not only reduce the visual appeal and marketability of the produce but also diminish its nutritional value and shelf life. In many cases, these disorders go unnoticed until after marketing or consumption stages, leading to hidden losses. Therefore, quality maintenance in fresh produce through proper postharvest handling, storage, and processing technologies is crucial to minimizing losses, maintaining food quality and safety, and ensuring sustainability in the horticultural value chain. Addressing physiological disorders through research, breeding, and technology adoption remains a key strategy in reducing postharvest losses and enhancing the economic and nutritional resilience of fruit and vegetable systems globally.

Postharvest physiological disorders in fruits and vegetables are complex phenomena resulting from a combination of preharvest and postharvest factors that disrupt normal metabolic and developmental processes, leading to visible or internal defects that compromise quality and marketability. Preharvest factors play a foundational role in the development of these disorders. Genetic makeup determines a crops susceptibility or tolerance to specific disorders, with certain cultivars being more prone to issues such as internal browning, cracking, or chilling injury. Environmental conditions during growth such as extreme temperatures, irregular irrigation, light intensity, and nutrient imbalances can significantly affect physiological development. Cultural practices, including pruning, fertilization, and irrigation scheduling, also influence the structural integrity, calcium distribution, and overall physiological status of the produce, thereby predisposing it to disorders. Postharvest factors are equally critical, with temperature and humidity being the most influential. Exposure to chilling or excessively high temperatures, fluctuations in relative humidity, and improper storage or packaging conditions can trigger physiological responses such as tissue breakdown, water loss, and oxidative stress. Rough handling and mechanical injuries during harvesting, packaging, or transportation further exacerbate vulnerability to disorders.

The interaction between biotic and abiotic stressors can intensify physiological damage. For example, fruits stressed by drought (abiotic) may exhibit weakened cell walls, making them more susceptible to pathogen attack (biotic), which can accelerate the development of disorders like water-soaked lesions or internal breakdown. At the mechanistic level, these disorders are often linked to physical and biochemical changes such as membrane degradation, imbalance in reactive oxygen species (ROS), enzymatic browning, and calcium deficiency. Such disruptions can impair cellular integrity, reduce firmness, and alter respiration and ethylene production rates, ultimately leading to visible symptoms like pitting, shriveling, cracking, or discoloration. Understanding the multifactorial causes and underlying mechanisms of physiological disorders is essential for developing effective pre- and postharvest management strategies aimed at minimizing losses and maintaining the quality and nutritional value of fruits and vegetables throughout the supply chain.

Effective management of postharvest physiological disorders in fruits and vegetables requires a comprehensive and integrated approach, addressing both preharvest and postharvest stages. In the field, preharvest strategies are fundamental in reducing the initial risk of disorder development. One of the most effective measures is the selection and cultivation of resistant or tolerant cultivars. These genetically improved varieties possess characteristics such as stronger cell walls, efficient nutrient uptake, and resilience to environmental stress, making them less prone to disorders like blossom end rot in tomatoes or bitter pit in apples. Proper nutrient management is another key factor, with calcium and boron being especially important for cell wall structure and membrane stability. Regular applications of calcium-based fertilizers or foliar sprays during critical growth stages help prevent calcium-related disorders. In addition, consistent irrigation practices are essential to avoid water stress, which can lead to fruit cracking or watercore. Employing drip irrigation systems and soil moisture monitoring can optimize water use and ensure steady nutrient transport. Canopy management practices, including strategic pruning and the use of shade nets, also help protect fruits from sunburn and temperature extremes, which often lead to physiological damage such as sunscald and heat injury.

Postharvest management is equally important in minimizing physiological disorders and maintaining the quality of fruits and vegetables. Controlling temperature and humidity during storage is critical, as improper conditions can trigger disorders like chilling injury, shriveling, and internal breakdown. For example, chilling-sensitive fruits like bananas, cucumbers, and papayas must be stored above their threshold temperatures to avoid cold-induced damage. Modified Atmosphere Packaging (MAP) and Controlled Atmosphere (CA) storage technologies provide regulated environments by adjusting oxygen, carbon dioxide, and ethylene levels, which slow down respiration and delay senescence. This helps reduce oxidative stress and the onset of disorders such as superficial scald and internal browning. Postharvest calcium treatments, such as dipping fruits in calcium chloride solutions, reinforcing cell wall strength and reducing the risk of calcium-deficiency disorders. The use of growth regulators like gibberellic acid and ethylene inhibitors like 1-methylcyclopropene (1-MCP) can also delay ripening and reduce senescence-related disorders.

Handling and transportation practices significantly influence the incidence of mechanical injuries and subsequent physiological disorders. Gentle harvesting techniques, cushioned packaging, and the use of ventilated crates can minimize bruising and compression injuries that often lead to water-soaked lesions or tissue breakdown. Establishing and maintaining an unbroken cold chain from the point of harvest through to retail, ensures that produce remains within optimal temperature and humidity ranges, preventing heat buildup and metabolic imbalance. Rapid cooling methods such as hydrocooling or forced-air cooling are particularly effective in removing field heat and stabilizing produce soon after harvest.

Innovative and non-chemical approaches are also gaining attention for their effectiveness in reducing postharvest physiological disorders without compromising food safety. Techniques such as UV-C light exposure, hot water dips, and the application of natural edible coatings like chitosan or aloe vera gel can enhance resistance to oxidative stress and prolong shelf life. Additionally, smart packaging technologies and biosensors now allow for real-time monitoring of temperature, humidity, and gas concentrations during storage and transport, enabling timely interventions to prevent disorder onset. The integration of digital tools and decision-support systems, which utilize weather data and crop modeling, further aids growers in making informed decisions to anticipate and mitigate stress conditions in the field.

The primary objective of this book, *Physiological Disorders of Fruits and Vegetables: Causes and Remedies*, is to provide a comprehensive understanding of nature, causes, and management of postharvest physiological disorders that limit the quality, storability, and marketability of fruits and vegetables. The book aims to bridge scientific knowledge with practical solutions by exploring the underlying physiological, biochemical, and environmental mechanisms that contribute to disorders, while also highlighting strategies for their prevention and management. By integrating theoretical foundations with real-world applications, this book serves as a valuable resource for horticulturists, researchers, extension professionals, industry personnel, postharvest technologists, students, and policymakers. It aims to bridge the gap between laboratory research and field-level implementation by presenting practical approaches, including improved cultural practices, physiological and molecular interventions, innovative postharvest technologies, and sustainable management strategies. Ultimately, the goal of this work is to support efforts in reducing postharvest losses, enhancing produce quality, and contributing to food system sustainability in a rapidly changing world.

The book begins with fundamental concepts of postharvest physiology and the extent of postharvest losses in fruits and vegetables, followed by detailed discussions on various physiological disorders affecting fresh produce. It examines both preharvest and postharvest factors contributing to disorder development, with dedicated chapters on temperature stress, low-temperature injury, calcium deficiency, water loss, and humidity-related problems. The role of storage atmosphere, and physiological and biochemical processes in disorder manifestation are comprehensively explained, alongside an in-depth analysis of their causes, mechanisms, and practical management strategies. The concluding chapters emphasize integrated approaches for disorder prevention and provide insights into future research directions to ensure sustainable postharvest handling of fruits and vegetables.

CHAPTER 1

Postharvest Physiology of Horticultural Crops

OSCAR W. MITALO[1*], PATRICK O. JUMA[2], and GHULAM KHALIQ[3]

[1]Tropical Agriculture Research Front, Japan International Research Center for Agricultural Sciences, Ishigaki, Okinawa, Japan

[2]Department of Horticulture and Food Security (HFS), Jomo Kenyatta University of Agriculture and Technology, Nairobi, Kenya

[3]Department of Horticulture, Faculty of Agriculture, Lasbela University of Agriculture, Water and Marine Sciences, Uthal, Pakistan

*Corresponding author

ABSTRACT

Horticultural crops are essentially living organisms that continue to exhibit key life processes even after harvest. Important biological processes that take place during postharvest handling include fruit ripening, senescence, respiration, transpiration, and sensitivity or response to both biotic and abiotic stimuli. These processes deplete the reservoirs of harvested horticultural crops since resultant losses can no longer be replenished, hence, the quality of the commodities deteriorates. Several factors including temperature, humidity, ethylene, and ambient atmospheric gas composition have been shown to significantly affect the rate of progression of the above-mentioned biological processes. Manipulation of these factors has enabled the development of various postharvest strategies aimed at reducing the rate of these deteriorative processes, and subsequently, prolonging the postharvest life of many horticultural crops. This chapter

Physiological Disorders of Fruits and Vegetables: Causes and Remedies.
Ghulam Khaliq, Sajid Ali, and Mohammed Wasim Siddiqui (Eds.)

DOI: 10.1201/9781998511334-1

summarizes our current understanding of the biological processes that operate during postharvest handling of horticultural crops with a special emphasis on current and emerging perspectives regarding their regulation by internal and external environmental factors such as ethylene, low temperature, CO_2/O_2 balance, and relative humidity.

1.1 INTRODUCTION

There is immense diversity among horticultural crops with regard to their genetic makeup, tissue and organ type, physiological state, and morphological structure (roots, stems, leaves, flowers, and fruits) and composition (Wu et al., 2018). Horticultural crops, particularly fruits and vegetables, are excellent sources of carbohydrates, fiber, vitamins, minerals, and phytochemicals which are essential for a healthy human diet (Slavin and Lloyd, 2012; Yahia and Carrillo-Lopez, 2018). Fruits and vegetables are also characterized by high water content, which can range between 61 and 95% by weight (Slavin and Lloyd, 2012). These compositional characteristics render horticultural crops highly susceptible to deterioration during postharvest handling. In fact, deterioration begins immediately after harvest, as harvested fruits and vegetables are living organisms which continue to maintain physiological and molecular systems and metabolic processes that existed before harvest. Major biological processes which contribute to postharvest deterioration include fruit ripening, senescence, respiration, and transpiration (DeEll et al., 2003). These processes lead to compositional changes and losses in food reserves, water content, and integrity of many horticultural crops. While attached to the parent plant, the flow of sap which contains water, photosynthates, and minerals replenishes losses from the above-mentioned biological processes (Yahia and Carrillo-Lopez, 2018). However, harvested fruits and vegetables are entirely on their own, and hence the usual losses of water and substrates caused by the above-mentioned biological processes can no longer be replenished (Khaliq et al., 2023; Khan et al., 2025; Lal et al., 2025). Moreover, harvested horticultural commodities are forced to cope with multiple biotic and abiotic stresses which originate from either the internal or external environment such as ethylene, temperature, humidity, pathogens, and ambient atmospheric composition; responses to these stresses greatly affect postharvest quality. Postharvest physiology of horticultural crops encompasses the study of biological, biochemical, and physiological processes that occur in fruits and vegetable plants after harvest, with the primary objective of maintaining quality, extending shelf life, and reducing postharvest losses. These crops, being highly perishable, undergo a series of metabolic changes, including respiration, transpiration, ethylene production, and enzymatic activities, which significantly influence their texture, color, flavor, nutritional value, and susceptibility to disorders and decay. Understanding these physiological processes is crucial for developing effective postharvest handling practices such as temperature management, controlled atmosphere storage, modified atmosphere packaging, and the application of chemical and biological treatments. Moreover, postharvest physiology plays a pivotal role in enhancing marketability, ensuring food security, and supporting the sustainability of horticultural supply chains. Integrating physiological knowledge with innovative technologies contributes to reducing postharvest losses, improving consumer satisfaction, and ensuring the economic viability of horticultural enterprises. This chapter will concentrate primarily on the physiological processes contributing to quality and quality loss during postharvest handling of horticultural crops, and the associated advancements in knowledge and technology development.

1.2 IMPORTANT BIOLOGICAL PROCESSES DURING POSTHARVEST HANDLING

Several physio-molecular activities occur simultaneously in fruits and vegetables after harvest and during postharvest handling. These processes are not entirely harmful, as some like fruit ripening contribute to edibility in certain commodities. The major biological processes that influence postharvest quality are discussed below.

1.3 WATER LOSS (TRANSPIRATION)

Transpiration refers to the process of losing water vapor from plants or plant organs. This process is important for the regulation of internal temperature. The moisture content at the time of harvest varies greatly among different horticultural crops (Table 1.1). During postharvest storage and handling, a significant percentage of the initial moisture content is lost due to transpiration. A certain degree of water loss may be necessary to prevent fungal growth and/or tissue browning in addition to regulating internal temperature (Ntuli et al., 2017; Ali et al., 2025). However, unlike when attached to the parent plant, the lost moisture cannot be replenished which leads to quality deterioration. As indicated in Table 1.1, different fruits and vegetables can withstand varying degrees of water loss. Excessive water loss beyond these acceptable levels results loss of saleable weight due to shriveling, as well as promoting tissue browning, loss in fruit texture and flavor, accelerated senescence, susceptibility to chilling injury, and membrane disintegration, and subsequently, loss of marketability (Singh et al., 2010; Lufu et al., 2020). In addition, moisture lost due to transpiration often accumulates on the surface of the horticultural products (Linke and Geyer, 2013), promoting the growth of postharvest pathogens. Therefore, appropriate treatments, packaging, and optimal storage conditions are essential to reduce water loss and maintain postharvest quality.

TABLE 1.1 Water Content and Acceptable Water Loss Percentage Among Some Common Horticultural Crops.

Commodity	Water Content (%)	References	Permissible Water Loss (%)	References
Lettuce	95	Vargas-Arcila et al. (2017)	3	Paull (1999)
Potato	63–83	Puttongsiri et al. (2012)	5	Chourasia et al. (2005)
Celery	95	Nguyễn et al. (2020)	10	Vina and Chaves (2003)
Cucumber	96	Manjunatha and Anurag (2014)	8–9	Moalemiyan and Ramaswamy (2012)
Apples	80–85	Rasooli Sharabiani et al. (2021)	7.5	Khan et al. (2017)
Grapefruits	91	Salehi and Kashaninejad (2018)	5	Khan et al. (2017)
Nectarines	91	Souza da Silva et al. (2019)	21	Khan et al. (2017)
Peaches	86	Golisz et al. (2013)	11–16.4	Khan et al. (2017)
Mangoes	56.3–86.1	Othman and Mbogo (2009)	7–9	Nunes et al. (2007)
Strawberries	90	Akhtar and Rab (2015)	6	Nunes et al. (1995)
Tomatoes	95	Sohail et al. (2011)	4–7	Khan et al. (2017)
Carrot	90	Becerra-Moreno et al. (2015)	8	Khan et al. (2017)
Cauliflower	92	Khan et al. (2017)	7	Khan et al. (2017)
Pepper fruit	90–92	Lownds et al. (1993)	4.5–7	Wills et al. (1981),
Asparagus	93	Khan et al. (2017)	8	Paull (1999)

1.3.1 ESTIMATION OF TRANSPIRATION RATES IN HORTICULTURAL CROPS

Water loss is most simply estimated by measuring weight loss. Here, the percentage of moisture loss is often equated to the weight loss (Bovi and Herppich, 2021). This method assumes that most of the mass reduction corresponds to the water lost through transpiration although a significant amount of water is also lost through respiration. Nevertheless, the below formula is widely used by postharvest physiologists and technologists around the world to provide a rough estimate of water lost during postharvest storage and handling.

$$\text{Mass loss}\,(\%) = \frac{M1 - M2}{M1} \times 100$$

where $M1$ is the initial mass of the product and $M2$ is the final mass.

The use of stable isotopes to assess water loss in postharvest horticultural is also a common practice in the horticulture industry. This approach is based on the fact that leaves are enriched with hydrogen and oxygen (O_2) isotopes as an after-effect of evapotranspiration. The isotope values can therefore be used as a biomarker for transpiration. Recently, Lufu et al. (2020) demonstrated that the $\delta^{18}O$ value correlated well with the weight decrease of plant samples. Therefore, projecting the $\delta^{18}O$ value of tissue water can provide a suitable parameter for checking postharvest alterations in tissue water of horticultural crops (Lufu et al., 2020).

Water loss induces and is strongly correlated with changes in the mechanical properties of horticultural crops. Hence, mechanical properties such as double compression (texture profile analysis) and puncture test have a potential for use as standard methods to track water loss rates especially in blueberries (Rivera et al., 2021) and apricot (Arafa, 2019). This method may not, however, be applicable to leafy vegetables since it is a big challenge to measure mechanical properties in leaves.

1.3.2 FACTORS AFFECTING TRANSPIRATION

Transpiration of produce during postharvest handling and storage can be influenced by both intrinsic and extrinsic factors.

1.3.2.1 PRODUCT SURFACE AREA

In apples, Maguire et al. (2010) observed that smaller fruits lost weight faster than larger fruits. Similarly, Díaz-Pérez et al. (2007) observed that in bell peppers, the transpiration rate decreased as the size of the product increased. Small-sized horticultural crops have a higher rate of moisture loss due to their higher surface area-to-volume ratio compared to larger fruits. Reporting the transpiration rate of fresh produce per unit surface area would be more practical than per fruit mass since water loss depends more on surface area.

1.3.2.2 MATURITY AT HARVEST

Immature fruits and vegetables tend to lose water faster than physiologically mature crops. Maturity allows for the proper development of peel structures that act as physical barriers to prevent water loss. Cutting and Wolstenholme (1992) showed that increasing maturity reduced the total amount of water lost in avocado during the ripening process. Similarly, Chomchalow et al. (2002) also showed that water loss rates were lower in nonethylene-treated mature tomatoes compared to medium and least mature tomatoes stored at 12.5°C for one day. Contrarily, in some fruits such as strawberries and persimmons, higher moisture loss rates were recorded in mature fruits compared to immature fruits (Ramin and Tabatabaie 2003; Rahman et al., 2016), probably due to structural changes in the outer layers of the product (Perkins-Veazie, 1995; Rahman et al. 2016). Kritzinger and Lötze (2019) reported that the cuticle permeability of plums was important in determining fruit moisture loss in fruit maturity-dependent manner. Therefore, harvesting fruits and vegetables at their optimal maturity is a critical precaution to minimize postharvest moisture loss during storage.

1.3.2.3 TEMPERATURE

The ambient temperature impacts the relative humidity (RH) of the surrounding, which in turn influences the water potential deficit between the product and its environment (Léchaudel et al., 2013; Zuo et al., 2025). An increase in product temperature facilitates water evaporation crop surfaces, while higher atmospheric temperatures raise the water-holding capacity of ambient air further facilitating more water loss in horticultural crops (Thompson et al., 2018). Theoretically, at 100% RH and constant temperature, transpiration does not occur. However, Rux et al. (2016) recently showed that water vapor saturation occurs even at substantial transpiration. These results suggest that heat generated from respiratory processes can significantly influence moisture evolution from horticultural crops under saturated conditions. It is, therefore, important to maintain optimum temperature and humidity throughout the postharvest value chain of horticultural crops.

1.3.2.4 RELATIVE HUMIDITY

Fluctuations in RH can also significantly influence the rate of water loss in fresh produce (Paull, 1999). Lower RH increases the water vapor pressure deficit across the fruit peel, increasing the moisture diffusion from the fruit to the surrounding air (Bovi et al., 2018). Storage of pomegranate fruit for 30 days of storage at 20°C and 65% RH resulted in higher water loss compared to fruits stored at the same temperature conditions and 95% RH (Mukama et al., 2019). High RH is generally important to prolong the shelf life of horticultural crops. However, high RH promotes spore germination and eventual growth of pathogenic fungi on crop surfaces (Oliveira et al., 2014). Therefore, a delicate balance must be made for the humidity to reduce water loss and disease incidence in postharvest horticultural crops.

1.4 RESPIRATION

Respiration provides energy for all living organisms, including horticultural crops. During respiration, plants break down carbohydrates or organic acids to release carbon dioxide (CO_2), water, and energy. In detached horticultural produce, respiration utilizes carbohydrates in the reservoirs with no options for replenishment. Thus, respiration results in losses of stored carbohydrates which contribute to losses in fresh weight and visual quality during postharvest handling.

1.4.1 ESTIMATION OF RESPIRATION

The respiration rate of fresh produce is expressed either as O_2 consumption rate and/or CO_2 production rate. There are three conventional methods for calculating these values: (a) closed/static system, (b) flowing /flushed system, and (c) permeable system. In a closed system, a gas-tight container of known volume is filled with a product of known weight leaving some head space. The container is closed with ambient air as the initial atmosphere in the head space. Changes O_2 and CO_2 concentration can be measured by obtaining the gas in the headspace using syringe over a certain period. The gas chromatograph is then used to quantify changes in O_2 and CO_2 amounts in the headspace gas. In the flowing system, horticultural produce is held in enclosed respiration chambers and a mixture of gas (CO_2, O_2, and nitrogen) is flown through at a constant rate using a precalibrated capillary flow meter. The composition of the gas mixture entering and leaving the respiration chambers is monitored regularly for O_2 and CO_2 concentrations. When the system reaches a steady state, the respiration rates are calculated from the absolute differences in gas concentrations between the outlet and the inlet. The permeable system uses a package of known dimensions and film permeability filled with a product of predetermined weight. The steady-state concentrations of O_2 and CO_2 are then determined and used to estimate respiration rates.

Apart from the above three, there is a fourth method for estimating respiration rates in horticultural crops which is based on modeling. Several respiration rate models have been proposed thus far including the Arrhenius, Michaelis–Menten, and Peleg models (Rahman et al., 2013; Maringgal et al., 2020). The Peleg model, for instance, has been applied to analyze experimental data on the dynamics of O_2, CO_2, and ethylene concentrations and describing the rate of gas composition function. The Peleg model has also been used to describe respiration rates of various fruits such as apples (Mahajan and Goswami 2001), papaya (Maringgal et al., 2021), lemon (Sultan et al., 2022), and mangoes (Ravindra and Goswami, 2008).

1.4.2 FACTORS INFLUENCING RESPIRATION RATES

There are multiple internal and external (environmental) factors that trigger stress in horticultural crops giving rise to a wide range of physiological, biochemical, and biophysical events that modify the respiration rate.

1.4.2.1 ATMOSPHERIC GAS COMPOSITION

It is essential to balance the O_2 and CO_2 levels throughout the storage period of horticultural crops since extremely high respiration rates result in deterioration. Low O_2 levels can slow down the rate of the respiration process, especially the aerobic part. However, a reduction of the O_2 concentration beyond a critical point can lead to increased anaerobic respiration which results in the production of alcohols and off-flavors (Schlie et al., 2020). For instance, storage of apples at O_2 contents of 0.5 kPa resulted in fermentation increased fermentation, overproduction of ethanol in the flesh (Park et al., 2022). A balance must be reached to significantly reduce the rate of aerobic respiration and avoid anaerobic respiration which is detrimental to the quality of horticultural crops.

1.4.2.2 TEMPERATURE

Temperature is the most important external factor influencing respiration rates in horticultural crops. Generally, biological reaction rates increase two- to three-fold for every 10°C rise in temperature within the optimal range of temperature (Saltveit, 2004; Campbell et al., 2007). Extremely high temperatures result in enzymatic denaturation halting the biological process including respiration (Knapp and Huang, 2022). Low temperatures can also deactivate most enzymes and cause a general decrease in the rate of most metabolic reactions (Rao, 2015). However, extended storage at suboptimal low temperatures can cause cold stress which results in chilling injury (Sevillano et al., 2009). Nevertheless, moderately lower temperatures can lower the respiration rate and have been shown to extend the postharvest life in broccoli (Parker, 2020), plums (Bal, 2013), pears (Ho et al., 2018), and dragon fruits (Ho et al., 2020).

1.4.2.3 WOUNDING

Wounding is one of the primary stresses horticultural crops are exposed to during harvesting or postharvest handling. Wounding is known to stimulate ethylene production rate, respiration rate, deterioration, senescence, and ripening of fruits and vegetables. A high respiration rate was observed in wounded red prickly pears compared to intact fruit (Ortega-Hernández et al., 2019). Wounding generally induced synthesis of enzymes involved in the respiration pathway, such as ATP-dependent phosphofructokinase involved in carbohydrate breakdown (Laties, 2019).

1.5 FRUIT RIPENING

Fruit ripening is a complex process that is synergistically controlled by both intrinsic developmental cues and various environmental factors (Chen et al., 2020). During ripening, multiple closely interlinked processes trigger changes in fruit color, texture, flavor, and aroma (Chen et al., 2020), and these alterations are closely related to the ultimate postharvest quality (Li et al., 2021). While fruit ripening-associated processes render fruit attractive, edible, and palatable for human consumption, they can also accelerate postharvest quality deterioration by increasing fruit susceptibility to pathogens, reducing postharvest life, and causing handling problems (Cantu et al., 2009; Chen et al., 2021). Given the biological and commercial importance of fruit ripening, significant progress has been made in understanding the regulatory mechanisms involved.

1.5.1 PHYSIOLOGY AND BIOCHEMISTRY

While the changes that occur during fruit ripening are both species- and cultivar-dependent, most fruits exhibit some or all of the changes described below.

1.5.1.1 FLESH SOFTENING

Some of the fruits that display obvious flesh softening during ripening include avocados, kiwifruit, European pears, tomatoes, and persimmons. Typically, fruit softening involves multiple adjustments and modifications to components of the cell wall, particularly polysaccharides and proteins (Wang et al., 2018). The primary cell walls of plants are generally comprised of cellulose microfibrils that are embedded in a matrix of hemicellulose and pectins (Posé et al., 2019). During fruit ripening, these components undergo solubilization, depolymerization, and transglycosylation which results in the loss of intercellular wall adhesion, and the loosening and disintegration of cell walls (Tucker et al., 2017; Wang et al., 2018; Posé et al., 2019). It has been shown that the above-mentioned cell wall modifications are rather complex, involving the coordinated and interdependent action of several families of enzymes and proteins such as polygalacturonases, pectate lyases, pectin methyltransferases, β-galactosidases, expansins, and xyloglucan endotransglucosylase hydrolases among others (Tucker et al., 2017; Forlani et al., 2019).

1.5.1.2 STARCH METABOLISM AND INCREASE IN SOLUBLE SUGARS

Another important change that occurs during ripening in most fruits is an increase in the overall content of soluble sugars. During ripening, starch, which is the major polysaccharide in most physiologically mature fruit, is actively converted into soluble sugars (mainly sucrose, fructose, and fructose) mainly by β- and α-amylases (Hu et al., 2016; Maria et al., 2016; Xiao et al., 2018) as well as sucrose synthases (Stein and Granot, 2019; Duan et al., 2021), sucrose phosphate synthases (Duan et al., 2021), and invertases (Duan et al., 2021). Additionally, the activity of various sugar carriers such as sucrose transporter (SUT), sugars will eventually be exported transporters (SWEET), and tonoplast monosaccharide transporters (TMTs) in sugar accumulation has been linked to increased sugar levels during fruit ripening in certain fruit species (Cheng et al., 2018; Shammai et al., 2018; Iqbal et al., 2020).

1.5.1.3 ACIDITY REDUCTION

Fruits contain a wide variety of organic acids with the major ones being citrate, malate, and ascorbate (Walker and Famiani, 2018). The content of these organic acids, particularly citrate and malate, decreases dramatically during fruit ripening (Batista-Silva et al., 2018; Ul Haq et al., 2024). There is little information regarding the molecular mechanisms that lead to the observed reduction in acidity during fruit ripening. However, it is hypothesized that increases in respiratory metabolism which occur during fruit ripening can result in decreased malate and citrate content (Centeno et al., 2011; Batista-Silva et al., 2018).

1.5.1.4 CHANGES IN COLOR

Fruit ripening-related color changes can occur in the peel (skin) and/or the flesh. These color changes result from the transformation of chloroplasts to chromoplasts (Rodrigo et al., 2013); a process that involves both chlorophyll degradation (Kuai et al., 2018) and synthesis or metabolism of several carotenoids (Sun et al., 2018). Chlorophylls are degraded sequentially to colorless compounds by several enzymes namely chlorophyll b reductase, chlorophyllase, pheophytinase, Mg dechelatase, pheophorbide a oxidase, and red chlorophyll catabolite reductase (Shimoda et al., 2016; Yin et al., 2016). Carotenoids, on the other hand, are synthesized and metabolized by multiple enzymes that include phytoene synthase, phytoene desaturase, ζ-carotene desaturase, lycopene ε-or β-cyclases, and ε-ring or β-ring hydroxylases (Kuai et al., 2018; Kapoor et al., 2022).

Besides chlorophylls and carotenoids, several other compounds contribute to the color of ripe fruit, including anthocyanins (Medina-Puche et al., 2014; Castellarin et al., 2007; Ravaglia et al., 2013) and flavonoids (Petroni and Tonelli, 2011).

1.5.1.5 RELEASE OF VOLATILE COMPOUNDS

Innumerable types of aroma volatiles are also produced during the fruit ripening process. The volatiles are biosynthesized from several substrates including carbohydrates, membrane lipids, and amino acids (El Hadi et al., 2013). While volatile profiles differ with fruit species and cultivar, the widespread ones are esters, alcohols, aldehydes, and terpenoids (Defilippi et al., 2009). The aroma of ripe kiwifruits, apples, pears, and peaches is largely due to esters, while that of ripe tomatoes, strawberries, and citrus fruits is contributed by terpenoids (Klee and Tieman, 2018). Esters produced by fruit can be either straight chain or branched chain. Straight chain esters are produced via the fatty acid degradation pathway through intermediates like alcohol and acyl coenzyme A (CoA) that contain three carbons or more (e.g., butanol, hexanol, and butyryl and hexanoyl CoA) (Cao et al., 2021). Conversely, branched chain esters are derived from amino acid precursors (e.g., isoleucine, valine, or leucine). Alcohol acyl transferases (AATs) which transfer an acyl group from a CoA donor to an alcohol acceptor catalyze the final step in ester biosynthesis (Souleyre et al., 2022).

1.5.2 HORMONAL CONTROL OF FRUIT RIPENING

Many hormones have thus far been implicated in the control of the fruit ripening process. The most studied hormone, however, is ethylene and mature fruits possess the ability to respond to both their own ethylene (endogenous) or exogenous ethylene from external sources (Li et al., 2019; Liu et al., 2020). Ethylene is produced in extremely low amounts by all plant tissues throughout growth and development (Liu et al., 2020), but the production rate increases during ripening and in response to biotic and abiotic stresses (Kou et al., 2021). As shown in Figure 1.1, two systems have been shown to be responsible for ethylene biosynthesis; System I is auto-inhibitory and produces the basal levels of ethylene required for plant growth and development, while System II is autocatalytic and is responsible for the massive burst of ethylene production rate during ripening and in response to stresses (Kou et al., 2021). Fruits are thus classified into either climacteric or nonclimacteric types depending on their ripening behavior (Kou and Wu, 2018; Table 1.2). Climacteric fruit display a sharp increase in ethylene production rate just before and during fruit ripening (Kou and Wu, 2018), while nonclimacteric fruit do not. Studies on several ethylene mutants have revealed that ethylene regulates the expression levels of a majority of ripening-associated genes, particularly in climacteric fruits (Liu et al., 2015). The general concept is that climacteric-type fruit ripening is largely controlled by ethylene, while nonclimacteric-type fruit ripening does not necessarily require ethylene. This is, however, not entirely true, as some climacteric fruit display ethylene-independent ripening processes (Asiche et al., 2018; Huang et al., 2022), while ethylene has been shown to regulate certain aspects of ripening in some nonclimacteric fruit (Fuentes et al., 2019). For instance, kiwifruits are generally classified as climacteric, yet the production of autocatalytic ethylene starts way after fruit softening has occurred (Richardson et al., 2011), while in the supposedly nonclimacteric citrus fruits, exogenous ethylene triggers changes in peel color (Mitalo et al., 2020, 2022; Zhu et al., 2021).

TABLE 1.2 A List of Common Climacteric and Nonclimacteric Fruits.

Climacteric Fruits	Nonclimacteric Fruits
Tomato, apple, mango, papaya, banana, pear, kiwifruit, persimmon, avocado, peach, plum, passion fruit, guava, durian, apricot, and cherimoya	Watermelons, citrus (orange, grapefruit, lemon, lime, etc.), grape, strawberry, pumpkin, squash, pineapple, loquat, olive, pomegranate, and jujube

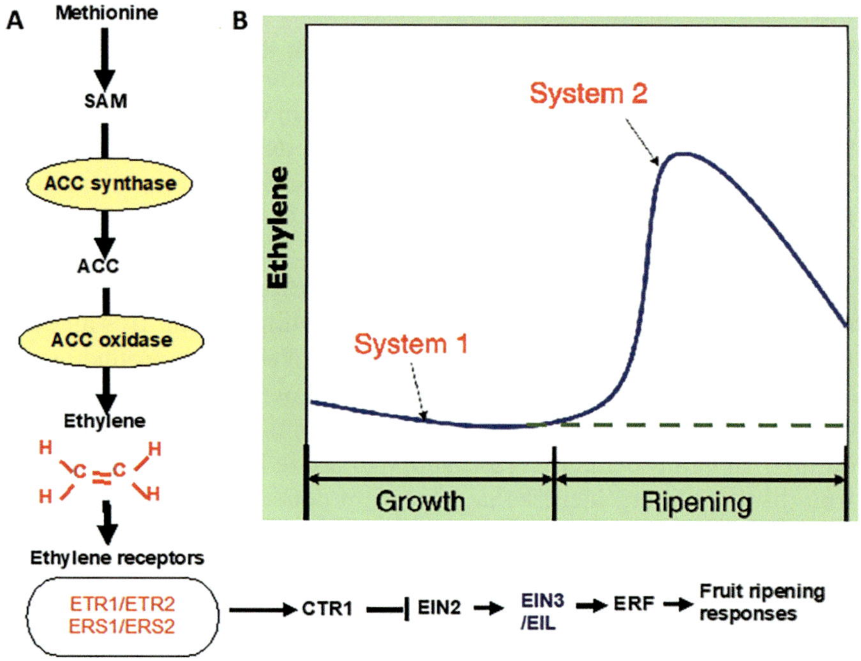

FIGURE 1.1 Ethylene production, perception and signal transduction pathways in ripening fruits. (A) Ethylene biosynthesis, perception, and signaling pathways. SAM—S-adensosylmethionine, ACC—1-Aminocyclopropane-1-carboxylic acid, ETR—Ethylene receptor, ERS—Ethylene response sensor, CTR—Constitutive triple response, EIN—Ethylene insensitive, EIL—EIN3-like, ERF—Ethylene response factor. (B) Two systems of ethylene production have been identified in ripening fruits. System 1 is present in all plant tissues and is responsible for production of basal amounts of ethylene. System II operates during fruit ripening and it can also be triggered by both biotic and abiotic stresses.

Apart from ethylene, other hormones have also been shown to regulate fruit ripening. It has been suggested that auxins mainly inhibit fruit ripening (Iqbal et al., 2017), as it is thought that low auxin concentration increases the sensitivity of fruit tissues to ethylene, while several auxin response factors negatively affect the ripening process (Zhang et al., 2020). Salicylic acid is another plant hormone that likely inhibits ethylene production and subsequently, the ripening process (Kumar et al., 2021). On the other hand, abscisic acid, brassinosteroids, and jasmonates were shown to promote ripening (Hu et al., 2020; Wu et al., 2020; Bai et al., 2021).

1.5.3 MOLECULAR REGULATION

Fruit ripening is regulated by a multitude of transcription factors (TF) that modulate the expression of various genes associated with ethylene biosynthesis, as well as specific ripening aspects described above. The identification of these TFs has been possible through studies of various spontaneous ripening mutants particularly in the model fruit, tomato (Wang et al., 2019; Li et al., 2022). The MADS-box family is perhaps the most studied of the TFs, particularly the master regulator, ripening inhibitor (RIN). Ripening inhibitor transcripts increase dramatically during ripening (Li et al., 2020), and the protein was shown to activate a large number of fruit ripening-associated genes (Fujisawa et al., 2013). In addition, *rin* is a spontaneous mutation that completely inhibits the biosynthesis of autocatalytic ethylene production and subsequently, fruit ripening (Ito et al., 2017). Other MADS-box domains that play critical roles in fruit ripening regulation are colorless nonripening (CNR) (Lai et al., 2020), tomato agamous-like (TAGL) (Itkin et al., 2009), and fruitfull (FUL) homologs, FUL1 and FUL2 (Bemer et al., 2012).

The NAC domain TFs have also been shown to play critical roles in the regulation of fruit ripening. Among them, nonripening (NOR) is the best characterized since *nor* tomato mutants also display a failed system II ethylene production and abnormal ripening phenotypes (Gao et al., 2019). Other NAC family TFs have also been shown to modulate ethylene biosynthesis as well as various fruit ripening aspects in numerous fruit species such as tomatoes (Kou et al., 2021), bananas (Li et al., 2020), and kiwifruits (Wang et al., 2022).

Ethylene response factors (ERFs) comprise a large multigene family that regulate fruit responses to ethylene and other hormones as well as responses to biotic and abiotic stresses (Gao et al., 2020). The role of ERFs in fruit ripening regulation has been extensively studied. In general, ERFs have been shown to act downstream of the ethylene signaling pathway, where they directly bind to and activate various ethylene-regulated ripening-associated genes (Gao et al., 2020). However, ERFs have also been shown to be involved in ethylene-independent fruit ripening regulation (Mitalo et al., 2020; 2022).

Several other TFs have also been shown to play minor but key roles in fruit ripening regulation. In tomatoes, the homeobox protein HB1 was shown to bind to the promoters of ethylene biosynthetic genes (Lin et al., 2008), while in bananas, Yang et al. (2021) demonstrated that four homeodomain leucine zipper (HD-ZIP) TFs such as MaHDZI.19, MaHDZI.26, MaHDZII.4 and MaHDZII.7 can activate genes encoding ethylene biosynthesis and cell wall modifying enzymes (Yang et al., 2021). Song et al. (2020) also reported that basic helix-loop-helix (bHLH) TFs may have a role in fruit ripening regulation, particularly anthocyanin accumulation in several fruit species (Song et al., 2020). The MYB family TFs have been shown to play important roles in the synthesis of flavonoids during fruit ripening in tomatoes (Jian et al., 2019), strawberries (Wang et al., 2020a), peaches (Khan et al., 2022), kiwifruits (Ampomah-Dwamena et al., 2019), and apples (Honda and Moriya, 2018). There is an enormous list of several other TF families that have been associated with fruit ripening including basic leucine-zippers (Wu et al., 2019), WRKYs (Gan et al., 2021), zinc finger proteins (Khaksar et al., 2019), GRAS (Liu et al., 2021), and TCP (Guo et al., 2018).

1.5.4 EPIGENETIC CONTROL OF FRUIT RIPENING

Recently, the role of epigenetic modifications in the control of the timing of fruit ripening comes to fruition. Epigenetic modifications such as DNA methylation dynamics, chromatin modeling, and histone acetylation and/or methylation are thought to govern the availability of genes for transcription (Li et al., 2021). For instance, the spontaneous tomato mutant *Cnr* results from hypermethylation in the promoter of squamosa binding protein (SPL)-CNR TF (Wang et al., 2020b), and this epigenetic change inhibits fruit ripening. Further, the generation of single-base methylomes has confirmed that a decrease in DNA methylation is required for fruit ripening to be initiated in tomatoes (Zhong et al., 2013). The enzyme DNA demethylase 2 (DML2) is responsible for the decrease in global DNA methylation levels (Lang et al., 2017), and *SlDML2* transcript levels increase dramatically during fruit ripening in tomatoes (Liu et al., 2015; Lang et al., 2017). In addition, *Sldml2* tomato mutants that were generated via CRISPR/Cas9 knockout failed to exhibit normal ripening (Lang et al., 2017). Interestingly, the reverse pattern was observed in citrus fruits and strawberries as DNA methylation levels increased during fruit ripening (Cheng et al., 2018; Huang et al., 2019).

1.5.5 REGULATORY FACTORS

1.5.5.1 THE EFFECT OF WATER LOSS

As mentioned earlier, horticultural crops have a high proportion of water which contributes to their succulent nature. A large amount of this water is lost during postharvest handling through transpiration (Yahia and Carrillo-Lopez, 2018), and this leads to water-deficit stress. It has been shown that water stress is a key contributor to ethylene production by the fruit (Figure 1.2), which then accelerates the ripening process. This was well demonstrated in persimmons (Nakano, 2002; Nakano et al., 2003) where packaging in perforated polyethylene bags or exposure to humidified airflow reduced the rate of water loss and noticeably delayed the initiation of ethylene production (Figure 1.2). In kiwifruits, Huang et al. (2021) recently demonstrated that water loss during storage induced premature ethylene production which accelerated fruit softening. Similar observations have been reported in jujubes (Dong et al., 2022) and strawberries (Jia et al., 2020). Furthermore, postharvest techniques that are widely known to reduce water loss in horticultural crops, such as modified atmosphere packaging (Park et al., 2018) and application of edible coatings/films (Thakur et al., 2019; Ali et al., 2022a, 2022b; Hira et al., 2022) also delay ethylene production and fruit ripening.

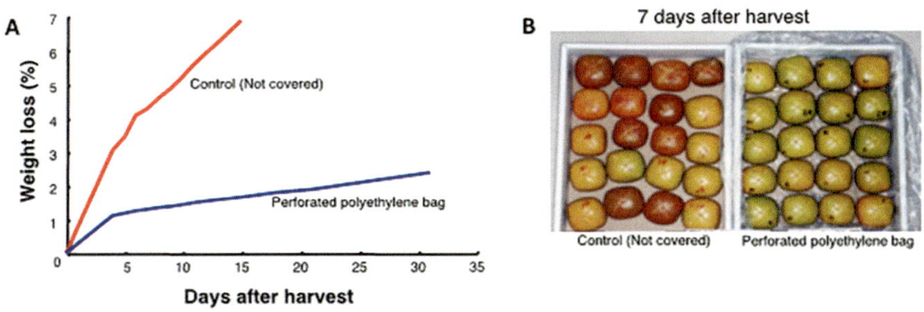

FIGURE 1.2 An illustration of the role of water stress in fruit ripening induction in persimmons. (A) Simple covering of persimmons with a perforated polyethylene bag noticeably reduced weight loss during storage. (B) Fruit that lost more weight (control noncovered) ripened faster than those covered with perforated polyethylene bag.

Source: Adapted from Nakano et al. (2003).

1.5.5.2 THE EFFECT OF CO_2 AND O_2 LEVELS

As horticultural crops are living organisms with functional respiratory systems, the concentration of CO_2 and O_2 in the ambient atmosphere affects postharvest quality. This is particularly important for fruit ripening since the respiration rate is known to increase during the ripening process (Khaliq, 2015; Kou and Wu, 2018). Indeed, high CO_2 and low O_2 concentrations were shown to inhibit fruit ripening (Kolati et al., 2020; Krupa and Tomala, 2021). These conditions (high CO_2 and low O_2) can be achieved by utilizing active modified atmosphere storage/packaging or controlled atmosphere storage (Belay et al., 2018). The use of edible coatings/films has also been shown to cause adjustments to internal CO_2 and O_2 concentrations, which in turn contribute to fruit ripening inhibition (Park, 1999). However, too much CO_2 and O_2 can damage the skin and flesh of fruit, in addition to promoting anaerobic respiration and the subsequent accumulation of alcohols and off-flavors (Schlie et al., 2020). There are also cases where high CO_2 levels can cause stress to the fruit which leads to accelerated softening, like is the case with persimmons after astringency removal procedure using CO_2 treatments (Fathi-Najafabadi et al., 2020).

1.5.5.3 POSTHARVEST DISEASES

During storage, fruits and vegetables are prone to attack by a myriad of pathogens that include mostly fungi and to some extent bacteria. The main postharvest pathogens that cause diseases in fruits across the world include *Penicillium* spp., *Botrytis* spp., *Colletotrichum* spp., *Fusarium* spp., *Aspergillus* spp., *Alternaria* spp., and *Pseudomonas* spp. (Huang et al., 2021). In addition to causing the well-known decay and rot in the affected fruits, these pathogens trigger a stress response in the fruits which then leads to activation of ethylene production and premature ripening (Figure 1.3). The effect of disease-induced ethylene is well exemplified by observations in kiwifruit (Asiche et al., 2018), where fruit stored in a group appeared to produce ethylene simultaneously within 1–3 weeks due to the presence of rot symptoms in a select few. When the kiwifruits were stored separately (about 10 cm apart from each other), a large proportion of healthy intact fruit did not produce detectable ethylene even after 56 days at 20°C.

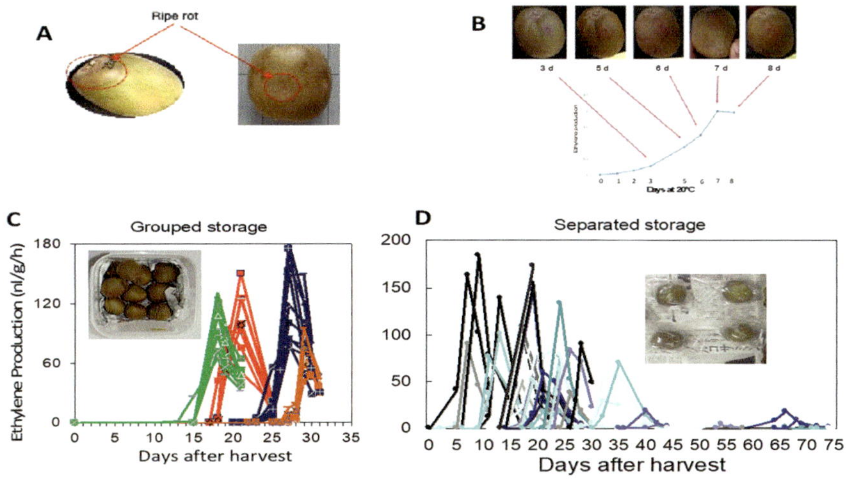

FIGURE 1.3 Postharvest diseases can trigger ethylene production, and subsequently, fruit ripening. (A) Postharvest kiwifruits are susceptible to ripe rot disease caused by *Botryosphaeria* spp., *Botrytis cinerea*, and *Phomopsis* spp. (B) These pathogens result in increased production of ethylene. (C) When kiwifruits are stored the conventional way (grouped), ethylene emanating from disease-infected fruits will cause healthy intact fruit to produce large amounts of ethylene leading to a seemingly synchronized ripening process. (D) Fruit that are separated from each other by about 10 cm start to produce ethylene haphazardly and a great percentage of healthy intact kiwifruits do not produce ethylene even after 75 days at 20°C.

Source: Reprinted with permission from Asiche et al. (2018).

1.5.5.4 LOW TEMPERATURE

In the horticulture industry, the most widely used technology to prolong postharvest life is cold storage, primarily due to the multifaceted effects of low temperature. It is widely recognized that low temperature slows most deteriorative metabolic processes in the cell especially processes associated with fruit ripening, respiration, and senescence (Figure 1.4; Rao, 2015). The retardatory effect of low temperature on fruit ripening can be thought of on two fronts. First, low temperature markedly delays the onset of fruit ripening-associated system II ethylene biosynthesis by inhibiting both the transcription of ethylene biosynthetic genes and the activities of associated enzymes (Gross et al., 2016). Second, low temperature restrains the growth of most pathogens that may damage the quality of horticultural crops (Vico et al., 2011; Manning et al., 2016), and this in turn alleviates the potentially disastrous effects of disease stress-induced ethylene.

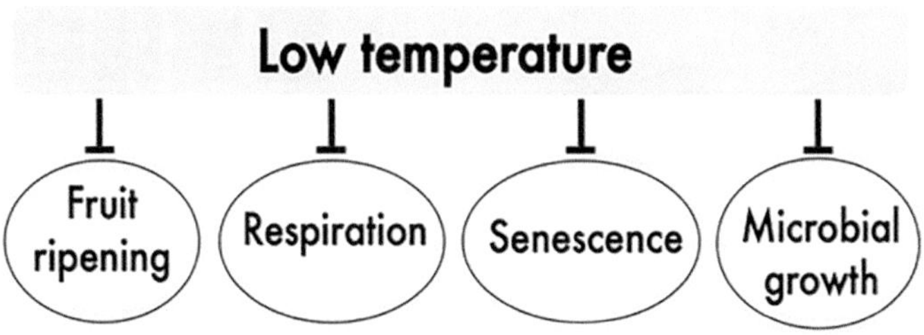

FIGURE 1.4 A simple illustration of the beneficial effects of cold storage on the postharvest life of horticultural crops. Low temperature (due to cold storage) inhibits most of the cell metabolic processes in addition to inhibiting the growth of postharvest pathogens.

Despite the above widely recognized inhibitory effects, it has been reported that low temperature can also stimulate or enhance the ripening process in some fruit species. For instance, most European pears, especially the winter cultivars lack the capacity to produce system II ethylene even when harvested at the appropriate maturity stage (Villalobolos-Acuña and Mitcham, 2008); low-temperature treatments are generally used to stimulate ethylene biosynthesis, and subsequently, fruit ripening (Nham et al., 2017; Mditshwa et al., 2023). In fact, late-maturing European pear cultivars like "Passe Crassane" rely on low temperature so heavily that relatively longer cold storage periods are required for them to initiate fruit ripening (El-Sharkawy et al., 2004). Years of extensive research have shown that low temperature activates the expression of various genes associated with ethylene biosynthesis, cell wall modifying enzymes and proteins, as well as some TF (Fonseca et al., 2005; Nham et al., 2017). Low-temperature exposure was also shown to activate the expression of ethylene biosynthetic genes, together with the activities of associated enzymes, in apples (Lara and Vendrell, 2003). However, the most unresolved question has been how low temperature can manage to trigger these changes in gene expression. Mitalo et al. (2019a) recently revealed that certain genes respond exclusively to low temperature during cold-induced ripening in "Passe Crassane" fruit, and this knowledge hopefully brings us one step closer to identifying the underlying regulatory mechanisms. It is, therefore, possible that low temperature first triggers the expression of certain unique genes that are then responsible for induction of ethylene biosynthesis and potentiating ripening-related ethylene responses in European pears.

Although classified as climacteric fruits, kiwifruits start system II ethylene production long after fruit softening has occurred (Richardson et al., 2011). Kiwifruits also soften extensively during cold storage even when ethylene levels are undetectable (Yin et al., 2009). It was previously thought that ethylene signaling is involved in this peculiar ripening behavior exhibited by kiwifruits, since prior studies had indicated their sensitivity to extremely low ethylene amounts (Jabbar and East, 2016). Moreover, a closely related theory suggested the possibility of low temperature triggering ethylene biosynthesis and increasing fruit sensitivity to ethylene (Murakami et al., 2014; Minas et al., 2016), similar to the case in European pears. By using potent inhibitors of ethylene signaling (1-methylcyclopropene, 1-MCP), together with elimination of the effect of ethylene from external sources including disease-induced, it was recently demonstrated that low temperature could regulate fruit ripening in kiwifruits independently of ethylene (Asiche et al., 2018; Mitalo et al., 2018, 2019b). Fruit stored at lower temperatures (5, 10, and 15°C) softened and reduced their acidity faster than fruit stored at room temperature (Mitalo et al., 2019b), and frequent treatments with 1-MCP to block ethylene signaling did not inhibit low-temperature-induced ripening changes. Further, RNA-Seq analysis revealed that the expression of certain genes increased in response to low temperature independently of ethylene. Interestingly, low temperature could not stimulate the release of aroma volatiles, which were shown to be strictly under ethylene regulation in kiwifruit (Mitalo et al., 2019c).

Several studies have also revealed that low temperature is involved in the peel degreening process in citrus fruits. Generally, peel degreening was enhanced both during cold storage and cold periods in the field in several citrus species including oranges, mandarins, lemons, and grapefruits (Matsumoto et al., 2009; Manera et al., 2012, 2013; Rodrigo et al., 2013; Conesa et al., 2019). In all cases, low-temperature promotion of peel color change was attributed to stimulation of carotenoid accumulation and activation of the expression of various carotenoid biosynthetic genes. It was initially thought that ethylene is involved in the low-temperature-induced changes, that is, basal levels of system I ethylene present in most tissues were physiologically active (Goldschmidt et al., 1993; Carmona et al., 2012) and considering the fact that exogenous ethylene treatments can trigger peel degreening in citrus fruits (Porat, 2008). However, recent studies in "Allen Eureka" lemons and "Aoshima" satsuma mandarins have shown that low temperature can trigger peel degreening independently of ethylene by activating the expression of genes associated with chlorophyll degradation, carotenoid biosynthesis, photosystem proteins, phytohormones, and TFs (Mitalo et al., 2020, 2022).

1.6 CONCLUSION

As living organisms, fresh horticultural crops are faced with a multitude of challenges that affect their postharvest quality. From a wealth of information generated from various studies, we now have a basic understanding of the biological processes taking place in horticultural crops during postharvest handling, as well as the various factors affecting their rates. This knowledge has increased our ability to develop appropriate and tailor-made postharvest strategies aimed at delaying the progress of these biological processes, thereby maintaining quality and extending postharvest life. However, there is a need to strike a balance between delaying the above biological processes and compromising the overall quality and usability of the commodities. For instance, we have seen that low temperature is arguably the most widely used technique to extend the postharvest life of most horticultural crops, because of its ability to slow down most of the metabolic activities. However, extended cold storage triggers chilling injury which instead leads to losses, especially in fruits and vegetable crops. This necessitates the development of novel strategies to either reduce the incidence of chilling injury or to increase innate cold tolerance in the susceptible crops. Further, there is still much to discover within the field of postharvest physiology as well as postharvest technology as they relate to the main goal of reducing food losses and improving food and nutrition security.

KEYWORDS

- **ethylene**
- **respiration**
- **ripening**
- **temperature**
- **transpiration**

REFERENCES

Akhtar, I., & Rab, A. (2015). Effect of fruit ripening stages on strawberry (*Fragaria x ananassa*. Duch) fruit quality for fresh consumption. *Journal of Agricultural Research, 53*, 413–422.

Ali, S., Anjum, M. A., Khan, A. S., Nawaz, A., Ejaz, S., Khaliq, G. et al. (2022a). Carboxymethyl cellulose coating delays ripening of harvested mango fruits by regulating softening enzymes activities. *Food Chemistry,* 380, 131804.

Ali, S., Khaliq, G., Iftikhar, A., Mir, S. A., & Iqbal, S. (2024). Nanoemulsions-based edible coatings for postharvest quality preservation of fruits and vegetables. In *Sustainable Postharvest Technologies for Fruits and Vegetables* (pp. 213-226). CRC Press.

Ali, S., Ullah, M. A., Nawaz, A., Naz, S., Shah, A. A., Gohari, G. et al. (2022b). Carboxymethyl cellulose coating regulates cell wall polysaccharides disassembly and delays ripening of harvested banana fruit. *Postharvest Biology and Technology, 191*, 111978.

Ampomah-Dwamena, C., Thrimawithana, A. H., Dejnoprat, S., Lewis, D., Espley, R. V., & Allan, A. C. (2019). A kiwifruit (*Actinidia deliciosa*) R2R3-MYB transcription factor modulates chlorophyll and carotenoid accumulation. *New Phytologist, 221*, 309–325.

Arafa, G. K. (2019). Effects of moisture content on some physical and mechanical properties of apricot fruits and kernels. *Misr Journal of Agricultural Engineering, 36*, 1209–1226.

Asiche, W. O., Mitalo, O. W., Kasahara, Y., Tosa, Y., Mworia, E. G., Owino, W. O. et al. (2018). Comparative transcriptome analysis reveals distinct ethylene–independent regulation of ripening in response to low temperature in kiwifruit. *BMC Plant Biology, 18*, 1–18.

Bai, Q., Huang, Y., & Shen, Y. (2021). The physiological and molecular mechanism of abscisic acid in regulation of fleshy fruit ripening. *Frontiers in Plant Science, 11*, 619953.

Bal, E. (2013). Postharvest application of chitosan and low temperature storage affect respiration rate and quality of plum fruits. *Journal of Agricultural Science and Technology, 15*, 1219–1230.

Batista-Silva, W., Nascimento, V. L., Medeiros, D. B., Nunes-Nesi, A., Ribeiro, D. M., Zsögön, A., & Araújo, W. L. (2018). Modifications in organic acid profiles during fruit development and ripening: Correlation or causation? *Frontiers in Plant Science, 9*, 1689.

Becerra-Moreno, A., Redondo-Gil, M., Benavides, J., Nair, V., Cisneros-Zevallos, L., & Jacobo-Velázquez, D. A. (2015). Combined effect of water loss and wounding stress on gene activation of metabolic pathways associated with phenolic biosynthesis in carrot. *Frontiers in Plant Science, 6*, 1–15.

Belay, Z. A., Caleb, O. J., Mahajan, P. V., & Opara, U. L. (2018). Design of active modified atmosphere and humidity packaging (MAHP) for 'wonderful' pomegranate arils. *Food and Bioprocess Technology, 11*, 1478–1494.

Bemer, M., Karlova, R., Ballester, A. R., Tikunov, Y. M., Bovy, A. G., Wolters-Arts, M. et al. (2012). The tomato fruitfull homologs TDR4/FUL1 and MBP7/FUL2 regulate ethylene-independent aspects of fruit ripening. *Plant Cell, 24*, 4437–4451.

Bovi, G. G., & Herppich, W. B. (2021). Keeping fruits and vegetables fresh by limiting respiration and transpiration. *Frontiers for Young Minds, 9*, 576906.

Bovi, G. G., Rux, G., Caleb, O. J., Herppich, W. B., Linke, M., Rauh, C., & Mahajan, P. V. (2018). Measurement and modelling of transpiration losses in packaged and unpackaged strawberries. *Biosystems Engineering, 174*, 1–9.

Campbell, C., Atkinson, L., Zaragoza-Castells, J., Lundmark, M., Atkin, O., & Hurry, V. (2007). Acclimation of photosynthesis and respiration is asynchronous in response to changes in temperature regardless of plant functional group. *New Phytologist, 176*, 375–389.

Cantu, D., Blanco-Ulate, B., Yang, L., Labavitch, J. M., Bennett, A. B., & Powell, A. L. (2009). Ripening-regulated susceptibility of tomato fruit to *Botrytis cinerea* requires *NOR* but not *RIN* or ethylene. *Plant Physiology, 150*, 1434–1449.

Cao, X., Wei, C., Duan, W., Gao, Y., Kuang, J., Liu, M. et al. (2021). Transcriptional and epigenetic analysis reveals that NAC transcription factors regulate fruit flavor ester biosynthesis. *Plant Journal, 106*, 785–800.

Carmona, L., Rodrigo, M. J., & Zacarías, L. (2012). Exploring the involvement of ethylene in the regulation of color changes in citrus fruit. *Acta Horticulture, 934*, 879–885.

Castellarin, S. D., Pfeiffer, A., Sivilotti, P., Degan, M., Peterlunger, E., & Di Gaspero, G. (2007). Transcriptional regulation of anthocyanin biosynthesis in ripening fruits of grapevine under seasonal water deficit. *Plant Cell Environment, 30*, 1381–1399.

Centeno, D. C., Osorio, S., Nunes-Nesi, A., Bertolo, A. L., Carneiro, R. T., Araújo, W. L. et al. (2011). Malate plays a crucial role in starch metabolism, ripening, and soluble solid content of tomato fruit and affects postharvest softening. *Plant Cell, 23*, 162–184.

Chen, T., Ji, D., Zhang, Z., Li, B., Qin, G., & Tian, S. (2021). Advances and strategies for controlling the quality and safety of postharvest fruit. *Engineering, 7*, 1177–1184.

Chen, T., Qin, G., & Tian, S. (2020). Regulatory network of fruit ripening: Current understanding and future challenges. *New Phytologist, 228*, 1219–1226.

Cheng, J., Niu, Q., Zhang, B., Chen, K., Yang, R., Zhu, J. K. et al. (2018). Downregulation of RdDM during strawberry fruit ripening. *Genome Biology, 19*, 1–14.

Chomchalow, S., El Assi, N. M., Sargent, S. A., & Brecht, J. K. (2002). Fruit maturity and timing of ethylene treatment affect storage performance of green tomatoes at chilling and nonchilling temperatures. *HortTechnology, 12*, 104–114.

Chourasia, M. K., Maji, P., Baskey, A., Goswami, T. K. (2005). Estimation of moisture loss from the cooling data of potatoes. *Journal of Food Process Engineering, 28*, 397–416.

Conesa, A., Manera, F. C., Brotons, J. M., Fernandez-Zapata, J. C., Simón, I., Simón-Grao, S., Alfosea-Simón, M., Nicolása, J. J. M., Valverde, J. M., & García-Sanchez, F. (2019). Changes in the content of chlorophylls and carotenoids in the rind of 'Fino 49' lemons during maturation and their relationship with parameters from the CIELAB color space. *Scientia Horticulturae, 243*, 252–260.

Cutting, J. G. M., & Wolstenholme, B. N. (1992). Maturity and water loss effects on avocado (*Persea americana* Mill.) postharvest physiology in cool environments. *Journal of Horticultural Sciences, 67*, 569–575.

DeEll, J. R., Prange, R. K., & Peppelenbos, H. W. (2003). Postharvest physiology of fresh fruits and vegetables. In *Handbook of postharvest technology: Cereals, fruits, vegetables, tea and spices* (pp. 455–483). Dekker.

Defilippi, B. G., Manriquez, D., Luengwilai, K., & Gonzalez-Agueero, M. (2009). Aroma volatiles: Biosynthesis and mechanisms of modulation during fruit ripening. *Advances in Botanical Research, 50*, 1–37.

Díaz-Pérez, J. C., Muy-Rangel, M. D., & Mascorro A. G. (2007). Fruit size and stage of ripeness affect postharvest water loss in bell pepper fruit (*Capsicum annuum* L.). *Journal of the Science of Food and Agriculture, 87*, 68–73.

Dong, X., Tang, H., Zhang, Q., Zhang, C., & Wang, Z. (2022). Transcriptomic analyses provide new insights into jujube fruit quality affected by water deficit stress. *Scientia Horticulturae,* 291, 110558.

Duan, Y., Yang, L., Zhu, H., Zhou, J., Sun, H., & Gong, H. (2021). Structure and expression analysis of sucrose phosphate synthase, sucrose synthase and invertase gene families in Solanum lycopersicum. *International Journal of Molecular Sciences, 22*, 4698.

El Hadi, M. A. M., Zhang, F. J., Wu, F. F., Zhou, C. H., & Tao, J. (2013). Advances in fruit aroma volatile research. *Molecules* 18, 8200–8229.

El-Sharkawy, I., Jones, B., Gentzbittel, L., Lelièvre, J. M., Pech, J. C., & Latché, A. (2004). Differential regulation of ACC synthase genes in cold-dependent and -independent ripening in pear fruit. *Plant, Cell Environment, 27*, 1197–1210.

Fathi-Najafabadi, A., Salvador, A., Navarro, P., Gil, R., & Besada, C. (2020). Effect of temperature during and immediately after CO_2-deastringency treatment on internal flesh browning after cold storage of persimmon fruit. *Scientia Horticulturae, 268*, 109363.

Fonseca, S., Monteiro, L., Barreiro, M. G., & Pais, M. S. (2005). Expression of genes encoding cell wall modifying enzymes is induced by cold storage and reflects changes in pear fruit texture. *Journal of Experimental Botany, 56*, 2029–2036.

Forlani, S., Masiero, S., & Mizzotti, C. (2019). Fruit ripening: The role of hormones, cell wall modifications, and their relationship with pathogens. *Journal of Experimental Botany, 70*, 2993–3006.

Fuentes, L., Figueroa, C. R., & Valdenegro, M. (2019). Recent advances in hormonal regulation and cross-talk during non-climacteric fruit development and ripening. *Horticulture, 5*, 45.

Fujisawa, M., Nakano, T., Shima, Y., & Ito, Y. (2013). A large-scale identification of direct targets of the tomato MADS box transcription factor ripening inhibitor reveals the regulation of fruit ripening. *Plant Cell, 25*, 371–386.

Gan, Z., Yuan, X., Shan, N., Wan, C., Chen, C., Xu, Y. et al. (2021). AcWRKY40 mediates ethylene biosynthesis during postharvest ripening in kiwifruit. *Plant Science, 309*, 110948.

Gao, J., Zhang, Y., Li, Z., & Liu, M. (2020). Role of ethylene response factors (ERFs) in fruit ripening. *Food Quality and Safety, 4*, 15–20.

Gao, Y., Zhu, N., Zhu, X., Wu, M., Jiang, C. Z., Grierson, D. et al. (2019). Diversity and redundancy of the ripening regulatory networks revealed by the fruit ENCODE and the new CRISPR/Cas9 CNR and NOR mutants. *Horticulture Research, 6*.

Goldschmidt, E. E., Huberman, M., & Goren, R. (1993). Probing the role of endogenous ethylene in the degreening of citrus fruit with ethylene antagonists. *Plant Growth Regulation, 12*, 325–329.

Golisz, E., Jaros, M., & Kalicka, M. (2013). Analysis of convectional drying process of peach. *Technical Sciences, 16*, 333–343.

Gross, K. C., Wang, C. Y., & Saltveit, M. E. (2016). *The commercial storage of fruits, vegetables, and florist and nursery stocks*. United States Department of Agriculture, Agricultural Research Service.

Guo, Z. H., Shu, W. S., Cheng, H. Y., Wang, G. M., Qi, K. J., Zhang, S. L., & Gu, C. (2018). Expression analysis of TCP genes in peach reveals an involvement of PpTCP. A2 in ethylene biosynthesis during fruit ripening. *Plant Molecular Biology Reporter, 36*, 588–595.

Hira, N., Mitalo, O. W., Okada, R., Sangawa, M., Masuda, K., & Fujita, N. (2022). The effect of layer-by-layer edible coating on the shelf life and transcriptome of 'Kosui' Japanese pear fruit. *Postharvest Biology and Technology, 185*, 111787.

Ho, P. L., Tran, D. T., Hertog, M. L., & Nicolaï, B. M. (2020). Modelling respiration rate of dragon fruit as a function of gas composition and temperature. *Scientia Horticulturae,* 263,109138

Ho, Q. T., Hertog, M. L., Verboven, P., Ambaw, A., Rogge, S., Verlinden, B. E., & Nicolaï, B. M. (2018). Down-regulation of respiration in pear fruit depends on temperature. *Journal of Experimental Botany, 69*, 2049–2060.

Honda, C., & Moriya, S. (2018). Anthocyanin biosynthesis in apple fruit. *Horticulture Journal, 87*, 305–314.

Hu, S., Liu, L., Li, S., Shao, Z., Meng, F., Liu, H. et al. (2020). Regulation of fruit ripening by the brassinosteroid biosynthetic gene SlCYP90B3 via an ethylene-dependent pathway in tomato. *Horticulture Research, 7.*

Hu, X., Kuang, S., Zhang, A. D., Zhang, W. S., Chen, M. J., Yin, X. R., & Chen, K. S. (2016). Characterization of starch degradation related genes in postharvest kiwifruit. *International Journal of Molecular Sciences, 17*, 2112.

Huang, X., Ren, J., Li, P., Feng, S., Dong, P., & Ren, M. (2021). Potential of microbial endophytes to enhance the resistance to postharvest diseases of fruit and vegetables. *Journal of the Science of Food and Agriculture, 101*(5), 1744–1757.

Huang, W., Hu, N., Xiao, Z., Qiu, Y., Yang, Y., Yang, J. et al. (2022). A molecular framework of ethylene-mediated fruit growth and ripening processes in tomato. *Plant Cell,* koac146. https://doi.org/10.1093/plcell/koac146.

Iqbal, N., Khan, N. A., Ferrante, A., Trivellini, A., Francini, A., & Khan, M. I. R. (2017). Ethylene role in plant growth, development and senescence: Interaction with other phytohormones. *Frontiers in Plant Science, 8*, 475.

Iqbal, S., Ni, X., Bilal, M. S., Shi, T., Khalil-ur-Rehman, M., Zhenpeng, P. et al. (2020). Identification and expression profiling of sugar transporter genes during sugar accumulation at different stages of fruit development in apricot. *Gene, 742*, 144584.

Itkin, M., Seybold, H., Breitel, D., Rogachev, I., Meir, S., & Aharoni, A. (200(). Tomato agamous-like 1 is a component of the fruit ripening regulatory network. *Plant Journal, 60*, 1081–1095.

Ito, Y., Nishizawa-Yokoi, A., Endo, M., Mikami, M., Shima, Y., Nakamura, N. et al. (2017). Re-evaluation of the rin mutation and the role of RIN in the induction of tomato ripening. *Nature Plants, 3*, 866–874.

Jabbar, A., & East, A. R. (2016). Quantifying the ethylene induced softening and low temperature breakdown of 'Hayward' kiwifruit in storage. *Postharvest Biology and Technology, 113*, 87–94.

Jia, K., Zhang, Q., Xing, Y., Yan, J., Liu, L., & Nie, K. (2020). A development-associated decrease in osmotic potential contributes to fruit ripening initiation in strawberry (*Fragaria ananassa*). *Frontiers in Plant Science, 11*, 1035.

Jian, W., Cao, H., Yuan, S., Liu, Y., Lu, J., Lu, W., Li, N., Wang, J., Zou, J., Tang, N., & Xu, C. (2019). SlMYB75, an MYB-type transcription factor, promotes anthocyanin accumulation and enhances volatile aroma production in tomato fruits. *Horticulture Research, 6.*

Kapoor, L., Simkin, A. J., George Priya Doss, C., & Siva, R. (2022). Fruit ripening: Dynamics and integrated analysis of carotenoids and anthocyanins. *BMC Plant Biology, 22*, 1–22.

Khaksar, G., Sangchay, W., Pinsorn, P., Sangpong, L., & Sirikantaramas, S. (2091). Genome-wide analysis of the Dof gene family in durian reveals fruit ripening-associated and cultivar-dependent Dof transcription factors. *Scientific Reports, 9*, 1–13.

Khaliq, G. H. U. L. A. M. (2015). Effects of edible coatings enriched with calcium chloride on physiological, biochemical and quality responses of mango (*Mangifera indica* L. cv. Choke Anan) fruit during cold storage. Universiti Putra Malaysia, Serdang, Selangor, Malaysia.

Khaliq, G., Ali, S., Gapper, N., & Nicola, S. (2023). Recent advances and approaches in the application of elicitors to enhance resistance mechanisms in fresh produce. *Frontiers in Sustainable Food Systems, 6*, 1061079.

Khan, D., Bhat, S., & Narayan, S. (2017). Storage methods for fruits and vegetables. *Missouri Botanical Garden,* 1–.9

Khan, I. A., Rahman, M. U., Sakhi, S., Nawaz, G., Khan, A. A., Ahmad, T. et al. (2022). PpMYB39 activates PpDFR to modulate anthocyanin biosynthesis during peach fruit maturation. *Horticulture, 8*, 332.

Khan, L., Khaliq, G., Ullah, E., Abdul-Rahaman, A., El-Mogy, M. M., Kesba, H. H., Ismail, A. M., & El-Beltagi, H. S. (2025). Guggul gum and clove essential oil synergistically reduced anthracnose disease and modulated physicochemical changes in tomato during storage. *Journal of Stored Products Research, 114*, 102713.

Klee, H. J., & Tieman, D. M. (2018). The genetics of fruit flavour preferences. *Nature Reviews Genetics, 19*, 347–356.

Knapp, B. D., & Huang, K. C. (2022). The effects of temperature on cellular physiology. *Annual Review of Biophysics, 51*, 499–526.

Kolati, S. B., Babu, D. R., Prasad, P. I., & Rao, K. N. (2020). Low oxygen and high carbon dioxide environment in controlled atmosphere cold storage: Recommendations for safe working conditions. *Plant Cell Biotechnology and Molecular Biology, 21*, 9–14.

Kou, X., & Wu, M. (2018). Characterization of climacteric and non-climacteric fruit ripening. In Y. Guo (Ed.), *Plant senescence* (Vol. 1744, pp. 89–102). Humana Press.

Kou, X., Zhou, J., Wu, C. E., Yang, S., Liu, Y., Chai, L., & Xue, Z. (2021). The interplay between ABA/ethylene and NAC TFs in tomato fruit ripening: A review. *Plant Molecular Biology, 106*, 223–238.

Kritzinger, I., & Lötze, E. (2019). Quantification of lenticels in Japanese plum cultivars and their effect on total fruit peel permeance. *Scientia Horticulturae, 254*, 35–39.

Krupa, T., & Tomala, K. (2021). Effect of oxygen and carbon dioxide concentration on the quality of minikiwi fruits after storage. *Agronomy 11*, 2251.

Kuai, B., Chen, J., & Hörtensteiner, S. (2018). The biochemistry and molecular biology of chlorophyll breakdown. *Journal of Experimental Botany, 69*, 751–767.

Kumar, N., Tokas, J., Raghavendra, M., & Singal, H. R. (2021). Impact of exogenous salicylic acid treatment on the cell wall metabolism and ripening process in postharvest tomato fruit stored at ambient temperature. *International Journal of Food Science and Technology, 56*, 2961–2972.

Lai, T., Wang, X., Ye, B., Jin, M., Chen, W., Wang, Y. et al. (2020). Molecular and functional characterization of the SBP-box transcription factor SPL-CNR in tomato fruit ripening and cell death. *Journal of Experimental Botany, 71,* 2995–3011.

Lal, S., Khaliq, G., Mastoi, A. H., Irshad, A., Jaffar, M. A., Khan, L., & Muhammad, N. (2025). Effect of *Fagonia indica* and garlic extract on *Alternaria alternata* and post-harvest quality of tomato fruit during storage. *Journal of Crop Health, 77*(6), 1–13.

Lang, Z., Wang, Y., Tang, K., Tang, D., Datsenka, T., Cheng, J. et al. (2017). Critical roles of DNA demethylation in the activation of ripening-induced genes and inhibition of ripening-repressed genes in tomato fruit. *Proceedings of the National Academy of Sciences, 114*, E4511–E4519.

Lara, I., & Vendrell, M. (2003). Cold-induced ethylene biosynthesis is differentially regulated in peel and pulp tissues of 'Granny Smith' apple fruit. *Postharvest Biology and Technology, 29*, 109–119.

Laties, G. G. (2019). The development and control of respiratory pathways in slices of plant storage organs. In *Biochemistry of wounded plant tissues* (Vol. 1, pp. 421–466).

Léchaudel, M., Lopez-Lauri, F., Vidal, V., Sallanon, H., & Joas, J. (2013). Response of the physiological parameters of mango fruit (transpiration, water relations and antioxidant system) to its light and temperature environment. *Journal of Plant Physiology, 170*, 567–576.

Li, B., Fan, R., Yang, Q., Hu, C., Sheng, O., Deng, G. et al. (2020). Genome-wide identification and characterization of the NAC transcription factor family in Musa acuminata and expression analysis during fruit ripening. *International Journal of Molecular Sciences, 21*, 634.

Li, S., Chen, K., & Grierson, D. (2019). A critical evaluation of the role of ethylene and MADS transcription factors in the network controlling fleshy fruit ripening. *New Phytologist, 221*, 1724–1741.

Li, S., Chen, K., & Grierson, D. (2021). Molecular and hormonal mechanisms regulating fleshy fruit ripening. *Cells, 10*, 1136.

Li, S., Zhu, B., Pirrello, J., Xu, C., Zhang, B., Bouzayen, M. et al. (2022). Roles of RIN and ethylene in tomato fruit ripening and ripening-associated traits. *New Phytologist, 226*, 460–475.

Li, X., Wang, X., Zhang, Y., Zhang, A., & You, C. X. (2022). Regulation of fleshy fruit ripening: From transcription factors to epigenetic modifications. *Horticulture Research, 9*.

Lin, Z., Hong, Y., Yin, M., Li, C., Zhang, K., & Grierson, D. (2008). A tomato HD-Zip homeobox protein, LeHB-1, plays an important role in floral organogenesis and ripening. *Plant Journal, 55*, 301–310.

Linke, M., & Geyer, M. (2013). Condensation dynamics in plastic film packaging of fruit and vegetables. *Journal of Food Engineering, 116*(1), 144–154.

Liu, M., Pirrello, J., Chervin, C., Roustan, J. P., & Bouzayen, M. (2015). Ethylene control of fruit ripening: Revisiting the complex network of transcriptional regulation. *Plant Physiology, 169*, 2380–2390.

Liu, Y., Shi, Y., Su, D., Lu, W., & Li, Z. (2021). SlGRAS4 accelerates fruit ripening by regulating ethylene biosynthesis genes and SlMADS1 in tomato. *Horticulture Research, 8*.

Liu, Y., Tang, M., Liu, M., Su, D., Chen, J., Gao, Y. et al. (2020). The molecular regulation of ethylene in fruit ripening. *Small Methods, 4*, 1900485.

Lownds, N.K, Banaras, M., & Bosland, P. W. (1993). Relationships between postharvest water loss and physical properties of pepper fruit (*Capsicum annuum* L.). *Horticultural Science, 28*, 182–184.

Lufu, R., Ambaw, A., & Opara, U. L. (2020). Water loss of fresh fruit, Influencing pre-harvest, harvest and postharvest factors. *Scientia Horticulturae, 272*, 109519.

Maguire, K. M., Banks, N. H., & Opara, L. U. (2010). Factors affecting weight loss of apples. *Horticultural Review: American Society for Horticultural Science, 25*, 197–234.

Mahajan, P. V., & Goswami, T. K. (2001). PH—Postharvest technology, enzyme kinetics based modelling of respiration rate for apple. *Journal of Agricultural Engineering, Res. 79*, 399–406.

Manera, F. J., Brotons, J. M., Conesa, A., & Porras, I. (2012). Influence of temperature on the beginning of degreening in lemon peel. *Scientia Horticulturae, 145*, 34–38.

Manera, F. J., Brotons, J. M., Conesa, A., & Porras, I. (2013). Relation between temperature and the beginning of peel color change in grapefruit (*Citrus paradisi* Macf.). *Scientia Horticulturae, 160*, 292–299.

Manjunatha, M., & Anurag, R. K. (2014). Effect of modified atmosphere packaging and storage conditions on quality characteristics of cucumber. *Journal of Food Science and Technology, 51*, 3470–3475.

Manning, M., Burdon, J., De Silva, N., Meier, X., Pidakala, P., Punter, M., & Billing, D. (2015). Maturity and postharvest temperature management affect rot expression in 'Hort16A' kiwifruit. *Postharvest Biology and Technology, 113*, 40–47.

Maria, T., Tsaniklidis, G., Delis, C., Nikolopoulou, A. E., Nikoloudakis, N., Karapanos, I., Aivalakis, G. (2016). Gene transcript accumulation and enzyme activity of β-amylases suggest involvement in the starch depletion during the ripening of cherry tomatoes. *Plant Gene, 5*, 8–12.

Maringgal, B., Hashim, N., Mohamed-Amin-Tawakkal, I. S., Mohamed, M. T. M., Hamzah, M. H., & Mohd-Ali, M. (2021). Effect of Kelulut honey nanoparticles coating on the changes of respiration rate, ascorbic acid, and total phenolic content of papaya (*Carica papaya* L.) during cold storage. *Foods, 10*, 432–448.

Maringgal, B., Hashim, N., Tawakkal, I. S. M. A., Mohamed, M. T. M., Hamzah, M. H. Ali, M. M., & Abd Razak, M. F. H. (2020). Kinetics of quality changes in papayas (*Carica papaya* L.) coated with Malaysian stingless bee honey. *Scientia Horticulturae, 267*, 109321.

Matsumoto, H., Ikoma, Y., Kato, M., Nakajima, N., & Hasegawa, Y. (2009). Effect of postharvest temperature and ethylene on carotenoid accumulation in the flavedo and juice sacs of Satsuma mandarin (*Citrus unshiu* Marc.) fruit. *Journal of Agricultural and Food Chemistry, 57*, 4724–4732.

Mditshwa, A., Khaliq, G., Hussein, Z., & Ejaz, S. (2023). Sustainable postharvest management practices for fresh produce. *Frontiers in Sustainable Food Systems, 7*, 1143759.

Medina-Puche, L., Cumplido-Laso, G., Amil-Ruiz, F., Hoffmann, T., Ring, L., Rodríguez-Franco, A. et al. (2014). MYB10 plays a major role in the regulation of flavonoid/phenylpropanoid metabolism during ripening of Fragaria× ananassa fruits. *Journal of Experimental Botany, 65*, 401–417.

Minas, I. S., Tanou, G., Karagiannis, E., Belghazi, M., & Molassiotis, A. (2016). Coupling of physiological and proteomic analysis to understand the ethylene and chilling induced kiwifruit ripening syndrome. *Frontiers in Plant Science, 7*, 120.

Mitalo, O. W., Asiche, W. O., Kang, S. W., Ezura, H., Akagi, T., Kubo, Y., & Ushijima, K. (2022). Examining the role of low temperature in satsuma mandarin fruit peel degreening via comparative physiological and transcriptomic analysis. *Frontiers in Plant Science, 13*.

Mitalo, O. W., Asiche, W. O., Kasahara, Y., Tosa, Y., Owino, W. O., Mworia, E. G., Ushijima, K., Nakano, R., & Kubo, Y. (2018). Characterization of ripening-related genes involved in ethylene-independent low temperature-modulated ripening in 'Rainbow Red' kiwifruit during storage and on-vine. *Horticulture Journal, 87*, 421–429.

Mitalo, O. W., Tosa, Y., Tokiwa, S., Kondo, Y., Azimi, A., Hojo, Y. et al. (2019a). 'Passe Crassane' pear fruit (*Pyrus communis* L.) ripening: Revisiting the role of low temperature via integrated physiological and transcriptome analysis. *Postharvest Biology and Technology, 158*, 110949.

Mitalo, O. W., Asiche, W. O., Kasahara, Y., Tosa, Y., Tokiwa, S., Ushijima, K., Nakano, R., & Kubo, Y. (2019b). Comparative analysis of fruit ripening and associated genes in two kiwifruit cultivars ('Sanuki Gold' and 'Hayward') at various storage temperatures. *Postharvest Biology and Technology, 147*, 20–28.

Mitalo, O. W., Tokiwa, S., Kondo, Y. et al. (2019c). Low temperature storage stimulates fruit softening and sugar accumulation without ethylene and aroma volatile production in kiwifruit. *Frontiers in Plant Science, 10*, 888.

Mitalo, O. W., Otsuki, T., Okada, R., Obitsu, S., Masuda, K., Hojo, Y. et al. (2020). Low temperature modulates natural peel degreening in lemon fruit independently of endogenous ethylene. *Journal of Experimental Botany, 71*, 4778–4796.

Moalemiyan, M., & Ramaswamy, H. S. (2012). Quality retention and shelf-life extension in mediterranean cucumbers coated with a pectin-based film. *Journal of Food Research, 1*, 159–168.

Mukama, M., Ambaw, A., Berry, T. M., & Opara, U. L. (2019). Analysing the dynamics of quality loss during precooling and ambient storage of pomegranate fruit. *Journal of Food Engineering, 245*,166–173.

Murakami, S., Ikoma, Y., & Yano, M. (2014). Low temperature increases ethylene sensitivity in *Actinidia chinensis* 'Rainbow Red' kiwifruit. *Journal of the Japanese Society for Horticultural Science, 83*, 322–326.

Nakano. (200). *Involvement of water stress-induced ethylene in postharvest softening in Japanese persimmon (Diospyros kaki Thunb.) fruit*. PhD thesis. Kyoto University Kyoto.

Nakano, R., Ogura, E., Kubo, Y., & Inaba, A. (2003). Ethylene biosynthesis in detached young persimmon fruit is initiated in calyx and modulated by water loss from the fruit. *Plant Physiology, 131*, 276–286.

Nguyễn, T. T. T., Huỳnh, T. Đ., Trần, Q. Đ., Lương, V. H. Đỗ, T. T. H., & Nguyễn, H. H. (2020). Effects of drying methods on color retention and chlorophyll of celery (*Apium graveolens* L.), spinach (*Spinacia oleracea* L.), Malabar spinach (Basella alba L.). *Tạp. chí Nông. nghiệp. và Phát triển. 19*, 53–61.

Nham, N. T., Macnish, A. J., Zakharov, F., & Mitcham, E. J. (2017). 'Bartlett' pear fruit (*Pyrus communis* L.) ripening regulation by low temperatures involves genes associated with jasmonic acid, cold response, and transcription factors. *Plant Science, 260*, 8–18.

Ntuli, V., Chatanga, P., Kwiri, R., Gadaga, H. T., Gere, J., Matsepo, T., & Potloane, R. P. (2017). Microbiological quality of selected dried fruits and vegetables in Maseru, Lesotho. *African Journal of Microbiology Research, 11*, 185–193.

Nunes, M. C. N., Brecht, J. K., Morais, A. M. M. B., & Sargent S. A. (1995). Physical and chemical quality characteristics of strawberries after storage are reduced by a short delay to cooling. *Postharvest Biology and Technology, 6*, 17–28.

Nunes, M. C. N., Emond, J. P., Brecht, J. K., Dea S., & Proulx, E. (2007). Quality curves for mango fruit (cv. Tommy Atkins and Palmer) stored at chilling and nonchilling temperatures. *Journal of Food Quality, 30*, 104–120.

Oliveira, M. J. de, Laranjeira, D., Câmara, M. P. S., Laranjeira, F. F., Armengol. J., & Michereff, S. J. (2014). Effects of wounding, humidity, temperature, and inoculum concentrations on the severity of corky dry rot caused by Fusarium semitectum in melon fruits. *Acta Scientiarum. Agronomy, 36*, 281–289.

Ortega-Hernández, E., Nair, V., Welti-Chanes, J., Cisneros-Zevallos, L., & Jacobo-Velázquez, D. A. (2019). Wounding and UVB light synergistically induce the biosynthesis of phenolic compounds and ascorbic acid in red prickly pears (Opuntia ficus-indica cv. Rojo Vigor). *International Journal of Molecular Sciences, 20*, 5327.

Othman, O., & Mbogo, G. (2009). Physico-chemical characteristics of storage-ripened Mango (*Mangifera indica* L.) fruits varieties of eastern Tanzania. *Tanzania Journal of Science,* 57–65.

Park, D., Al Shoffe, Y., Algul, B. E., & Watkins C. B. (2022). Fermentative metabolism of three apple cultivars during storage under low partial pressures of oxygen. *Postharvest Biology and Technology, 3*, 112037.

Park, H. J. (1999). Development of advanced edible coatings for fruits. *Trends Journal of Food Science and Technology, 10*, 254–260.

Park, M. H., Sangwanangkul, P., & Choi, J. W. (2018). Reduced chilling injury and delayed fruit ripening in tomatoes with modified atmosphere and humidity packaging. *Scientia Horticulturae, 231*, 66–72.

Parker, S. (2002, Dec). *Effects of cooling and postharvest storage methods on broccoli quality*. The University of Tennessee, pp. 1–209.

Paull, R. (1999). Effect of temperature and relative humidity on fresh commodity quality. *Postharvest Biology and Technology, 15*, 263–277.

Perkins-Veazie, P. (1995). Growth and ripening of strawberry fruit. *Horticultural Review: American Society for Horticultural Science,* 267–297.

Petroni, K., & Tonelli, C. (2011). Recent advances on the regulation of anthocyanin synthesis in reproductive organs. *Plant Science, 181*, 219–229.

Porat, R. (2008). Degreening of citrus fruit. *Tree and Forestry Science and Biotechnology, 2*, 71–76.

Posé, S., Paniagua, C., Matas, A. J., Gunning, A. P., Morris, V. J., Quesada, M. A., & Mercado, J. A. (2019). A nanostructural view of the cell wall disassembly process during fruit ripening and postharvest storage by atomic force microscopy. *Trends Journal of Food Science and Technology, 87*, 47–58.

Puttongsiri, T., Choosakul, N., & Sakulwilaingam, D. (2012). Moisture content and physical properties of instant mashed potato. In *International conference on nutrition and food science* (pp. 92–95).

Rahman, E. A. A., Talib, R. A., Aziz, M. G., & Yusof, Y. A. (2013). Modelling the effect of temperature on respiration rate of fresh cut papaya (*Carica papaya* L.) fruits. *Food Science and Biotechnology, 22*, 1581–1588.

Rahman, M. M., Moniruzzaman, M., Ahmad, M. R., Sarker, B. C., & Alam, M. K. (2016). Maturity stages affect the postharvest quality and shelf-life of fruits of strawberry genotypes growing in subtropical regions. *Journal of the Saudi Society of Agricultural Sciences, 25*, 28–37.

Ramin, A., & Tabatabaie, F. (2003). Effect of various maturity stages at harvest on storability of persimmon fruits (*Diospyros Kaki* L.). *Journal of Agricultural Science and Technology, 5*,113–123

Rao, C. G. (2015). *Engineering for storage of fruits and vegetables: Cold storage, controlled atmosphere storage, modified atmosphere storage*. Academic Press.

Rasooli Sharabiani, V., Kaveh, M., Abdi, R., Szymanek, M., & Tanaś, W. (2021). Estimation of moisture ratio for apple drying by convective and microwave methods using artificial neural network modeling. *Scientific Reports, 11*, 1–12.

Ravaglia, D., Espley, R. V., Henry-Kirk, R. A., Andreotti, C., Ziosi, V., Hellens, R. P. et al. (2013). Transcriptional regulation of flavonoid biosynthesis in nectarine (*Prunus persica*) by a set of R2R3 MYB transcription factors. *BMC Plant Biology, 13*, 1–14.

Ravindra, M. R., & Goswami, T. K. (2008). Modelling the respiration rate of green mature mango under aerobic conditions. *Biosystems Engineering, 99*, 239–248.

Richardson, A. C., Boldingh, H. L., McAtee, P. A., Gunaseelan, K., Luo, Z., Atkinson, R. G. et al. (2011). Fruit development of the diploid kiwifruit, Actinidia chinensis 'Hort16A'. *BMC Plant Biology, 11*, 1–14.

Rivera, S., Kerckhoffs, H., Sofkova-Bobcheva, S., Hutchins, D., & East, A. (2021). Influence of water loss on mechanical properties of stored blueberries. *Postharvest Biology and Technology, 176*, 111498.

Rodrigo, M. J., Alquézar, B., Alós, E., Medina, V., Carmona, L., Bruno, M., Al-Babili, S., & Zacarías, L. (2013). A novel carotenoid cleavage activity involved in the biosynthesis of Citrus fruit-specific apocarotenoid pigments. *Journal of Experimental Botany, 64*, 4461–4478.

Rux, G., Mahajan, P. V. Linke, M. Pant, A. Sängerlaub, S. Caleb, O. J., & Geyer, M. (2016). Humidity-regulating trays, moisture absorption kinetics and applications for fresh produce packaging. *Food and Bioprocess Technology, 9*, 709–716.

Salehi, F., & Kashaninejad, M. (2018). Mass transfer and color changes kinetics of infrared-vacuum drying of grapefruit slices. International Journal of Fruit Science, 18(4), 394-409.

Saltveit, M. E. (2004). Respiratory metabolism. In K. C. Gross, C. Y. Wang, and M. Saltveit (Eds.), *The commercial storage fruits, vegetables, florist and nursery stock* (p. 68). Agriculture Handbook Number 66, United States Department of Agriculture.

Schlie, T. P., Köpcke, D., Rath, T., & Dierend, W. (2020). Importance of individual apple fruit characteristics for DCA-CF storage on the variety 'Elstar, PCP'. *Erwerbsobstbau, 62*, 57–60.

Sevillano, L., Sanchez-Ballesta, M. T., Romojaro, F., & Flores, F. B. (2009). Physiological, hormonal and molecular mechanisms regulating chilling injury in horticultural species. Postharvest technologies applied to reduce its impact. *Journal of the Science of Food and Agriculture*, *89*(4), 555–573.

Shammai, A., Petreikov, M., Yeselson, Y., Faigenboim, A., Moy-Komemi, M., Cohen, S. et al. (2018). Natural genetic variation for expression of a SWEET transporter among wild species of *Solanum lycopersicum* (tomato) determines the hexose composition of ripening tomato fruit. *Plant Journal, 96*, 343–357.

Shimoda, Y., Ito, H., & Tanaka, A. (2016). Arabidopsis STAY-GREEN, Mendel's green cotyledon gene, encodes magnesium-dechelatase. *Plant Cell, 28*, 2147–2160.

Singh, P., Langowski, H., Wani, A. A., & Saengerlaub, S. (2010). Recent advances in extending the shelf life of fresh Agaricus mushrooms, a review. *Journal of the Science of Food and Agriculture, 90*, 1393–1402.

Slavin, J. L., & Lloyd, B. (2012). Health benefits of fruits and vegetables. *Advances in Nutrition, 3*, 506–516.

Sohail, M., Ayub, M., Ahmad, I., Ali, B., & Dad, F. (2011). Physicochemical and microbiological evaluation of sun dried tomatoes in comparison with fresh tomatoes. *Pakistan Journal of Biochemistry and Molecular Biology, 44*, 106–109.

Song, C. B., Shan, W., Kuang, J. F., Chen, J. Y., & Lu, W. J. (2020). The basic helix-loop-helix transcription factor MabHLH7 positively regulates cell wall-modifying-related genes during banana fruit ripening. *Postharvest Biology and Technology, 161*, 111068.

Souleyre, E. J., Nieuwenhuizen, N. J., Wang, M. Y., Winz, R. A., Matich, A. J., Ileperuma, N. R. et al. (2022). Alcohol acyl transferase genes at a high flavor intensity locus contribute to ester biosynthesis in kiwifruit. *Plant Physiology,* kiac316. https://doi.org/10.1093/plphys/kiac316.

Souza da Silva, E., Rupert Brandão, S. C., Lopes da Silva, A., Fernandes da Silva, J. H., Duarte Coêlho, A. C., & Azoubel, P. M. (2019). Ultrasound-assisted vacuum drying of nectarine. *Journal of Food Engineering, 246*, 119–124.

Stein, O., & Granot, D. (2019). An overview of sucrose synthases in plants. *Frontiers in Plant Science, 10*, 95.

Sultan, M., Hafez, O. M., & Saleh, M. A. (2022). Quality assessment of lemon (Citrus aurantifolia, swingle) coated with self-healed multilayer films based on chitosan/carboxymethyl cellulose under cold storage conditions. *International Journal of Biological Macromolecules, 200*, 12–24.

Sun, T., Yuan, H., Cao, H., Yazdani, M., Tadmor, Y., & Li, L. (2018). Carotenoid metabolism in plants: The role of plastids. *Molecular Plant, 11*, 58–74.

Thakur, R., Pristijono, P., Bowyer, M., Singh, S. P., Scarlett, C. J., Stathopoulos, C. E., & Vuong, Q. V. (2019). A starch edible surface coating delays banana fruit ripening. *LWT, 100*, 341–347.

Thompson, R., Hornigold, R., Page, L., & Waite, T. (2018). Associations between high ambient temperatures and heat waves with mental health outcomes: a systematic review. *Public health*, *161*, 171–191.

Tucker, G., Yin, X., Zhang, A., Wang, M., Zhu, Q., Liu, X., Xie, X., Chen, K., & Grierson, D. (2017). Ethylene and fruit softening. *Food Quality and Safety, 1*, 253–267.

Vico, I., Jurick, W. M., Camp, M. J., Janisiewicz, W. J., & Conway, W. S. (2011). Temperature suppresses decay on apple fruit by affecting *Penicillium solitum* conidial germination, mycelial growth and polygalacturonase activity. *The Plant Pathology Journal, 9*, 144–148.

Villalobos-Acuña, M., & Mitcham, E. J. (2008). Ripening of European pears: The chilling dilemma. *Postharvest Biology and Technology, 49*, 187–200.

Vina, S. Z., & Chaves, A. R. (2003). Texture changes in fresh cut celery during refrigerated storage. *Journal of the Science of Food and Agriculture, 83*, 1308–1314.

Walker, R. P., & Famiani, F. (2018). Organic acids in fruits: Metabolism, functions and contents. *Horticultural Review, 45*, 371–430.

Wang, D., Yeats, T. H., Uluisik, S., Rose, J. K., & Seymour, G. B. (2018). Fruit softening: Revisiting the role of pectin. *Trends Plant Science, 23*, 302–310.

Wang, H., Zhang, H., Yang, Y., Li, M., Zhang, Y., Liu, J. et al. (2020a). The control of red colour by a family of MYB transcription factors in octoploid strawberry (Fragaria× ananassa) fruits. *Plant Biotechnology Journal, 18*, 1169–1184.

Wang, R., Angenent, G. C., Seymour, G., & de Maagd, R. A. (2020b). Revisiting the role of master regulators in tomato ripening. *Trends Plant Science,, 25*, 291–301.

Wang, R., Shu, P., Zhang, C., Zhang, J., Chen, Y., Zhang, Y. et al. (2022). Integrative analyses of metabolome and genome-wide transcriptome reveal the regulatory network governing flavor formation in kiwifruit (*Actinidia chinensis*). *New Phytologist, 233*, 373–389.

Wang, R., Tavano, E. C. D. R., Lammers, M., Martinelli, A. P., Angenent, G. C., & de Maagd, R. A. (2019). Re-evaluation of transcription factor function in tomato fruit development and ripening with CRISPR/Cas9-mutagenesis. *Scientific Reports, 9*, 1–10.

Wills, R. H.H; Lee, T. H., Graham, D., McGlasson, W. B., & Hall, E. G. (1981). *Postharvest: An introduction to the physiology and handling of fruit and vegetables*. Granada.

Wu, C., Shan, W., Liang, S., Zhu, L., Guo, Y., Chen, J. et al. (2019). MaMPK2 enhances MabZIP93-mediated transcriptional activation of cell wall modifying genes during banana fruit ripening. *Plant Molecular Biology, 101*, 113–127.

Wu, S., Zhang, B., Keyhaninejad, N., Rodríguez, G. R., Kim, H. J., Chakrabarti, M. et al. (2018). A common genetic mechanism underlies morphological diversity in fruits and other plant organs. *Nature Communications, 9*, 1–12.

Wu, Y. Y., Liu, X. F., Fu, B. L., Zhang, Q. Y., Tong, Y., Wang, J. et al. (2020). Methyl jasmonate enhances ethylene synthesis in kiwifruit by inducing NAC genes that activate ACS1. *Journal of Agricultural and Food Chemistry, 68*, 3267–3276.

Xiao, Y. Y., Kuang, J. F., Qi, X. N., Ye, Y. J., Wu, Z. X., Chen, J. Y., & Lu, W. J. (2018). A comprehensive investigation of starch degradation process and identification of a transcriptional activator MabHLH6 during banana fruit ripening. *Plant Biotechnology Journal, 16*, 151–164.

Yahia, E. M., & Carrillo-Lopez, A. (Eds.). (2018). *Postharvest physiology and biochemistry of fruits and vegetables*. Woodhead Publishing.

Yang, Y. Y., Shan, W., Kuang, J. F., Chen, J. Y., & Lu, W. J. (2021). Four HD-ZIPs are involved in banana fruit ripening by activating the transcription of ethylene biosynthetic and cell wall-modifying genes. *Plant Cell Report, 39*, 351–362.

Yarkwan, B., & Oketunde, O. (2016). A study of the nutritional composition of freshly squeezed and processed orange juices. *Food Science and Quality Management, 48*,126–132.

Yin, X. R., Allan, A. C., Zhang, B., Wu, R. M., Burdon, J., Wang, P., Ferguson, I. B., & Chen, K. S. (2009). Ethylene-related genes show a differential response to low temperature during 'Hayward' kiwifruit ripening. *Postharvest Biology and Technology, 52*, 9–15.

Yin, X. R., Xie, X. L., Xia, X. J., Yu, J. Q., Ferguson, I. B., Giovannoni, J. J., & Chen, K. S. (2016). Involvement of an ethylene response factor in chlorophyll degradation during citrus fruit degreening. *Plant Journal, 86*, 403–412.

Zhang, T., Li, W., Xie, R., Xu, L., Zhou, Y., Li, H. et al. (2020). CpARF2 and CpEIL1 interact to mediate auxin–ethylene interaction and regulate fruit ripening in papaya. *Plant Journal, 103*, 1318–1337.

Zhong, S., Fei, Z., Chen, Y. R., Zheng, Y., Huang, M., Vrebalov, J. et al. (2013). Single-base resolution methylomes of tomato fruit development reveal epigenome modifications associated with ripening. *Nature Biotechnology, 31*, 154–159.

Zhu, K., Sun, Q., Chen, H., Mei, X., Lu, S., Ye, J. et al. (2021). Ethylene activation of carotenoid biosynthesis by a novel transcription factor CsERF061. *Journal of Experimental Botany, 72*, 3137–3154.

Zuo, X., Wang, J., Li, Y., Zhang, J., Wu, Z., Jin, P., Cao, P., & Zheng, Y. (2025). Recent advances in high relative humidity strategy for preservation of postharvest fruits and vegetables: A comprehensive review. *Food Chemistry,* 144130.

CHAPTER 2

Postharvest Losses in Fruits and Vegetables

PHEBE DING[1*] and GHULAM KHALIQ[2]

[1]*Department of Crop Science, Faculty of Agriculture, Universiti Putra Malaysia, UPM Serdang, Selangor, Malaysia*

[2]*Department of Horticulture, Faculty of Agriculture, Lasbela University of Agriculture, Water and Marine Sciences, Uthal, Pakistan*

*Corresponding author

ABSTRACT

Postharvest losses have been a hot topic in intergovernmental meetings as it is closely related to food security especially when global population increases tremendously. The causes of postharvest losses are multiple, complicated, and interconnected. The significant losses in fresh fruit and vegetables are mainly due to difficulty in accessing markets, insufficient knowledge, nonadaptability to new technologies, lack of infrastructure, limited resources, and government policies. To mitigate postharvest losses, holistic and comprehensive strategies by integrating effective policies, education, training, infrastructure, and research are needed.

2.1 INTRODUCTION

It is projected that by 2050 the global population may reach 9 billion which would create a great pressure upon agriculturists and governments to produce sufficient food for everyone. It would be possible either by increasing food production or preventing food losses in order to feed the

Physiological Disorders of Fruits and Vegetables: Causes and Remedies.
Ghulam Khaliq, Sajid Ali, and Mohammed Wasim Siddiqui (Eds.)

DOI: 10.1201/9781998511334-2

world population. Scientists believe that preventing food losses is cheaper than increasing food production. This is because food loss impacts the environment, and it contributes about 8% of annual greenhouse gas emissions and consumes 25% of all water used in food production (Flanagan et al., 2019). As a result, the United Nations has set Sustainable Development Goal target 12.3 with the hope that by 2030 the food losses along production and supply chains, including postharvest losses (PHL), can be reduced. It seems proper postharvest handling and technologies are needed so that the global population can be sufficiently fed without bringing additional hectares under production or without changing present agricultural practices to avoid further impact to the earth. Thus, initiatives to secure food by reducing food waste and loss have been undertaken globally. In 2019, the global food loss index (FLI) as estimated by the Food and Agriculture Organization (FAO) is 14% after measuring the changes in percentage losses for a basket of 10 main commodities by a country in comparison with a base period (FAO, 2019). The FLI measures PHL along the food supply chain starting from the farm up to, but not including, the retail level (FAO, 2018). Postharvest system comprises of activities from the moment of harvest, packing house operation, storage, distribution, marketing, and finally to the end users. The system is a chain that interconnects each activity and a break at any point of the chain may lead to low consumer acceptability, deterioration of nutrients, and financial loss to the producers, which is regarded as PHL.

The main cause of the losses is limited shelf life of commodities where the inherent biological behavior makes fresh commodities susceptible to water loss and postharvest disease (Shahidah and Ding, 2020). It is estimated that one-third of food produced on this earth is wasted or undergoes loss (Gustavsson et al., 2011). Among the losses, fruits and vegetables encounter the highest losses, about 40%. Generally, the magnitude of PHL in fresh fruit and vegetables is estimated to be 5–25% in developed countries and 25–50% in developing countries (FAO, 2011; FAO, 2021). However, these losses may vary depending on the nature of products, regions, and quality standards. In Australia, the PHL of tomato along supply chains was between 40.3% and 55.9% depending on the market distance (McKenzie et al., 2017). In Ethiopia, a country in Africa, the losses of fruit and vegetables for retailers range from 5% to 83% depending on the nature of commodities (Yeshiwas and Tadele, 2021). For Indian litchi, the fruit encountered 35.3–43.8% losses along the supply chain (Kumar et al., 2016).

In Sri Lanka, the PHL range from 20 to 40% for vegetables and 30% to 40% for fruits (Rajapaksha et al., 2021), while in China, fruits and vegetables recorded 13.2% and 27.7% losses, respectively (Lu et al., 2022). In Southeast Asia, it was reported that the PHL for fruit and vegetables in Indonesia, the Philippines, Thailand, and Vietnam are estimated to be 20–50%, 27–42%, 17–35%, and 20–25%, respectively (Gunasekera et al., 2017). Quality standard of produces also have an impact on PHL in certain countries, especially developed and industrialized countries. In Australia, due to product specifications, 68.6–86.7% of undamaged and edible tomato were rejected and discarded as outgrades (McKenzie et al., 2017). During harvesting and grading stages (including field and packing shed), Australian tomatoes experienced overall losses of 90.3–97.5%, while after the farm-gate, losses were between 2.5 and 9.7% of overall losses.

2.2 FACTORS CONTRIBUTING TO POSTHARVEST LOSSES IN FRUITS AND VEGETABLES

The factors causing PHL are complicated and involve socioeconomic, business, natural environment, and political context of a country. Market availability, insufficient knowledge, nonadaptability to new technologies, lack of infrastructure, limited resources, and government policies are among the common factors of PHL globally. These factors either directly or indirectly contribute to PHL.

2.2.1 MARKET AVAILABILITY

In Africa, the mismatch and miscommunication between producers and available markets has caused losses to tomato (Arah et al., 2015). In Ghana, difficulties in market accessibility have caused growers huge fresh commodities loss at postharvest (Dovlo et al., 2021). Losses of 36.6% in onion, 27.3% in tomato, 27.1% in watermelon, and 26.5% in sweet potato were reported in Ghana due to lack of ready market at retail centers (Sugri et al., 2021). As a result, producers have to sell their harvest at very low cost to prevent total loss. Similarly, in India, lack of markets has contributed to PHL in their fruits and vegetables (Singh et al., 2022). However, niche markets that demand premium grade produces have promoted the implementation of good postharvest handling.

Consumers are now demanding high-quality produce because the awareness, education level, and purchasing power increase has changed their consumption pattern and lifestyle. These consumers are expecting more specific terms for their food requirements, and this has led food dealers to respond to these niche markets. In Malaysia, a survey of vegetable producers revealed that market availability explained 13.211% of the total variance. The most influential factor was buyers' concern about product quality, with the highest factor loading (0.707), followed by the availability of sufficient marketing channels (0.698), the necessity to meet market demand (0.659), and the presence of appropriate marketing channels (0.592) (Ruslan et al., 2013). As such, producers have to follow the demands of consumers by implementing good postharvest handling practices to meet consumers' needs. It seems market availability is an important motivator for growers and handlers to implement good postharvest handling, and hence reduce PHL.

2.2.2 KNOWLEDGE

Knowledge is the new experiences, information, understanding, and skills that is gained through education or experience (Mohajan, 2016). Food production is a dynamic system involving the interplay between human and ecosystems with high degrees of complexity, uncertainty, and adaptiveness (Burdock and Ampt, 2018). Thus, knowledge plays an important role in postharvest activities so that the losses of fresh fruits and vegetables can be minimized. Studies have also revealed growers who have knowledge about technology, attain high awareness, and tend to adopt postharvest practices (Mwaijande, 2017). This is because knowledge affects human behavior and gives impetus for growers and handlers to adopt a technology (Singh and Singh, 2018).

In Malaysia, insufficient knowledge of good handling practices contributed to 36.7% PHL (Osman et al., 2009). This is followed by insufficient knowledge in supply and demand, and off-season or off-grade produce diversification which contribute to 35.7 and 27.6% losses in fruit and vegetables, respectively. In Nigeria, the huge losses of fruit and vegetables in the country were due to insufficient postharvest handling knowledge among growers and handlers (Muhammad et al., 2012). It was found out that 58.2% of the Nigerian vegetable growers have lack of knowledge in packaging (Egwuonwu, 2020). Similarly, lack of knowledge in postharvest handling among growers contributes to PHL in Ghana (Dovlo et al., 2021). In Ethiopia, 100% of respondents expressed that they do not have any formal knowledge on handling practices and this may be the reason retailers experienced 5–83% losses depending on type of commodities (Yeshiwas and Tadele, 2021).

In Tanzania, it was revealed that improving educational knowledge helped reduce PHL among fruit growers and handlers (Kereth et al., 2013). Very often the magnitude of PHL is correlated with seller's educational status and selling experience (Yeshiwas and Tadele, 2021). When the educational status and selling experience of the retailers increase, the loss of fruit and vegetables decreases. Educational status does have impact on effort expectancy in informational technology of a person that leads to his behavioral intention (Chiemeke and Eveiwkpaefe, 2011). It seems educational knowledge of growers and handlers can affect their behavior, which eventually impacts PHL.

2.2.3 ATTITUDE

Attitude shows a bodily state of readiness to respond in a characteristic way to a stimulus such as a situation, concept, or an object and it affects behavior (Chuang et al., 2020). Once an attitude is adopted, it can provide a long-term effect (Olgyaiova et al., 2005). In adopting good postharvest handling practices and technologies, attitudes of growers, handlers, and retailers are determinants of success. It is possible to improve their performance in adoption and commercialization of fresh fruit and vegetables once their attitudes have changed (Meena et al., 2009). It is reported that growers' attitude can contribute to PHL of fruit and vegetables as it can influence the adoption of agriculture technologies by growers (Rose et al., 2016). In Malaysia, attitude toward postharvest activities due to workers' apathy (49.7%) and inefficient/improper handling systems (50.3%) were the contributing factors to fruit and vegetables PHL (Osman et al., 2009). A study showed that 57.6% and 35.6% Nigerian plantain growers and wholesalers, respectively, have unfavorable attitude toward postharvest handling (Ladapo and Oladele, 2011). This may be due to poor educational background and inadequate information on the value addition potentials of plantain. It seems growers and handlers' attitude is related to their educational status that is able to determine their adoption of good postharvest practices.

2.2.4 INFRASTRUCTURE

Countries such as Ethiopia (Bateno and Buke, 2018; Yeshiwas and Tadele, 2021), Ghana (Dovlo et al., 2021), and Nigeria (Egwuonwu, 2020) have lack of infrastructure and this leads to PHL in their fruit and vegetables. In Ethiopia, the long transportation journey from production site to market place has caused 18% lettuce and 14.8% tomato damage, improper packaging material leads to 22% banana and 21% tomato physically damaged while the high light intensity with elevated temperatures in the market has caused 34.4% lettuce and 33.6% mango damage (Bateno and Buke, 2018). In India, poor infrastructure and logistics are the cause of PHL in fruit and vegetables (Singh et al., 2022). However, there are cases where growers and handlers refuse to implement the appropriate postharvest practices if the government does not provide financial incentives to build the infrastructure (Kitinoja et al., 2011). In countries like Malaysia, the fresh vegetable producers are ready to implement postharvest handling practices on their fresh produce due to the availability of proper infrastructure (Ruslan et al., 2013). The constraints in having postharvest handling infrastructure in developing countries are poverty of the farming community and lack of capital to acquire improved technologies (Rajapaksha et al., 2021). It seems financial support is pivotal in mitigating PHL in these countries.

2.3 MITIGATING POSTHARVEST LOSSES IN FOOD SUPPLY CHAIN

A food supply chain comprises all parties involved in food production and distribution for human being consumption. The chain may start from nurseries, farms, harvesting, sorting, trimming, washing, grading, storing, distributing, marketing, and waste management. It is a dynamic and market-driven activity that includes unforeseen circumstances such as Covid-19 pandemic, which contributed to food losses due to labor shortage, lockdown, import/export restrictions, distribution, and consumer food-buying behavior (Kahramanoğlu et al., 2021).

For postharvest, food losses occur at any point along the chain from harvesting to end users. Reducing PHL should always be priority for addressing food insecurity issues which requires integration of plant physiology, postharvest science and technology, engineering, agribusiness, and government policies. Careful and proper postharvest handling has a significant impact in fruit and vegetables PHL as any changes in the chain give irreversible effect to fruit and vegetables quality as the chain progresses.

2.3.1 HARVESTING

Harvesting is an activity to separate the fruit or vegetable from its mother plant where they were supplied with water and nutrients. PHL during harvesting may be due to improper harvesting maturity, methods, or techniques and time of harvesting during the day. The maturity stage at harvest is a crucial factor affecting postharvest life and eating quality of fresh produce (Mijin et al., 2021; Ali et al., 2022). If fruit is harvested before achieving its physiological maturity, it would encounter erratic ripening and poor eating quality although the fruit may have extended postharvest life. While delayed harvesting may cause the fruit prone to pathogen attack and vulnerable to mechanical injury during handling, distribution, and marketing that eventually leads to quality deterioration and short postharvest life.

In addition to harvesting stage, fruit can be classified as climacteric or nonclimacteric fruit depending on their respiration rate during ripening. For climacteric fruit, there is a surge in respiration rate during ripening and fruit can continue ripening if it is harvested at physiological maturity stage. Contrary to climacteric fruits, nonclimacteric fruits exhibit slow or nonexistent ripening even the fruit harvested at physiological maturity stage. The ripening of nonclimacteric fruit takes place on-tree for optimal eating quality. This classification is able to affect harvesting stage and postharvest handling of a fruit.

Jackfruit exhibits a climacteric respiration pattern; therefore the fruit can undergo ripening when harvesting carried out at its optimal maturity stage. Jackfruit harvested at 11 weeks after anthesis (WAA) fail to ripen, but those harvested at 12 WAA emitted its typical sweet aroma within 8–11 days indicating the fruit has ripened and fixed for fresh consumption by 12 WAA (Mijin et al., 2021). The days taken for the fruit to emit aroma decrease with harvesting stage and by 15 WAA, jackfruit has undergone on-tree ripening with aroma emission before harvesting. The firmness of jackfruit harvested at 12 WAA was significantly firmer than those harvested at delayed stage of 15 WAA. In contrast, fructose, glucose, and flavonoids content of jackfruit aril that harvested at 12 WAA was significantly lower than those harvested at 15 WAA. This study also revealed harvesting maturity is essential in ensuring optimum postharvest life and eating quality of a fruit.

Harvesting method is another cause of PHL in fruit and vegetables. Improper harvesting methods either manually or mechanically results in mechanical damage such as bruises, cuts, puncture, split, and abrasion to harvested crops. Manual harvesting or hand picking is a labor extensive and time-consuming operation as compared to mechanized harvesting. However, mechanized harvesting causes more wounding to the crops. Robotic picking using three-finger gripper induced higher grasping force and pressure in apple than those hand picking with undetectable bruise damage (Li et al., 2016).

Bruising fruit contributes to downgrading and fruit rejection thereby contributing to PHL (Zaharan et al., 2020). The external bruising of fruit is obvious where the damaged fruit can be discarded during selection. Very often the internal bruise injury is potentially overlooked during selection and grading but the bruise damage manifests along the supply chain. This has been illustrated in Carabao mango fruit where the internal bruise injury due to careless harvesting extent throughout supply chain and fruit ripening stages (Lacap et al., 2021). Besides, increasing level of induced injury due to harvest could also contribute to higher water loss and accelerate wilting of commodity (Lufu et al., 2020). Careless harvesting caused bruising in strawberry, hence, induced ethylene production that accelerated decay and deterioration of fruit (Chang and Brecht, 2020). Thus, careful harvesting is essential in reducing PHL.

The severity of produce bruising is also affected by time of harvesting during the day. Harvesting fruit at hotter time of the day induce faster senescence in bruising fruit (Lufu et al., 2020). Higher temperature during harvesting incurred higher field heat in the fruit and accelerates its respiratory activity that could increase bruise damage susceptibility and shriveling of fruit. Study on kiwifruit revealed fruit with higher temperature resulting in more absorbed energy which tended to increase bruising (Ahmadi, 2012). In apple, fruit harvested later in the day were less susceptible to bruising than those harvested in the morning hours due to reduced cells turgor (Abbott et al., 2009). For avocado, harvesting at dawn could cause fruit to be susceptible to handling damage at harvest as found in fruit harvested after rain that leads to higher risk of fungal infection than the fruit harvested in the mid-afternoon (Elmsly et al., 2007).

2.3.2 TEMPERATURE AND RELATIVE HUMIDITY MANAGEMENT

Since harvested fresh commodities are living organs, respiration actively takes place throughout their postharvest life. During respiration process, the stored food of complex organic substances is oxidized to release carbon dioxide, water vapor, and energy for running cellular processes to sustain cells and organism alive. Along the oxidation processes, water is transpired from commodity and causes them to shrivel and wilt with development of undesirable texture. Loss of 5–10% water from the initial weight of fresh fruit and vegetables could cause them become unsaleable on top of browning, loss in flavor, accelerating senescence, susceptibility to chilling injury, and membrane disintegration (Lufu et al., 2020). In papaya, 8% weight loss could cause the fruit unmarketable due to rubbery texture and low-gloss skin appearance (Azene et al., 2014). Besides, the heat generated from respiratory activity tends to increase commodity temperature, thus leading to an increase in transpiration and water loss (Becker and Fricke, 1996).

The loss of water from cell tissues to atmosphere is affected by both plant characteristics (e.g., surface-to-volume ratio, peel surface structure, ripening stage) and climatic conditions (e.g., temperature, relative humidity (RH)) (Khaliq, 2015; Pereira et al., 2018). Water is lost in the form of water vapor through opening on epidermal tissue of commodities such as stomata, lenticels, and microcracks depending on species (Shahidah and Ding, 2020; Ul Haq et al., 2024). Under different temperature and RH conditions, commodity encounters different levels of water loss. Temperature affects respiration rate of commodity where every 10°C increase in temperature, the respiratory activity will be doubled from the minimum and reduce the overall shelf life by half (Alam et al., 2021). While RH affects moisture diffusion from commodity to surrounding atmosphere due to water vapor pressure deficit (Lufu et al., 2020; Mditshwa et al., 2023).

The water loss of fresh fig variety Ipoh Blue Giant stored for 12 days at 5 and 10°C was about 3.2% but significant increase of 6% water loss was recorded when fig fruit stored at 15°C (Ding and Choon, 2021). Passion fruit showed about 9.5% water loss when stored at 5 and 10°C for 21 days but water loss increased significantly to 15.31% when fruit stored at 15°C (Shahidah et al., 2021). When storing passion fruit at 20 and 25°C, the fruit experienced even more water loss with 19.51 and 23.20%, respectively. Pomegranate stored at 65% RH showed significantly higher water loss of 29.1% as compared to 5.8% loss for fruit stored at 95% after 30 days of storage at 20°C (Mukama et al., 2019). Zainal et al. (2019) found out water loss in rock melon was associated with cell turgor loss that related to cell wall turgidity in the pericarp tissue of fruit.

Temperature fluctuations along the supply chain have been shown to impact PHL. A study in Queensland, Australia revealed the temperature of tomato increased from 10.2 to 12°C after 392 km of transportation from farm to Brisbane market in November (McKenzie et al., 2017). The core temperature of tomato gradually increased to a peak of 25°C after removing fruit from refrigerated storage to display and eventually 5.4% of the fruit was deemed unsaleable by fifth day of display. During a 4-day trip from Florida to California, strawberries experienced temperature fluctuation by a minimum of 0.8°C and a maximum of 5°C (Pelletier et al., 2011). Using modelling PHL, it was estimated 2% losses in Florida tomato when average temperature increases by 3°C during postharvest operations which include transportation (Suthar et al., 2019). Generally, 30–49% PHL occurred during transit due to temperature fluctuations which affect quality and shelf life of commodities (Al-Dairi et al., 2022). It seems temperature fluctuations along the supply chain needs to be addressed during combating PHL.

2.3.3 PRECOOLING

Precooling, the most critical operation in postharvest handling, is used to remove field heat of harvested produces so that cellular metabolism can be decelerated and thus prolong products postharvest life (Ding and Raja Norhidayu, 2021). Precooling can be carried out using ice, water, or air depending produce characteristics, produce quantity, cooling medium temperature and flow rate, packing configuration and stacking arrangement of the produce, energy efficiency of the method, the availability of skilled labor and the economic viability (capital and running costs) (Makule et al., 2022). Delay in precooling reduces the quality and postharvest life of fresh fruit and vegetables. It has been noted that strawberries without precooling treatment were not fit for shipment 1 day after harvest (Pelletier et al., 2011). The partial and delayed precooled strawberries held at 21°C were inferior to that of fruit stored at 0 or 6°C. These researchers also revealed the prompt precooling strawberries to 1.7°C had acceptable visual quality even after 3 days at 0°C but the delayed or partial precooling treatments fruit ranged from slightly poor to poor visual quality after the same period of time.

The effectiveness of precooling methods depends on commodity type and cooling time. Highbush blueberry cultivar Farthing can be precooled using hydrocooling without skin split while cultivar Emerald experienced skin split after 7 days of storage (Carnelossi et al., 2019). Thus, forced-air cooling is more suitable to precool Emerald blueberry than using hydrocooling. For broccoli, it was demonstrated that both vacuum cooling and package icing can be used to extend its shelf life and retain quality during supply chain (Dirapan et al., 2021). Precooled sweet corn ears to 7.25°C by submerging the ears in cold water could retain significant higher soluble solids concentration and pH with lower water loss compared to other cooling time and control (Ding and Raja Norhidayu, 2021). For rock melon, it was found out hydrocooling at half-cooling time with fruit core temperature cooled down to 14.8°C was better in preserving cell wall structure and maintaining fruit quality than other cooling times (Zainal et al., 2019).

2.3.4 COLD STORAGE FACILITIES

Cold storage facilities are part of cold chain systems in minimizing PHL of fresh fruit and vegetables. Stressful environmental conditions at marketplace due to improper cold storage facilities and shaded structure or building for selling fruit and vegetables is another aspect to consider in addressing PHL. The facilities are needed to slow down the metabolism of commodities, hence, to delay commodities quality loss and senescence. A nonconducive environment exposes produces to high temperature and light intensity. In the USA, rot incidence of apple increases while its firmness losses when the fruit kept in farm markets without cooling facilities or at elevated temperatures (McArtney et al., 2011). For tomato, a temperature increases from 18 to 20 to 26°C reduced average fruit shelf life by 4 ± 1 day while fungal susceptibility increased by 11 ± 5% (Thole et al., 2021). Sweet corn ears that were precooled to 7.25°C and then stored at 12 ± 2°C has better visual quality than those stored at 25 ± 2C after 8 days of storage (Ding and Raja Norhidayu, 2021).

Mechanical refrigeration system is energy consuming and capital-intensive technology. Refrigeration system with off-grid power, low-cost, zero or renewable energy and natural refrigerants is more feasible in developing countries and area without electricity. Study revealed a zero-energy cooling chamber in which the internal temperature is 10–15°C lower than the external ambient temperature, was able to prolong the post-harvest life of fresh fruit and vegetables for short-term storage (Ashitha et al., 2021). Similar positive findings were also reported in Nepal (Mishra et al., 2020) and Pakistan (Khalid et al., 2020) where this zero energy cooling chamber is able to prolong the postharvest life of fresh vegetables significantly. The low-cost and eco-friendly zero energy cool chamber, by using solar-driven adsorption refrigerator and an evaporative cooling system, was able to reduce internal temperature of chamber to 12.07°C while the external ambient temperature was 31.5°C for 23 days (Islam and Morimoto, 2014).

CoolBot is a device coupled to room air conditioner capable that developed by Store It Cold. This device is capable to trick air conditioner while preventing components from freezing and eventually to create a small-scale with well-insulated room become a cold storage room with temperature as low as 2°C. The CoolBot cold storage technology is a cheaper alternative for small scale growers and it is also eco-friendly by having low carbon emission and low operation costs (Sharma et al., 2019). Onion stored under CoolBot cold storage showed reduction in losses by 25% (Saran et al., 2012). In Cambodia, PHL of tomato along supply chain was significantly lower after the fruit was treated with ice box, CoolBot chamber or evaporative cooler as compared to control (Socheath et al., 2020). In India, CoolBot cool room successfully increased the marketable life of fruit and vegetables by preventing distress sale (Dubey and Rahman, 2016).

2.3.5 PHYSICAL TREATMENTS

Fresh fruit and vegetables are vulnerable to bruise damages during handling. Hence, packaging provides physical protection to commodity by enclosing product to avoid or minimize mechanical injuries, tampering, and even contamination from physical, chemical, and biological sources using materials made from paper, plastics, paperboard, and fiberboard (Prasad and Kochhar, 2014). Besides, packaging provides barrier against airflow, minimize the flow rate of moisture carried away from around the commodity and minimize diffusion across commodity epidermal tissue, hence reduce water loss and wilting (Lufu et al., 2020). However, while minimizing the flow rate of moisture carried away from commodity, the packaging is also preventing gases exchange between inside and outside a package. The elevated carbon dioxide in packaging may inhibit normal ripening of fruit (Krupa and Tomala, 2021). The accumulation of carbon dioxide in polyvinyl chloride packaging inhibited the production of ethylene in plums hence retard fruit ripening (Wang et al., 2021). On top of carbon dioxide, the accumulation of ethylene produced by fruit and vegetables causes undesirable quality changes to the commodities such as color, taste, odor, texture, or induce microbial growth (Gaikwad et al., 2020). As such, the traditional packaging by having permeability and breathability is no longer able to fulfil industry expectations. Ethylene-removing packaging has been developed to scavenge ethylene and reduce its concentration surrounding commodities (Awalgaonkar et al., 2020). To date, the packaging technologies have evolved from active packaging to intelligent packaging and currently to smart packaging (Alam et al., 2021). Embedded biosensors are able to monitor the freshness, spoilage and ripeness degree of commodity, temperature and humidity, and atmosphere composition inside the package along food value chain, thus reducing PHL.

Besides packaging, heat treatments using hot water and vapor heat are effective treatments for tropical and subtropical fruits. Heat treatments can be used to combat microorganism on the surface of fruit or eggs and larva of insects in the fruit flesh (Marouf and Khali, 2020). Dipping Chok Anan mango fruit in hot water of 55°C for 5 min was able to reduce disease incidence significantly while physico-chemical quality characteristics of fruit was not altered by the hot water dipping treatment (Ding and Salumiah, 2013). The ripening of Berangan banana fruit was delayed by dipping the fruit in hot water of 50°C for 10 min indicating hot water dip treatment is able to prolong the postharvest life of banana (Mirshekari et al., 2015). By retaining core temperature of guava at 47.5°C for 25 min under vapor heat treatment, the fruit displayed higher marketability index, cosmetic and eating quality with lower disease incidence, ion leakage, and weight loss during ambient storage conditions as compared to control (Malik et al., 2021). White-fleshed dragon fruit that treated with vapor heat at 49°C for 70 min was able to control *Planococcus lilacinus* Cockerell (Hemiptera: Pseudococcidae), an invasive pest, with minimal impacts on fruit quality (Ren et al., 2022).

Ultra-violet C (UV-C) is a nonionizing and nonthermal physical treatment in postharvest. The irradiation triggers natural disease resistances of plant tissues to enhance the storability of commodities where it retains the cell wall integrity of Berangan banana's crown (Nuratika et al., 2017). Thus, crown rot diseases of Berangan banana were suppressed by 0.01 kJ/m^2 UV-C irradiation. It was also revealed that phenolic contents and antioxidant capacity of Berangan banana was enhanced by the same amount of UV-C (Ding and Nur Ayuni, 2015). *In vitro* study revealed UV-C was able to inhibit conidial germination and sporulation of *Colletotrichum gloesporioides* on mango, indicating it is a potential treatment to reduce chemical treatment in controlling postharvest pathogen and thus reducing PHL (Gunasegaran et al., 2018).

Coating regulates moisture and gaseous exchange between internal and external atmosphere of the coated commodities by slowing down commodities respiration rate, conserving store energy, retarding microbial growth, and hence extending their postharvest life (Shahidah and Ding, 2020; Khan et al., 2025). Combining 1.5% chitosan and 15 mM vanillin is able to retain postharvest quality of tomato fruit during ambient temperature storage by delaying weight loss up to 90% and reducing changes in firmness, soluble solids concentration and color (Zahir et al., 2021). These researchers also found out similar combination of coating formulation was able to control *Fusarium oxysporum* fruit rot incidence in tomato fruit by 70.84% (Zahir et al., 2020). With 10% gum Arabic + 2% sodium bicarbonate coating, the progression disease rate of *Collectotrichum* sp. in red-fleshed dragon fruit was reduced by 50% as compared to control (Ding and Lee, 2018).

2.3.6 CHEMICAL TREATMENTS

Fresh fruit and vegetables are nutritious; thus, they are susceptible to pathogenic fungi infection and then decay, thus shortening their postharvest life and quality (Khaliq et al., 2023; Lal et al., 2025). The fungi that cause postharvest diseases infect the fruit either in the field and remain latent until fruit ripened or through injuries, wounds or bruises inflicted along postharvest activities (Palou, 2018; Ul Haq et al., 2024). Postharvest diseases have caused huge economic losses, thus synthetic fungicides have been widely used as a cost-effective measurement to control the pathogenic fungi. Synthetic fungicides with active ingredients imazalil (IMZ) and thiabendazole (TBZ) are employed to control postharvest decay in citrus fruit caused by *Penicillium digitatum* (Schirra et al., 2011). To control anthracnose in mango, TBZ has been selected while benomyl is widely used to control brown rot and rhizopus rot in stone fruits caused by *Rhizopus stolonifera.* Although synthetic fungicides have been proven effective in controlling postharvest decay effectively, prolong use of synthetic chemicals leads to significant problems to human health, environmental contamination, reduced efficacy of fungicides due to the proliferation of resistant fungal biotypes, restricted access to organic markets and importing countries legislation (Palou, 2018). Hence, alternative approaches using generally recognized as safe (GRAS) to control postharvest diseases and regulate metabolic activity of commodities have attracted researchers' attention. Inorganic or organic salts such as carbonates, sorbates, benzoates and silicates have been studied for its efficacy in controlling postharvest pathogens (Palou et al., 2018). Both sodium carbonate and potassium sorbate are the GRAS salts and have shown good ability to control fungi that cause world largest decay losses in citrus, *P. digitatum* (Pers.: Fr.) Sacc. (commonly known as green mold) and *Penicillium italicum* Wehmer (commonly known as blue mold).

Unlike synthetic fungicides that target pathogenic filamentous fungi, the causal agents of postharvest decay, ethylene evolved by fresh fruit and vegetables accelerates deterioration of commodities quality and hence shortens their postharvest life. As such ethylene inhibitor, scavenger, and scrubber are often used to inhibit ethylene action and retain commodities quality. By exposing mature green Berangan banana to 1 μL/L 1-methylcyclopropene (1-MCP) for 4 h, the ripening of fruit was delayed as compared to control (Ding and Khairul, 2009). In another banana study, immersing fruit in 50 or 100 μL/L ethephon and subsequently treated with 400 nL/L 1-MCP had extended green life during cold storage and fruit underwent normal ripening after ripening initiation (Satekge and Magwaza, 2020). For Kampuchea guava, treating fruit with 600 nL/L 1-MCP for 6 h was able to delay the development of fruit color, flesh firmness and disease occurrence (Ding and Ong, 2010). Mango fruit treated with 1-MCP displayed significant inhibition on cell wall enzymes activity and maintain fruit firmness, hence extended fruit postharvest life (Li et al., 2020).

In addition to 1-MCP, nitric oxide (NO) is also being used to prolong postharvest life of fruit. NO is a signaling molecule commonly found in plant and released as gaseous bioactive molecule by plant during development or under stress conditions (Buet et al., 2021). Winter jujube fumigated with NO showed higher firmness while activities of cell wall enzymes decreased during storage at 0°C for 75 days (Zhao et al., 2019a). Recently, sodium nitroprusside (SNP), the donor of NO in aqueous solution, is also extensively being studied. SNP is not only delaying quality loss in pear, apple, peaches and plums (Adhikary et al., 2021), it also suppressed post-harvest anthracnose decay in mango fruit (Ren et al., 2020). Melatonin is a signaling molecule too where it may involve in seed germination, circadian rhythm and photo-response regulation, development, flowering, and defense (Wang et al., 2020). Melatonin, known as N-acetyl-5-methoxytryptamine, is derived from tryptophan, an amino acid (Zhao et al., 2019b). Recently, the exogenous melatonin application on fruit and vegetables has gained attention as it can enhance disease resistance and chilling tolerance, and delay ripening and senescence of commodities (Xu et al., 2019). However, the efficacy of melatonin treatment varies greatly depending on concentration, application method, and time.

Amino acids such as alanine, aspartic acid, arginine, cysteine, glycine, methionine, ornithine, phenylalanine, serine, tryptophan, tyrosine, and valine are believed to provide energy and play roles in catabolism of harvested fruit and vegetables. Since amino acids have GRAS status, their impacts in addressing PHL are worth investigation. Green life of broccoli was extended while its ethylene production and ion leakage decreased after treating with arginine, alanine, aspartic acid, glutamic acid, glycine, ornithine, phenylalanine, serine, tyrosine, and tryptophan, but not valine (Sohail et al., 2021). For Sandhuri guava kept at 5°C for 24 days, the application of arginine successfully maintained fruit quality by retaining higher firmness, marketability, anthocyanins and phenolic with reduced chilling injury as compared to control (Ali et al., 2022). L-cysteine showed positive impact on Stanley plum fruit that stored at 1°C with higher ascorbic acid, anthocyanin, flavonoids, and phenol contents, and total antioxidant activity as compared to control fruit (Sogvar et al., 2020).

2.3.7 TRANSPORTATION AND DISTRIBUTION

Transporting harvested commodities from production sites to collection center and then to marketplace is an important process in food supply chain. It can be carried out by truck, rail, ship or airplane. As such, inadequate transport infrastructure and facility such as road and vehicle condition, temperature control in shipping containers and at entry and exit points, during transit and voyage do have impact on fresh commodities losses. During land shipping, road condition, vehicle travel speed and suspension characteristics, duration of vibration exposure, packaging, stacking methods, and temperature management of load and containers are factors contributing to mechanical damage of fresh commodities (Al-Dairi et al., 2022). The vibration and compression along distribution causes high respiratory activity in fresh commodities which eventually leads to wilting and shorten their postharvest life. A well-organized service of transportation is vital to distribute the harvested commodities to destination as quickly as possible with minimum damage (Yeshiwas and Tadele, 2021).

In the hill areas of Bangladesh, the poor road and transportation facilities contributed to 8.8% losses in banana (Hossain et al., 2017). In Oman, weight loss and firmness of tomato were highly affected by transportation distance, storage temperature, and storage period with highest weight loss and softening occurred in tomatoes transported from the farthest distance and stored at 22°C (Al-Dairi et al., 2021). A study in Ethiopia revealed bad road access and transportation facility (even in 20 km distance) caused external mechanical damages and internal injuries to potato tubers, hence enhance degradation rate of tubers at later stages of storage (Kuyu et al., 2019). It was found out high transit temperatures coupled with poor road network contributed to PHL in the fresh commodities (Sugri et al., 2021). In India, litchi encounters weight loss of 7.07–9.42% after 1000 km of road transit (Kumar et al., 2016). The total soluble solids, respiration, and ethylene production rate of litchi increased while acidity and color decreased significantly after the transit.

Besides having good road condition and reducing vibration during shipping, advanced technology is needed as well in addressing PHL. The time-temperature indicator (TTI) is one of the intelligent packaging technologies. It is developed based on chemical and enzymatic reactions (Kalaïtzis et al., 2016). It is a small self-adhesive label that can be placed on individual packages or shipping containers. The color of TTI changed when it experiences abusive conditions. Digital twins, a real-time monitoring technology, has significantly to quantify the quality evolution of fresh mango during shipment and predict the remaining shelf life of the fruit (Defraeye et al., 2019). Besides, the Internet of Things (IoT) has also been applied as smart logistics in transporting fresh commodity as this technology is able to monitor and control the process conditions of fresh produce, hence stakeholders are able to take proper decisions to rectify the conditions and reduce PHL (Onwude et al., 2020).

2.4 POSTHARVEST LOSSES MITIGATION STRATEGIES

Strategies used to address PHL needs to integrate and consider socioeconomic, business, natural environment, and political context of a country. For Algeria, one of the factors of PHL was due to the stubbornness of farmers to continue to apply traditional methods during harvest although the country has implemented a lot of policies (Tria et al., 2020). A study revealed the main reasons for Malaysian vegetable growers rejected postharvest practice due to lack of government incentives (Taufiq, 2017). However, rigorous enforcement by the authority has improved postharvest handling among Malaysian vegetables producers (Ruslan et al., 2013). In Ghana, the growers lack technologies to store their commodities due to small-scale nature of production (Sugri et al., 2021). As such, appropriate technologies to manage harvest surpluses and maintain quality have been included in their government policies to address PHL in the country. The major challenge in Ethiopia is lack of agro-processing to value-adding its agricultural commodities to combat PHL (Teferra, 2022). Thus, Ethiopian government organizes Industrial Parks Development Agency to follow up and support agro-processing investments. It seems the strategies and interventions to mitigate PHL in all countries vary from each other.

Many countries expressed the lack of postharvest activities knowledge among growers and handlers are positively correlated to PHL incidence. On top of traditional education, knowledge can be gained through training. Training is one of the most popular strategies to develop cognitive and behavioral changes of an individual through knowledge and skills (Alzate, 2015). Training is able to bridge the knowledge and skills gap. A survey revealed Ethiopian growers and handlers did not receive any training on postharvest handling practices of fruit and vegetables except for the traditional knowledge that they have (Yeshiwas and Tadele, 2021). This could explain high PHL in the country. For Romania, vocational training for the sector workers is essential in reducing PHL (Iordachescu et al., 2019). Sri Lankan expressed that due to inadequacy of trained manpower in postharvest handling has caused PHL along food supply chain (Rajapaksha et al., 2021). In Tanzania, a pilot Postharvest Training and Services Center funded by the United States Agency for International Development (USAID) has successfully helped more than 13,000 African extension staffs, growers, food processors, and marketers to reduce PHL where 79% attendees using improved postharvest practices in their operation (Kitinoja and Barrett, 2015). This implies a good integration of skilled labor, postharvest knowledge, and techniques would increase productivity, save resources, time, and effort, hence secure more food.

Other strategies to mitigate PHL are to increase agricultural extension staffs. It was reported that the deficiency of agricultural extension service has affected the improvement of marketing channels of fruit and vegetables in Sri Lanka (Rajapaksha et al., 2021). In Egypt, tomato growers obtained their information of production and supply mainly from market agencies (36%) as compared to only 5.5% from extension staffs (Ali et al., 2021) indicating either lack of extension staffs or knowledge to assist growers. As country policy, the government of Ghana included capacity building of postharvest technologies extension staffs to give training to growers and handlers (Sugri et al., 2021).

Extension staffs who interact and transfer knowledge and skills to growers and handlers received their education in vocational training centers, colleges, and universities. However, it was revealed postharvest science and management courses are lacking in the curriculum of training institutions. Less than 10% of agricultural institutions in the Philippines are teaching postharvest science courses such as Postharvest Handling of Horticultural Crops as a subject (Nuevo and Apaga, 2010). Although agricultural vocational training colleges of Lesotho directly supports postharvest management, the nonagriculture vocational training colleges offered postharvest management course as under subtopics, thus limiting the number of extension staffs equipped with postharvest knowledge (Muroyiwa et al., 2020).

Since PHL occurs along the food supply chain, the training and knowledge should not limit to extension staff, but should extend to other players along the chain. A short course and refresher course to create awareness among industry players such as growers, handlers, supermarkets, exporters, and entrepreneurs are necessary too. The intervention of government and international research organizations to set up a central site for conducting postharvest research and offering local extension is recommended especially for developing countries as these countries encountered substantial PHL as compared to developed countries. In addition, the establishment of country-level Postharvest Working Groups is a good move to connect researchers, extension staffs, growers, handlers, marketers, transporters, exporters, and importers (Kitinoja et al., 2011). The information, training, shared learning, issues, and problems around PHL reduction can be facilitated and exchanged through the group. Regionally, a link with the various Postharvest Working Groups would further facilitate the collaboration and PHL reduction. Through such group targeted researches and immediate action can be carried out to address issues faced by food supply chain actors at both national and regional levels.

On top of postharvest human capital, appropriate postharvest innovation is also important in addressing PHL. However, the innovation is highly depending on available funding mechanism. It has been reported that agricultural funding that allocated for postharvest research areas in most countries is less than 5% of the funding research as compared to research funding on the improvement of agricultural production (Kitinoja et al., 2011). It has been proven that the magnitude of postharvest research correlate negatively with food losses. In Africa, except in South Africa, research on postharvest is very limited, thus contributing to about 18% of global postharvest food losses (Alamar et al., 2017). Contrary, the extensive postharvest research in Europe contributes relatively low input to global food loss (ca. 7.7%). In addition to the funding for postharvest research, finance schemes such as microcredit institutions and Credit Agricole is essential in reducing PHL too. In Kenya, the funding by Rockefeller Foundation's YieldWise Initiative to create awareness and provide proven technologies such as CoolBot to reduce PHL among mango growers was worthwhile with positive outcomes (Mujuka et al., 2020).

The success of a PHL prevention program needs good program designs and effective implementation management (McNamara and Tata, 2015). Postharvest management is handling of fresh produces from farm to consumer plate so that produces quality are retained with competitive market price and producers make profit (Bekele, 2021; Delgado et al., 2021). A good governance in the management is essential as PHL prevention program involves mobilizing multiple actors, significant institutional capacity, human resources, and appropriate technologies that suit to socioeconomic, business, natural environment, and political context of a country. In Lesotho, the reason of high food loses was due to poor integration of postharvest management in the country agricultural policy and strategies to minimize PHL (Muroyiwa et al., 2020). A direct policy is needed to guide postharvest management activities in Lesotho besides issues of poor planting materials, lack of infrastructures and facilities, lack of knowledge and skills, norms, and gender dimensions.

Wrong stage of harvesting contributes to substantial PHL in fruits and the stage varies among species. It is important to prepare harvest maturity guidelines for growers. Improper harvesting maturity is one of the causes of quality loss in mango (Sivakumar et al., 2011). A significant PHL occurred in Sri Lanka vegetables due to harvesting of over-matured okra, beans and bitter gourd which cause tough texture to the produces (Rajapaksha et al., 2021). Quality standards such as size, weight, shape, color, and so on are requirement established by the markets so that only produces fulfilled the standards are allowed to enter the markets (Hailu and Derbew, 2015). Quality standards vary greatly among countries and cultures where consumer preferences and purchasing power are the major determinant. The standards may either give positive or negative impact to PHL. Produces that meet the standards may lead to increased opportunities for exports and access into international and niche markets, and the growers are rewarded with higher price (Muroyiwa et al., 2020). Contrary, the quality standards may lead to losses at the agricultural production stage due to rejection of unqualified produces especially quality standards set by retailers of developed countries. However, very often the rejected produce does not pose food safety risk and is edible, but just simply it fails to meet cosmetic and physical criteria of markets (Yahia et al., 2019). Strategies are needed to vindicate this type of produces in order to reduce PHL especially when considering responsible consumption and production as one of the sustainable development goals target.

2.5 CONCLUSION

PHL occurs rampantly and the magnitude is threatening food security worldwide. The causes of PHL are mainly connected to financial, managerial, knowledge and technical limitations in harvesting, handling, infrastructure, and facilities to prolong commodities quality and postharvest shelf life. Interventions strategies from government and nongovernment agencies as well as private sectors have a vital role in addressing PHL as it is a complicated and difficult task involving multiple parties along the entire food supply chain. Regional involvement and collaboration are also needed to facilitate exchange of information. A holistic and comprehensive approach integrating effective policies, education, training, infrastructure, and research should be part of the strategies in combating PHL.

KEYWORDS

- **attitude**
- **coating**
- **food security**
- **harvesting**
- **knowledge**
- **maturity**

REFERENCES

Abbott, B., Holford, P., & Golding J. B. (2009). Comparison of "Cripps Pink" apple bruising. *Acta Horticulturae, 880*, 223–230.

Adhikary, T., Gill, P. P. S., Jawandha, S. K., Bhardwaj, R. D., & Anurag, R. K. (2021). Efficacy of postharvest sodium nitroprusside application to extend storability by regulating physico-chemical quality of pear fruit. *Food Chemistry, 346*, 128934.

Ahmadi, E. (2012). Bruise susceptibilities of kiwifruit as affected by impact and fruit properties. *Research in Agricultural Engineering, 58*(3), 107–113.

Alam, A. U., Rathi, P., Beshai, H., Sarabha, G. K., & Deen, M. J. (2021). Fruit quality monitoring with smart packaging. *Sensors, 21*, 1509.

Alamar, M. del C., Falagán, N., Aktas, E., & Terry, L. A. (2017). Minimizing food waste: A call for multidisciplinary research. *Journal of the Science of Food and Agriculture, 98*(1), 8–11.

Al-Dairi, M., Pathare, P. B., & Al-Yahyai, R. (2021). Effect of postharvest transport and storage on color and firmness quality of tomato. *Horticulturae, 7*, 163.

Al-Dairi, M., Pathare, P. B., Al-Yahyai, R., & Opara, U. L. (2022). Mechanical damage of fresh produce in postharvest transportation: Current status and future prospects. *Trends in Food Science and Technology, 124*, 195–207.

Ali, S., Khan, A. S., Anjum, M. A., Ejaz, S., Khaliq, G., Hussain, S., Ahmad, R., & Saleem, M. S. (2022). Pre-and postharvest physiological disorders of citrus fruit. In Citrus Production (pp. 371-390). CRC Press.

Ali, A., Xia, C., Ouattara, N., Mahmod, I., & Faisal, M. (2021). Economic and environmental consequences' of postharvest loss across food supply chain in the developing countries. *Journal of Cleaner Production, 323*, 129146.

Ali, M. H., Khan, A. S., Jaskani, M. J., Anwar, R., Ali, S., Malik, A. U., Hasan, M. U., Rehman, R. N. U., & Ayyub, S. (2022). Pre-storage application of L-arginine mitigates chilling injury and maintains quality of Sandhuri guava fruit. *Journal of Food Processing and Preservation, 46*(3), e16405.

Alzate, J. V. (2015). Demonstration with lecture and guided tour: Effective technology transfer methods on mango postharvest technologies. *International Journal of Novel Research in Education and Learning, 2*(6), 1–6.

Arah, I. K., Kumah, E. K., Anku, E. K., & Amaglo, H. (2015). An overview of postharvest losses in tomato production in Africa: Causes and possible prevention strategies. *Journal of Biology, Agriculture and Healthcare, 5*(16), 78–88.

Ashitha, G., Sonaraj, P. R., Sooraj K. P. M., & Thejus S. (2021). Zero energy cooling chamber and zero energy cooling system in building wall. *International Journal of Engineering Research & Technology, 9*(6), 100–106.

Awalgaonkar, G., Beaudry, R., & Almenar, E. (2020). Ethylene-removing packaging: Basis for development and latest advances. *Comprehensive Reviews in Food Science and Food Safety, 19*(6), 3980–4007.

Azene, M., Workneh, T. S., & Woldetsadik, K. (2014). Effect of packaging materials and storage environment on postharvest quality of papaya fruit. *Journal of Food Science and Technology, 51*(6), 1041–1055.

Bateno, T., & Buke, T. (2018). Postharvest losses and management assessment of commercial fruit and vegetable at Wolaitazone, at Sodo Market. *International Journal of Research and Innovations in Earth Science, 5*(3), 99–105.

Becker, B. R., & Fricke, B. A. (1996). Transpiration and respiration of fruits and vegetables. In *Proceedings of the meeting of Commission C2 with Commissions B2, D1 and D2–3, of the International Institute of Refrigeration*, October 2–4, pp. 110–121.

Bekele, D. (2021). Role of postharvest management for food security: A review. *Advances in Crop Science and Technology, 9*(7), 475.

Burdock, R. P., & Ampt, P. (2018). The characteristics of five food production systems and their implications for sustainable landscapes. *Journal of Agricultural Science, 10*(2), 23–47.

Buet, A., Steelheart, C., Perini, M. A., Galatro, A., Simontacchi, M., & Grozef, G. E. G. (2021). Nitric oxide as a key gasotransmitter in fruit postharvest: An overview. *Journal of Plant Growth Regulation, 40*, 2286–2302.

Carnelossi, M. A. G., Sena, E. O. A., Berry, A. D., & Sargent, S. A. (2019). Effect of forced-air cooling, hydrocooling, or their combination on fruit quality of two Southern Highbush blueberry cultivars. *HortScience, 54*(1), 136–142.

Chang, L. Y., & Brecht, J. K. (2020). Variation among strawberry cultivars in bruising susceptibility related to wound ethylene production and sensitivity. *HortScience, 55*(4), 444–448.

Chiemeke, S. C., & Evwiekpaefe, A. E. (2011). A conceptual framework of a modified unified theory of acceptance and use of technology (UTAUT) model with Nigerian factors in E-commerce adoption. *Educational Research, 2*(12), 1719–1726.

Chuang, J. H., Wang, J. H., & Liou, Y. C. (2020). Farmers' knowledge, attitude and adoption of smart technology in Taiwan. *International Journal of Environmental Research and Public Health, 17*(19), 7236.

Defraeye, T., Tagliavini, G., Wu, W., Prawiranto, K., Schudel, S., Kerisima, M. A., Verboven, P., & Bühlmann, A. (2019). Digital twins probe into food cooling and biochemical quality changes for reducing losses in refrigerated supply chains. *Resources, Conservation and Recycling, 149*, 778–794.

Delgado, L., Schuster, M., & Torero, M. (2021). Quantity and quality food losses across the value chain: A comparative analysis. *Food Policy, 98*, 101958.

Ding, P., & Choon, S. Y. (2021). Effects of storage temperature on postharvest quality of Malaysian grown fig (*Ficus carica* L.) cv. Ipoh Blue Giant. *Mindanao Journal of Science and Technology, 19*(2), 91–106.

Ding, P., & Khairul, B. D. (2009). Responses of *Musa* AAA Berangan to 1-methycyclopropene. *Pertanika Journal of Tropical Agricultural Science, 32*(2), 125–132.

Ding, P., & Lee P. E. (2018). Effect of gum Arabic and sodium bicarbonate coating on disease development of red-fleshed pitaya fruit (*Hylocereus polyrhizus*). In *Proceedings at the International Symposium on Horticulture 2018—Emerging Challenges and Opportunities in Horticulture Supporting Sustainable Development Goals*, November 27–30, pp. 151–160.

Ding, P., & Nur Ayuni, M. (2015). UV-C irradiation enhanced total phenolic contents and antioxidant capacity of *Musa* AAA "Berangan". *Acta Horticulturae,* 213–216.

Ding, P., & Ong, P. T. (2010). Extending Kampuchea guava shelf life at 27°C using 1-methylcyclopropene. *International Food Research Journal, 17*(1), 63–69.

Ding, P., & Raja Norhidayu, R. M. (2021). Hydro-cooling as means to retain fresh sweet corn ears quality. *Sains Malaysiana 50*(12), 3593–3602.

Ding, P., & Salumiah, M. (2013). Physico-chemical characteristics of Chok Anan mango fruit after hot water treatment. *Pertanika Journal of Tropical Agricultural Science, 36*(4), 359–372.

Dirapan, P., Boonyakiat, D., & Poonlarp, P. (2021). Improving shelf life, maintaining quality, and delaying microbial growth of broccoli in supply chain using commercial vacuum cooling and package icing. *Horticulturae 7*(11), 506.

Dovlo, M., Kunu, E. K., & Rudin, N. A. (2021). Statistical analysis on factors that contribute to postharvest losses of crops in rural area at Nkwanta Districs, Volta Regions, Ghana. *Jambura Agribusiness Journal, 2*(2), 46–54.

Dubey, N., & Rahman, N. M. (2016). Effect of CoolBot cool room on shelf-life of cabbage and cauliflower. *American International Journal of Research in Formal, Applied and Natural Sciences, 13*(1), 66–69.

Egwuonwu, H. A. (2020). Factors influencing postharvest losses among vegetable farmers in Mbaitoli local government area in Imo State. *Asian Journal of Agricultural and Horticultural Research, 7*(1), 22–28.

Elmsly, T. A., Mandemaker, A. J., Dixon, E. M., Smith, D. B., & Dixon, J. (2007). Influence of time of day of harvest on "Hass" avocado ripe fruit quality. *New Zealand Avocado Growers' Association Annual Research Report, 7*, 91–96.

FAO. (2018). *Food losses and waste reduction and value chain development for food security in Egypt and Tunisia*. FAO.

FAO. (2019). *The state of food and agriculture 2019: Moving forward on food loss and waste reduction.* Rome License, CC BY-NC-SA 3.0 IGO. FAO.

FAO. (2011). *Global Food Losses and Food Waste–Extent, Causes and Prevention*. FAO.

FAO, IFAD, UNICEF, WFP, & WHO. (2021). *The state of food security and nutrition in the world 2021. Transforming food systems for food security, improved nutrition and affordable healthy diets for all*. FAO.

Flanagan, K., Robertson, K., & Hanson, C. (2019). *Reducing food loss and waste: Setting a global action agenda*. World Resources Institute.

Gaikwad, K. K., Singh, S., & Negi, Y. S. (2020). Ethylene scavengers for active packaging of fresh food produce. *Environmental Chemistry Letters, 18*, 269–284.
Gunasegaran, B., Ding, P., & Kadir, J. (2018). Morphological identification and *in vitro* evaluation of *Colletotrichum gloesporioides* in "Chok Anan" mango using UV-C irradiation. *Acta Horticulturae,* 599–602.
Gunasekera, D., Parsons, H., & Smith, M. (2017). Postharvest loss reduction in Asia-Pacific developing economies. *Journal of Agribusiness in Developing and Emerging Economies, 7*(3), 303–317.
Gustavsson, J., Cederberg, C., Sonesson, U., van Otterdijk, R., & Meybeck, A. (2011). *Global food losses and food waste: Extent, causes and prevention.* FAO.
Hailu, G., & Derbew, B. (2015). Extent, causes and reduction strategies of postharvest losses of fresh fruits and vegetables—A review. *Journal of Biology, Agriculture and Healthcare, 5*(5), 49–64.
Hossain, M. A., Khatun, M., Matin, M. A., & Dewan, M. F. (2017). Postharvest loss assessment of major fruits grown in hill regions of Bangladesh. *Bangladesh Journal of Agricultural Research, 42*(1), 171–184.
Islam, M. P., & Morimoto, T. (2014). A new zero energy cool chamber with a solar-driven adsorption refrigerator. *Renewable Energy, 72*, 367–376.
Iordachescu, G., Ploscutanu, G., Pricop E. M., Baston, O., & Barna, O. (2019). Postharvest losses in transportation and storage for fresh fruits and vegetables sector. *Journal of International Scientific Publications, 7*, 244–251.
Kahramanoğlu, I., Rengasamy, K. R. R., Usanmaz, S., Alas, T., Halvaci, M., Okatan, V., Askin, M. A., & Wan, C. (2021). Improving the safety and security of fruits and vegetables during COVID-19 pandemic with postharvest handling. *Critical Reviews in Food Science and Nutrition,* 1–11.
Kalaïtzis, P., Bita, E. C., & Hilmi, M. (2016). *Innovative postharvest technologies for sustainable value chain.* https://www.researchgate.net/publication/331207145_INNOVATIVE_POSTHARVEST_TECHNOLOGIES_FOR_SUSTAINABLE_VALUE_CHAIN. Accessed on Aug 20, 2022.
Kereth G. A., Lyimo, M., Mbwana, H. A., Mongi, R. J., & Ruhembe, C. C. (2013). Assessment of postharvest handling practices: Knowledge and losses of fruits in Bagamoyo district of Tanzania. *Food Science and Quality Management, 11*, 8–15.
Kitinoja, L., Saran, S., Roy, S. K., & Kader, A. A. (2011). Postharvest technology for developing countries: Challenges and opportunities in research, outreach and advocacy. *Journal of the Science of Food and Agriculture, 91*(4), 597–603.
Kitinoja, L., & Barrett, D. M. (2015). Extension of small-scale postharvest horticulture technologies-A model training and services center. *Agriculture, 5*(3), 441–455.
Kumar, V., Purbey, S. K., & Anal, A. K. D. (2016). Losses in litchi at various stages of supply chain and changes in fruit quality parameters. *Crop Protection, 79*, 97–104.
Khalid, M., Riaz, M., Mahmood, M. H., Miyazaki, T., Sultan, M., Ashraf, M. N., & Khan, Z. M. (2020). Study on zero energy evaporative cooling chamber for agricultural products storage. In *Proceeding of International Exchange and Innovation Conference on Engineering & Sciences (IEICES)* (Vol. 6, pp. 104–109).
Khaliq, G. H. U. L. A. M. (2015). Effects of edible coatings enriched with calcium chloride on physiological, biochemical and quality responses of mango (*Mangifera indica* L. cv.

Choke Anan) fruit during cold storage. Universiti Putra Malaysia, Serdang, Selangor, Malaysia.

Khaliq, G., Ali, S., Gapper, N., & Nicola, S. (2023). Recent advances and approaches in the application of elicitors to enhance resistance mechanisms in fresh produce. Frontiers in Sustainable Food Systems, 6, 1061079.

Krupa, T., & Tomala, K. (2021). Effect of oxygen and carbon dioxide concentration on the quality of minikiwi fruits after storage. *Agronomy, 11,* 2251.

Khan, L., Khaliq, G., Ullah, E., Abdul-Rahaman, A., El-Mogy, M. M., Kesba, H. H., Ismail, A.M., & El-Beltagi, H. S. (2025). Guggul gum and clove essential oil synergistically reduced anthracnose disease and modulated physicochemical changes in tomato during storage. *Journal of Stored Products Research, 114,* 102713.

Kuyu, C. G., Tola, Y. B., & Abdi, G. G. (2019). Study on postharvest quantitative and qualitative losses of potato tubers from two different road access districts of Jimma zone, South West Ethiopia. *Heliyon, 5*(8), e02272.

Lacap, A. T., Bayogan, E. R. V., Secretaria, L. B., Joyce, D. C., Ekman, J. H., & Goldwater, A. D. (2021). Bruise injury and its effect on Carabao mango fruit quality. *Philippine Journal of Science, 150* (6B), 1689–1701.

Ladapo M. A., & Oladele, O. I. (2011). Effect of knowledge, attitude and constraints on postharvest losses among plantain farmers and wholesalers in south-western Nigeria. *Life Sciences Journal, 8*(2), 476–482.

Lal, S., Khaliq, G., Mastoi, A. H., Irshad, A., Jaffar, M. A., Khan, L., & Muhammad, N. (2025). Effect of *Fagonia indica* and garlic extract on *Alternaria alternata* and post-harvest quality of tomato fruit during storage. *Journal of Crop Health, 77*(6), 1–13.

Li, L., Li, C., Sun, J., Sheng, J., Zhou, S., Xing, M., Yi, P., He, X., Zheng, F., Tang, Y., Li, J., & Tang, J. (2020). The effects of 1-methylcyclopropene in the regulation of antioxidative system and softening of mango fruit during storage. *Hindawi Journal of Food Quality, 6090354*, 1–11.

Li, J., Karkee, M., Zhang, Q., Xiao, K., & Feng, T. (2016). Characterizing apple picking patterns for robotic harvesting. *Computers and Electronics in Agriculture, 127*, 633–640.

Lu, S., Cheng, G., Li, T., Xue, L., Liu, X., Huang, J., & Liu, G. (2022). Quantifying supply chain food loss in China with primary data: A large-scale, field-survey based analysis for staple food, vegetables, and fruits. *Resources, Conservation and Recycling, 177*, 106006.

Lufu, R., Ambaw, A., & Opara, U. L. (2020). Water loss of fresh fruit: Influencing pre-harvest, harvest and postharvest factors. *Scientia Horticulturae, 272*, 109519.

Makule, E., Dimoso, N., & Tassou, S. A. (2022). Precooling and cold storage methods for fruits and vegetables in sub-Saharan Africa—A review. *Horticulturae, 8*, 776.

Malik, U. A., Hasan, M. U., Hassan, W. U., Khan, A. S., Shah, M. S., Rajwana, I. A., Latif, M., & Anwar, R. (2021). Postharvest quarantine vapour heat treatment attenuates disease incidence, maintains eating quality and improves bioactive compounds of "Gola" and "Surahi" guava fruits. *Journal of Food Measurement and Characterization, 15*, 1666–1679.

Marouf, A. M., & Khali M. (2020). Study the effect of postharvest heat treatment on infestation rate of fruit date palm (*Phoenix dactylifera* L.) cultivars grown in Algeria. *Journal of Nutritional Science and Healthy Diet, 1*(1), 37–40.

McArtney, S., Parker, M., Obermiller, J., & Hoyt, T. (2011). Effects of 1-methycyclopropene on firmness loss and the development of rots in apple fruit kept in farm markets or at elevated temperatures. *HortTechnology, 21*(4), 494–499.
McKenzie, T. J., Singh-Peterson, L., & Underhill, S. J. R. (2017). Quantifying postharvest loss and the implication of market-based decisions: A case study of two commercial domestic tomato supply chains in Queensland, Australia. *Horticulturae, 3*(3), 44.
McNamara, P. E., & Tata, J. S. (2015). Principles of designing and implementing agricultural extension programs for reducing postharvest loss. *Agriculture, 5*(4), 1035–1046.
Mditshwa, A., Khaliq, G., Hussein, Z., & Ejaz, S. (2023). Sustainable postharvest management practices for fresh produce. *Frontiers in Sustainable Food Systems, 7*, 1143759.
Meena, M. S., Kumar, A., Singh, K. M., & Meena, R. (2009). Farmers' attitude towards postharvest issues of horticultural crops. *Indian Journal of Extension Education, 9*(3), 15–19.
Mijin, S., Ding, P., Saari, N., & Ramlee, S. I. (2021). Effects of pollination techniques and harvesting stage on the physico-chemical characteristics of jackfruit. *Scientia Horticulturae, 265*, 110199.
Mirshekari, A., Ding, P., & Ghazali, H. M. (2015). Enzymatic activity and microstructural changes of hot water treated banana during ripening. *Journal of Agricultural Science and Technology, 17*, 949–962.
Mishra, A., Jha, S. K., & Ojha, P. (2020). Study on zero energy cool chamber (ZECC) for storage of vegetables. *International Journal of Scientific and Research Publications, 10*(1), 427–433.
Mohajan, H. K. (2016). Knowledge is an essential element at present world. *International Journal of Publication and Social Studies*, *1*(1), 31–53.
Mukama, M., Ambaw, A., Michael, T., & Linus, U. (2019). Analysing the dynamics of quality loss during precooling and ambient storage of pomegranate fruit. *Journal of Food Engineering, 245*, 166–173.
Muhammad, R. H., Hionu, G. C., & Olayemi, F. F. (2012). Assessment of the postharvest knowledge of fruits and vegetable farmers in Garun Mallam L. G.A of Kano, Nigeria. *International Journal of Development and Sustainability, 1*(2), 510–515.
Mujuka, E., Mburu, J., Ogutu, A., & Ambuko, J. (2020). Returns to investment in postharvest loss reduction technologies among mango farmers in Embu County, Kenya. *Food Energy Security, 9*, e195.
Muroyiwa, B., Sholopa, L., Liloetla, P., & Rantlo, M. (2020). Integration of postharvest management in agricultural policy and strategies to minimise postharvest losses in Lesotho. *Journal of Development and Agricultural Economics, 12*(2), 84–94.
Mwaijande, F. (2017). Farmers' adoption and willingness to pay for postharvest technologies in Tanzania: Policy implication for enhancing food security. *Journal of Postharvest Technology, 5*(1), 1–6.
Nuevo, P., & Apaga, A. R. (2010). Technology reducing postharvest losses and maintaining quality of fruits and vegetables (Philippines). In *Proceedings of 2010 AARDO Workshop on Technology on Reducing Post-harvest Losses and Maintaining Quality of Fruits and Vegetables*, October 3–9, pp 154–167.
Nuratika, T. S. M., Ding, P., Jugah, K., & Hasanah, M. G. (2017). Biochemical and cell wall ultrastructural changes in crown tissue of banana (*Musa* AAA "Berangan") fruit as

mediated by UVC irradiation against crown rot fungal infection. *Postharvest Biology and Technology, 128*, 144–152.

Olgyaiova, K., Pongrácz, E., Mikkola, T., Radoslav, S. K., & Keiski, R. L. (2005). Attitudes toward waste minimization in Finland and Czech Republic—Barriers and drivers. In *Proceedings of the RESOPT Closing Seminar "Waste Minimization and Utilization in Oulu Region: Drivers and Constraints"* (pp. 85–109).

Onwude, D. I., Chen, G., Eke-emezie, N., Kabutey, A., Khaled, A. Y., & Sturm, B. (2020). Recent advances in reducing food loses in the supply chain of fresh agricultural produce. *Processes, 8*, 1431.

Osman, A., Saari, N., Saleh, R., Bakar, J., Zainal, N. D., & Yacob, M. (2009). Postharvest handling practices on selected local fruits and vegetables at different levels of the distribution chain. *Journal of Agribusiness Marketing, 2*, 39–53.

Palou, L. (2018). Postharvest treatments with GRAS salts to control fresh fruit decay. *Horticulturae, 4*(4), 46.

Palou, L., Moscoso-Ramírez, P. A., & Montesinos-Herrero, C. (2018). Assessment of optimal postharvest treatment conditions to control green mold of oranges with sodium benzoate. *Acta Horticulturae, 1194*, 221–226.

Pelletier, W., Brecht, J. K., Nunes, M. C. N., & Emond, J. P. (2011). Quality of strawberries shipped by truck from California to Florida as influenced by postharvest temperature management practices. *HortTechnology, 21*(4), 482–493.

Pereira, E., Silva, R. G. B., Spagnol, W. A., & Junior, V. S. (2018). Water loss in table grapes: Model development and validation under dynamic storage conditions. *Food Science and Technology (Campinas), 38*(3), 473–479.

Prasad, P., & Kochhar, A. (2014). Active packaging in food industry: A review. *Journal of Environmental Science, Toxicology and Food Technology, 8*(5), 1–7.

Rajapaksha, L., Gunathilake, D. M. C. C., Pathirana, S. M., & Fernando, T. N. (2021). Reducing postharvest losses in fruits and vegetables for ensuring food security—Case of Sri Lanka. *MOJ Food Processing & Technology, 9*(1), 7–16.

Ren, L., Qian, L., Xue, M., Peng, C., Chen, N., & Zhan, G. (2022). Vapor heat treatment against *Planococcus lilacinus* Cockerell (Hemiptera: Pseudococcidae) on dragon fruit. *Pest Management Science, 78*(1), 150–158.

Ren, Y., Xue, Y., Tian, D., Zhang, L., Xiao, G., & He, J. (2020). Improvement of postharvest anthracnose resistance in mango fruit by nitric oxide and the possible mechanisms involved. *Journal of Agricultural and Food Chemistry, 68*(52), 15460–15467.

Rose, D. C., Sutherland, W. J., Parker, C., Lobley, M., Winter, M., Morris, C., & Dicks, L. V. (2016). Decision support tools for agriculture: Towards effective design and delivery. *Agricultural Systems, 149*, 165–174.

Ruslan, N. Z., Man, N., Nawi, N. M., & Ding, P. (2013). Factors that influence the implementation of postharvest handling practices among fresh vegetable producers in selected states in Malaysia. *Journal of International Food & Agribusiness Marketing, 25*, 87–97.

Saran, S., Roy, S. K., & Kitinoja, L. (2012). Appropriate postharvest technologies for small scale horticultural farmers and marketers in Sub-Saharan Africa and South Asia-Part 2. Field trial results and identification of research needs for selected crops. *Acta Horticulturae, 934*, 41–52.

Satekge, T. K., & Magwaza, L. S. (2020). The combined effect of 1-methylcyclopropene (1-MCP) and ethylene on green life and postharvest quality of banana fruit. *International Journal of Fruit Science, 20*(Sup 3), S1539–S1551.

Schirra, M., D'Aquino, S., Cabras, P., & Angioni, A. (2011). Control of postharvest diseases of fruit by heat and fungicides: Efficacy, residue levels, and residue persistence—A review. *Journal of Agricultural and Food Chemistry, 59*, 8531–8542.

Shahidah, M. N., Siti, N. F., & Ding, P. (2021). Effects of temperature and storage duration on keeping quality of Malaysian grown passion fruit (*Passiflora edulis* Sims f. flavicarpa Deg.). *Acta Horticulturae, 1327*, 557–562.

Shahidah, M. D., & Ding, P. (2020). Trends and advances in edible biopolymer coating for tropical fruit: A review. *Food Research International, 134*, 109208.

Sharma, M. K., Kumar, S., Kumar, A., & Kumar, S. (2019). Cold chain management in perishable commodities for minimum postharvest losses: A review. *International Journal of Chemical Studies, SP6*, 405–411.

Shinga, M. H., Silue, Y., & Fawole, O. A. (2025). Recent Advancements and Trends in Postharvest Application of Edible Coatings on Bananas: A Comprehensive Review. Plants, 14(4), 581.

Singh, S. K., & Singh, A. K. (2018). Factors affecting the extent of knowledge about mango postharvest management practices. *Journal of Pharmacognosy and Phytochemistry, SP1*, 1041–1043.

Singh, A., Vaidya, G., Jagota, V., Darko, D. A., Agarwal, R. K., Debnath, S., & Potrich, E. (2022). Recent advancement in postharvest loss mitigation and quality management of fruits and vegetables using machine learning frameworks. *Hindawi Journal of Food Quality,* 6447282.

Sivakumar, D., Jiang, Y., & Yahia, E. M. (2011). Maintaining mango (*Mangifera indica* L.) fruit quality during the export chain. *Food Research International, 44*(5), 1254–1263.

Socheath, T., Buntong, B., & Acedo, A. Jr. (2020). Postharvest management options to improve tomato value chain in Cambodia. *International Journal of. Environmental and Rural Development, 11*(2), 154–162.

Sogvar, O. B., Razavi, F., Rabiei, V., & Gohari, G. (2020). Postharvest application of L-cysteine to prevent enzymatic browning of Stanley plum fruit during cold storage. *Journal of Food Processing and Preservation, 44*(10), e14788.

Sohail, M., Wills, R. B. H., Bowyer, M. C., & Pristijono, P. (2021). Multiple amino acids inhibit postharvest senescence of broccoli. *Horticulturae, 7*(4), 71.

Sugri, I., Abubakari, M., Owusu, R. K., & Bidzakin, J. K. (2021). Postharvest losses and mitigating technologies: Evidence from Upper East Region of Ghana. *Sustainable Futures, 3*, 100048.

Suthar, R. G., Barrera, J. I., Judge, J., Brecht, J. K., Pelettier, W., & Muneepeerakul, R. (2019). Modeling postharvest loss and water and energy use in Florida tomato operations. *Postharvest Biology and Technology, 153*, 61–68.

Taufiq, S. A. (2017). *Postharvest handling practices on vegetable crops at different stages of the distribution chain: A case study in Selangor, Malaysia*. MSc Thesis Management Studies, Wageningen University, 63 p.

Teferra, T. F. (2022). The cost of postharvest losses in Ethiopia: Economic and food security implications. *Heliyon, 8*, e09077.

Thole, V., Vain, P., & Martin, C. (2021). Effect of elevated temperature on tomato postharvest properties. *Plants, 10*(11), 2359.

Tria, D., Alghorbany, A., Muhamad, A. I., & Alam, M. M. (2020). Government policies, financial scopes and technological usages for agricultural development and postharvest loss reduction in Algeria. *International Journal of Postharvest Technology and Innovations, 7*(4), 335–352.

Ul Haq, Z., Khaliq, G., Irshad, A., El-Beltagi, H. S., Mahmoud Ismail, A., & El-Mogy, M. M. (2024). Effect of beeswax and lemon grass oil on *Alternaria alternata* and post-harvest quality of bitter gourd during storage. *New Zealand Journal of Crop and Horticultural Science,* 1–21.

Wang, L., Hong K., Xu, R., Zhao, Z., & Cao, J. (2021). The alleviation of cold-stimulated flesh reddening in "Friar" plum fruit by the elevated CO_2 with polyvinyl chloride (PVC) packaging. *Scientia Horticulturae, 281*, 109997.

Wang, S. Y., Shi, X. C., Wang, R., Wang, H. L., Liu, F., & Laborda, P. (2020). Melatonin in fruit production and postharvest preservation: A review. *Food Chemistry, 320*, 126642.

Xu, T., Chen, Y., & Kang, H. (2019). Melatonin is a potential target for improving postharvest preservation of fruits and vegetables. *Frontiers in Plant Sciences, 10*, 1388.

Yahia, E. M., Fonseca, J. M., & Kitinoja, L. (2019). Chapter 2—Postharvest losses and waste. In M. Y. Elhadi (Ed.), Postharvest technology of perishable horticultural commodities (pp. 43–69). Woodhead Publishing.

Yeshiwas, Y., & Tadele, E. (2021). An investigation into major causes for postharvest losses of horticultural crops and their handling practice in Debre Markos, North-Western Ethiopia. *Hindawi Advances in Agriculture, 1985303, 1*–10.

Zaharan, H., Fawole, O. A., & Opara, U. L. (2020). Harvest and postharvest factors affecting bruise damage of fresh fruits. *Horticultural Plant Journal, 6*(1), 1–13.

Zahir, S. S., Ding, P., Sabir, A. A., Atif, A., Yaqubi, A., & Yusoff, S. F. (2021). Maintaining antioxidants in tomato fruit using chitosan and vanillin coating during ambient storage. *Food Research, 5*(5), 274–286.

Zahir, S. S., Ding, P., Nakasha, J. J., & Yusoff, S. F. (2020). Combining chitosan and vanillin to retain postharvest quality of tomato fruit during ambient temperature storage. *Coatings, 10*(12), 1222.

Zainal, B., Ding, P., Ismail, I. S., & Saari, N. (2019). Physico-chemical and microstructural characteristics during postharvest storage of hydro-cooled rockmelon (*Cucumis melo* L. reticulatus cv. Glamour). *Postharvest Biology and Technology, 152*, 89–99.

Zhao, Y., Zhu, X., Hou, Y., Wang, X., & Li, X. (2019a). Effects of nitric oxide fumigation treatment on retarding cell wall degradation and delaying softening of winter jujube (*Ziziphus jujuba* Mill. cv. Dongzao) fruit during storage. *Postharvest Biology and Technology, 156*, 110954.

Zhao, D., Yu, Y., Shen, Y., Liu, Q., Zhao, Z., Sharma, R., & Reiter, R. J. (2019b). Melatonin synthesis and function: Evolutionary history in animals and plants. *Frontiers in Endocrinology, 10*, 249.

CHAPTER 3

Postharvest Physiological Disorders in Fruits and Vegetables

NIRMAL KUMAR MEENA[1*], K. CHOUDHARY[2], SAJEEL AHMAD[3], MENAKA M.[3], and VIVEK KUMAR SAURABH[3]

[1]Department of Fruit Science, College of Horticulture and Forestry, Jhalawar Agriculture University, Kota, India

[2]Agriculture University, Jodhpur, India

[3]Division of Food Science and Postharvest Technology, ICAR-Indian Agricultural Research Institute, New Delhi, India

*Corresponding author

ABSTRACT

Postharvest physiological disorders are non-pathogenic ailments that significantly reduce the quality and shelf life of the produce. Certain environmental factors such as high- and low-temperature stress, nutritional deficiencies, heat injury, and storage atmospheric imbalance are responsible factors for such physiological disorders. Detection and correct identification of symptoms are needed to implement correct management strategies. Application of pre- and postharvest nutrient solutions, hormones, along with cultural management practices, are the main approaches to control these incidences. This chapter describes postharvest physiological disorders, symptoms, mechanism, and their possible management.

3.1 INTRODUCTION

Fruits are considered as "protective food" due to their enormous health benefits since they are rich in phytonutrients. Certain preharvest factors

Physiological Disorders of Fruits and Vegetables: Causes and Remedies.
Ghulam Khaliq, Sajid Ali, and Mohammed Wasim Siddiqui (Eds.)

DOI: 10.1201/9781998511334-3

such as irrigation, temperature, mineral nutrition, canopy management, rootstock, and postharvest factors like treatments and storage environment are crucial in determining the postharvest quality of fresh produce. The biochemical composition of fruits varies greatly with preharvest conditions, and after harvest there is no scope for further improvement. Some recent studies showed that climate change has negatively affected the food value of fruits (Hertel et al., 2010).

Physiological disorders are maladies caused due to non-pathogenic organisms or abiotic factors such as weather, nutritional deficiency, or storage-related problems (Table 3.1). Preharvest management of mineral nutrition is necessary to maintain the quality and normal metabolic process of fruit and vegetables. Postharvest metabolic processes are mainly regulated by storage temperature and that determines the shelf life of the produce (Madebo et al., 2021; Mditshwa et al., 2023). Fruits exposed to low temperatures (0–13°C) often develop chilling injury, characterized by brown discoloration, surface pitting, off-flavor, leakage, surface streaks, and susceptibility to microbial attack (Barman et al., 2015; Mishra et al., 2016; Ali et al., 2021). Freezing temperature (below −1 degree Celsius) causes tissue disruption mainly in temperate fruits. Apart from this, high temperatures cause scald on the surface, especially in pomegranate, citrus, and karonda. Mineral deficiencies are more prevalent in fruit orchards and most postharvest disorders are caused due to deficiency of Ca, K, boron, and Zn. Bitter pit of apple is a well-known example of Ca deficiency-related disorder (Mishra et al., 2016; Zhang et al., 2020). Granulation in citrus is sometimes caused due to rootstock and fertilizer imbalances. Black tip, senescence, and premature ripening are mainly caused by harmful gases and ethylene injury, respectively. Moreover, many physiological disorders such as oleocellosis, rind breakdown, black tip, etc. are results of complex interactions between weather stresses and mineral deficiencies.

These physiological disorders are characterized by several typical symptoms such as surface injury, discoloration, browning, pitting, and tissue breakdown. However, these symptoms vary from fruit to fruit, and the severity of the disorder (Zhang et al., 2020). Most chilling-injured fruits show symptoms after removal from cold storage and can be characterized by pitting and surface discoloration. The jelly seed of mango is difficult to identify from the surface rather it can be seen only after cutting the fruits. Jelly seed can be identified through nondestructive methods like X-ray imaging, CT-scan and electron microscopy. Sap burn injury may appear on the surface as dark brown streaks and burnt spots.

There are several management strategies to control these disorders. Both pre- and postharvest approaches need to be implemented in order to reduce the incidence of the disorder. Use of pre-harvest spray of minerals, and hormones, use of balanced fertilizers, proper harvesting at optimum maturity stage, and maintaining ideal composition inside cold storage including temperature are major approaches (Lafuente and Zacarias, 2006; Zheng et al., 2011; Mishra et al., 2016). Recently, the use of novel phytohormones such as melatonins, Brassinosteroids, GABA, and plant extracts gained attention.

3.2 CLASSIFICATION OF POSTHARVEST DISORDERS

There are several determining factors that cause problems in postharvest produce. Several pre-harvest factors such as temperature, mineral nutrition, mechanical bruising, water, rootstock, fruit position, crop load, etc., can cause disorder that remains up to the supply chain (Hewett, 2006; Ferguson et al., 1999). However, postharvest factors are more prevalent, especially during storage which affects the overall quality and shelf life of the produce. These postharvest factors include storage temperature, ethylene, gaseous combination, and postharvest treatments (Khaliq et al., 2023; Khan et al., 2025; Lal et al., 2025). Overall, postharvest disorders can be classified into several categories based on their causes. Nutritional deficiency related, environmental factors related, storage related, temperature related, low oxygen/elevated CO_2 related, and others are the major broad categories (Table 3.1).

3.3 CALCIUM-RELATED DISORDERS

Fruit contains a significant amount of calcium (Ca), which is also used in numerous physiological processes as an intracellular signal transduction molecule. Through the maintenance of cell membranes and appropriate photosynthetic function, calcium is essential for plant stress tolerance (Qu et al., 2012; De Freitas et al., 2016). It has long been a tradition to dip various fruits and vegetables in calcium salts to extend their shelf life. Calcium is a crucial element for maintaining a number of physiological processes in plants. It is very common practice to apply calcium salts, such as calcium chloride, calcium phosphate, calcium propionate, and calcium gluconate, both before and after harvest to minimize numerous physiological disorders with fruits and vegetables (Aguayo et al., 2008; Manganaris et al., 2005). The occurrence of physiological disorders in apple fruit after harvest, such as bitter pit (BP), has been linked to low calcium (Ca) content and high amounts of magnesium (Mg), potassium (K), and nitrogen (N) (de Freitas et al., 2012; de Freitas et al., 2015). Calcium ions (Ca^{2+}), a vital component for plant cells, function as second messengers and are essential for the regulation of plant physiology, especially in postharvest fruits and vegetables. Numerous studies have been conducted on the functions of calcium and calcium-sensing receptors in plants, and it has been discovered that certain of these receptors also contribute to the ripening of fruit (Khaliq, 2015; Gao et al., 2019). It is well-recognized that calcium plays a crucial role in many physiological processes including membrane formation and function as well as enzyme activity in plants. The loss of cell compartmentalization and the enzyme processes that result in these disorder symptoms may be prevented or delayed by strengthening the cell components (Wills and Mahendra, 1989).

3.3.1 MECHANISM

Calcium ions create a cross-link or bridge between the free carboxyl groups of these pectin chains, strengthening the cell and preserving the produce's textural quality.

Numerous fruits and vegetables experience calcium-related physiological disorders that are typically accompanied by membrane integrity loss, loss of cell compartmentalization, and finally cell death in their tissues. For maintaining proper cell-wall structure and lowering the frequency of internal breakdown diseases in horticultural crops, sufficient Ca^{2+} is thought to be crucial. The stabilization of membrane systems and production of calcium pectates, which strengthen the rigidity of the central region and cell wall of the fruit, have been credited with maintaining firmness and resistance to softening as a result of calcium administration. Calcium (Ca), which is required to develop strong fruit tissue, superior quality, and long shelf life, is dependent on xylem translocation. Ca travels with the transpiration stream and bonds with polysaccharides to strengthen cell walls. Calcium stops the deterioration of cell walls, leaky membranes, and early senescence. Plant disease resistance has been linked to calcium's involvement in preserving firmness. Applying calcium to fruits may increase their resistance to postharvest pathogens or decrease their susceptibility to postharvest diseases and disorders. By forming a bond with pectin and improving the cell wall's cohesion, calcium ions improve

TABLE 3.1 Major Causal Factor of Postharvest Physiological Disorders in Fruits.

Major Causes	Disorders	Examples
Temperature	Low-temperature breakdown	Superficial scald, Flesh browning, core browning, and soggy breakdown in apple.
	Chilling injury	Most of the tropical and subtropical fruits such as mango, banana, papaya, and guava.
	Freezing injury	Peach, pear, apple, apricot, plum, etc.
	Sunscald /heat injury	Apple, pomegranate, karonda, fig, and citrus.
Nutritional deficiency	Ca related	Bitter pit in apple, blossom-end rot in tomato, melons, cavity spots, soft nose and jelly seed in mango, and end spot of avocado
	Boron related	Internal necrosis, internal breakdown, fruit splitting, and cracking.
Storage related	Excess ethylene	Ethylene injury/senescence in most of the tropical and subtropical fruits.
	CO_2 injury	Blackheart in potato, off-flavor development, and anaerobiosis in many tropical and subtropical produce.
Others	Granulation	Due to poor fertilization and rootstock problems.
	Mechanical injury	Pinhole injury in citrus fruits, pitting, oleocellosis, and browning in custard apple.
	Polyphenolic oxidation	Pericarp browning in litchi and lenticel browning in mango and apple.
	Accumulation of sorbitol	Water core in apple.
	Low temperature	Flesh browning in apple.

the structure of the cell wall and slow down the aging and ripening of fruit (Laster and Grusak 2004). Calcium concentration in the extracellular calcium pool rises as a result of the exogenous application of calcium salts, maintaining cell membrane stability, and cell wall rigidity, and reducing fruit softening (Gao et al., 2019) resistance. Calcium salts participate in the signaling pathways of hormones like ethylene, abscisic acid (ABA), and auxin, which help them play a significant role in fruit physiology. Due to calcium's inhibitory influence on ethylene synthesis, calcium has a favorable effect on the buildup of total phenols and antioxidant activity as well as slowing down the degradation of pigments like chlorophyll and carotenoids (Aghdam et al., 2013; Barzegar et al., 2018).

3.3.2 EFFECT ON POSTHARVEST QUALITY

In order to strengthen the cell wall and create cross-links or bridges between the free carboxyl groups of the pectin chains, calcium is necessary to maintain the textural quality of product (García et al., 1996). Calcium, both naturally occurring and artificially given, can stiffen plant tissue by attaching to the pectin carboxyl groups produced by PME (Stanley et al., 1995; Manganaris et al., 2007). In fresh-cut musk melon, Luna-Guzmán and Barrett (2000) added calcium chloride and discovered an unfavorable bitterness; this bitterness was eliminated when calcium lactate was added in the same quantity. Calcium treatment could successfully retain fruit firmness and postpone fruit ripening and softening because calcium plays a crucial function in preserving the structure of cell walls in fruits (Shiri et al., 2016). Ethylene-induced senescence and related disorders can be significantly delayed by exogenous calcium spray, and also helpful in maintaining fruit quality (Zhao and Wang, 2015). Some of the important physiological disorders due to Ca deficiency are being discussed herewith.

3.4 BITTER PIT

Bitter pit in apple is the foremost physiological disorder that appears both in pre- and postharvest conditions. Typically, the bitter pit is characterized by flesh breakdown and flesh pressing (Watkins et al., 2004; Amarante et al., 2012). It has been reported that this disorder is more prevalent in acidic soils along with year and location (Volz et al., 2006). Bitter pit-affected fruits are very low in Ca content; hence Ca content is directly correlated with the occurrence of this disorder. Amarante et al. (2012) reported that reduced Ca content below 5 mg/100 g, increased susceptibility to bitter pit. Specifically in the distal part of the fruit, this condition is characterized by the disintegration of cells in the flesh just beneath the peel, resulting in tiny, dark depressions (Amarante et al., 2012). Low calcium (Ca) content, high magnesium (Mg), potassium (K), and along with high nitrogen (N) content are associated with bitter pit. There are only a few studies about the detection of bitter pit. Lötze and Theron (2005) distinguished bitter pit-affected fruits of "Braeburn" using fluorescence imaging and obtained an accuracy of 75–100%. Moreover, near-infrared technology (NIR) spectroscopy has been frequently used to detect bitter pit symptoms in many apple cultivars (Nicolaï et al.,2006; Kafle et al., 2016; Jarolmasjed et al., 2017).

The use of Ca^{2+} sprays and resistant plant cultivars is increased, especially during hot, dry summers (Miqueloto et al., 2014). Calvo and Candan (2005) found that 1-methylcyclopropene (1-MCP) and diphenylamine (DPA) are effective in reducing symptoms of bitter pit. DPA is considered a synthetic antioxidant that reduces the bitter pit incidence but due to health concerns is not being used now anymore. Postharvest dipping and spraying with $CaCl_2$ solution have been found beneficial in reducing the incidence of bitter pit (Ait-Oubahou et al., 1995).

3.5 SUPERFICIAL SCALD

A physiological disorder of apples and pears is known as superficial scald, sometimes known as storage scald. The term "scald" is a catch-all term for a variety of skin conditions that affect apples and pears. It appears as brown or black spots on the fruit's skin, although it also goes by other names including rugose and lenticel scald. Although scald development is thought to be an oxidative process, it can be controlled by the application of antioxidants, especially the frequently used DPA, as well as low oxygen storage. It has been suggested that the amount of α-farnesene may be less significant than the activity of antioxidant enzymes that may influence its oxidation (Lurie and Watkins, 2012).

3.6 GRANULATION

Granulation, a serious internal physiological disorder that frequently develops in citrus fruit during pre- and postharvest storage. The disorder reduces the fruit's organoleptic quality, which results in a loss of marketable and edible value due to drying taste, flavor loss, and stiffened texture. Segment drying, a serious internal physiological disorder of citrus fruit, frequently happens before and after harvest. During segment drying development, granulated vesicles show drying taste, flavor loss, and texture deterioration (e.g., hardening), which are connected to numerous physiochemical changes (Huang et al., 2023) Treatment with calcium after harvest may be useful in reducing the likelihood of granulation. Granulation was significantly reduced in "Huyou" fruit after treatment with $CaCl_2$, with the granulation index dropping by about 90% (Chen et al., 2005).

3.7 SOFT NOSE

Mango is one of the most important crops in tropical and subtropical regions. There are a few important disorders in mango prevalent during handling which impair its quality and acceptability. Some of the internal disorders are characterized by soft pulp and loosening of pulp integrity. These symptoms are referred to as jelly seed, soft nose, and internal breakdown (Raymond et al., 1998). Shivashankar et al. (2016) established a correlation of jelly seed with premature seed germination mediated through very long chain fatty acids. Burdon et al. (1991) already well-established the role of Ca content and reported that jelly seed disorder is caused by the deficiency of Ca. Generally, it is observed that jelly seed is related to genotype. A soft nose is also an important disorder that impairs the quality of fruits. Sometimes it is also referred to as "jelly seed." Nitrogen deficiency and excessive calcium are the main causes of soft nose (Burdon et al., 1991). Early signs of senescence were distinguished by a distinct yellow color of the mesocarp around the stone (jelly seed), at the distal end, or at the sinus of the fruit (soft nose). Advanced jelly seed and soft nose signs included advanced degradation and/or yellowing of the mesocarp (Raymond et al., 1998). Therefore, a potential contributing cause to the onset of the condition may be the low calcium content of fruit from the soft nose-prone region. Before the fruit ripens, the common sign is a breakdown of the flesh toward the fruit's apex. The fruits' mesocarp cells exhibit noticeable cell wall breakdown and cell separation. Harvest the fruit when it is fully mature, fertilize with calcium, such as calcium ammonium nitrate, and employ sod culture as a preharvest management technique (Wainwright and Burbage, 1989).

3.8 BLOSSOM-END ROT

Tomatoes, eggplant, squashes, green peppers, and melons are the main sources of it. It results from the fruit's lack of calcium. The fact that the blossom end of tomato fruits always has the lowest level of Ca^{2+} has led researchers to conclude that a lack of calcium is the main cause of BER (Saure, 2001). Brown, leathery, and decaying sores are visible on the blossom end of the infected fruits, which are also deformed. At the bloom end, a black sunken spot forms that later spreads to the surrounding water-soaked area (Saure, 2014).

3.9 INTERNAL BROWNING

Internal browning (IB) is a low-temperature disorder that affects many fruits, with flesh tissue browning after being removed from cold storage and brought to room temperature. By slowing down protein hydrolysis and oxidative metabolism, the increased degree of calcium treatment may increase the shelf life of Mauritius pineapple fruits and avert the degeneration brought on by IB. Calcium can lessen cell membrane damage brought on by the cold (Selvarajah et al., 1998). Calcium spray applied to pineapples of the Smooth Cyanne cultivar prior to harvest minimizes IB during storage (Pusittigul et al., 2011).

3.10 GLASSINESS

The water-soaked appearance of the watermelon's flesh is a disorder, and this disorder is linked to the low calcium content of the fruit tissue (Madrid et al., 2004). Another study supported the theory that this disorder resulted from a decrease in calcium in nutrient solution, which accelerated fruit softening and glossiness (Serrano et al., 2002).

3.11 PULP SPOT

It results from a calcium deficiency in the avocado during the time of fruit maturity. The fruit is chopped, and soon after, the first pulp spot symptoms appear. The 1.0 mm-diameter patches, which are initially smooth and glassy, develop along the vascular bundle. After being exposed to the atmosphere, the spots change color, usually turning from brown to dark brown after 30 minutes. The severity of the pulp spot varies from season to season, and as the season goes on, the incidence quickly declines. Fruit with spots on the pulp contain less calcium than fruit with sound tissue. This suggests that pulp spot susceptibility is significantly influenced by the calcium content. In the controlled atmosphere (CA) environment of 10.0% CO_2 and 2.0% O_2, pulp spot was eliminated after 33 days of storage at 5°C. Sprays of calcium can lower the frequency of pulp spots (Koen et al., 1989).

3.12 BORON-RELATED DISORDERS

Boron is an essential and one of the most complex micronutrients in fruit which is constantly needed throughout plant life due to its function as a structural element of the plant cell wall. Boron participates in many physiological reactions, particularly in cell wall formation and metabolism of sugars and phenolic compounds (Khalaj et al., 2016). Its deficiency causes various physiological disorders in fruits not only when these are attached to plants but also after their harvesting. Some major boron deficiency disorders in the storage of fruits are: "Corky core" or "internal/external cork" or "drought spot" of apples; internal necrosis of aonla; brown heart, core browning, and cavities in pears (Garcia and Streif, 1993).

Internal cork of apple often does not show any external symptoms. When fruits are cut open small, round water-soaked spots appear throughout the flesh. The brown tissue normally occurs closer to the core than the skin. As the corky tissue dies, the fruit becomes misshapen.

External cork shows irregular depressions to appear as fruit matures. Raised brown or reddish spots may occur. These later cracks and the skin dries up and scales off, leaving a rough corky layer. Dark brown to greenish-black lesions with a green border occur on or just beneath the skin, mostly in the depressions. When the affected fruit is cut open, corky lesions are scattered throughout the flesh. Adequate boron availability in apple trees is also associated with Ca uptake by fruits; however, its deficiency can cause serious physiological disorders such as the bitter pit, which justifies that foliar applications and soil fertilization with boron are common in pomiculture (Botelho et al., 2022). The preharvest application of boron reduces flesh breakdown in apple (Brackmann et al., 2016). Dart (2004) stated that internal/external cork can be controlled through foliar spray of polyborate powder (20.5% soluble boron) and it should be sprayed in late spring when there is ample foliage to absorb the nutrients. Boron can also be applied after harvesting fruits but must be applied when leaves are still green.

3.13 INTERNAL NECROSIS

This particular disorder of aonla fruit has been observed very often in the case of Banarasi and Francis varieties. The disorder starts with browning of the innermost part of the mesocarp at the time of fruit maturity and extends towards the epicarp resulting into brown–black appearance of the flesh. In case of severe incidence, these black spots become corky and gummy pockets develop. Although internal necrosis is a preharvest disorder, it remains after the harvesting of fruits. The disorder is caused due to deficiency of the boron element which has a significant role in the mobilization of food material from source to sink as a result accumulates the photosynthates (Singh et al., 2001). It was observed that internal necrosis can be minimized by foliar by the application of boron 0.50% at the pin head and pea stage of fruits (Patel et al., 2017; Dadheech et al., 2021). A combined spray of $ZnSO_4$ @ 0.4% with $CuSO_4$ @ 0.4% and borax (0.4%) during September–October has been found effective. Spray of 0.5–0.6% borax in the month of September–October (Mishra et al., 2016).

3.14 CORE BROWNING

Browning disorders are a typical postharvest problem, induced by adverse storage conditions, however, the development of disorders during post-harvest ripening and storage of fruit also depends on a range of preharvest factors. Browning disorders of pears are observed in different forms such as asymmetrical browning, brown core, radial browning, brown and/or dry spots, cavities, etc. (Franck et al., 2007). The disorder is characterized by the breakdown of flesh up to 1 cm from the peel (Larrigaudiere et al., 2004). Cavities can be apparent in different shapes such as small spots in a star pattern in between the five carpels as well as randomly localized dried lesions or localized cavities, usually of a bigger size. Lammertyn et al. (2000) proposed that cavities arise from the brown tissue due to the time course of IB and the appearance of cavities. Franck et al. (2007) reported that browning patterns in pears did not evolve spatially over time but became more severe during storage. The application of boron has been shown to reduce browning incidence in "Conference pears" (Xuan et al., 2001).

3.15 NITROGEN AND PHOSPHORUS-RELATED DISORDERS

Phosphorus and nitrogen have a significant role in the postharvest life of fruits hence the low concentration of phosphorus and nitrogen was associated with biochemical browning reactions in fruit flesh at harvest and with fruit (Olivos et al., 2012). Flesh browning is a common storage disorder of pears, which reduces their market life and lowers consumer quality. Enzymatic flesh browning is a major component of postharvest problems in stone fruit which is also a commercial problem since it can affect fruit quality in fresh as well as canned forms. Peach and nectarine cultivars from California breeding programs are very susceptible to flesh browning, reducing their storage potential (Crisosto et al., 2008). Olivos et al. (2012) reported in Grand Pearl Nectarine that fruit with low N and K had higher flesh browning incidence. Fruit from deficient treatments of N and K had 3.0 times more fruit with flesh browning (50–60%) than fruit from the control treatment (20%) after storage for 11 days at 5°C. High levels of nitrogen may increase the chance of water core in susceptible varieties such as Braeburn, Red Delicious, and Fuji (Dart and Newman, 2005).

3.16 STORAGE-RELATED DISORDERS

Storage practices are also crucial for the safe transportation and marketing of fresh produce. Several factors such as low storage temperature, elevated levels of CO_2 or O_2, ethylene stress, and so forth determine the produce's quality as well as shelf life.

3.16.1 SUPERFICIAL SCALD

Under storage, superficial scald of temperate fruits is the most common and devastating disorder which reduces the quality and marketability (Gong et al., 2021; Papoutsis, 2024). Emongor et al. (1994) stated that scald reduces appearance and utility causing significant wastage in apple fruit industries. The disorder is characterized by brown discoloration over the fruit surface and cell damage (Lurie and Watkins, 2012). The surface becomes rough and changes its surface color. It is reported that in slight incidence, it changes the color slightly and it can be up to a depth of 0.5 cm (Emongor et al., 1994). Generally, storage scald does not appear in storage, but it appears when fruit is taken out of storage. Sometimes, it has been observed that warmer storage conditions favored the disorder incidence. It is reported that prolonged cold storage of apple fruits is one of the main causes of superficial scald (Watkin et al., 1995; Lurie and Watkins, 2012). Further, the imbalance in antioxidants and oxidation of alpha-farnesene (a natural volatile found in the wax of fruit) triggers peel browning as a result of tissue damage. Emongor et al. (1994) reported that the severity of scald is proportional to peel anti-oxidants and the extent of c~-farnesene oxidation. In addition to that, several other hypotheses have been given by various researchers among prolonged exposure to cold stress and oxidative stress in "Granny Smith" associated with releasing chlorogenic acid in the vacuole, which reacts with polyphenol oxidase (PPO) enzyme, accelerated peel scald (Busatto et al., 2018). In another study, ozone-induced storage scald has been reported (Karagiannis et al., 2020). A recent study of global protein analysis by Du et al. (2017) elucidated that ethylene biosynthesis, phenylpropanoid metabolism, allergens, and sulfur amino acids-containing proteins actively participate in scald development. Gong et al. (2021) found that some apple cultivars such as Granny Smith, Red Delicious, Cortland, and Law Rome are more sensitive to develop superficial scald. Dällenbach et al. (2020) reported low flesh and peel calcium content in scald-affected fruits thereby low calcium may be one of the associated factors responsible for scald development. Early harvested fruits, bigger size, and immature fruits are more prone to scald development.

There are several pre- and postharvest management strategies to control the superficial scald. Padfield (1977) reported that the preharvest application of ethephon (500 ppm) significantly eliminated scald. Post-harvest application of DPA, 1-MCP, intermittent warming, temperature conditioning, etc. has been commercially applied worldwide. It has been reported that postharvest application of DPA and 1-MCP on "Granny Smith" apple induced resistance against scald (Karagiannis et al., 2020). Dällenbach et al. (2020) obtained low scald and significant control upon 1-MCP treatment followed by dynamic controlled storage.

Others: Likewise, soggy breakdown and soft scald are other cold storage-related disorders causing significant losses in apple fruit. It has been reported that soft scald can affect the flesh along with the skin of the fruit. The soggy breakdown shows symptoms like a complete ring of soft, brown, and spongy tissue when severe incidence occurs. Soggy break-down can also be controlled by the use of DPA, intermittent warming, and temperature conditioning.

3.16.2 BROWNING

Inside the storage structure, gaseous balance is crucial which determines the shelf life of the fruits. An imbalance of respiratory gases may cause certain disorders and causes significant deterioration. Browning is an important disorder in the storage environment and causes significant losses, especially in temperate fruits such as pear, peach, and apple. Internal breakdown, browning, brown core, core breakdown, brown heart, etc. are collectively known as complex browning and are more prevalent during storage due to abiotic factors. Internal breakdown is another form of chilling injury caused by prolonged storage at low temperatures and causes flesh browning, red pigmentation, and off-flavor development in stone fruits (Peace et al., 2005). IB in pears can be seen only after cutting the fruits (Franck et al., 2007). Preharvest factors such as growing temperature, lights, and harvesting stage substantially affect the degree of incidence. Late-harvested fruits are more susceptible to browning. Asymmetric browning, radial browning, cavities, brown spots, brown core, etc. are major forms of the browning disorder in pear fruit as previously reported (Lammertyn et al., 2000; Giraud et al., 2001; Franck et al., 2007). Conference pear has been reported to be more susceptible to most browning disorders with large cavities and lesions (Zerbini et al., 2002; Lammertyn et al., 2003). Core breakdown symptoms are more associated with senescence and can be accelerated by higher CO_2 inside the storage. Researchers tried to distinguish core breakdown and brown heart as both are related to senescence except in response to elevated CO_2. It has been further observed that core breakdown easily appears when more mature fruits are stored at high temperatures, low O_2, and elevated CO_2 conditions (Lammertyn et al., 2003). Browning in fruits is more commonly associated with the oxidation of phenolic compounds due to PPO enzyme activities.

3.17 MANAGEMENT

There are several storage interventions that can help in reducing the incidence of browning. Li et al. (2017) reported that "Yali" pear fruit when exposed to 0.025 MPa hypobaric treatment significantly reduced the incidence of core browning by 37.3%. Similarly, Veltman et al. (2003) reported that storage containing 0 kPa O_2 with or without CO_2 had no water core browning symptoms and protected membrane damage in pear fruits. Modified storage conditions have also proven better for browning management. It has been found that modified atmospheric storage (12.3% O_2 + 5.6% CO_2) yielded browning-free fruits and significantly extended the shelf life of "Bartlett" pear up to 4 months at −1.1°C (Wang and Sugar, 2013). In addition, various chemical agents such as calcium-based formulations, L-cystine, antioxidant-like compounds (ALCs), and ascorbic acid have been used to reduce browning. The application of ethylene scrubbers and plant hormones also contributes to browning reduction by decreasing the activity of polyphenol oxidase (PPO).

3.18 OXYGEN AND CO_2 INJURY

During the storage of fresh commodity, it is necessary to maintain the level of oxygen and carbon dioxide. Exposure to excess CO_2 may create injury to the fresh fruit and sometimes may develop off-flavor. In many commodities, especially in potatoes, excessive CO_2 and less oxygen create the blackening of the central core commonly called as blackheart of potato. Although this gaseous imbalance usually causes anaerobic respiration and leads to oxidation of fatty acids. Sometimes, excessive ethylene also triggers uneven ripening, premature ripening, and bitterness in many commodities. In many cucumber-like fruits, exposure to ethylene causes brown-yellowish streaks on the surface. Ethylene injury is more common in non-climacteric fruits. It has been reported that excess ethylene caused a white core in kiwifruits. In many tropical and subtropical fruits, brown discoloration is commonly seen as a result of ethylene (Watkins, 2017).

3.19 HEAT INJURY

Sun light is an important factor in developing quality, pigments including volatiles. But up to a certain level solar radiation is beneficial. Due to climate change, elevated temperatures altered fruit physiology and transformed fruit production (Morales-Quintana et al., 2020). Photo-oxidation, excess solar radiation, and air temperature affect fruit quality negatively (Lal and Sahu, 2017). They also reported that the losses due to heat vary from 6 to 30% depending upon the season, crop, and intensity. Heat injury occurs during the fruit developmental stage in preharvest condition, but it leads to deterioration in quality and consumer acceptability of the produce (Figure 3.1C). Heat shock/ radiant heat causes scorching and scalding in many fruit crops such as tomatoes, apples, pomegranates, and berries. Sunscald is a very common heat injury in apple and pomegranate. It is reported that rising temperatures of 45°C caused sunburn injury in pomegranate (Melgarejo et al., 2004). They also reported a total loss of up to 40% in pomegranate yield in Spain. Singh et al. (2012) reported the damage due to heat injury is around 0.9–19.13% in different varieties. In the case of apples, sunburn is characterized by white to brown patches over the fruit surface (Morales-Quintana et al., 2020). They also show lenticel marking and postharvest scald (Martinez-Hernandez et al., 2017). Fruit epidermis layer (cuticle), cultural operations, and cultivar are some of the protective phenomena in fruits (Wünsche et al., 2002). Felicetti and Schrader (2009) reported that internal antioxidant compounds and photoprotective pigments give some protection against sunburn injury in apples and other fruits. It is suggested that the application of foliar spray of NAA, GA3, and salicylic acid may enhance drought tolerance in plants thus heat injury may be avoided by a certain degree. As per the advisory of the National Research Centre on Pomegranate, Solapur (India), it is recommended that Kaolin (5%) is very effective in reducing sunburn in pomegranate. In summer and heat-prevailing areas, shade management is practiced.

3.20 LOW-TEMPERATURE DISORDERS

After harvesting, ideal storage conditions are the foremost requirements of any fruit or vegetable. Combinations of temperature and relative humidity are needed. The fluctuation in storage temperature is harmful as it leads to rapid metabolic processes and deterioration in quality. The requirement of temperature varies from commodity to commodity, stage of harvesting, and purpose of storage. There are two most common temperature injuries, that is, freezing injury and chilling injury which affect most of the fruits. Fruits react to temperature changes differently depending on the places they were grown (tropical, subtropical, or temperate). Deviation from optimum storage temperatures can be a source of low-temperature injury namely, freezing injury and chilling injury.

3.21 FREEZING INJURY

Fruits of tropical and subtropical origin can suffer from freezing damage if they are kept at temperatures below the freezing point for an extended period of time, which can cause the tissues to collapse (Yahia et al., 2019). The freezing point of fruits is slightly lower than the freezing point of pure water due to the "freezing point depression" (Table 3.1). The concentration of dissolved solutes (such as sugars) in the tissue and cells determines the temperature at which produce freezes (below 0°C). Pure water typically freezes at 0°C, whereas products with high solute concentrations typically freeze at lower temperatures (Kader, 1985; Yahia et al., 2019). Frosty conditions in the orchard can cause freezing before harvest, or incorrect storage practices can cause freezing after harvest. The degree of damage depends on a number of variables, including temperature and exposure time. Fruit that is frozen for a longer time suffers more freezing damage when it is thawed, which might take days or weeks. Fruits that are stored closest to an air vent or those that are exposed to ambient air are more prone to damage. The pulp is affected by freezing more negatively than the peel, in contrast to chilling injury. Without causing any damage, the development of ice crystals in the intercellular spaces gives tissues the appearance of being water-soaked. However, such fruit is likely to have a faster rate of respiration, be more prone to rot, soften more quickly than typical fruit, and turn mealy. The membrane and juice sacs are especially prone to damage, and there may be significant internal distortion. Even if this is prevented, moisture evaporation through the peel is increased, leading to dryness and shrivelling of the damaged pulp.

3.21.1 SYMPTOMS

Freeze damage begins with brown discoloration of the skin surface and extends to the vascular bundles during subsequent refrigerated storage. During storage, lesions may remain intact or sunken, depending on the degree of damage to the sub-epidermal tissue. Affected fruits fail to ripen, become disorganized and weakened, and have a short shelf life. When the product freezes, water present inside and between the cells forms into ice crystals and the cells dehydrate themselves. Frozen fruit volume is reduced by up to 10% and the skin is noticeably wrinkled. In extreme freezing, ice crystals can penetrate cell walls. Thawing kills the cells, releasing liquid and causing the fruit to wilt and often discolor. Mass reduction during thawing forms the basis for the separation of damaged citrus fruit by flotation. Frozen apples soften and become mushy. In tangerines, crystals form in the pulp, and in grapes, the pulp is milky white. Frost damage to stone fruit is characterized by the translucent, watery appearance of the flesh. Severely frozen fruit tends to decompose when thawed, with damaged tissue becoming soft, brown, and watery.

3.21.2 CONTROL

The ideal way to prevent frost damage is to keep fresh fruits at higher freezing temperatures at all times. When no apparent damage is seen in frozen fruit, damaged fruit seems to be less dense than healthy fruit. Hence, it is based on the fact that it tends to be separated by flotation in a specially prepared solution. Accurate and well-positioned thermostats are important in a fruit store to prevent the supply air from getting too cold. Pears are especially vulnerable as they are only stored below freezing. Care must also be taken when transporting. Fruit suspected to be frozen should be moved carefully, as bruises are amplified in frozen tissue.

3.22 CHILLING INJURY

Chilling injury (CI) of fruits is very unique from freezing damage, as it occurs at temperatures above freezing and below their critical safe temperature. The critical temperature is the temperature below which chilling injury symptoms appear, usually between 5 and 15°C (Table 3.2). Most fruits of tropical and subtropical origin are sensitive to freezing temperatures (Gross et al., 2002). The term "chilling injury" is synonymous with chilling disorder, chilling damage, chilling disorder, low-temperature breakdown, low-temperature injury, low-temperature breakdown, and low-temperature disorder (Morris, 1982; Ali et al., 2023).

3.22.1 SYMPTOMS

There are several symptoms common to chilling and often indicate the severity of the injury. These symptoms rarely become apparent after transfer to cold temperatures. The first visible sign of chilling injury is the appearance of dark spots on the skin (Fan et al., 2022). As shown in Figure 3.1A, the most common symptoms such as surface browning, discoloration, pitting, flesh browning, water browning, membrane leakage, etc. have been reported in most of the tropical and subtropical fruits (Chaudhary et al., 2014; Li et al., 2014; Lado et al., 2016; Habibi et al., 2021; Fan et al., 2022). Moreover, these symptoms vary from fruit to fruit and the intensity of chilling injury. In the case of mango, surface discoloration and pitting are the first and most common symptoms of chilling injury (Cantre et al., 2017). Likewise in banana, a brown-blackish streak appears over the surface and fruits fail to ripen (Khademi et al., 2019).

3.22.2 SENSITIVITY OF FRUITS TO CHILLING TEMPERATURE

Based on sensitivity to chilled conditions, fruits are classified in the following ways (Kader, 2002; Kader and Yahia, 2011). (1) *Highly chilling sensitive fruits*: Bananas, pineapples, mangoes, papayas, sweet potatoes, pomegranates, olives, grapefruits, lemons and limes, avocadoes, bread-fruit, carambola, cherimoya, citrus fruits, durian, feijoa, guava, jackfruit, jujube, lanzone, longan, litchi, mangosteen, passion fruit, pepino, plantain, persimmon (Fuyu), prickly (cactus) pear, rambutan, sapodilla, sapote, and tamarillo. (2) *Low chilling sensitive fruits*: Few temperate fruits like apples and pears. (3) *Chilling insensitive fruits*: Date, fig, kiwifruit, loquat, persimmon (Hachiya), and oranges.

3.22.3 ALLEVIATION OF CHILLING INJURY

Reduction of chilling damage can be achieved either by increasing the resistance of products to chilling conditions or by delaying the occurrence of chilling injury symptoms in commodities that are chilling-sensitive. Several methods have been shown to reduce chilling injury, including high- or low-temperature conditioning, intermittent warming, controlled atmospheres, chemicals and growth regulators application, waxes and other surface coatings, and packaging (Wang, 2004; Lafuente and Zacarias, 2006; Kader and Yahia, 2011). The first three approaches manipulate and modify the surrounding environment of the produce, while the others involve the direct handling of commodities.

3.22.4 TEMPERATURE CONDITIONING

Exposure of chilling-sensitive tropical and subtropical fruits to temperatures slightly above the critical chilling range increases the tolerance of these commodities to chilling during subsequent low-temperature storage and delays the development of symptoms of injury (Chaudhary et al., 2014; Wang et al., 2017). It appears that low-temperature conditioning causes numerous physiological and biochemical changes in fruits that provide an adaptive response to chilling stress. These modifications include reducing the loss of membrane phospholipids. Increased sugar, starch, and proline content; maintains high levels of polyamines, squalene, and long-chain aldehydes. Increase the ratio of unsaturated and saturated fatty acids. Similarly, short-term treatment at temperatures between 35 and 60°C can reduce the chilling damage of some commodities (Wang, 1994). Heat treatment mainly refers to pre-storage exposure of commodities to hot water, hot air, or steam for a specified period. High temperature, a low-cost method accelerates the healing of cuts and bruises and increases anti-fungal compounds that protect against invading pathogens, ultimately enhancing overall appearance (Wang et al., 2001).

3.22.5 INTERMITTENT WARMING

The process of interrupting low temperature by one or more brief heatings. This short-term heat treatment approach can extend the shelf life of some chilling-sensitive commodities, but must be applied before damage is irreversible. This helps in the repair of damage to membranes, organelles, or metabolic pathways (Lyons and Breidenbach, 1987). The timing and frequency of intermittent heating depend on the post-harvest shelf life of the commodity.

3.22.6 CONTROLLED ATMOSPHERES

Precise control of the atmospheric gases surrounding the product, including high levels of CO_2 and low levels of oxygen, can delay the onset of chilling injury symptoms. However, their effectiveness in reducing CI varies with the product and can be either beneficial, nil, or harmful. An elevated CO_2 level could significantly reduce the CI index in postharvest guava fruit (Alba-Jimenez et al., 2018).

3.22.7 GROWTH REGULATORS

Plant growth regulators such as ABA, ethylene, polyamines, and some triazole-derived compounds such as paclobutrazol and uniconazole have been shown to alter the chilling tolerance of plant tissues. ABA stabilizes microtubule networks, inhibits ion leakage, and prevents the loss of glutathione and membrane phospholipids (Carvajal et al., 2017). Treatment of chilling-sensitive crops with methyl jasmonate (MeJA) induces the synthesis of stress proteins including heat shock proteins, pathogenesis-related proteins and alternative oxidases (Ding et al., 2002; Fung et al., 2004). Exogenous treatment with polyamines increases internal polyamine levels and reduces chilling-induced damage through antioxidant activity and membrane-stabilizing effects (Habibi and Ramezanian, 2017).

FIGURE 3.1 (A) Chilling injury symptoms in mango. (B) Sap burn injury in mango. (C) Heat injury in Karonda. (D) Browning on custard apple.

3.22.8 CHEMICAL TREATMENTS

Various compounds with different chemical structures are known to be effective in reducing CI. These compounds include calcium, ethoxyquin, sodium benzoate, imazalil, thiabendazole, mineral oil, safflower oil, vegetable oil, and squalene. Tissue calcium levels correlate with the susceptibility of fruit to chilling damage. Lime fruits with the lowest calcium content in the juice developed the highest rates of chilling injury (Slutzky et al., 1982). Antioxidants such as ethoxyquin and sodium benzoate have been reported to prevent peroxidation of membrane lipids and ultimately reduce chilling damage. Post-harvest treatments with dimethylpolysiloxane, safflower oil, or mineral oil prevented chilling-induced discoloration of the banana peel. In oranges and grapes, the fungicides imazalil and thiabendazole have been reported to suppress latent infections, slow skin senescence, and reduce pitting. The development of CI in grapefruit is delayed by vegetable oil. Oil is thought to be able to reduce CI by reducing water loss and oxidative stress.

3.22.9 PACKAGING

Wrapping fruit in plastic film helps prevent cold damage by maintaining high relative humidity and altering the concentration of oxygen and carbon dioxide in the atmosphere surrounding the product. The reduction in water loss from the tissue appears to inhibit the breakdown of the epidermis and underlying cells, thus preventing pitting. Individually wrapping the fruit in heat-shrink film also significantly reduced pitting and burns in chilled grapefruits (Miller et al., 1990). Packaging has also been shown to delay chilling damage in bananas, Japanese apricots, pineapples, and lemons.

3.22.10 WAXING AND OTHER COATINGS

Similar to film packaging, the edible coating creates a microclimate around the fruit, limits gas exchange, slows down respiration and transpiration rates, and improves appearance (Zhang et al., 2019).

3.22.11 UV RADIATION

Recently, several studies have reported that residue-free UV treatment can reduce CI in horticultural products. It was observed that UV-B-treated (5 kJ/m^2 for 4 h) mango fruits had reduced CI index, malondialdehyde (MDA) content, and ion leakage compared with control fruits (Ruan et al., 2015).

3.22.12 GENETIC MODIFICATION

Large genetic variation in plants provides an opportunity to modify susceptibility to chilling by transferring resistance genes to sensitive species.

Structural information from cDNA clones of mRNAs encoding resistant genes, or the use of classical genetic mapping, is required to narrow down the chromosomal regions and genomes that carry resistant traits. Genes can be characterized using immunological screens or antisense RNA technology.

TABLE 3.2 Average Critical Safe Temperature and Common Chilling Injury Symptoms in Some Important Fruits.

Fruit	Critical Safe Temperature (°C)	Symptoms	References
Apple	−1 to 4	Pitting and flesh browning	Saba and Watkins (2020)
Apricot	7	Flesh browning	Koushesh Saba et al. (2012), Wang et al. (2016)
Avocado	13	Grayish brown discoloration of flesh, pitting, and off-flavor development	Woolf et al. (2020)
Banana	13	Brown discoloration on the surface, dark water-soaked areas, failure to ripen, hardening of the central placenta, turn black under severe chilled conditions and off-taste	Murata (1969), Nguyen et al. (2003), Khademi et al. (2019)
Grapefruit	10–13	Internal discoloration, browning in albedo and flavedo, pitting, yellow rind, absence of red color, scald, and watery breakdown	Lado et al. (2015, 2016), Chaudhary et al. (2014)
Kiwifruit	5	Pitting, water-soaked lesion, peel turns brown and black, lignified tissues, and grainy appearance	Suo et al. (2018), Yang et al. (2016)
Lemon and lime	10–13	Pitting, red blotch, albedo discoloration, and sunken depressions	Martinez-Hernandez et al. (2017), Wongsheree et al. (2009), Patel et al. (2016)
Mango	10–13	Surface discoloration, uneven ripening or failure to ripen, poor flavor, increased decay, pitting, color change of mesocarp, and susceptibility to rot	Nair and Singh (2004), Li et al. (2014), Cantre et al. (2017)
Olive	5–8	Shrinkage, flesh browning, pit, and fruit browning	Kiritsakis et al. (1998), Fan et al. (2022)
Orange	3–5	Pitting, brown stain, superficial scald in flavedo, and watery breakdown	Habibi et al. (2020, 2021), Rasouli et al. (2019)

TABLE 3.2 *(Continued)*

Fruit	Critical Safe Temperature (°C)	Symptoms	References
Papaya	8	Skin pitting, failure to ripen, hard lumps in the pulp, water-soaked flesh, off-flavor development, and sensitivity to decay	Shadmani et al. (2015), Pan et al. (2017)
Peach	−1 to 0	Mealiness and flesh browning	Zhang et al. (2020)
Pineapple	10	Uneven ripening, dull color, water-soaked flesh, off-flavor, and endogenous brown spot	Nukuntornprakit et al. (2020)
Pomegranate	5	Scald, pitting, colorless arils, browning of internal white segments, firmness loss, and increased decay	Pareek et al. (2015)
Rambutan	8	Browning of peel and thorns	Kondo et al. (2001)

3.23 OTHER DISORDERS

Besides all above-mentioned factors, certain inherent factors are also associated with postharvest disorders. For example, lenticel browning in mango is associated with biochemical changes in the peel (Prasad et al., 2016, 2020). Furthermore, they also reported that cultivar susceptibility is one of the dominating factors and lenticel discoloration (LD) is highly dependent on cultivar. Similarly, Ali et al. (2018) reported that pericarp browning in litchi is also associated with the action of enzymes and phenolic oxidation. In the case of custard apple, this oxidative browning is very common and frequent (Figure 3.1D). This injury can take place by touching the fruits frequently. These problems can be addressed through certain postharvest applications of anti-browning agents, salicylic acid, hot water treatments, and proper handling (Ali et al., 2018; Yun et al., 2021). Sap burn injury in mango is also one of the major problems during harvest and poses an obstacle, a bad appearance on the surface of the fruit (Barman et al., 2015) (Figure 3.1B). Maqbool and Malik (2008) recommended some anti-sapping agents, and they found that the use of Ca $(OH)_2$ could minimize the problem very effectively. Likewise, Barman et al. (2015) reported that sodium hydroxide (1%) has significantly reduced the incidence of sap burn in mango.

3.24 DETECTION OF SYMPTOMS OF PHYSIOLOGICAL DISORDERS

It is very crucial to detect proper symptoms in order to apply suitable management strategies. Besides several manual and physical methods, certain non-destructive techniques need to be implemented. It has been reported that machine vision cell load technology can be used to assess external visual symptoms (Walsh et al., 2020). The use of near-infrared spectroscopy (NIRS) has come up with certain statistical models and validations (Walsh et al., 2020). It has been reported that using vis-NIR (550 and 650 nm) in "Keitt" mango could easily detect jelly seed and black flesh with an accuracy of 65% (Mogollón et al., 2020). In the case of apple bitter pit, Torres et al. (2021) used a handheld vis-NIR to detect symptoms and obtained significant results with an accuracy of 75–81%. However, these instruments need to be standardized and calibrated according to cultivar and types of symptoms.

3.25 CONCLUSION

Postharvest physiological disorders are a great challenge to the fresh industry. Proper identification and management are essential to reduce fresh produce losses and maintain the nutritional quality of fruits and vegetables. Various pre- and postharvest causes are responsible for postharvest physiological disorders. Proper detection and various techniques to reduce the postharvest physiological disorders are discussed in this chapter.

KEYWORDS

- **bitter pit**
- **browning**
- **low temperature**
- **pitting**
- **rootstock**
- **sunburn**
- **watercore**

REFERENCES

Aghdam, M. S., Dokhanieh, A. Y., Hassanpour, H., & Rezapour, F. J. (2013). Enhancement of antioxidant capacity of cornelian cherry (Cornus mas) fruit by postharvest calcium treatment. *Scientia Horticulturae, 161*, 160–164.

Aguayo, E., Escalona, V. H., & Artes, F. (2008). Effect of hot water treatment and various calcium salts on quality of fresh-cut "Amarillo" melon. *Postharvest Biology and Technology, 47*, 397–406.

Ait-Oubahou, A., El-Otmani, M., Charhabaili, Y., Fethi. J., & Ben dada, M. (1995). Effect of harvesting date and postharvest calcium treatment on stored apple quality. In *Postharvest physiology, pathology and technologies for horticultural commodities: Recent advances* (pp. 57–64). Agadir, 16–21 Jan 1994.

Alba-Jimenez, J. E., Benito-Bautista, P., Nava, G. M., Rivera-Pastrana, D. M., Vazquez Barrios, M. E., & Mercado-Silva, E. M. (2018). Chilling injury is associated with changes in microsomal membrane lipids in guava fruit (*Psidium guajava* L.) and the use of controlled atmospheres reduce these effects. *Scientia Horticulturae, 240*, 94–101.

Ali, S., Khan, A. S., Malik, A. U., Anwar, R., Anjum, M. A., Nawaz, A. et al. (2021). Combined application of ascorbic and oxalic acids delays postharvest browning of litchi fruits under controlled atmosphere conditions. *Food Chemistry, 350*, 129277.

Ali, S., Anjum, M. A., Ullah, S., Nawaz, A., Ejaz, S., & Khaliq, G. (2023). Salicylic acid mediated postharvest chilling and disease stress tolerance in horticultural crops. In Plant stress mitigators (pp. 69-85). Academic Press.

Ali, S., Khan, A. S., Malik, A. U., Shaheen, T., & Shahid, M. (2018). Pre-storage methionine treatment inhibits postharvest enzymatic browning of cold stored "Gola" litchi fruit. *Postharvest Biology and Technology, 140*, 100–106.

Amarante, C. D., Miqueloto, A., Steffens, C. A., Dos Santos, A., & Argenta, L. C. (2012, June). Changes in xylem functionality during apple fruit development: Implications on calcium concentration and incidence of bitter pit. In *VII International Postharvest Symposium 1012* (pp. 135–140).

Barman, K., Asrey, R., Pal, R. K., Jha, S. K., & Sharma, S. (2015). Influence of different desapping agents on the incidence of sapburn, ripening behaviour and quality of mango. *Journal of Food Science and Technology, 52*(1), 161–170.

Barzegar, T., Fateh, M., & Razavi, F. (2018). Enhancement of postharvest sensory quality and antioxidant capacity of sweet pepper fruit by foliar applying calcium lactate and ascorbic acid. *Scientia Horticulturae, 241*, 293–303.

Botelho, R. V., Muller, M. M. L., Umburanas, R. C., Laconski, J. M. O., & Terra, M. M. (2022). Boron in fruit crops: Plant physiology, deficiency, toxicity, and sources for fertilization. In *Boron in Plants and Agriculture* (pp. 29–50). Academic Press.

Brackmann, A., Thewes, F. R., Anese, R. D. O., & Linke Junior, W. (2016). Preharvest boron application and its relation with the quality of "Galaxy" apples after harvest and controlled atmosphere storage. *Ciência Rural, 46*, 585–589.

Burdon, J. N., Moore, K. G., & Wainwright, H. (1991). Mineral distribution in mango fruit susceptible to the physiological disorder soft-nose. *Scientia Horticulturae, 48*(3–4), 329–336.

Busatto, N., Farneti, B., Commisso, M., Bianconi, M., Iadarola, B., Zago, E., Ruperti, B., Spinelli, F., Zanella, A., Velasco, R., & Ferrarini, A. (2018). Apple fruit superficial scald resistance mediated by ethylene inhibition is associated with diverse metabolic processes. *The Plant Journal, 93*(2), 270–285.

Calvo, G., & Candan, A. P. (2005, July). 1-Methylcyclopropene (1-MCP) affects physiological disorders in "Granny Smith" apples depending on maturity stage. In *IX International Controlled Atmosphere Research Conference* (Vol. 857, pp. 63–70).

Cantre, D., Herremans, E., Verboven, P., Ampofo-Asiama, J., Hertog, M. L., & Nicolai B. M. (2017). Tissue breakdown of mango (*Mangifera indica* L. cv. Carabao) due to chilling injury. *Postharvest Biology and Technology, 125*, 99–111.

Carvajal, F., Palma, F., Jiménez-Muñoz, R., Jamilena, M., Pulido, A., & Garrido, D. (2017). Unravelling the role of abscisic acid in chilling tolerance of zucchini during postharvest cold storage. *Postharvest Biology and Technology, 133*, 26–35.

Chaudhary, P. R., Jayaprakasha, G. K., Porat, R., & Patil, B. S. (2014). Low temperature conditioning reduces chilling injury while maintaining quality and certain bioactive compounds of "Star Ruby" grapefruit. *Food Chemistry, 153*, 243–249.

Chen, K., Xu, C., Li, F., & Zhang, S. (2005). Postharvest granulation of "Huyou" (*Citrus Changshanensis*) fruit in response to calcium. *Israel Journal of Plant Sciences, 53*(1), 35–40.

Crisosto, C. H., Crisosto, G. M., & Day, K. R. (2008). Market life update for peach, nectarine, and plum cultivars grown in California. *Advances in Horticultural Sciences, 22*, 201–204.

Dadheech, S., Khandelwal, R., & Sharma, R. (2021). Effect of foliar spray of nutrients and plant growth regulators on yield, quality and physiochemical characteristics in Aonla cv NA 10.

Dällenbach, L. J., Eppler, T., Bühlmann-Schütz, S., Kellerhals, M., & Bühlmann, A. (2020). Pre-and postharvest factors control the disease incidence of superficial scald in the new fire blight tolerant apple variety "Ladina". *Agronomy, 10*(4), 464.

Dart, J. (2004). Boron deficiency (cork) in pome fruits. Ag Fact H 4.

Dart, J. A., & Newman, S. M. (2005). Watercore of apples. *Profitable & Sustainable Primary Industries—Primefact, 49*, 1–2.

De Freitas, S. T., Amarante, C. D., & Mitcham, E. J. (2016). Calcium deficiency disorders in plants. In *Postharvest ripening physiology of crops* (pp. 477–502).

de Freitas, S. T., do Amarante, C. V. T., & Mitcham, E. J. (2015). Mechanisms regulating apple cultivar susceptibility to bitter pit. *Scientia Horticulturae, 186*, 54–60.

de Freitas, S. T., Handa, A. K., Wu, Q., Park, S., & Mitcham, E. J. (2012). Role of pectin methylesterases in cellular calcium distribution and blossom-end rot development in tomato fruit. *The Plant Journal, 71*(5), 824–835.

Ding, C. K., Wang, C., Gross, K. C., & Smith, D. L. (2002). Jasmonate and salicylate induce the expression of pathogenesis-related-protein genes and increase resistance to chilling injury in tomato fruit. *Planta, 214*(6), 895–901.

Du, L., Song, J., Campbell Palmer, L., Fillmore, S., & Zhang, Z. Q. (2017). Quantitative proteomic changes in development of superficial scald disorder and its response to diphenylamine and 1-MCP treatments in apple fruit. *Postharvest Biology and Technology, 123*, 33–50.

Emongor, V. E., Murr, D. P., & Lougheed, E. C. (1994). Preharvest factors that predispose apples to superficial scald. *Postharvest Biology and Technology, 4*(4), 289–300.

Fan, Z., Lin, B., Lin, H., Lin, M., Chen, J., & Lin, Y. (2022). γ-Aminobutyric acid treatment reduces chilling injury and improves quality maintenance of cold-stored Chinese olive fruit. *Food Chemistry: X, 13*, 100208.

Felicetti, D. A., & Schrader, L. E. (2009). Changes in pigment concentrations associated with sunburn browning of five apple cultivars. II. Phenolics. *Plant Science, 176*(1), 84–89.

Ferguson, I., Volz, R., & Woolf, A. (1999). Preharvest factors affecting physiological disorders of fruit. *Postharvest Biology and Technology, 15*(3), 255–262.

Franck, C., Lammertyn, J., Ho, Q. T., Verboven, P., Verlinden, B., & Nicolaï, B. M. (2007). Browning disorders in pear fruit. *Postharvest Biology and Technology, 43*(1), 1–13.

Fung, R. W., Wang, C. Y., Smith, D. L., Gross, K. C., & Tian, M. (2004). MeSA and MeJA increase steady-state transcript levels of alternative oxidase and resistance against chilling injury in sweet peppers (*Capsicum annuum* L.). *Plant Science, 166*(3), 711–719.

Gao, Q., Xiong, T., Lia, X., Chena, W., & Zhua, X. (2019). Calcium and calcium sensors in fruit development and ripening. *Scientia Horticulturae, 253*, 412–421.

Garcia, J. M., & Streif, J. (1993). *Quality and storage potential of pear, pt. 1: Influence of CA-and ULO-storage*. Gartenbauwissenschaft (Germany).

García, J. M., Herrera, S., & Morilla, A. (1996). Effects of postharvest dips in calcium chloride on strawberry. *Journal of Agricultural and Food Chemistry, 44*(1), 30–33.

Giraud, M., Westercamp, P., Coureau, C., Chapon, J. F., & Berrie, A. (2001). *Recognizing Postharvest Diseases of Apple and Pear*. Centre Technique Interprofessionel des Fruits et Legumes.

Gong, Y., Song, J., Palmer, L. C., Vinqvist-Tymchuk, M., Fillmore, S., Toivonen, P., & Zhang, Z. (2021). Tracking the development of the superficial scald disorder and effects of treatments with diphenylamine and 1-MCP using an untargeted metabolomic approach in apple fruit. *Food Chemistry: Molecular Sciences, 2*, 100022.

Gross , K. , Wang , C. Y., & Saltveit , M. E. (Eds.). (2002). The commercial storage of fruit, vegetables, and florist and nursery stocks. USDA Agr. Hbk. 66. http://www.ba.ars.usda.gov/hb66/index.html. Accessed on Jan 3, 2023.

Habibi, F., & Ramezanian, A. (2017). Vacuum infiltration of putrescine enhances bioactive compounds and maintains quality of blood orange during cold storage. *Food Chemistry, 227*, 1–8.

Habibi, F., Ramezanian, A., Guillén, F., Martínez-Romero, D., Serrano, M., & Valero, D. (2020). Susceptibility of blood orange cultivars to chilling injury based on antioxidant system and physiological and biochemical responses at different storage temperatures. *Foods, 9*(11), 1609.

Habibi, F., Serrano, M., Zacarías, L., Valero, D., & Guillén, F. (2021). Postharvest application of 24-epibrassinolide reduces chilling injury symptoms and enhances bioactive compounds content and antioxidant activity of blood orange fruit. *Frontiers in Plant Science, 12*, 629733.

Hertel, T. W., Burke, M. B., & Lobell, D. B. (2010). The poverty implications of climate-induced crop yield changes by 2030. *Global Environmental Change, 20*(4), 577–585.

Hewett, E. W. (2006). An overview of preharvest factors influencing postharvest quality of horticultural products. *International Journal of Postharvest Technology and Innovation, 1*(1), 4–15.

Huang, C., Hou, J., Huang, M., Hu, M., Deng, L., Zeng, K., & Yao, S. (2023). A comprehensive review of segment drying (vesicle granulation and collapse) in citrus fruit: Current state and future directions. *Scientia Horticulturae, 309*, 111683.

Jarolmasjed, S., Zúñiga Espinoza, C., & Sankaran, S. (2017). Near infrared spectroscopy to predict bitter pit development in different varieties of apples. *Journal of Food Measurement and Characterization, 11*(3), 987–993.

Kader, A. A. (1985). Ethylene-induced senescence and physiological disorders in harvested horticultural crops. *HortScience, 20*(1), 54–57.

Kader, A. A. (2002). *Postharvest technology of horticultural crops* (Vol. 3311, pp. 39–47). University of California Agriculture and Natural Resources, Publication.

Kader, A. A., & Yahia, E. M. (2011). Postharvest biology of tropical and subtropical fruits. In *Postharvest biology and technology of tropical and subtropical fruits* (pp. 79–111). Woodhead Publishing.

Kafle, G. K., Khot, L. R., Jarolmasjed, S., Yongsheng, S., & Lewis, K. (2016). Robustness of near infrared spectroscopy based spectral features for non-destructive bitter pit detection in honeycrisp apples. *Postharvest Biology and Technology, 120*, 188–192.

Karagiannis, E., Tanou, G., Scossa, F., Samiotaki, M., Michailidis, M., Manioudaki, M., Laurens, F., Job, D., Fernie, A. R., Orsel, M., & Molassiotis, A. (2020) Systems-based approaches to unravel networks and individual elements involved in apple superficial scald. *Frontiers Plant Sciences, 11*, 8. DOI: 10.3389/fpls.2020.00008

Khademi, O., Ashtari, M., & Razavi, F. (2019). Effects of salicylic acid and ultrasound treatments on chilling injury control and quality preservation in banana fruit during cold storage. *Scientia Horticulturae, 249*, 334–339.

Khalaj, K., Ahmadi, N., & Souri, M. K. (2016). Improvement of postharvest quality of Asian pear fruits by foliar application of boron and calcium. *Horticulturae, 3*(1), 15.

Khaliq, G. H. U. L. A. M. (2015). Effects of edible coatings enriched with calcium chloride on physiological, biochemical and quality responses of mango (*Mangifera indica* L. cv. Choke Anan) fruit during cold storage. Universiti Putra Malaysia, Serdang, Selangor, Malaysia.

Khaliq, G., Ali, S., Gapper, N., & Nicola, S. (2023). Recent advances and approaches in the application of elicitors to enhance resistance mechanisms in fresh produce. *Frontiers in Sustainable Food Systems, 6,* 1061079.

Khan, L., Khaliq, G., Ullah, E., Abdul-Rahaman, A., El-Mogy, M. M., Kesba, H. H., Ismail, A. M., & El-Beltagi, H. S. (2025). Guggul gum and clove essential oil synergistically reduced anthracnose disease and modulated physicochemical changes in tomato during storage. *Journal of Stored Products Research, 114,* 102713.

Kiritsakis, A., Nanos, G. D., Polymenopulos, Z., Thomai, T., & Sfakiotakis, E. M. (1998). Effect of fruit storage conditions on olive oil quality. *Journal of the American Oil Chemists' Society, 75*(6), 721–724.

Koen, T. J., Du Plessis, S. F., & Terblanche, J. H. (1989, Nov). Nutritional factors involved in physiological post-harvest fruit disorders of avocados (cv. Fuerte). In *International symposium on the culture of subtropical and tropical fruits and crops* (Vol. 275, pp. 543–550).

Kondo, S., Posuya, P., & Kanlayanarat, S. (2001). Changes in physical characteristics and polyamines during maturation and storage of rambutans. *Scientia Horticulturae, 91*(1–2), 101–109.

Koushesh Saba, M., Arzani, K., & Barzegar, M. (2012). Postharvest polyamine application alleviates chilling injury and affects apricot storage ability. *Journal of Agricultural and Food Chemistry, 60*(36), 8947–8953.

Lado, J., Rodrigo, M. J., Cronje, P., & Zacarías, L. (2015). Involvement of lycopene in the induction of tolerance to chilling injury in grapefruit. *Postharvest Biology and Technology, 100*, 176–186.

Lado, J., Rodrigo, M. J., Lopez-Climent, M., Gomez-Cadenas, A., & Zacarías, L. (2016). Implication of the antioxidant system in chilling injury tolerance in the red peel of grapefruit. *Postharvest Biology and Technology, 111*, 214–223.

Lafuente, M. T., & Zacarias, L. (2006). Postharvest physiological disorders in citrus fruit. *Stewart Postharvest Review, 1*(2).

Lal N., & Sahu, N. (2017). Management Strategies of sun burn in fruit crops—A review. *International Journal of Current Microbiology and Applied Sciences, 6*(6), 1126–1138.

Lal, S., Khaliq, G., Mastoi, A. H., Irshad, A., Jaffar, M. A., Khan, L., & Muhammad, N. (2025). Effect of *Fagonia indica* and garlic extract on *Alternaria alternata* and post-harvest quality of tomato fruit during storage. *Journal of Crop Health, 77*(6), 1–13.

Lammertyn, J., Aerts, M., Verlinden, B. E., Schotsmans, W., & Nicolaı, B. M. (2000). Logistic regression analysis of factors influencing core breakdown in "Conference" pears. *Postharvest Biology and Technology, 20*(1), 25–37.

Lammertyn, J., Scheerlinck, N., Jancsok, P., Verlinden, B. E., Nicolaı, B. M. (2003). A respiration–diffusion model for Conference pears II: Simulation and relation to core breakdown. *Postharvest Biology and Technology, 30*, 43–55.

Larrigaudiere, C., Lentheric, I., Puy, J., & Pinto, E. (2004). Biochemical characterisation of core browning and brown heart disorders in pear by multivariate analysis. *Postharvest Biology and Technology, 31*(1), 29–39.

Laster, G. E., & Grusak, M. A. (2004). Field application of chelated calcium: Post-harvest effects on cantaloupe and honeydew fruit quality. *Horticulture Technology, 14*, 2–38.

Li, J., Bao, X., Xu, Y., Zhang, M., Cai, Q., Li, L., & Wang, Y. (2017). Hypobaric storage reduced core browning of Yali pear fruits. Scientia Horticulturae, 225, 547-552.

Li, P., Zheng, X., Liu, Y., & Zhu, Y. (2014). Pre-storage application of oxalic acid alleviates chilling injury in mango fruit by modulating proline metabolism and energy status under chilling stress. *Food Chemistry, 142*, 72–78.

Lötze, E., & Theron, K. I. (2005, Jan). Dynamics of calcium uptake with pre-harvest sprays to reduce bitter pit in "golden delicious". In *V International Symposium on Mineral Nutrition of Fruit Plants* (Vol. 721, pp. 313–320).

Luna-Guzmán, I., & Barrett, D. M. (2000). Comparison of calcium chloride and calcium lactate effectiveness in maintaining shelf stability and quality of fresh-cut cantaloupes. *Postharvest Biology and Technology, 19*(1), 61–72.

Lurie, S., & Watkins, C. B. (2012). Superficial scald, its etiology and control. *Postharvest Biology and Technology, 65*, 44–60.

Lyons, J. M., & Breidenbach, R. W. (1987). Chilling injury. In J. Weichmann (Ed.), *Postharvest Physiology of Vegetables* (pp. 305–326). Marcel Dekker.

Madebo, M. P., Hu, S., Zheng, Y., & Jin, P. (2021). Mechanisms of chilling tolerance in melatonin treated postharvest fruits and vegetables: A review. *Journal of Future Foods, 1*(2), 156–167.

Madrid, R., Valverde, M., Alcolea, V., & Romojaro, F. (2004). Influence of calcium nutrition on water soaking disorder during ripening of Cantaloupe melon. *Scientia Horticulturae, 101*(1–2), 69–79.

Manganaris, G. A., Vasilakakis, M., Diamantidis, G., & Mignani, I. (2005) Effect of calcium additives on physicochemical aspects of cell wall pectin and sensory attributes of canned peach (*Prunus persica* (L.) Batsch cv. Andross. *Journal of the Science of Food and Agriculture, 85*, 1773–1778.

Manganaris, G. A., Vasilakakis, M., Diamantidis, G., & Mignani, I. J. F. C. (2007). The effect of postharvest calcium application on tissue calcium concentration, quality attributes, incidence of flesh browning and cell wall physicochemical aspects of peach fruits. *Food Chemistry, 100*(4), 1385–1392.

Maqbool, M. E. H. D. I., & Malik, A. U. (2008). Anti-sap chemicals reduce sapburn injury and improve fruit quality in commercial mango cultivars. *International Journal of Agriculture and Biology, 10*, 1–8.

Martinez-Hernandez, G. B., Artes-Hernandez, F., Gomez, P. A., Breto, J., Orihuel Iranzo, B., & Artes, F. (2017). Postharvest treatments to control physiological and pathological disorders in lemon fruit. *Food Packaging and Shelf Life, 14*, 34–39.

Melgarejo, P., Martınez, J. J., Hernández, F. C. A., Martınez-Font, R., Barrows, P., & Erez, A. (2004). Kaolin treatment to reduce pomegranate sunburn. *Scientia Horticulturae, 100*(1–4), 349–353.

Mditshwa, A., Khaliq, G., Hussein, Z., & Ejaz, S. (2023). Sustainable postharvest management practices for fresh produce. *Frontiers in Sustainable Food Systems, 7,* 1143759.

Miller, W. R., Chun, D., Risse, L. A., Hatton, T. T., & Hinsch, R. T. (1990). Conditioning of Florida grapefruit to reduce peel stress during low-temperature storage. *HortScience, 25*(2), 209–211.

Miqueloto, A., do Amarante, C. V. T., Steffens, C. A., dos Santos, A., & Mitcham, E. (2014). Relationship between xylem functionality, calcium content and the incidence of bitter pit in apple fruit. *Scientia Horticulturae, 165*, 319–323.

Mishra, D. S., Tripathi, A., & Nimbolkar, P. K. (2016). Review on physiological disorders of tropical and subtropical fruits: Causes and management approach. *International Journal of Agriculture Environment and Biotechnology, 9*(6), 925.

Mogollón, R., Contreras, C., da Silva Neta, M. L., Marques, E. J. N., Zoffoli, J. P., & de Freitas, S. T. (2020). Non-destructive prediction and detection of internal physiological disorders in "Keitt" mango using a hand-held Vis-NIR spectrometer. *Postharvest Biology and Technology, 167*, 111251.

Morales-Quintana, L., Waite, J. M., Kalcsits, L., Torres, C. A., & Ramos, P. (2020). Sun injury on apple fruit: Physiological, biochemical and molecular advances, and future challenges. *Scientia Horticulturae, 260*, 108866.

Morris, L. L. (1982). Chilling injury of horticultural crops: An overview. *HortScience, 17*(2), 161–162.

Murata, T. (1969). Physiological and biochemical studies of chilling injury in bananas. *Physiologia Plantarum, 22*(2), 401–411.

Nair, S., & Singh, Z. (2004). Chilling injury in mango fruit in relation to biosynthesis of free polyamines. *The Journal of Horticultural Science and Biotechnology, 79*(4), 515–522.

Nguyen, T. B. T., Ketsa, S., & Van Doorn, W. G. (2003). Relationship between browning and the activities of polyphenoloxidase and phenylalanine ammonia lyase in banana peel during low temperature storage. *Postharvest Biology and Technology, 30*(2), 187–193.

Nicolaï, B. M., Lötze, E., Peirs, A., Scheerlinck, N., & Theron, K. I. (2006). Non-destructive measurement of bitter pit in apple fruit using NIR hyperspectral imaging. *Postharvest Biology and Technology, 40*(1), 1–6.

Nukuntornprakit, O. A., Luengwilai, K., & Siriphanich, J. (2020). Chilling injury in pineapple fruit is related to mitochondrial antioxidative metabolism. *Postharvest Biology and Technology, 170*, 111330.

Olivos, A., Johnson, S., Xiaoqiong, Q., & Crisosto, C. H. (2012). Fruit phosphorous and nitrogen deficiencies affect "Grand Pearl" nectarine flesh browning. *Hortscience, 47*(3), 391–394.

Padfield, C. A. S. (1977). Pre-harvest ethephon treatment, a possible control of apple scald. *New Zealand Journal of Experimental Agriculture, 5*(2), 183–184. DOI: 10.1080/03015521.1977.10425961

Pan, Y. G., Yuan, M. Q., Zhang, W. M., & Zhang, Z. K. (2017). Effect of low temperatures on chilling injury in relation to energy status in papaya fruit during storage. *Postharvest Biology and Technology, 125*, 181–187.

Papoutsis, K. (2024). Alternatives to DPA and ethoxyquin for preventing the development of superficial scald in apples: A review. Food Chemistry: X, 101730.

Pareek, S., Valero, D., & Serrano, M. (2015). Postharvest biology and technology of pomegranate. *Journal of the Science of Food and Agriculture, 95*(12), 2360–2379.

Patel, B., Tandel, Y. N., Patel, A. H., & Patel, B. L. (2016). Chilling injury in tropical and subtropical fruits: A cold storage problem and its remedies: A review. *International Journal of Science, Environment and Technology, 5*(4), 1882–1887.

Patel, J. S., Dhruve, J. J., Motka, G. N., & Patel, A. D. (2017). Influence of plant growth regulators and boron on nutritional quality and shelf life of aonla fruit. *International Journal of Current Microbiology and Applied Sciences, 6*(4), 2533–2540.

Peace, C. P., Crisosto, C. H., Garner, D. T., Dandekar, A. M., Gradziel, T. M., & Bliss, F. A. (2005, Jan). Genetic control of internal breakdown in peach. In *VI International Peach Symposium* (Vol. 713, pp. 489–496).

Prasad, K., Sharma, R. R., & Srivastav, M. (2016). Postharvest treatment of antioxidant reduces lenticel browning and improves cosmetic appeal of mango (*Mangifera indica* L.) fruits without impairing quality. *Journal of Food Science and Technology, 53*(7), 2995–3001.

Prasad, K., Sharma, R. R., Srivastav, M., & Asrey, R. (2020). Relationship between lenticel discoloration and biochemical and quality attributes in mango (*Mangifera indica* L.) fruit. *Acta Physiologiae Plantarum, 42*(12), 1–12.

Pusittigul, I., Siriphanich, J., & Juntee, C. (2011, Dec). Role of calcium on internal browning of pineapples. In *International Symposium on Tropical and Subtropical Fruits* (Vol. 1024, pp. 329–338).

Qu C, Liu C, Gong X, Li C, Hong M, Wang L, Hong F. (2012). Impairment of maize seedling photosynthesis caused by a combination of potassium deficiency and salt stress. *Environmental and Experimental Botany, 75*, 134–141.

Rasouli, M., Saba, M. K., & Ramezanian, A. (2019). Inhibitory effect of salicylic acid and Aloe vera gel edible coating on microbial load and chilling injury of orange fruit. *Scientia Horticulturae, 247*, 27–34.

Raymond, L., Schaffer, B., Brecht, J. K., & Crane, J. H. (1998). Internal breakdown in mango fruit: symptomology and histology of jelly seed, soft nose and stem-end cavity. *Postharvest Biology and Technology, 13*(1), 59–70.

Ruan, J., Li, M., Jin, H., Sun, L., Zhu, Y., Xu, M., & Dong, J. (2015). UV-B irradiation alleviates the deterioration of cold-stored mangoes by enhancing endogenous nitric oxide levels. *Food Chemistry, 169*, 417–423.

Saba, M. K., & Watkins, C. B. (2020). Flesh browning development of "Empire" apple during a shelf life period after 1-methylcyclopropene (1-MCP) treatment and controlled atmosphere storage. *Scientia Horticulturae, 261*, 108938.

Saure, M. C. (2001). Blossom-end rot of tomato (*Lycopersicon esculentum* Mill.)—A calcium-or a stress-related disorder? *Scientia Horticulturae, 90*(3–4), 193–208.

Saure, M. C. (2014). Why calcium deficiency is not the cause of blossom-end rot in tomato and pepper fruit—A reappraisal. *Scientia Horticulturae, 174*, 151–154.

Selvarajah, S., Herath, H. M. W., Bandara, D. C., & Abeysinghe Banda, D. M. G. (1998). Effect of pre-harvest calcium treatment on post-harvest quality of pineapple.

Serrano, M., Amorós, A., Pretel, M. T., Martínez-Madrid, M. C., Madrid, R., & Romojaro, F. (2002). Effect of calcium deficiency on melon (*Cucumis melo* L.) texture and glassiness incidence during ripening. *Food Science and Technology International, 8*(3), 147–154.

Shadmani, N., Ahmad, S. H., Saari, N., Ding, P., & Tajidin, N. E. (2015). Chilling injury incidence and antioxidant enzyme activities of *Carica papaya* L. "Frangi" as influenced by postharvest hot water treatment and storage temperature. *Postharvest Biology and Technology, 99*, 114–119.

Shiri, M. A., Ghasemnezhad, M., Fatahi Moghadam, J., & Ebrahimi, R. (2016). Effect of $CaCl_2$ sprays at different fruit development stages on postharvest keeping quality of "Hayward" kiwifruit. *Journal of Food Processing and Preservation, 40*(4), 624–635.

Shivashankar, S., Sumathi, M., & Singh, H. S. (2016). Premature seed germination induced by very-long-chain fatty acids causes jelly seed disorder in the mango (*Mangifera indica* L.) cultivar "Amrapali" in India. *The Journal of Horticultural Science and Biotechnology, 91*(2), 138–147.

Singh, G., Nath, V., Pandery, S. D., Ray, P. K., & Singh, H. S. (2012). *The litchi* (p. 181). Food and Agriculture Organization of the United Nations.

Singh, H. K., Srivastava, A. K., Dwivedi, R., & Kumar, P. (2001). Effect of foliar feeding of micronutrients on plant growth, fruit quality, yield and internal fruit necrosis of aonla (*Emblica officinalis* Gaertn.) cv. Francis. *Progressive Horticulture, 33*(1), 80–83.

Slutzky, B., Gonzalez-Abreu, A., & Berdam, I. (1982). Chilling injury related to mineral composition of grapefruit and limes during cold storage. In *Proceedings of the International Society of Citriculture/[International Citrus Congress, November 9–12, 1981, Tokyo, Japan; K. Matsumoto, editor]*. International Society of Citriculture, 1982–1983.

Stanley, D. W., Bourne, M. C., Stone, A. P., & Wismer, W. V. (1995). Low temperature blanching effects on chemistry, firmness and structure of canned green beans and carrots. *Journal of Food Science, 60*(2), 327–333.

Suo, J., Li, H., Ban, Q., Han, Y., Meng, K., Jin, M., Zhang, Z., & Rao, J. (2018). Characteristics of chilling injury-induced lignification in kiwifruit with different sensitivities to low temperatures. *Postharvest Biology and Technology, 135*, 8–18.

Torres, E., Guinjuan, I. R., & Castellví, S. A. (2021). Potential of VIS/NIR spectroscopy to detect and predict bitter pit in "Golden Smoothee" apples. *Spanish Journal of Agricultural Research, 19*(1), 1001.

Veltman, R. H., Lentheric, I., Van der Plas, L. H. W., & Peppelenbos, H. W. (2003). Internal browning in pear fruit (*Pyrus communis* L. cv Conference) may be a result of a limited availability of energy and antioxidants. *Postharvest Biology and Technology, 28*(2), 295–302.

Volz, R., Alspach, P., Fletcher, D., & Ferguson, I. (2006). Genetic variation in bitter pit and fruit calcium concentrations within a diverse apple germplasm collection. *Euphytica, 149*(1–2), 1–10.

Wainwright, H., & Burbage, M. B. (1989). Physiological disorders in mango (*Mangifera indica* L.) fruit. *Journal of Horticultural Science, 64*(2), 125–135.

Walsh, K. B., McGlone, V. A., & Han, D. H. (2020). The uses of near infra-red spectroscopy in postharvest decision support: A review. *Postharvest Biology and Technology, 163*, 111139.

Wang, C. Y. (1994). Chilling injury of tropical horticultural commodities. *HortScience, 29*(9), 986–988.

Wang, C. Y. (2004, Sept). Alleviation of chilling injury in tropical and subtropical fruits. In *III International Symposium on Tropical and Subtropical Fruits* (Vol. 864, pp. 267–273).

Wang, C. Y., Bowen, J. H., Weir, I. E., Allan, A. C., & Ferguson, I. B. (2001). Heat-induced protection against death of suspension-cultured apple fruit cells exposed to low temperature. *Plant, Cell & Environment, 24*(11), 1199–1207.

Wang, J., Zhou, X., Zhou, Q., Cheng, S., Wei, B., & Ji, S. (2017). Low temperature conditioning alleviates peel browning by modulating energy and lipid metabolisms of "Nanguo" pears during shelf life after cold storage. *Postharvest Biology and Technology, 131*, 10–15.

Wang, Y., & Sugar, D. (2013). Internal browning disorder and fruit quality in modified atmosphere packaged "Bartlett" pears during storage and transit. *Postharvest Biology and Technology, 83*, 72–82.

Wang, Z., Cao, J., & Jiang, W. (2016). Changes in sugar metabolism caused by exogenous oxalic acid related to chilling tolerance of apricot fruit. *Postharvest Biology and Technology, 114*, 10–16.

Watkins, C. B., Nock, J. F., Weis, S. A., Jayanty, S., & Beaudry, R. M. (2004). Storage temperature, diphenylamine, and pre-storage delay effects on soft scald, soggy breakdown andbitter pit of "Honeycrisp" apples. *Postharvest Biology and Technology, 32*(2), 213–221.

Watkins, C. B. (2017). Postharvest physiological disorders of fresh crops. *Encyclopedia of Applied Plant Sciences,* 315–322.

Watkins, C. B., Bramlage, W. J. & Cregoe, B. A. (1995). Superficial scald of "granny smith" apples is expressed as a typical chilling injury. *Journal of American Society of Horticultural Science, 120*, 88–94 (1995).

Wills, R. B. H., & Mahendra, M. S. (1989). Effect of postharvest application of calcium on ripening of peach. *Australian Journal of Experimental Agriculture, 29*(5), 751–753.

Wongsheree, T., Ketsa, S., & van Doorn, W. G. (2009). The relationship between chilling injury and membrane damage in lemon basil (Ocimum× citriodourum) leaves. *Postharvest Biology and Technology, 51*(1), 91–96.

Woolf, A. B., Arpaia, M. L., Defilippi, B. G., & Bower, J. P. (2020). Subtropical fruits: Avocados. In *Controlled and modified atmospheres for fresh and fresh-cut produce* (pp. 389–397). Academic Press.

Wünsche, J. N., Bowen, J., Ferguson, I., Woolf, A., & McGhie, T. (2002, August). Sunburn on apples-causes and control mechanisms. In *XXVI International Horticultural Congress: Key Processes in the Growth and Cropping of Deciduous Fruit and Nut Trees* (Vol. 636, pp. 631–636).

Xuan, H., Streif, J., Pfeffer, H., Dannel, F., Römheld, H., & Bangerth, F. (2001). Effect of pre-harvest boron application on the incidence of CA-storage related disorders in "Conference" pears. *The Journal of Horticultural Science and Biotechnology, 76*(2), 133–137.

Yahia, E. M., Carrillo-López, A., & Sañudo, A. (2019). Physiological disorders and their control. In *Postharvest technology of perishable horticultural commodities* (pp. 499–527). Woodhead Publishing.

Yang, Q., Wang, F., & Rao, J. (2016). Effect of putrescine treatment on chilling injury, fatty acid composition and antioxidant system in kiwifruit. *PLoS One, 11*(9), e0162159.

Yun, Z., Gao, H., Chen, X., Chen, Z., Zhang, Z., Li, T., Qu, H., & Jiang, Y. (2021). Effects of hydrogen water treatment on antioxidant system of litchi fruit during the pericarp browning. *Food Chemistry, 336*, 127618.

Zerbini, P. E., Rizzolo, A., Brambilla, A., & Grassi, M. (2002). Loss of ascorbic acid during storage of Conference pears in relation to the appearance of brown heart. *Journal of the Science of Food and Agriculture, 82*, 1–7.

Zhang, W., Zhao, H., Jiang, H., Xu, Y., Cao, J., & Jiang, W. (2020). Multiple 1-MCP treatment more effectively alleviated postharvest nectarine chilling injury than conventional one-time 1-MCP treatment by regulating ROS and energy metabolism. *Food Chemistry, 330*, 127256.

Zhang, W., Zhao, H., Zhang, J., Sheng, Z., Cao, J., & Jiang, W. (2019). Different molecular weights chitosan coatings delay the senescence of postharvest nectarine fruit in relation to changes of redox state and respiratory pathway metabolism. *Food Chemistry, 289*, 160–168.

Zhao, Y., & Wang, C. (2015). Effect of calcium chloride in combination with salicylic acid on post-harvest freshness of apples. *Food Science and Biotechnology, 24*(3), 1139–1146.

Zheng, Y. Q., Deng, L., He, S. L., Zhou, Z. Q., Yi, S. L., Zhao, X. Y., & Wang, L. A. (2011). Rootstocks influence fruit oleocellosis in "Hamlin" sweet orange (*Citrus sinensis* L. Osbeck). *Scientia Horticulturae, 128*, 108–114.

CHAPTER 4

Preharvest Factors Affecting Postharvest Physiological Disorders

GHULAM KHALIQ[1*], SAJID ALI[2], SHAGHEF EJAZ[2], OSCAR W. MITALO[3], and PHEBE DING[4]

[1]*Department of Horticulture, Faculty of Agriculture, Lasbela University of Agriculture, Water and Marine Sciences, Uthal, Pakistan*

[2]*Department of Horticulture, Faculty of Agricultural Sciences and Technology, Bahauddin Zakariya University, Multan, Pakistan*

[3]*Tropical Agriculture Research Front, Japan International Research Center for Agricultural Sciences, Ishigaki, Okinawa, Japan*

[4]*Department of Crop Science, Faculty of Agriculture, Universiti Putra Malaysia, Selangor, Malaysia*

*Corresponding author

ABSTRACT

Postharvest losses in fresh produce are global issues which affect growers, industry, and consumers. Fresh produce such as fruits and vegetables is highly perishable and subject to rapid quality loss after harvest. Postharvest losses in fresh produce can be either qualitative or quantitative. Postharvest quality deterioration of fresh produce starts from the field and continues until it reaches the customer. Maintaining quality is one of the major challenges in handling fresh produce. Postharvest physiological disorders of fruits and vegetables occur due to abiotic stresses and unfavorable biochemical changes. The development of postharvest physiological disorders of fresh produce in part depends on some preharvest practices during production.

Physiological Disorders of Fruits and Vegetables: Causes and Remedies.
Ghulam Khaliq, Sajid Ali, and Mohammed Wasim Siddiqui (Eds.)

DOI: 10.1201/9781998511334-4

4.1 INTRODUCTION

Fresh produce, such as fruits and vegetables, is among the most widely consumed food items in the world. Fresh produce contains higher amounts of lycopene, a type of carotenoid with antioxidant properties that is beneficial in reducing the incidence of some chronic diseases like cancer and many other cardiovascular disorders (Freeman and Reimers, 2011). Fruits and vegetables have been proven to have extensive beneficial effects against various diseases and disorders, as well as, anti-microbial, antioxidant, anti-inflammatory, anti-aging, and anti-diabetic properties (Moretti et al., 2010). Fruits and vegetables contain a high amount of bioactive compounds, such as phenolic acids, vitamin C, vitamin E, flavonoids, ellagitannins, and carotenoids. Fruits and vegetables are also a good source of vitamins A, B1, B2, and B6. The nutritional value of fruits and vegetables depends on the variety, growing conditions, and ripening stage (Meera et al., 2017). Postharvest quality losses in fresh produce are a major concern for growers, industry, and consumers (Khaliq et al., 2020; Mditshwa et al., 2023). There are limited advances in the quality maintenance of fresh produce, hence resulting in a high number of qualitative losses. Postharvest physiological disorders have been recognized as one of the main causes of quality deterioration of fruits and vegetables. Qualitative losses in fresh produce can negatively affect several aspects, including grower income, nutritional value, and consumer acceptability.

Physiological disorders are superficial from the outside and these damaged tissues are a key factor in pathogen invasion. The occurrence of postharvest physiological disorders reduces the shelf life and decreases the quality of fresh produce (Ferguson et al., 1999; Ali et al., 2022). Symptoms of postharvest physiological disorders are easily noticeable to consumers and sellers, who therefore avoid purchasing such products. There are various symptoms of postharvest physiological disorders like peel browning, peel cracking, senescent breakdown, watercore breakdown, and brown core (Watkins, 2017). Peel splitting results in open wounds, promoting rapid moisture loss and quality deterioration. Moreover, cracking fruit is not attractive to consumers, and also decreases its quality and marketability.

The development of postharvest physiological disorders of fruits and vegetables begins very early in the farm planning process (Hewett, 2006). The effects of preharvest factors on postharvest physiological disorders are often underestimated and overlooked. However, many of the decisions that we make during crop production can greatly influence the postharvest physiological disorders of fresh produce. It is critical to remember that fresh produce quality is only maintained after harvesting. It is not improved during storage and marketing channels. Therefore, it is of the utmost importance to consider the preharvest factors that allow us to minimize the occurrence of postharvest physiological disorders of fruits and vegetables.

Many postharvest physiological disorders are the direct result of many preharvest factors. Fruits and vegetables that are inappropriately fertilized and irrigated are of poor quality and can never be improved in quality by any postharvest treatment methods (Kathayat and Rawat, 2019). Postharvest quality of fruit cannot be improved after harvest, but can only be maintained (Khaliq et al., 2021). The best quality of fresh produce is desirable during the whole process like harvesting, storage, transport, ripening, and marketing (Léchaudel and Joas, 2007). Two types of quality must be satisfied: functional and sensorial quality. Functional qualities include both physiological disorders and susceptibility to postharvest diseases. Sensorial quality reflects the consumers' expectations in terms of the final product (Tyagi et al., 2017). The postharvest physiological disorders depend not only on postharvest factors, but also on several preharvest factors such as cultivar, irrigation, genetics, nutrition, and environmental conditions (Mishra et al., 2016). Therefore, it is important to know the effect of preharvest factors on the occurrence of postharvest physiological disorders in fresh produce during storage. The purpose of this chapter is to investigate and identify the preharvest factors that affect the postharvest physiological disorders associated with an environmental condition, cultivar type, irrigation, sunlight, and nutrition of fresh produce.

4.2 PHYSIOLOGICAL DISORDER

Physiological disorders are defined as plant tissue disturbances caused by abnormal external or internal environmental conditions (Antunes et al., 2022). These physiological disorders are the result of failure of the physiological process of tissues. Physiological disorders are neither caused by an infection nor a mechanical factor. Pre- and postharvest factors are the main causes of physiological disorders, which result in abnormal physiological processes. Physiological disorders are caused by abiotic stresses such as unfavorable environmental conditions (temperature, water stress, air pollutants, and humidity), poor soil conditions (pH, texture, structure, and drainage), chemical exposure (weedicides, herbicides, pesticides), imbalanced concentrations of plant growth regulators, and nutrient deficiencies or excesses (Gangadhara et al., 2021). Physiological disorders are commonly started in the field before harvesting, however, symptoms appear during storage and marketing (Li et al., 2021). Physiological disorders also occur due to genetic susceptibility, enzymatic activities, over-ripeness, aging, or due to cultural practices (Zheng et al., 2020). Physiological disorders cause pitting, off-flavors, off-odors, sunken regions, and generally, internal or external browning (Mishra et al., 2016). Physiological disorders are thought to be the main cause of fruit quality deterioration, shelf life, appearance, organoleptic, and multifunctional nutritional levels (Ramírez-Gil et al., 2021).

Postharvest physiological disorders appear after harvesting the fresh produce resulting from abnormal metabolism, which affects susceptible plant tissues and finally cell death (Schrader et al., 2008). Physiological disorders are different from pathogenic infection and other unfavorable postharvest quality changes, like chlorophyll loss, water loss, ripening, and other softening-related processes related to normal plant senescence (Moretti et al., 2010). The definition of postharvest physiological disorders also eliminates a number of direct postharvest factors such as mechanical damage, bruising, handling damage, and exposure to chemical solutions (Anastasiadi et al., 2021). Generally, postharvest physiological disorders affect both marketability and shelf life. Postharvest physiological disorders, such as peel browning, cracking, chilling injury, and watercore affect fruit quality and reduce the market value (Luyckx et al., 2016). Postharvest (temperature, relative humidity, gas combination) and preharvest conditions are the main factors affecting the postharvest physiological disorders in fresh produce (Watkins, 2009). Some disorders are primarily associated with one factor, while the majority of disorders occur due to multiple factors, such as genetics, nutrition, and environment (Zheng et al., 2020). Therefore, before overcoming these physiological disorders with a specific management approach, there is a need to know the reason behind the particular cause of the disorders. Physiological disorders have a stronger impact on the quality and consumers' acceptability of fruits and vegetables. Depending on the stage of occurrence, physiological disorders can arise from preharvest and postharvest factors (Lafuente and Zacarias, 2006; Lufu et al., 2020).

4.3 PREHARVEST FACTORS

The preharvest factors lead to the occurrence and development of postharvest physiological disorders in fruits and vegetables are complex and multifaceted, involving disruptions in cellular structure, metabolic balance, and stress-response pathways (De Wit et al., 2024). Preharvest conditions such as nutrient imbalances, water stress, extreme temperatures, and inappropriate cultural practices can compromise the physiological and biochemical integrity of developing fruits and vegetables. For instance, calcium deficiency during fruit development weakens cell wall and membrane stability, predisposing tissues to disorders like bitter pit in apples or blossom end rot in tomatoes. Similarly, excessive nitrogen fertilization promotes rapid vegetative growth and delays maturity, resulting in soft-textured produce with low resistance to mechanical damage and senescence-related disorders. Environmental stresses such as high temperatures or water deficits generate oxidative stress in plant tissues, leading to the accumulation of reactive oxygen species (ROS), which damage lipids, proteins, and cellular organelles (Anastasiadi et al., 2021). This oxidative damage may remain latent during harvest but manifests as visible disorders during storage. Furthermore, irregular irrigation or sudden changes in soil moisture can disrupt calcium transport and turgor pressure, causing internal breakdown, cracking, or shriveling. The hormonal imbalances induced by preharvest stresses, particularly involving ethylene, abscisic acid, and auxins can also alter ripening physiology and accelerate senescence, making produce more susceptible to chilling injury, internal browning, or uneven ripening during postharvest handling (Tipu et al., 2024). Thus, preharvest factors set the physiological foundation of produce quality, and any stress or imbalance during crop growth stages can predispose harvested commodities to the development of physiological disorders under postharvest conditions.

The preharvest factors that affect the postharvest physiological disorders of fresh produce include genetic makeup, cultivar, orchard management, cultural practices, growing location, environmental stress, plant water status, weather conditions, fruit maturity, and mineral nutrition in the field (Ferguson et al., 1999; Tyagi et al., 2017). Some physiological disorders initially develop while fruits are still in the field, and visible symptoms appear during storage and ripening (Younmoon et al., 1997). Preharvest factors have an impact on the development of postharvest physiological disorders of fresh produce (Hernandez et al., 2016; Pedreschi et al., 2016; Ramírez-Gil et al., 2021). There is relatively little understanding of the origins of preharvest factors that may be associated with the development of postharvest physiological disorders. It is necessary to identify preharvest factors that may be related to the prevalence of postharvest physiological disorders based on prior knowledge and current demands to produce quality fresh produce. Postharvest physiological disorders development depends on a variety of preharvest factors, including planting age, cultural practices, crop density, nutrient levels in the fruit, canopy density, tree vigor, and optimal water availability throughout fruit development (Willingham et al., 2006). Below is a detailed discussion of some of the preharvest factors that can affect the postharvest physiological disorders of fresh produce.

4.3.1 TEMPERATURE

The development of postharvest physiological disorders in fruits and vegetables is frequently associated with preharvest low and high temperatures (Ali et al., 2024; Khan et al., 2025). The majority of fruits and vegetables have specific temperature needs for the best development and quality characteristics (Khaliq et al., 2023; Lal et al., 2025). Preharvest temperature is one of the most significant environmental factors affecting the growth, production, quality, and postharvest physiological disorders of horticultural commodities (Beverly et al., 1993). Preharvest temperature affects photosynthesis, maturity, ripening, color, aroma, and carbohydrate biosynthesis (Yang and Barth, 1997; Woolf and Ferguson, 2000). It is well recognized that high preharvest temperatures might affect a commodity's resistance to high postharvest temperature treatments or low postharvest storage temperatures. The metabolism of nutrients and mineral absorption by plants is influenced by temperature (Watkins, 2009). Lower temperatures affect the flower sex and fruit setting, while higher temperatures enhance transpiration. Variations in temperature can have an impact on plant hormone levels, photosynthesis, respiration, aqueous interactions, membrane stability, and fruit development (Kader et al., 1974; Park et al., 2024). High temperatures can have an impact on mineral accretion by speeding up biological reactions mediated by various enzymes. When grown at high temperatures, fruits like grapes and apples have higher sugar content and reduced acid content (Yamada et al., 1994). High temperatures during the olive fruit growth stage affect the accumulation of tannins and subsequent turnover of tannins causing low flavor development (Rallo et al., 2018). Higher temperatures can increase the air's ability to absorb water vapor, there may be a greater need for water. A reduced or depleted soil water reservoir could result from higher transpiration, which would cause water stress in plants during dry seasons. For instance, exposure to high temperatures can alter the morphology, anatomy, physiology, and ultimately the biochemistry of plant tissues, which in turn can impact the growth and development of fruits (Schrader et al., 2008). These rapid biochemical changes due to temperature fluctuation may result in dramatic declines in the quality of fresh produce (Khaliq, 2015).

Postharvest physiological disorders like chilling injury in avocados are associated with high preharvest temperatures (Table 4.1). Similarly, apple scald and watercore could be the effect of late harvest and low temperature in the field (Yamada et al., 1994). These high- and low-temperature effects on postharvest physiological disorders, may be due to long-term effects with temperature fluctuation, where fruits are exposed in the field (Ferguson et al., 1999). Kiwi planted at higher temperatures reached maturity sooner than the same crops grown at lower temperatures. High temperature is causing more sunburn and fruit cracking in apples, apricots, cherries, and litchi. The presence of mild temperatures during the three days prior to the harvest was linked to the thermo-sensitivity of papaya fruit. When papaya fruit was exposed to minimum temperatures above 22.4°C for 3 days in the field, it caused the fruit to develop thermotolerance, which prevented the disruption of the softening process in the pulp caused by conventional hydrothermal treatment (42°C for 30 min followed by 20 min at 49°C). Several studies confirmed that preharvest temperature exposure during fruit development affects the postharvest chilling injury of many fruits and vegetables (Woolf and Ferguson, 2000). During three months prior to harvest, pineapple is

TABLE 4.1 Preharvest Factors Affecting Postharvest Physiological Disorders.

Preharvest Factors	Physiological Disorders	Fruits/ Vegetables	References
Rainfall	Skin freckle	Papaya	De Oliveira and Vitória (2011)
Radiation	Sunburn	Muskmelon	Moretti et al. (2010)
Cultural practices	Russet spotting	Lettuce	Kader et al. (1974)
Calcium deficiency	Chilling injury	Avocado	Ramírez-Gil et al. (2021)
Potassium deficiency	Blotchy ripening	Tomato	Gangadhara et al. (2021)
Irrigation	Granulation	Citrus	Mishra et al. (2016)
Nutrition	Bitter pit	Apple	Watkins (2017)
Climate	Superficial-scald	Guava	Nxumalo et al. (2019)
Temperature	Translucence	Pineapple	Ferguson et al. (1999)
Cultivar type	Finger drop	Banana	Luyckx et al. (2016)
Genetic factors	Brown heart	Radish	Manzoor et al. (2021)
Poor soil drainage	Fruit cracking	Pomegranate	Galindo et al. (2014)
Thinning	Purple spot	Loquat	Gariglio et al. (2008)
Rootstock	Cork spot	Pear	Yahia et al. (2019)
Maturity	Internal browning	Apricot	Jing et al. (2018)
Canopy position	Mealiness	Peach	Crisosto et al. (1997)

sensitive to a physiological disorder called water-soaking flesh at a preharvest temperature between 15 and 23°C. Low preharvest temperatures in pitahayas inhibit photosynthesis, but high temperatures promote it. These variations in photosynthesis affect the postharvest bioactive compounds, total soluble solids, and phenolics in pitahayas.

When fruits like bananas, tomatoes, apples, and pears are exposed to high temperatures, there is a disruption of fruit ripening with abnormal pigmentation, minimal softening, and a considerable drop in ethylene production. Radish grown in the autumn–winter season has lower possibilities of developing pith tissue incidence than radish cultivated in the spring–summer. High temperature in the soil caused the incidence of internal breakdown disorder in radish (Manzoor et al., 2021). A hard lump in pulp is a physiological disorder of papaya fruit that is caused by the inactivation of cell wall degrading enzymes in response to high-temperature stress (De Oliveira and Vitória, 2011). The mRNAs associated with softening and ripening may be inhibited or reduced when temperatures are high. Paull and Chen (2000) found that a commodity degree of sensitivity or tolerance to heat stress is correlated with the quantity of heat protection proteins present during harvest and the postharvest development of heat shock proteins. The injury lesion may be connected to transcription, translation, and the capacity of the cell to recover once an injury threshold has been reached. Preharvest peel pitting may be caused by cold, dry winds, low temperatures, and low relative humidity. The physiological characteristics of membranes and cuticles can be altered by high-temperature conditions.

4.3.2 RADIATION

Variations in the radiation interception in the field have been commonly associated with the development of physiological disorders in fruits and vegetables. Fruits and vegetables that are exposed to the sunlight have more phenolic and vitamin C than those that have been shaded. Direct sunlight causes green fruits to ripen unevenly, resulting in the appearance of yellow patches on the sun-exposed side of the fruit (Emongor et al., 1994). Affected areas have a leathery and harder texture than the surrounding tissues. The biosynthesis of metabolites, including phytohormones, carbohydrates, lipids, and phenolics depends on light-dependent processes of photosynthesis (Chen et al., 2008). The most crucial factor is the availability of solar radiation in areas with a high plant density since it controls photosynthesis and other biochemical activities. The ideal irradiance levels are needed to optimize the accumulation of the various nutritional and antioxidant components in fruits (Kays, 1999). Crop load and fruiting position appear to have a greater impact on a variety of storage physiological disorders than any environmental factor (Tsouvaltzis et al., 2023). Fruiting position and crop load have a typical effect on the development of physiological disorders in apple fruit (Younmoon et al., 1997). Peach fruit from trees with low crop loads and shaded positions have been observed to have higher incidences of mealiness and flesh browning. High-density planting increases plant competition, limits light availability, and may consequently result in reducing fruit quality (Everett et al., 2007). The average light intensity and temperature during the fruit season have an influence on the physiological disorders of strawberry after harvest. In muskmelon, total soluble solids decreased as plant density increased (Mendlinger, 1994). Insufficient light can promote physiological disorders in fruit crops.

Sunburn is a physiological disorder of fruits and vegetables growing in areas with hot climate and high (visible and ultraviolet) light intensity. Sunburn is caused by solar radiation, whereas the term "sunscald" is a physiological disorder of the bark caused by freezing. The symptoms of sunburn browning are yellow, brown, or dark spots on the irradiated side of the fruit (Munné-Bosch and Vincent, 2019). Sunburn disorder deteriorates fruit quality and reduces the shelf life. The fruit's position in the canopy may severely affect sunburn disorder (Ferguson et al., 1999). When watermelon is exposed to sunlight at a temperature of 41.6°C, sunburn appears on the fruit's upper surface. The symptoms are a gray region on the fruit rind which results in a pigment degradation process. Pepper and cucumber fruits also exhibit symptoms of sunburn when exposed to hot temperature and high-intensity sunlight. Sunburn has been identified in tomatoes, which causes discoloration due to chlorophyll degradation. Sunburn browning occurs in the field due to high light radiation on the sun-exposed side of the apple (Streif and Saquet, 2003). High light intensity can cause harder texture and browning in guava fruit (Figure 4.1).

The canopy of fruit trees is important in terms of light exposure. Fruit with a smaller canopy is more exposed to the sun, developing the characteristic ripe fruit color quickly. Many fruits, including olives, peaches, grapes, mangoes, grapes, and apples, have shown the effects of increased sunlight exposure. The synthesis of esters, which are associated with fruit

FIGURE 4.1 Guava fruit peel discoloration.

aroma, is considerably altered by the increased exposure of apples to sunlight caused by canopy management (Moggia et al., 2015). Avocados grown in the sun on the tree revealed that sun fruit had a higher dry matter, calcium, magnesium, and higher oil content than shady fruit (Woolf et al., 1999). Rind breakdown is the physiological disorder of orange fruit, and the cause of this disorder is the intensity and position of the fruit to sunlight. The occurrence of rind breakdown varies among orchards and from year to year. The position of the fruit on the tree has been demonstrated to play a significant role in the development of rind breakdown (Jayasekara et al., 2021). The fruits exposed to sunlight are the most susceptible. Apple fruit quality parameters, including fruit weight, color, soluble solids, sucrose, glucose, fructose, sorbitol, malic, and citric acids, declined with increased planting density (Argenta et al., 2022). The outer face of the fruit is more vulnerable than the inside face. The influence of fruiting position and crop load characteristics in apple fruit has been shown physiological disorders (Chen et al., 2008).

4.3.3 NUTRITION

The production of fruits and vegetables depends heavily on the nutrients present in the field. Both the quality and quantity of fruits and vegetables are significantly affected by nutrient availability in the field. A lot of research has been conducted to investigate the influence of plant nutrition on the postharvest physiological disorder development of fruits and vegetables. Various kinds of physiological disorders originate from preharvest nutritional imbalances (do Amarante et al., 2020). Nutrients affect the appearance, flavor, susceptibility to abiotic stress, juice composition, and the emergence of physiological disorders (Singh et al., 2013; Ahmed et al., 2024). Fruits and vegetables often show numerous browning symptoms that have been attributed to deficiencies in some mineral constituents of the produce (Meera et al., 2017). A balanced mineral intake is necessary for the growth and development of fruits and vegetables. Therefore, a deficiency or excess of any important mineral will result in the occurrence of postharvest physiological disorders. These physiological disorders are prevented by balancing the specified mineral during growth and development.

Physiological disorders can be caused by increasing the nitrogen input to the plant and may diminish fruit quality by lowering the fruit sugar content (Passam et al., 2007). Some significant fruit quality characteristics, including total soluble solids, glucose, fructose, and pH can be negatively impacted by the high nitrogen application (Willingham et al., 2006). Fruit flavors can be enhanced by adding decreased nitrogen sources like ammonium. Increased incidence of scald has also been associated with high nitrogen nutrition (Emongor et al., 1994). The application of potassium boosted photosynthesis, which resulted in the accumulation of carbohydrates and improved fruit quality and yield (Arah et al., 2015). Potassium is a key nutrient for plants' growth and development, and both high and low levels of potassium have been related to abnormal metabolism (Zheng et al., 2020). High potassium levels are involved with the development of bitter pits in apples. Low potassium is associated with changes in the ripening tomato and delays the development of a full red color by inhibiting lycopene biosynthesis (Gangadhara et al., 2021). There may be roles for other minerals in the development of other physiological disorders. In the production of tomatoes, an adequate supply of potassium fertilizer enhances the titratable acidity of the fruit, while also improving fruit color and reducing the occurrence of yellow shoulder (Passam et al., 2007). A physiological disorder in tomatoes known as "yellow shoulder" is characterized by discolored areas of the fruit (Masarirambi et al., 2009). However, quality traits such as total soluble solids pH, acidity, juice, or fruit color attributes are not significantly impacted by variations in the availability of phosphorus in crop-growing soils. Fruit with high phosphate, low calcium, and high potassium contents have been linked to an increase in scald incidence.

Boron deficiency leads to a physiological disorder known as internal cork in apple fruit. This condition is marked by pitting of the flesh and is often indistinguishable from bitter pit. The differences between the two disorders are that internal cork is prevented by the application of boron sprays and develops only on the tree, while bitter pit responds to calcium treatment and can develop after harvest. When using trace elements, the amount of boron utilized has the greatest impact on the quality of the fruit, however, other micronutrients may have an impact on the postharvest physiological disorders only when the plants exhibit severe deficient symptoms (Zheng et al., 2020). Fruit firmness is greatly affected by boron availability, which is a major storage concern (Smit and Combrink, 2004). The application of iron, copper, and cobalt has induced similar symptoms to low-temperature breakdown and superficial scald in apples. However, this does not necessarily suggest that these elements played a part in the emergence of natural disorders. The enzymatic browning, or the browning of cut or damaged tissues when exposed to air, is caused by heavy metals, particularly copper (Prange and DeEll, 1997). The physiological disorder creasing has been linked to endogenous variables, cultural practices, and climatic conditions. Molybdenum plays a crucial role in the creasing physiological disorder. The most prevalent zinc deficiency was treated with a foliar spray of zinc sulfate, which enhanced plant health overall and reduced leaf chlorosis and dieback in kinnow. The enzymes involved in converting polysaccharides into simple sugars, which raise the total soluble contents of fruits, would have been stimulated by zinc application (Eman et al., 2007).

Preharvest application of calcium combined with other minerals has a positive effect on the reduction of postharvest physiological disorders. Calcium is an important element of plant cell walls and membrane structure (Khaliq et al., 2016). Deficiencies in soil calcium have been associated with a number of postharvest disorders, including a bitter pit of apple, blackheart in celery, soft nose in mango, blossom end rot of watermelon, and tipburn of lettuce, cauliflower, and cabbage. Compared to other minerals, calcium deficiencies have been related to a more physiological disorder (Barrientos-Priego et al., 2016). A torrential rainfall and low calcium level increase pulp softening in papaya fruit (Figure 4.2). When calcium salts are used as a preharvest spray, some of these disorders, like tomato blossom-end rot, can be easily controlled, but others, like apple bitter pit, can only be partially controlled. However, this variation in the level of control is most likely associated with how much amount of calcium is absorbed by the fruit. Spongy tissue is a physiological disorder of mango associated with the lower content of calcium and magnesium in the leaves (Shivashankar, 2014). A significant number of proteins are involved in cell signaling associated with calcium, which is a crucial component of cell walls and the middle lamella (Buchanan et al., 2013). In addition to a number of physiological responses, calcium has also been investigated for its part in how plants respond to adapt to biotic and abiotic stresses (Ramírez-Gil et al., 2021). The majority of fruits and vegetables require calcium for healthy growth, hardness, and sufficient storage life. Any impact of calcium content on the quality of fresh produce appears to occur earlier in fruit development than right before harvest. Applications of calcium before harvest have reduced the incidence of postharvest physiological disorders of fresh produce after harvest (Barrientos-Priego et al., 2016). The prevalence of postharvest physiological disorders correlated with calcium/potassium and calcium + magnesium/potassium in the soil. Evidence reveals that calcium is a nutrient closely associated with the occurrence of postharvest physiological disorders in fresh produce (Arah et al., 2015). Low calcium levels have been related to vascular bundle darkening, low postharvest shelf life, and chilling injury (Everett et al., 2007).

FIGURE 4.2 A heavy rainfall before the end of the final growth stage and low calcium enhances pulp softening in papaya fruit.

Calcium assimilation is connected to soil levels of magnesium and potassium, due to an antagonism between them. Fruits and vegetables textural quality and senescence are influenced by calcium levels (Gangadhara et al., 2021). The preharvest calcium application can lower susceptibility to postharvest physiological disorders. Increasing the ratio of calcium + magnesium/potassium in fresh food is generally advised for growers to manage postharvest physiological disorders better (Everett et al., 2007).

4.3.4 IRRIGATION

Irrigation water is essential for plant growth, development, and various metabolic processes. Adequate soil moisture during the growing and development of fruits and vegetables is crucial for the maintenance of postharvest quality. Fruits and vegetables often contain 90% or more water in their cells and tissues. Water is necessary for the plants' growth, flowering, fruit development, and sprouting. Rainfall, irrigation water, and relative humidity in the environment are the primary sources of water for plants. The quality and physiological disorders of the crops are significantly influenced by the relative humidity and soil moisture contents (Li, 2019). However, other factors like temperature, light, and nutrient availability also affect the development of physiological disorders. Water stress affects commodity size and physiological disorder during the growing season and fruit development. The physiological disorders such as tomato blossom end rot and apple bitter pit are controlled by deficit irrigation water.

Fruit quality normally increases when soil moisture and nutrient levels arise from inadequate to optimal levels resulting in the largest yield and the greatest ability to retain quality. Improper irrigation and nutritional deficiency in certain areas can cause physiological disorders in fruit and vegetable crops (Taghavi et al., 2019). The yields of several fruit crops were significantly reduced due to water shortage during the production period (Anastasiadi et al., 2021). The growth and development of fruit depend on the timing and number of irrigations in crop production. Water is a limited resource in most production areas, so growers have come up with more effective water management strategies that maintain crop yield while putting areas because proper irrigation is essential throughout the growing season (Nangare et al., 2016). Water stress decreases crop output and hastens the ripening of fruit. Preharvest water stress is associated with the development of postharvest physiological disorders in fruits and vegetables. Moisture stress is causing sunburn and fruit cracking in apples, apricots, cherries, and litchi.

Deficit irrigation during fruit development boosted fruit total soluble solids levels and reduced the occurrence of physiological disorders (Behboudian and Mills, 1997). The moisture content of the fruits was somewhat reduced when irrigated with saline water, however, the overall fruit yield was not significantly affected. Irrigation in the early morning every three days produced larger yields than irrigation every day. The addition of microelements to irrigation water in a solution form can enable the maximum utilization of trace elements. The availability of soil water during the growth and development of fruits and vegetables had a great effect on the incidence of physiological disorders (Alikhani-Koupaei et al., 2018). Preharvest water stress during the growing season can have an influence on the harvested plant organ and result in soft or dehydrated fruit that is more susceptible to physiological disorders during storage (Arah et al., 2015). The flavor and nutritional value of fruits and vegetables can be negatively affected by an excess of water during the growing season due to the dilution of soluble solids and acids (Shewfelt and Prussia, 1993). One of the most important causes of radish root cracking is inadequate irrigation. The quantity and irrigation frequency affect the water content, which leads to the appearance of cracks on radish roots during active secondary growth (Manzoor et al., 2021). The cracking/splitting incidence was decreased in radish that were irrigated properly than in those that were irrigated inadequately (Luengwilai et al., 2012).

4.3.5 RAINFALL

The climatic factors such as rainfall, solar intensity, irrigation, and transpiration capacity have affected the frequency of physiological disorders in fresh produce (De Oliveira and Vitória, 2011). Rainfall affects the water supply to the plant, which results in changes in the biochemical composition of the fruits (Lara et al., 2022). Fruits and vegetables are more susceptible to heavy rain. Fruit and vegetable quality may be damaged over extended periods of rain. High rainfall and its effects on fruit growth also led to an increase in the incidence of physiological disorders, like in apples and cherries. Fruit cracking is primarily a physiological disorder, that is, most related to excessive air humidity after significant rainfall or intensive irrigation in hot weather (Papoutsis and Edelenbos, 2021). Transverse and radial splitting have been observed in some fruits during the rainy season. The extended period of rain caused the fruit splitting of watermelon. When the watermelon fruit is cut open, water-soaked flesh edge areas have been observed. The symptom might be related to rapid water uptake or increased stress levels that affect fruit ripening. Rainfall increased the susceptibility of pomegranate plants to physiological disorder that had previously been affected by water stress, which led to a rise in fruit turgor pressure (Galindo et al., 2014). Pomegranate peel cracking may be associated with improper environmental conditions, heavy rainfall, nutritional deficiency, and water imbalance (Figure 4.3). Because aril turgor increased considerably more than peel turgor, thus the pressure of the arils on the peel favored peel cracking.

FIGURE 4.3 Pomegranate peel cracking due to water stress.

4.3.6 *GENETIC FACTORS, CULTIVAR, AND VARIETY*

Genetic composition, cultivar, and variety of fruits and vegetables are the major components in the sensitivity to postharvest physiological disorders (De Oliveira and Vitória, 2011). The cultivar and variety susceptibility to various abiotic stresses depends on their biochemical and genetic composition such as bioactive compounds, enzyme activity, and expression of genes (Hanna, 2009). Various postharvest quality attributes are directly related to genetic composition (Penna and Jain, 2023). The resistance mechanism patterns of fruit and vegetable varieties differ from each other grown even in the same agro-climatic conditions, and these differences are possibly due to the diverse genetic makeup of the variety (Zhu et al., 2022).

The chemical composition and bioactive compounds of each cultivar are more or less controlled by genetic factors (Hernandez et al., 2016). Variations among cultivar susceptibility to develop chilling injury of tomato could be explained by the ability to biosynthesize heat shock protein, metabolism of several bioactive compounds, and cell wall degradation. The response of cultivars to chilling injury has been associated with the activity of antioxidant enzymes, hormone metabolism, carbohydrate metabolism, and the variable gene expression involved in the cell membrane system (Kathayat and Rawat, 2019). Each species and variety has a particular susceptibility to chilling injury. For example, lemons and limes are typically more susceptible to chilling injury than mandarins and oranges. Similarly, the Navel oranges cultivar Navelina fruit has a higher tolerance to chilling injury than the Navelate and Roberts cultivars which are extremely sensitive to chilling injury. Several studies have confirmed that fruit cracking is associated with a number of genetic, environmental, physiological, biochemical, and cultural factors. Cell wall structure, turgor pressure, and relative humidity are believed to have a major effect on peel splitting. Banana fruit postharvest quality is greatly affected by peel splitting (Figure 4.4).

FIGURE 4.4 Banana fruit peel splitting.

Cultivar and variety selection is, therefore, critical to manage the postharvest physiological disorders of fruits and vegetables. Watercore is a genetic disorder in some apple cultivars; however, it does not occur in other apple cultivars (Harker et al., 1999). Producers use cultivars with improved quality attributes and long postharvest shelf life for numerous commodities. Plant geneticists are combining plant breeding techniques with molecular biology methods to produce novel genotypes with improved flavor, firmness maintenance, resistance to biotic and abiotic stresses, and other desirable traits. The occurrence of postharvest physiological disorders could be decreased by selecting an appropriate genotype, cultivar, and variety (Prange and DeEll, 1997). For instance, apples and pears are highly sensitive to superficial scald, a physiological disorder that causes black and brown blemishes on the fruit skin during storage. The susceptibility to superficial scald disorder depends on the genotype, cultivar, and variety (Marc et al., 2020). Golden Delicious, Empire, Pink Lady, Gala, and Braeburn apple cultivars are more resistant to superficial scald, whereas Fuji, Granny Smith, and Law Rome are sensitive to it. The navel orange cultivar is more susceptible to stem-end rind breakdown disorder than the other orange cultivars/varieties. Internal browning is a symptom of chilling injury in peach, and this varies among cultivars, depending on the capacity of each cultivar to metabolize sucrose. The pineapple cultivar Queen type is more susceptible to internal browning disorder than the Smooth Cayenne type, possibly due to variation in gene expression.

The choice of the best cultivar and variety is a crucial factor for consumer acceptability and the subsequent development of postharvest physiological disorders (Shewfelt and Prussia, 1993). Numerous cultivars were shown to be susceptible to only one physiological disorder, while several cultivars were affected by more than one specific physiological disorder, which resulted in a significant decline in market value (Manzoor et al., 2021). Apple cultivars York Imperial, Marigold, Prima Starking, and Starkrimson are more susceptible to the bitter pit disorder than the Fuji, Red Gravenstein, Gala, and Stonetosh cultivars. The cultivar type affects the postharvest physiological disorder because different cultivars exhibit various quality traits, some are more appealing to growers and consumers than others. It is crucial for a producer to choose a cultivar with acceptable yields, desired fruit quality, and a longer shelf life (Hanna, 2009). A poor cultivar choice could result in fruits of low quality and fruits that are less marketable. Different cultivars vary in fruit size, color, texture, flavor, and storage capacity. Fruit cultivars react differently when stored under the same conditions to some physiological disorders. Therefore, cultivar and variety selection is crucial in order to respond to physiological disorders during the postharvest storage life.

4.3.7 ROOTSTOCK SCION RELATIONSHIP

Rootstock and scion affect the postharvest quality and physiological disorders. Numerous horticultural commodities' productivity, quality, and physiological disorders are significantly affected by the rootstock. The rootstock has a significant impact on physiological disorders, probably as a result of rootstock-mediated changes in tolerance to harmful factors (such as drought and soil), and increased nutrient and water uptake (Rallo et al., 2018). The rootstocks affect floral bud toughness, flowering initiation, and fruit ripening time (Willingham et al., 2006). Compared to self-rooted plants, grafted plants typically exhibit more water and mineral uptake, which increases plant vigor and the availability of nutrients for fruit growth, development, and final size (Rouphael et al., 2010). For instance, the Seckel pear grafted onto *Pyrus betulifolia* rootstock produces good quality fruit, however, the same rootstock for Anjou pear exhibits cork spot in the flesh areas (Yahia et al., 2019). Some plum and prune rootstocks may favor fruit that is more susceptible to gum stains, stem end shrivel, and interior browning. For many crops, rootstock and scion have a significant influence on fruit yield, quality, and physiological disorders. Higher fruit firmness, total phenolic, and vitamin C were observed when Flame seedless grapes were grafted onto Paulson rootstock compared with those grafted on Flame, Ruggeri, and Freedom rootstocks. Granulation is a common physiological disorder due to crystallization in citrus fruit (Jayasekara et al., 2021). Rootstock and scion relationship may reduce or increase the development of physiological disorder. When compared to fruits from older plants, the incidence of physiological disorder is considerably higher in those from younger plants (Yahia et al., 2019). Vigorous rootstocks like rough lemon have a higher incidence of granulation than the less vigorous rootstocks. The incidence of granulation has been reported to rise with late maturity and prolonged cold conditions during the entire period of maturity.

4.3.8 PRUNING, THINNING, AND FRUIT SIZE

Pruning and thinning affect the fruit size, while the occurrence of physiological disorders in some fruits and vegetables is related to fruit size.

A good strategy to lessen competition between fruits is to manage the amount of flowers, fruits, or fruit trusses. Therefore, pruning makes sure nutrients are directed to fewer fruit which can result in increased fruit size and boosting fruit sugar content (Gautier et al., 2001). The incidence of postharvest physiological disorders is increased by increased leaf area, which may be related to microclimate with higher humidity and aging tissues (Everett et al., 2007). Tree density may be related to pruning, which would cause a decline in the canopy index and tree age. The impact of pruning on fruit quality attributes is influenced by a variety of variables, including the fruit-to-leaf ratio, truss position, and genetic background (Gautier et al., 2001). Under the correct growing conditions, pruning can enhance fruit size in most cultivars, which can be utilized to promote the marketability of fruits, which often have smaller sizes.

Pruning improves fruit quality by promoting light penetration into the canopy. For instance, removing internal water sprouts can greatly boost light penetration and enhance the fruit's size, color, and soluble solid concentration of lower position fruit (Takishita et al., 2021). Fruit at the top of the tree is always of higher quality than fruit at the bottom, where the canopy is shaded. One of the most crucial methods for adjusting fruit size is fruit thinning by regulating fruit number. Fruit thinning has been done using a standard based on fruit and leaf ratio number per canopy volume. Small fruit removal from trees could increase fruit size; however, improper thinning can reduce fruit quality in a number of ways. A number of factors influence the size of fruits and vegetables. High plantation densities and an extensive canopy limit root growth and reduce plant photosynthetic activity by decreasing light penetration and photosynthesis, which compromises fruit size (Hanna, 2009). The growth of fruit tissues depends on the turgor exerted by the water in fruit cells during fruit development, which substantially diminishes the fruit size at harvest. Larger fruits have a significantly higher occurrence of physiological disorder than little fruits. Fruit size and environmental factors affect the occurrence of physiological disorders. Citrus cultivar Harumi small fruit indicated severe peel wrinkle; however, dehydration of flesh and rind puffing were observed in large fruit during storage (Takishita et al., 2021).

4.3.9 MATURITY STAGE

Numerous physiological disorders are significantly influenced by the maturity stage at harvest. Fruits and vegetables can be harvested at various stages of maturity, such as the mature green, half-ripe, full red, physiological, and commercial maturity stage (Toivonen, 2007; Busatto and Herrera, 2025). Each stage of harvest has a characteristic that the fruit will display after harvest. All fruits and vegetable varieties showed the longest shelf life when harvested at the green mature stage (Moneruzzaman et al., 2009). Fruits and vegetables that are harvested too early or too late display postharvest physiological disorders. The maturity stage at harvest has a significant effect on quality attributes like texture, fiber, taste, aroma, and uniformity (Bowen and Watkins, 1997; Latt et al., 2025). Immature harvested fruits and vegetables are very susceptible to shriveling and mechanical harm. Overripe fruits and vegetables frequently have poor texture, flavor, and greater chances of physiological disorder. Zacarias et al. (2001) stated that oleocellosis increased with fruit maturity, sudden changes in relative humidity, water stress, and climate change from rain to wind (Figure 4.5). The proper harvesting stage of maturity decreases susceptibility to postharvest physiological disorders of fresh produce. Apple harvested at the proper maturity stage is more resistant to bitter pits than fruit harvested at an immature stage. Similarly, some particular cultivars of citrus fruit harvested at the advanced stage of maturity showed peel cracking around the calyx or on the peel surface (Lafuente and Zacarias, 2006; Jayasekara et al., 2021).

FIGURE 4.5 Oleocellosis due to the breakdown of the oil glands in the flavedo of Kinnow.

Preharvest factors like maturity stage and geographic/climatic may predispose the development of physiological disorders (Lo'ay and Dawoo, 2019). For instance, sugar transport to vine-ripened fruits appears to increase in the latter stages of maturity. When fruits are harvested early or in a green state, sugar import to fruits will be interrupted, which is undesirable and leads to postharvest degradation of starch contents (Yamada et al., 1994). Harvesting at a later stage also encourages higher sugar accumulation in the ripe fruits, which have a shorter shelf life and are more prone to physiological disorders (Toivonen, 2007; Zhu et al., 2022). Fresh produce's pH is a crucial factor in the fresh produce processing industry. Because fruits and vegetables are processed as high-acid foods, the more acidic they are, the easier they will be to process. Therefore, high-pH cultivars might not be suited for processing. Fruit and vegetable acidity is maximum at the pink stage of maturity. Fruits can be collected when they are mature green to provide growers ample time for long-distance distribution, but for local marketing, harvesting at the fully ripe stage is preferred to maximize nutritional content (Moneruzzaman et al., 2009).

The maturity stage and environmental conditions affect the occurrence of chilling injury in various fruits and vegetables. For example, fruit picked earlier and later in the season is more prone to chilling injury than fruit picked in the middle of the season. Fruits that are harvested before they are fully mature and ripe may become more susceptible to chilling injury or may never reach their full ripening stage (Jing et al., 2018). Tropical and sub-tropical fruits and vegetables that are not harvested at the proper stage of maturity showed chilling injury symptoms. Maturity stage and climatic conditions may have varied effects on the development of the storage physiological disorder, such as flesh browning, watercore, and cavity development in apples (Al Shoffe et al., 2021). The incidence of watercore also increased as fruit maturity increased, hastening the cavity development and flesh browning (Harker et al., 1999). The occurrence of watercore is associated with flesh browning during storage in the late-harvested apple fruit (Kweon et al. 2012). Ethylene evolution has been used to determine the physiological maturity of climacteric fruits, such as apples, mangoes, and bananas. After harvest, apple with watercore exhibited higher ethylene evolution than unaffected fruit. The link between watercore development and maturation has been observed in numerous fruits and vegetables.

4.4 CONCLUSION

Postharvest physiological disorders reduce the quality and consumer acceptability of fresh produce. This chapter suggests that numerous preharvest factors, rather than one single component, may be responsible for the development of postharvest physiological disorders in fresh produce. It is essential for fruits and vegetable growers to identify and know the preharvest factors such as cultural practices, irrigation frequency, planting density, temperature, radiation, nutrition, and proper selection of specific cultivars for specific conditions in order to reduce the occurrence of postharvest physiological disorders. These conditions combined define the level of postharvest physiological disorders incidence and severity in fresh produce. A systematic approach to managing plant nutrition, climatic conditions, and cultural practices is essential to reduce the occurrence of postharvest physiological disorders, most of which are linked to preharvest factors. There are still many shortcomings due to insufficient research on the mechanism of interaction between these postharvest physiological disorders and preharvest factors. For the implementation of integrated management programs in sustainable techniques, more investigation of various environmental factors that influence the emergence and development of postharvest physiological disorder are necessary.

KEYWORDS

- **environmental factors**
- **perishable**
- **postharvest physiological disorders**
- **preharvest factors**
- **quality loss**

REFERENCES

Al Shoffe, Y., Nock, J. F., Zhang, Y., & Watkins, C. B. (2021). Physiological disorder development of 'Honeycrisp'apples after pre-and post-harvest 1-methycyclopropene (1-MCP) treatments. *Postharvest Biology and Technology, 182*, 111703.

Ahmed, N., Zhang, B., Chachar, Z., Li, J., Xiao, G., Wang, Q., et al. (2024). Micronutrients and their effects on horticultural crop quality, productivity and sustainability. Scientia Horticulturae, 323, 112512.

Alikhani-Koupaei, M., Fatahi, R., Zamani, Z., & Salimi, S. (2018). Effects of deficit irrigation on some physiological traits, production and fruit quality of 'Mazafati'date palm and the fruit wilting and dr Sankat opping disorder. *Agriculture Water Management, 209*, 219–227.

Ali, S., Khaliq, G., Iftikhar, A., Mir, S. A., & Iqbal, S. (2024). Nanoemulsions-based edible coatings for postharvest quality preservation of fruits and vegetables. In *Sustainable postharvest technologies for fruits and vegetables*, (pp. 213–226). CRC Press.

Ali, S., Khan, A. S., Anjum, M. A., Ejaz, S., Khaliq, G., Hussain, S., Ahmad, R., & Saleem, M. S. (2022). Pre-and postharvest physiological disorders of citrus fruit. In *Citrus Production*, (pp. 371–390). CRC Press.

Anastasiadi, M., Falagán, N., Rossi, S., & Terry, L. A. (2021). A comprehensive study of factors affecting postharvest disorder development in celery. *Postharvest Biology and Technology, 172*, 111384.

Antunes, M. D., Guimarães, A. C., Gago, C., Guerreiro, A., Panagopoulos, J., Vilas Boas, E., & Miguel, M. G. (2022). Membrane fatty acids and physiological disorders in cold-stored 'Golden Delicious' apples treated with 1-MCP and calcium chloride. *Horticulturae, 8*(2), 162.

Arah, I. K., Amaglo, H., Kumah, E. K., & Ofori, H. (2015). Preharvest and postharvest factors affecting the quality and shelf life of harvested tomatoes: a mini review. *International Journal of Agronomy, 2015*

Argenta, L. C., do Amarante, C. V. T., de Freitas, S. T., Brancher, T. L., Nesi, C. N., & Mattheis, J. P. (2022). Fruit quality of 'Gala' and 'Fuji' apples cultivated under different environmental conditions. *Scientia Horticulturae, 303*, 111195.

Barrientos-Priego, A. F., Martínez-Damian, M. T., Vargas-Madríz, H., & Lazaro-Dzul, M. O. (2016). Effect of preharvest calcium spraying on ripening and chilling injury in 'Hass' (*Persea americana* Mill.) avocado. *Revista Chapingo Serie Horticultura, 22*, 145–159.

Behboudian, M. H., & Mills, T. M. (1997). Deficit irrigation in deciduous orchards. *Horticultural Reviews, 21*, 105–131.

Beverly, R. B., Latimer, J. G., & Smittle, D. A. (1993). Preharvest physiological and cultural effects on postharvest quality. In *Postharvest handling* (pp. 73–98). Academic Press.

Bieleski, R. L., & Bartley. T. (1999). Maturity and regional influences on watercore development and its postharvest disappearance in 'Fuji' apples. *Journal of the American Society for Horticultural Science, 124*, 166–172.

Bowen, J. H., & Watkins, C. B. (1997). Fruit maturity, carbohydrate and mineral content relationships with watercore in 'Fuji' apples. *Postharvest Biology and Technology, 11*(1), 31–38.

Busatto, N., & Herrera, R. (2025). Fruit Development and Ripening—A Molecular and Physiological View Modulating and Enhancing Fruit Quality. Journal of Plant Growth Regulation, 1-3.

Buchanan, B., Wilhelm, G., & Rossell, L. (2013). In *Biochemistry and molecular biology of plants* (2nd Ed.). John Wiley & Sons, West Sussex, UK.

Chen, L. S., Li, P., & Chen, L. (2008). Effects of high temperature coupled with high light on the balance between photo oxidation and photoprotection in the sun-exposed peel of apple. *Planta, 228*, 745–756.

Crisosto, C. H., Johnson, R. S., DeJong, T., & Day, K. R. (1997). Orchard factors affecting postharvest stone fruit quality. *HortScience, 32*, 820–823.

De Oliveira, J. G., & Vitória, A. P. (2011). Papaya: Nutritional and pharmacological characterization, and quality loss due to physiological disorders. An overview. *International Food Research Journal, 44*(5), 1306–1313.
do Amarante, C. V., Silveira, J. P. G., Steffens, C. A., de Freitas, S. T., Mitcham, E. J., & Miqueloto, A. (2020). Post-bloom and preharvest treatment of 'Braeburn' apple trees with prohexadione-calcium and GA4+ 7 affects vegetative growth and postharvest incidence of calcium-related physiological disorders and decay in the fruit. *Scientia Horticulturae, 261*, 108919.
Eman, A. A., Abdou, M. M. M., Omayma, M. M. I. (2007). GA3 and zinc spray for improving yield and fruit quality of 'Washington Naval' Orange tree growing under sandy soil conditions. *Research Journal of Biological Science, 3*, 498–503.
Emongor, V. E., Murr, D. P., & Lougheed, E. C. (1994). Preharvest factors that predispose apples to superficial scald. *Postharvest Biology and Technology, 4*(4), 289–300.
Everett, K. R., Boyd, L. M., Pak, H. A., & Cutting, J. G. M. (2007). Calcium, fungicide sprays and canopy density influence postharvest rots of avocado. *Australasian Plant Pathology, 36*, 22–31.
Ferguson, I., Volz, R., & Woolf, A. (1999). Preharvest factors affecting physiological disorders of fruit. *Postharvest Biology and Technology, 15*(3), 255–262.
Freeman, B. B., & Reimers, K. (2011). Tomato consumption and health: emerging benefits. *American Journal of Lifestyle Medicine, 5*(2), 182–191.
Galindo, A., Rodríguez, P., Collado-González, J., Cruz, Z. N., Torrecillas, E., Ondoño, S., Corell, M., Moriana, A., & Torrecillas, A. (2014). Rainfall intensifies fruit peel cracking in water stressed pomegranate trees. *Agricultural and Forest Meteorology, 194*, 29–35.
Gangadhara, K., Selvakumar, R., Singh, P. K., Manjunathagowda, D. C., & Kumar, J. (2021). Tomato: Physiological disorders and their management. *Biotica Research Today, 3*(10), 944–948.
Gariglio, N., Reig, C., Martinez-Fuentes, A., Mesejo, C., Agustı, M. (2008). Purple spot in loquat (*Eriobotrya japonica* Lindl.) is associated to changes in flesh-rind water relations during fruit development. *Scientia Horticulturae, 119*, 55–58.
Gautier, H., Guichard, S., & Tchamitchian, M. (2001). Modulation of competition between fruits and leaves by flower pruning and water fogging, and consequences on tomato leaf and fruit growth. *Annals of Botany, 88*(4), 645–652.
Hanna, H. Y. (2009). Influence of cultivar, growing media, and cluster pruning on greenhouse tomato yield and fruit quality. *HortTechnology, 19*(2), 395–399.
Harker, F. R., Watkins, C. B., Brookfield, P. L., Miller, M. J., Reid, S., Jackson, P. J.,
Hernandez, I., Fuentealba, C., Olaeta, J. A., Lurie, S., Defilippi, B. G., Campos-Vargas, R., & Pedreschi, R. (2016). Factors associated with postharvest ripening heterogeneity of 'Hass' avocados (*Persea americana* Mill). *Fruits, 71*, 259–268.
Hewett, E. W. (2006). An overview of preharvest factors influencing postharvest quality of horticultural products. *International Journal of Postharvest Technology and Innovation, 1*(1), 4–15.
Jayasekara, A., Abeywickrama, K., Daranagama, A., & Kodituwakku, T. (2021). Physiological disorders of selected Citrus fruit species in Sri Lanka and their effect on fruit quality. *Journal of Horticulture and Postharvest Research, 4*(3), 385–398.
Jing, Y., Ma, X., Jin, P., & Zhu, X. (2018). Effects of harvest maturity on chilling injury and storage quality of apricots. *Journal of Food Quality, 2018*(1), 1–7.

Kader, A. A., Lyons, J. M., & Morris, L. L. (1974). Postharvest responses of vegetables to preharvest field temperature. *HortScience*, *9*(6), 523–527.

Kathayat, K., & Rawat, M. (2019). Physiological disorders in vegetable crops. In S. K. Singh & S. Kaur (Eds), *Advances in horticultural crop management and value* (pp. 313–314). *addition*; Laxmi Publications.

Kays, S. J. (1999). Preharvest factors affecting appearance. *Postharvest Biology and Technology, 15*(3), 233–247.

Khaliq, G. (2015). Effects of edible coatings enriched with calcium chloride on physiological, biochemical and quality responses of mango (*Mangifera indica* L. cv. Choke Anan) fruit during cold storage. Doctoral thesis, Universiti Putra Malaysia.

Khaliq, G., Ali, S., Gapper, N., & Nicola, S. (2023). Recent advances and approaches in the application of elicitors to enhance resistance mechanisms in fresh produce. *Frontiers in Sustainable Food Systems, 6*, 1061079.

Khaliq, G., Mohamed, M. T. M., Ghazali, H. M., Ding, P., & Ali, A. (2016). Influence of gum arabic coating enriched with calcium chloride on physiological, biochemical and quality responses of mango (*Mangifera indica* L.) fruit stored under low temperature stress. *Postharvest Biology and Technology, 111*, 362–369.

Khaliq, G., Saleh A., Bugti G. A., & Hakeem K. R. (2020). Guggul gum incorporated with basil essential oil improves quality and modulates cell wall-degrading enzymes of jamun fruit during storage. *Scientia Horticulturae, 273*, 109608.

Khaliq, G., Ullah, M., Memon, S. A., Ali, A., & Rashid, M. (2021). Exogenous nitric oxide reduces postharvest anthracnose disease and maintains quality of custard apple (*Annona squamosa* L.) fruit during ripening. *Journal of Food Measurement and Characterization, 15*(1), 707–716.

Khan, L., Khaliq, G., Ullah, E., Abdul-Rahaman, A., El-Mogy, M. M., Kesba, H. H., Ismail, A. M., & El-Beltagi, H. S. (2025). Guggul gum and clove essential oil synergistically reduced anthracnose disease and modulated physicochemical changes in tomato during storage. *Journal of Stored Products Research, 114,* 102713.

Kweon, H. J., Kim, M. J., Moon, Y. S., Lee, J. W., Choi, C., Choi, D. G., Lee, D. H., & Kang, I. K. (2012). Relationship between preharvest factors and the incidence of storage disorders in 'Fuji' apples during CA storage. *Horticultural Science and Technology, 30*(1), 50–55.

Lafuente, M. T., & Zacarias, L. (2006). Postharvest physiological disorders in citrus fruit. *Stewart Postharvest Review, 2*(1), 1–9.

Lal, S., Khaliq, G., Mastoi, A. H., Irshad, A., Jaffar, M. A., Khan, L., & Muhammad, N. (2025). Effect of *Fagonia indica* and garlic extract on *Alternaria alternata* and post-harvest quality of tomato fruit during storage. *Journal of Crop Health, 77*(6), 1–13.

Lara, I., Bustamante, C. A., & Villarreal, N. M. (2022). Fruit responses to biotic and abiotic stressors during postharvest. *Frontiers in Plant Science, 13*, 914841.

Latt, T. T., Lee, J., Park, H., Byeon, S. E., Han, S. Y., Lwin, H. P., et al. (2025). Pre-storage conditioning treatment reduces physiological disorders by mediating the targeted primary metabolites in 'Chuhwangbae'pears during shelf life after cold storage. Horticulture, Environment, and Biotechnology, 1-16.

Léchaudel, M., & Joas, J. (2007). An overview of preharvest factors influencing mango fruit growth, quality and postharvest behaviour. *Brazilian Journal of Plant Physiology, 19*, 287–298.

Li, H. (2019). Pre-harvest cultural practices affecting postharvest physiological disorders, minerals, and titratable acidity in cucurbit melon. In *Postharvest physiological disorders in fruits and vegetables* (pp. 15–28).

Li, Y., Zhang, S., & Dong, Y. (2021). Improving storability, physiological disorders, and antioxidant properties of 'Bartlett' and 'd'Anjou' pears (*Pyrus communis* L.) by pre-harvest 1-methylcyclopropene spraying. *International Journal of Food Science, 56*(1), 115–125.

Lo'ay, A. A., & Dawood, H. D. (2019). Chilling injury, fruit color maturity stages, and antioxidant enzyme activities of lemon 'baladi CV' fruits under cold storage stress. *Scientia Horticulturae, 257*, 108676.

Luengwilai, K., Beckles, D. M., & Saltveit, M. E. (2012). Chilling injury of harvested tomato (*Solanum lycopersicum* L.) cv. Micro-Tom fruit is reduced by temperature pre-treatments. *Postharvest Biology and Technology, 63*, 123–128.

Lufu, R., Ambaw, A., & Opara, U. L. (2020). Water loss of fresh fruit: Influencing pre-harvest, harvest and postharvest factors. *Scientia Horticulturae, 272*, 109519.

Luyckx, A., Lechaudel, M., Hubert, O., Salmon, F., & Brat, P. (2016). Banana peel physiological post-harvest disorders: A review. *MOJ Food Process Technology, 3*(1), 226–231.

Manzoor, A., Bashir, M. A., Naveed, M. S., Cheema, K. L., & Cardarelli, M. (2021). Role of Different abiotic factors in inducing pre-harvest physiological disorders in radish (*Raphanus sativus*). *Plants, 10*(10), 2003.

Marc, M., Cournol, M., Hanteville, S., Poisson, A. S., Guillou, M. C., Pelletier, S., Laurens, F., Tessier, C., Coureau, C., Renou, J. P., & Delaire, M. (2020). Pre-harvest climate and post-harvest acclimation to cold prevent from superficial scald development in Granny Smith apples. *Scientific Reports, 10*(1), 1–15.

Masarirambi, M. T., Mhazo, N. O. R. M. A. N., Oseni, T. O., & Shongwe, V. D. (2009). Common physiological disorders of tomato (*Lycopersicon esculentum*) fruit found in Swaziland. *The Journal of Agricultural Society Science, 5*, 123–127.

Mditshwa, A., Khaliq, G., Hussein, Z., & Ejaz, S. (2023). Sustainable postharvest management practices for fresh produce. Frontiers in Sustainable Food Systems, 7, 1143759.

Meera, S., Farheen, H., Jignasa, R., Preeti, G., Adodaria, B. K., Patel, H. N., Malam, V. R., & Kanzaria, D. R. (2017). Physiological disorders and quality improvement by appropriate remedies in fruit crops. *Bioscience Trends, 10*(32), 6979–6981.

Mendlinger, S. (1994). Effect of increasing plant density and salinity on yield and fruit quality in muskmelon. *Scientia Horticulturae, 57*(1–2), 41–49.

Mishra, D. S., Tripathi, A., & Nimbolkar, P. K. (2016). Review on physiological disorders of tropical and subtropical fruits: Causes and management approach. *International Journal of Agriculture Environment and Biotechnology, 9*(6), 925.

Moggia, C., Pereira, M., Yuri, J. A., Torres, C. A., Hernández, O., Icaza, M. G., & Lobos, G. A. (2015). Preharvest factors that affect the development of internal browning in apples cv. Cripp's Pink: Six-years compiled data. *Postharvest Biology and Technology, 101*, 49–57.

Moneruzzaman, K. M., Hossain, A. B. M. S., Sani, W., Saifuddin, M., & Alenazi, M. (2009). Effect of harvesting and storage conditions on the postharvest quality of tomato. *Australian Journal of Crop Science, 3*(2), 113–121.

Moretti, C. L., Mattos, L. M., Calbo, A. G., & Sargent, S. A. (2010). Climate changes and potential impacts on postharvest quality of fruit and vegetable crops: A review. *Food Research International, 43*(7), 1824–1832.

Munné-Bosch, S., & Vincent, C. (2019). Physiological mechanisms underlying fruit sunburn. *Critical Reviews in Plant. Science, 38*(2), 140–157.

Nangare, D. D., Singh, Y., Kumar, P. S., & Minhas, P. S. (2016). Growth, fruit yield and quality of tomato (*Lycopersicon esculentum* Mill.) as affected by deficit irrigation regulated on phenological basis. *Agricultural Water Management, 171*, 73–79.

Nxumalo, K. A., Matsuane, C., & Masarirambi, M. T. (2019). Calcium-related post-harvest physiological disorders of fruits and vegetables in Eswatini: A review. *Current Journal of Applied Science and Technology, 33*(6), 1–10.

Park, H., Eo, H. J., Kim, C. W., Stewart, J. E., Lee, U., & Lee, J. (2024). Physiological disorders in cold-stored 'Autumn Sense'hardy kiwifruit depend on the storage temperature and the modulation of targeted metabolites. Food Chemistry, 460, 140730.

Papoutsis, K., & Edelenbos, M. (2021). Postharvest environmentally and human-friendly pre-treatments to minimize carrot waste in the supply chain caused by physiological disorders and fungi. *Trends in Food Science and Technology, 112*, 88–98.

Passam, H. C., Karapanos, I. C., Bebeli, P. J., & Savvas, D. (2007). A review of recent research on tomato nutrition, breeding and post-harvest technology with reference to fruit quality. The *European Journal of Plant Science and Biotechnology, 1*, 1–21.

Paull, R. E., & Chen, N. J. (2000). Heat treatment and fruit ripening. Postharvest biology and technology, 21(1), 21-37.

Pedreschi, R., Hollak, S., Harkema, H., Otma, E., Robledo, P., Westra, E., Somhorst, D., Ferreyra, R., & Defilippi, B. G. (2016). Impact of postharvest ripening strategies on 'Hass'avocado fatty acid profiles. *South African Journal of Botany, 103*, 32–35.

Penna, S., & Jain, S. M. (2023). Fruit crop improvement with genome editing, in vitro and transgenic approaches. *Horticulturae, 9*(1), 58.

Prange, R. K., & DeEll, J. R. (1997). Preharvest factors affecting postharvest quality of berry crops. *HortScience, 32*(5), 824–830.

Rallo, L., Díez, C. M., Morales-Sillero, A., Miho, H., Priego-Capote, F., & Rallo, P. (2018). Quality of olives: A focus on agricultural preharvest factors. *Scientia Horticulturae, 233*, 491–509.

Ramírez-Gil, J. G., Henao-Rojas, J. C., & Morales-Osorio, J. G. (2021). Postharvest diseases and disorders in avocado cv. Hass and their relationship to preharvest management practices. *Heliyon, 7*(1), 05905.

Rouphael, Y., Schwarz, D., Krumbein, A., & Colla, G. (2010). Impact of grafting on product quality of fruit vegetables. *Scientia Horticulturae, 127*(2), 172–179.

Schrader, L. E., Sun, J., Zhang, J., Felicetti, D., & Tian, J. (2008). Heat and light-induced apple skin disorders: Causes and prevention. *Acta Horticulturae, 772*, 51–58.

Shewfelt, R. L., & Prussia, S. E. (1993). In *Postharvest handling: A systems approach*. Academic Press Inc.

Shivashankar, S. (2014). Physiological disorders of mango fruit. *Horticultural Reviews, 42*, 313–348.

Singh, S. K., Singh, R. S., & Awasthi, O. P. (2013). Influence of pre and post-harvest treatments on shelf life and quality attributes of ber fruits. *Indian Journal of Horticulture, 70*(4), 610–613.

Smit, J. N., & Combrink, N. J. J. (2004). The effect of boron levels in nutrient solutions on fruit production and quality of greenhouse tomatoes. *South African Journal of Plant and Soil, 21*(3), 188–191.

Streif, J., & Saquet, A. A. (2003). Internal flesh browning of 'Elstar' apples as influenced by pre-and postharvest factors. *Acta Horticulturae, 599*, 523–527.

Taghavi, T., Siddiqui, R., & Rutto, L. K. (2019). The effect of preharvest factors on fruit and nutritional quality in strawberry. In T. Asao & M. Asaduzzaman (Eds.), *Strawberry pre-and post-harvest management techniques for higher fruit quality*. IntechOpen.

Takishita, F., Nishikawa, F., Matsumoto, H., & Kato, M. (2021). Fruit thinning and physiological disorders in citrus variety 'harumi'. *Reviews in Agricultural Science, 9*, 20–31.

Toivonen, P. M. A. (2007). Fruit maturation and ripening and their relationship to quality. *Stewart Postharvest Review, 3*(2), 1–5.

Tsouvaltzis, P., Gkountina, S., & Siomos, A. S. S. (2023). Quality traits and nutritional components of cherry tomato in relation to the harvesting period, storage duration and fruit position in the truss. *Plants, 12*, 315.

Tyagi, S., Sahay, S., Imran, M., Rashmi, K., & Mahesh, S. S. (2017). Pre-harvest factors influencing the postharvest quality of fruits: A review. *Current Journal of Applied Science and Technology, 23*(1), 12.

Watkins, C. B. (2009). Postharvest physiological disorders and mineral nutrients. *New York, Fruit Quarterly, 17*(3), 17–20.

Watkins, C. B. (2017). Postharvest physiological disorders of fresh crops. In *Encyclopedia of applied plant sciences* (pp. 315–322).

Willingham, S. L., Pegg, K. G., Anderson, J. M., Cooke, A. W., Dean, J. R., Giblin, F. R., & Coates, L. M. (2006). Effects of rootstock and nitrogen fertiliser on postharvest anthracnose development in Hass avocado. *Australasian Plant Pathology, 35*, 619–629.

Woolf, A. B., Bowen, J. H., & Ferguson, I. B. (1999). Preharvest exposure to the sun influences postharvest responses of 'Hass' avocado fruit. *Postharvest Biology and Technology, 15*(2), 143–153.

Woolf, A. B., & Ferguson, I. B. (2000). Postharvest responses to high fruit temperatures in the field. *Postharvest Biology and Technology, 21*(1), 7–20.

Yahia, E. M., Gardea-Béjar, A., de Jesús Ornelas-Paz, J., Maya-Meraz, I. O., Rodríguez-Roque, M. J., Rios-Velasco, C., Ornelas-Paz, J., & Salas-Marina, M. A. (2019). Preharvest factors affecting postharvest quality. *Postharvest Technology of Perishable Horticultural Commodities*, 99–128.

Yamada, H., Ohmura, H., Arai, C., & Terui, M. (1994). Effect of preharvest fruit temperature on ripening, sugars, and watercore occurrence in apples. *Journal of the American Society for Horticultual Science, 119*(6), 1208–1214.

Yang, L. A., & Barth, M. M. (1997). Preharvest factors affecting postharvest quality of vegetables. *HortScience, 32*(5), 812–816.

Younmoon, P., Jung, K. H., Yul, K. H., & Ha, R. O. (1997). Preharvest factors affecting the incidence of physiological disorders during CA storage of 'Fuji' apples. *Horticulture Environment and Biotechnology, 38*(6), 725–729.

Zacarias, L., Alferez, F., Gariglio, N., Almela, V., & Agustı´, M. (2001). Rind breakdown in Navelate oranges: Influence of rootstock. In *Proceedings of the International Society on Citriculture, Orlando, Florida, USA*.

Zheng, Y., Ma, Y. Y., Liu, W., & Qiu, F. Plant nutrition and physiological disorders in fruit crops. In A. K. Srivastava & C. Hu (Eds), *Fruit crops: Diagnosis and management of nutrient constraints* (pp. 47–50). Elsevier Publishers.

Zhu, Y., Huang, Q., Pan, Y., Zhang, Z., Yuan, R., & Nie, Y. (2022). Abnormal behavior of chilling injury in postharvest papaya fruit is associated with sugar metabolism. *Journal of Food Science, 87*(3), 919–928.

CHAPTER 5

Postharvest Factors Affecting Physiological Disorders

POE NANDAR KYAW[1*], EI HAY MAN[1], WAI WAI LWIN[1], KHIN THIDA MYINT[1], and ZORA SINGH[2]

[1]*Department of Horticulture, Yezin Agricultural University (YAU), Myanmar*

[2]*Horticulture, School of Science, Edith Cowan University, Joondalup, Western Australia*

*Corresponding author

ABSTRACT

Fresh fruit and vegetables are mainly in demand for their nutritive value, appetizing nature and, in modern days, for their esthetic value to garnish the dishes. Preserving quality and freshness is of prime importance in trading fresh produce during off-seasons or across continents and is achieved by the advanced postharvest technologies and handling practices. However, certain postharvest storage conditions may lead to the development of physiological disorders which ultimately diminish marketability and consumer acceptance of the fresh produces. In addition, mismanagement of postharvest storage conditions leads to a reduction in key quality attributes such as antioxidant properties, sweetness, flavor, and freshness. Moreover, even decent postharvest technologies, either it may be under low temperature or controlled atmosphere or modified atmosphere storage conditions, can negatively impact the quality of fresh produce during extended storage. Among several factors which govern the quality deterioration of fresh produces, this chapter will mainly discuss the

Physiological Disorders of Fruits and Vegetables: Causes and Remedies.
Ghulam Khaliq, Sajid Ali, and Mohammed Wasim Siddiqui (Eds.)

DOI: 10.1201/9781998511334-5

important postharvest aspects for incidence of physiological disorders and quality reduction of fresh produces during postharvest handling.

5.1 INTRODUCTION

Physiological disorders have been regarded as one of the major root causes for reduction in quality, esthetic value, and economic value of horticultural crops (Babellahi et al., 2020; Strano et al., 2022; Li et al., 2022) and are triggered by several factors ranging from preharvest to postharvest aspects (Freitas and Pareek, 2019). Preharvest cultural practices such as irrigation and over- or under-application of fertilizers, and environmental factors during growth and development such as temperature fluctuations, rainfall, and light distributions influence the occurrence of physiological disorders (Freitas and Pareek, 2019; Drogoudi et al., 2021). Maturity at harvest is also an important preharvest determinant for the tolerance of physiological disorders in fruit during storage (Jing et al., 2018). For example, over-matured "Fuji" apples had incidence of radial browning during storage for 6 months under CA or normal air at –0.5°C, although the degree of incidence was higher in CA stored fruit as the result of CO_2 stress induced browning (Chigwaya, 2020).

Storage environment as such temperature, relative humidity, atmospheric oxygen, and carbon dioxide compositions, including ethylene, are the key postharvest factors governing the occurrence of physiological disorders in fruits and vegetables during and after storage (Freitas and Pareek, 2019). In addition, special postharvest treatments such as 1-MCP (Larrigaudière and Giné-Bordonabaand, 2019) and heat treatments (Fallik and Ilic, 2019) may sometimes aggravate physiological disorders if they were erroneously applied. Genetic factors, which can either be pre- and postharvest trigger, are also crucial for the response of fruit and vegetables to the factors eliciting the physiological disorder symptoms (Freitas and Pareek, 2019).

This chapter will mainly focus on the factors affecting physiological disorders during postharvest handling of fruit and vegetables. Moreover, how do the postharvest handling practices and technologies negatively impact, when they are mismanaged, the quality of fresh products will also be discussed.

5.2 POSTHARVEST FACTORS REGULATING PHYSIOLOGICAL DISORDERS

Fresh produces have been successfully traded during off-seasons or across the continents for centuries with the advancement in postharvest storage facilities and technologies. The shelf life of fruit and vegetables has doubled or tripled longer depending on the type of storage technology applied. However, mishandling of these technologies may sometimes induce the occurrence of disorders during storage or on shelf displays. The most common storage environmental factors which influence the storage disorders are temperature, elevated CO_2 composition, and ethylene. Physiological disorders in certain fruits and vegetables under different storage systems are listed in Table 5.1.

5.2.1 TEMPERATURE

Despite maximizing the marketability and minimizing the losses of most horticultural products, low temperature storage may induce injuries in chilling sensitive tropical and subtropical crops. Low-temperature storage at 5 ± 1°C and 90–95% RH for 60 days induced chilling injury (CI) symptoms in 'Tarocco' sweet orange; however, the degree of CI incidence varied depending on the cultivars (Strano et al., 2021). Bell pepper stored at 4°C with 70% RH were infected with CI such as internal browning, fruit surface shriveling, and surface pitting (Babellahi et al., 2020). "Conference" pears stored at −1°C for 180 days were infected with internal browning and senescent scald (Łysiak et al., 2021). These chilling injuries, for instance, internal browning in pineapples, surface pitting in banana, water-soaked flesh and scalding skin in papaya, and dehydration and sunken surface in cucumber, negatively impact the quality and reduce the net profit returns (Liang et al., 2020; Tarmizi and Dolhaji, 2022).

CI refers to exposure to low but non-freezing temperatures, which causes damage to cold-sensitive plants. Chilling temperatures (1−10°C) disturb the cell processes leading to numerous physiological changes, results in CI and ultimately death of the harvested produces (Watkins, 2016). The chilling stress induces membrane lipid disorganization which results in malfunctioning of membrane lipid such as ion leakage, insufficient ATP supply, and excessive generation of reactive oxygen species

(ROS), in severe cases, which may end up in irreversible cell death (Li et al., 2019; Liang et al., 2020). The extent of membrane lipid peroxidation determines the ability to recover from CI in fruit and vegetables. According to Vazquez-Hernandez et al. (2020), resistance to chilling would be associated with the ratio of unsaturated to saturated fatty acids (UFA/SFA). The higher the UFA, the higher the membrane stability and thus, the lower is the incidence of CI (Jing et al., 2018). However, Antunes et al. (2022) observed no clear association between the skin chilling disorders with the UFA/SFA and any of the membrane fatty acids, such as palmitic, linoleic, oleic and stearic, in Golden Delicious apples. Further clarification is needed to understand the changes in properties of membrane lipids and the influence of postharvest treatments on CI development.

Tolerance to chilling injury in fruits and vegetables is also associated with maturity at harvest (Khaliq et al., 2023). For example, "Saimaiti" apricot harvested at 50–80% fruit color development stage were resistant to CI as the fruit contain more soluble solids which increase the tolerance to CI by lowering the freezing point of intercellular constituents (Jing et al., 2018). In addition, genetic factor also determines chilling tolerance. Habibi et al. (2020) proved that the blood orange cultivars "Sanguinello" and "Sanguine" containing high levels of catalase, ascorbate peroxidase, and superoxide dismutase were more tolerant to chilling injuries. The authors postulated that these compounds might have a role in maintaining membrane integrity.

In recent years, many researchers have studied on physical and chemical treatments to alleviate CI in postharvest horticultural products: common physical treatments include temperature management (Zhao et al., 2019), controlled atmosphere (Alba-Jim´ enez et al., 2018; Park et al., 2018), modified atmosphere packaging and UV irradiation (Ruan et al., 2015), and so forth. Chemical treatments are also used and they mainly include plant hormones and natural elicitors, such as oxalic acid (Li et al, 2018; Liang et al., 2017), nitric oxide (NO) (Venkatachalam, 2018), hydrogen sulfide (Aghdam et al., 2018), calcium ions (Zhang et al., 2021), ethylene (Chen et al., 2021), 1-MCP (Zhang et al., 2020), melatonin (Kong et al., 2020), methyl jasmonate (Ma et al., 2020), salicylic acid (Ge et al., 2020), glycine betaine (Luo et al., 2020), and abscisic acid (Carvajal et al., 2017).

Playing important roles in fighting against abiotic stress, application of phytohormones are also alternative approach to avoid postharvest storage disorders. The expression of *SlCYP90B3 (SlCYP90B3-OE)*, a gene responsible for brassinosteroid biosynthesis, is linked to chilling tolerance in tomato by increasing the activities of ascorbate peroxidase, catalase, peroxidase, and superoxide dismutase, which are related to plant defense antioxidant systems (Hu et al. 2022). Intermittent warming (20°C with 70% RH) for 24 h at early stage (15th day) of 70 days cold storage (2 ± 0.5C and 90 ± 5% RH) alleviate chilling injuries, that is, dehydration, browning, and pitting in "Rabab-e-Neyriz" pomegranate by maintaining high membrane integrity and preserving high membrane safety-related antioxidants, spermine and putrescine.

Another common disorder that occurs during extended low temperature storage is superficial scald and it occurs mostly in pome fruit. Superficial scald is known as black or brown patches on fruit skin during cold storage and are associated with cell death or necrosis in hypodermal cortical tissue (Lurie and Watkins, 2012). Although the internal quality is not affected, development of superficial scald reduced appearance quality and consequently unmarketable (Mditshwa et al., 2016). The levels of tolerance to superficial scald development vary depending on the cultivars (Caracciolo et al., 2020). Superficial scald generally originates from oxidation of α-farnesen, an important secondary metabolite in plant defense system, which generates highly reactive oxidative products leading to cell membrane disintegration. This disintegration of membrane lipids and proteins cause necrosis symptoms on fruit skin (Dias et al., 2020).

Enhancing antioxidant systems or restricting oxidation by limiting oxygen supply can effectively control superficial scald. Kawhena et al. (2021) reported that the development of superficial scald on "Granny Smith" apples can be inhibited by storing the fruit under low oxygen-controlled atmosphere technologies, repeated low oxygen stress (RLOS), and dynamic controlled atmosphere-chlorophyll florescence (DCA-CF) at 0°C in a season of low superficial scald potential. Internal quality parameters were also maintained for up to 10 months of storage at 0C and after a simulated 6 weeks of shipment and handling period plus 7 days shelf life (20°C) by applying ultra-low oxygen stress (RLOS) and DCA-CF. Restricting oxygen concentration to less than 1% in the storage atmosphere suppress the development of superficial scald, senescent scald, and internal browning in "Granny Smith" apple, "Bartlett" pear, and "Spadona" pear by reducing the levels of α-farnesene and the byproduct of its oxidation, 6-methyl-5-hepten-2-one (MHO) (Pesis et al., 2014).

Golden Delicious apples stored for 60 days at 0.5°C, 90–96% RH plus 7 days at room temperature (≈22°C) were affected with storage physiological disorders such as diffuse skin browning, superficial scalds, and bitter pit (Antunes et al., 2022). Core browning symptoms were observed in "Whangkeumbae" pears stored for 180 days at 1–3°C with 90–95 % RH, which was evidenced by the progressive increased in production of ROS species and expressions of enzymes responsible for browning (Zhou et al., 2020). Similarly, different pear cultivars stored at 4°C for 1 month developed water soaking and core browning symptoms (Zhai et al., 2018). Dipping in methyl jasmonate (1 mM) or calcium chloride (4% wt/vol) solutions for 2 min as postharvest treatment eliminate peel browning by improving antioxidants accumulation and cell membrane integrity in "Grand Nain" banana stored at 10 ± 2°C with 85–90% RH for 20 days (Elbagoury et al., 2020). Soaking the fruit in melatonin (100 μM) solution for 12 h at room temperature (26°C) prevent the occurrence of water soaking and core browning symptoms in pear cultivars stored at 4°C for 1 month (Zhai et al., 2018).

Mismanagement of storage temperature may cause freezing injury in fruits and vegetables, resulting in the formation of ice crystals in the tissues. The ice crystals are capable of rupturing the cell wall and cellular membranes, thereby allowing cell fluids to escape and resulting in moisture loss, quality degradation, or flavor changes. Normally, the freezing point in fruit ranges from −4 to −1.1°C. However, most fruits and vegetables can be freezed usually by one or more degrees below their freezing point before they actually freeze.

Ice crystal formation causes both physical and chemical damage. There are two types of damage in cellular structures that lead to loss of firmness. The first type causes crystal punctures in cell membranes, leading to turgor loss. In the second type, ice crystals break up the cell wall structure, then the tissues breakdown, and after that degradative enzymes are released. The freezing tissues generally lose rigidity, resulting in thawing, and water-soaking appears. In general, fresh fruits and vegetables are sensitive to freezing injury. Some commodities may be frozen without injury for a long time, but others are freezed within a short period. Fruits and vegetables can be divided into three groups based on freezing sensitivity: (i) most susceptible (cucumbers, summer squash); (ii) moderately susceptible (cantaloupe, watermelon, winter squash); and (iii) least susceptible (none of the cucurbits) (Wang and Wallace, 2004). The freezing can be controlled by adjusting the temperature.

5.2.2 CONTROLLED ATMOSPHERIC GASES

Controlled atmosphere storage is a beneficial mean for long-term storage which allows trading of fruit and vegetables during off seasons and across continents (Bodbodak and Moshfeghifar, 2016); however, elevated CO_2 in the storage environment may sometimes injure the stored products. The accumulation of CO_2 restricts O_2 absorption for normal cell metabolism leading to development of browning in plant cells (Li et al., 2022). Storage at 0°C with high CO_2 (4% along with 21% O_2 and 75% N) concentration induced core browning in pears which was evidenced with enhanced accumulation of H_2O_2, a ROS involved in oxidation process and expression of polyphenol oxidase (PPO), an enzyme responsible for oxidation of phenolic compounds (Li et al., 2022).

The increased CO_2 resulting from intercellular space disorganization restrict O_2 absorption for normal cell metabolism causing energy (ATP) deficiency (Chigwaya, 2020). Low porosity and low pore connectivity regions in fruit tissue restrict the respiratory gas exchange leading to accumulation of CO_2 resulting in increased susceptibility to internal browning (Chigwaya, 2020). In addition, the reduction of antioxidant activity in fruit stored under CA may also contribute to oxidative browning (Łysiak et al., 2021).

The high incidence of internal browning was observed in "Fuji" apples stored under CA (1.5% O_2 and 0.5% CO_2) for 6 months and delayed CA (3 weeks at –0.5°C in normal air plus 5 months at –0.5°C in CA 1.5% O_2 and 0.5% CO_2) (Chingwaya, 2020). Similarly, "Laetitia" plum stored under CA (1 kPa of O_2, < 0.5 kPa of CO_2, 1.5 ± 0.2°C, 92 ± 2% RH) for 55 days were affected with internal browning (Steffens et al., 2022) and "Conference" pears stored under CA (2% O_2, 1% CO_2, and –1°C) for 180 days were also infected with internal browning and senescent scald (Łysiak et al., 2021).

NO minimizes the oxidative browning incidence in plum, through modulation of oxidative enzymes such as PPO. However, the effectiveness may vary depending on the amount applied, exposure time, and tolerance of the cultivar to NO. Significant reduction of browning incidence was attained by multiple applications of low dose NO (2 $\mu L\ L^{-1}$) at 5 days interval in "Laetitia" plum stored under CA (1 kPa of O_2, < 0.5 kPa of CO_2, 1.5 ± 0.2°C, 92 ± 2% RH) for 55 days (Steffens et al., 2022). In contrast, severe browning incidence was observed in plums treated with 5 and 10 $\mu L\ L^{-1}$ NO, indicating the toxic effect of NO at high dose.

Application of 0.05 gm^{-3} 1-MCP for 24 h significantly reduced internal browning and senescent scald in "Conference" pears either stored under CA (2% O_2, 1% CO_2, and −1°C) for 180 days or stored under normal air at −1°C for 180 days (Łysiak et al., 2021). Fumigation with 1-MCP (1 $\mu L \cdot L^{-1}$) at 25°C for 16 h or in combination with preconditioning treatment at 22°C for 1 week reduced the severity of flesh browning, core browning, water soaking, and cavity in "Wonhwang" pear stored at 1°C with ≈ 90% RH (Seo et al., 2019). In contrast, "Golden Delicious" apples treated with 625 nL L^{-1} of 1-MCP for 20 h at 0.5°C showed higher incidence (27.21%) of bitter pit compared to control (18.61%) (Antunes et al., 2022). Immersion of Golden Delicious apples for 2 min in a solution containing 1.5% $CaCl_2$ reduced the occurrence of bitter pit escalated by 1-MCP treatment (Antunes et al., 2022). Zoffoli et al. (2018) also revealed that 1-MCP inhibited superficial scald development in "Granny Smith" apple; however, the incidence of bitter pit and lenticel blotch pit disorder were even advanced in 1-MCP treated fruit. Lee et al. (2022) proposed that the vulnerability of fruit to storage disorders was provoked by 1-MCP through regulations of metabolites such as volatile aromatic compounds, and amino acids in different types of fruit tissues.

5.2.3 ETHYLENE

The majority of postharvest physiological disorders are influenced directly or indirectly by ethylene in horticultural crops, especially in climacteric fruits. The negative effects of ethylene during postharvest senescence are overripening, accelerated quality loss, and increased susceptibility of physiological disorders and fruit pathogens (Martínez-Romero et al., 2007). Ethylene also aggravates chilling disorders such as pulp water soaking in papaya (Zou et al., 2014), pitting in zucchini (García et al., 2020), and internal browning in plum (Steffens et al., 2022), which usually are initiated by low temperature storge. Ethylene, however, alleviates internal browning in peach stored at 5°C for 28 days by downregulating the expression of enzymes related to chilling disorders such as PPO, peroxidase, and lipoxygenases (Zhu et al., 2019). Ethylene also enhances the cold tolerance at seedling stage which was confirmed by the overexpression of the genes related to ethylene signal induction in transgenic apple seedlings (Wang et al., 2021). The role of ethylene in chilling response is yet to be explored more.

According to Watkins (2016), the low concentrations of ethylene have also caused physiological disorders as below:

1. Russet spotting of lettuce: localized spot-like lesions occur in the epidermis or mesophyll, and increasing severity, vascular tissue discoloration, and mesophyll cell collapse.
2. In rolling of petals, fading, wilting, and abscission of many flowers: closing of open flowers ("sleepiness").
3. Gummosis, bud necrosis, and flower blasting in flowering bulbs: inhibition of shoot and root elongation and abscission of flower buds and leaves.
4. Epinastic responses (downward curvature of leaves) in ornamental plants.
5. Development of white inclusions in core kiwifruit (Actinidia deliciosa var. deliciosa): exposed to ethylene under CA conditions
6. Rusty brown discoloration and carbon dioxide-induced brown stain in lettuce.
7. Bitterness in carrots: isocoumarin accumulation.

5.2.4 RELATIVE HUMIDITY

Relative humidity of storage environment influences the physiological loss of weight in fruit and vegetables by retaining their surface moisture. However, high humidity during storage encourages development of fruit cracks in Royal Gala apples stored in perforated polyethylene bags at 0.5°C for 6 months (Lee et al., 2019). According to the authors, the perforated bags gradually maintained relative humidity higher than 95% during the 7-days shelf-life period and the fruit inside the bags showed ≈ 33% higher incidence of cracking compared to control.

5.3 POSTHARVEST FACTORS RESPONSIBLE FOR QUALITY LOSS

Physiological disorders of fruit and vegetables are triggered by abnormal changes in cell metabolism that highly influence quality (Ali et al., 2024; Lal et al., 2025). Physiological disorders can affect the internal and external appearance of produces, resulting in changes in flavor and nutritional quality and reducing the storage life and marketability of fruit and vegetables. All these factors, thus, increase quality deterioration of produces and postharvest losses. In addition, postharvest loss and quality deterioration of horticultural crops may sometimes occur due to improper care, inappropriate harvesting equipment and materials, and lack of motivation to improve and upgrade the harvesting and handling techniques from time to time (Kasso and Bekele, 2018).

Microclimate of the stored products, such as temperature, when they are not managed in a proper way, also indispensably hastens the rate of quality deterioration during postharvest handling of fresh produces (Khan et al., 2025). Storing the products at higher temperature than optimum range or at low temperature for long time impairs the quality of many fruit and vegetables. High storage temperature (22°C) also negatively affects the peel redness and color intensity in pomegranates and they last only 4 weeks, while storage temperature of 5°C maintained better peel color and higher TSS/TA ratio up to 16 weeks (Fawole and Opara, 2013). Low temperature storage of pears at −1°C maintained moisture content of the fruit for 60 days; however, the moisture loss and firmness reduction were severe alongside with higher transpiration percentage following 180 days storage at −1°C (Łysiak et al., 2021). The mass loss and the incidence of postharvest diseases and disorders were found most visibly at −1°C. Unless the storage atmospheric gases are controlled to the optimum concentration, the fruit cannot be stored for 180 days economically, even if they are subjected to postharvest treatments such as 1-MCP (Łysiak et al., 2021).

According to Al-Dairi et al. (2021), storage time and storage temperature highly influence color and texture, which greatly affect consumer's acceptance in markets of fresh tomato. Storing the fruit at 10°C retained these quality parameters. The authors concluded that storage temperature was one of the vital factors that required intensive monitoring along postharvest supply chain and marketing (Al-Dairi et al., 2021). In blood orange, a marked reduction of anthocyanin was observed during subsequent shelf-storage at room temperature (22°C and 75% RH) although the anthocyanin content in fruit was generally increased during low temperature storage for 60 days at 5 ± 1°C and 90–95% RH (Strano et al., 2021). Moreover, low temperature storage alone could not control the development of fungal diseases such as gray mold, blue mold, bitter rot, and brown rot.

Prestorage temperature or atmospheric gas conditioning treatments were regarded as effective postharvest storage methods to prevent postharvest disorder in apples (Zoffoli et al., 2018; Riaño et al., 2022). However, temperature conditioning at 20°C for 10 days in combination with ultra-low oxygen treatments increased the accumulations of ethanol and acetaldehyde through anaerobic respiration. In the long run, the acceptability, firmness, and color development of the treated fruit were reduced after storing at 0°C for 150 days (Zoffoli et al., 2018). Similarly, sensory quality of "Fuji" apples were reduced as a consequent of internal ethanol and acetaldehyde accumulations resulting off-flavor development with ultra-low oxygen treatments at 0 or 5°C for 15 or 30 days, although corky physiological disorders were repressed in fruit stored at 5°C (Riaño et al., 2022).

Management of atmospheric gas compositions in the storage environment supplements the effect of temperature control in maintaining produces quality. Nonetheless, ill-treatments in modifying or controlling storage atmospheric gas concentrations may precede negative effects on quality. Nonperforated or poorly permeable film liners create anoxic condition inside the packages with higher condensation percentage resulting in more decay incident, though, packing film liners with better permeability of CO_2, O_2, and moisture maintained better fruit quality by creating modified atmosphere (Lufu et al., 2021). According to D'Aquino et al. (2022), "Simeto" mandarins aged rapidly after 7 days at 5°C with 85–90% RH, plus 7 days at 20C and 55–60% RH, resulting in higher respiratory activities and quicker loss of fruit firmness and overall eating quality. Modified atmospheric packing films slowed down the metabolic processes related to aging of "Simeto" mandarins by restricting oxygen in-diffusion and by improving carbon dioxide accumulation inside the package and preserved the eating quality. However, the incidence of pathogenic infection was greater due to the condensation inside the packages in the films with lower moisture permeability. In these poorly permeable films, the increase in off-flavor and off-odor were also observed alongside with the increased level of acetaldehyde and ethanol.

Thewes et al. (2021) observed a higher loss of flesh firmness in apples stored under 0.15 kPa O_2 + 0 kPa CO_2, which is likely due to the advanced ripening stage of these apples and proved by the high ethylene production and respiratory rate. The excessive anaerobic metabolic compounds produced by the fruit stored under higher pO_2 became harmful and may have degraded part of the middle lamella, reducing its firmness. However, the storage condition of 0.4 kPa O_2 + 1.0 kPa CO_2 maintained flesh firmness and exhibited beneficial effect of increased CO_2. The harmful effect of anaerobic metabolic compounds on fruit quality was also proved by Weber et al. (2020). The authors observed that the application of exogenous ethanol concentration higher than 500 $\mu L\ L^{-1}$ lead to decreased flesh firmness in "Braeburn" apples after 6 months of storage. Physiological disorders initiated by ethylene such as cell weakening and membrane leakage, weight loss, flavor loss, surface pitting, internal browning, textural changes, softening, mealiness, and flavor loss not only adversely change the quality but also reduce consumer attraction by affecting the physical appearance of the produces. In addition, ethylene production in produce increases rate of respiration, maturation, ripening, and senescence process resulting in reduced shelf life and marketability (Ansari and Tuteja, 2014).

TABLE 5.1 Physiological Disorders in Fruits and Vegetables Under Different Storage Systems.

Crop	Symptoms of Disorders	Storage Conditions	Researcher
Mushroom	Stem elongation and cap expansion	RA-storage	Thompson (2016)
Kiwi and persimmon	Softening	RA-storage	Thompson (2016)
Lettuce	Russet spotting	RA-storage	Schotsman et al. (2009)
Carrot	Bitter taste	RA-storage	Wills and Golding (2016)
Asparagus	Shriveling or lignification of pericyclic fibers	RA-storage	Techavuthiporn and Boonyaritthongchai (2016)
Lettuce	Browning of midrib	RA-storage	Wills and Golding (2016)
Apple	Skin burning	RA-storage	Wills and Golding (2016)
Banana	Crown rot	MA-storage	Bastiaanse et al. (2010)
Potato	Blackheart and sprouting	MA-storage	Kader (2002)
Lettuce	Brown stain	MA-storage	Kader (2002)
Apple and pear	Internal browning	MA-storage	Kader (2002)
Apple and pear	Internal breakdown or soft scald	Hypoberic storage	Burg (2004)
Apple	Sclad	Hypoberic storage	Wang and Dilley (2000)
Mushroom	Browning	Hypoberic storage	Robitaille and Badenhop (1981)
Peach and nectarine	Skin browning, skin pitting, and internal flesh discoloration	CA-storage	Wang et al. (1990)
Banana	Skin browning	CA-storage	Klieber et al. (2002)

TABLE 5.1 *(Continued)*

Crop	Symptoms of Disorders	Storage Conditions	Researcher
Kiwifruit	Hard core	CA-storage	Arpaia (1990)
Cauliflower	Discoloration and off-flavors	CA-storage	Lipton (1990)
Cabbage	Meristem browning or red heart	CA-storage	Berard (1994)
Brussel sprout	Meristem browning or Redheart	CA-storage	Lipton and Mackey (1987)
Apple	Alcoholic tastes	CA-storage	Thompson (1998)
Cucumber	Chilling injury	CA-storage	Eaks (1956)
Tomato	Chilling injury	CA-storage	Kader and Morris (1975)
Grapefruit	Chilling injury	CA-storage	Purvis and Cubbedge (1990)
Lettuce	Brown heart, brown stain, and heart leaf injury	CA-storage	Schotsman et al. (2009)
Melon and tomato	Irregular ripening	CA-storage	Bartz and Brecht (2003)
Zucchini	Chilling injury	CA-storage	Mencarelli (1987)
Chili pepper	Chilling injury	CA-storage	Kader and Morris (1975)
Apple	Corking	Long-term storage	Schotsman et al. (2009)
Cabbage	Black midrib and necrotic spot	CA-storage	Schotsman et al. (2009)
Mushrooms	Internal and external browning	CA-storage	Schotsman et al. (2009)
Citrus	Peel collapse and pitting	CA-storage	Schotsman et al. (2009)
Broccoli	Off-flavor	CA-storage	Bartz and Brecht (2003)

5.4 CONCLUSION

The use of storage facilities and technologies is essential for the successful trade of fresh produce. For profit maximization, however, it is important to alleviate or protect the shortcomings of these facilities and technologies such as occurrence of storage disorders or quality impairment during post-harvest handling. Comprehensively understanding the factors governing the development of disorders and quality deterioration along the postharvest supply chain and the underlying mechanisms of storage disorders are essential to develop satisfactory solutions for eliminating the storage disorders in fresh produces.

KEYWORDS

- **fresh produces**
- **gaseous composition**
- **physiological disorders**
- **postharvest**
- **quality**
- **temperature**

REFERENCES

Aghdam, M. S., Mahmoudi, R., Razavi, F., Rabiei, V., & Soleimani, A. (2018). Hydrogen sulfide treatment confers chilling tolerance in hawthorn fruit during cold storage by triggering endogenous H_2S accumulation, enhancing antioxidant enzymes activity and promoting phenols accumulation. *Scientia Horticulturae, 238*, 264–271.

Alba-Jim´ enez, J. E., Benito-Bautista, P., Nava, G. M., Rivera-Pastrana, D. M., V´ azquezBarrios, M. E., & Mercado-Silva, E. M. (2018). Chilling injury is associated with changes in microsomal membrane lipids in guava fruit (*Psidium guajava* L.) and the use of controlled atmospheres reduce these effects. *Scientia Horticulturae, 240*, 94–101.

Al-Dairi, M., Pathare, P. B., & Al-Mahdouri, A. (2021). Effect of storage conditions on postharvest quality of tomatoes: A case study at market-level. *Journal of Agricultural and Marine Sciences, 26*(1), 13–20.

Ali, S., Khaliq, G., Iftikhar, A., Mir, S. A., & Iqbal, S. (2024). Nanoemulsions-based edible coatings for postharvest quality preservation of fruits and vegetables. In *Sustainable postharvest technologies for fruits and vegetables,* (pp. 213–226). CRC Press.

Ansari, M. W., & Tuteja, N. (2014). Post-harvest quality risks by stress/ethylene: Management to mitigate. *Protoplasma, 252*(1), 21–32.

Antunes, M. D., Guimarães, A. C., Gago, C., Guerreiro, A., Panagopoulos, J., Vilas Boas, E., & Miguel, M. G. (2022). Membrane fatty acids and physiological disorders in cold-stored 'golden delicious' apples treated with 1-MCP and calcium chloride. *Horticulturae, 8*(2), 162.

Arpaia, M. L. (1990). Kiwi fruit. In P. D. Lister, G. D. Blanpied & R. K. Prange (Eds.), *Controlled atmosphere disorders of commercial fruits and vegetables* (Publication 1847/E, pp. 41–44). Agriculture and Agri-Food Canada, Kentville, Canada.

Babellahi, F., Paliwal, J., Erkinbaev, C., Amodio, M. L., Chaudhry, M. M. A., & Colelli, G. (2020). Early detection of chilling injury in green bell peppers by hyperspectral imaging and chemometrics. *Postharvest Biology and Technology, 162*, 111100.

Bartz, J. A., & Brecht, J. K. (2003). In *Postharvest physiology and pathology of vegetables*. Marcel Dekker, New York.

Bastiaanse, H., De Lapeyre de Bellaire, L., Lassois, L., Misson, C., & Jijakli, M. H. (2010). Integrated control of crown rot of banana with Candida oleophila strain O, calcium chloride and modified atmosphere packaging. *Biological Control, 53*(1), 100–107.

Berard, L.S. (1994). Storage disorders of cabbage. In R. J. Howard, J. A. Garland & W. L. Seaman (Eds), *Diseases and pests of vegetable crops in Canada* (pp. 110–112). Canadian Phytopathology Society and Entomology Society of Canada, Ottawa, Canada.

Bodbodak, S., & Moshfeghifar, M. (2016). Advances in controlled atmosphere storage of fruits and vegetables. In *Eco-friendly technology for postharvest produce quality* (pp. 39–76). Academic Press.

Burg, S. P. (2004). In *Postharvest physiology and hypobaric storage of fresh produce*. CAB International. Oxford, UK.

Caracciolo, G., Magri, A., Petriccione, M., Maltoni, M. L., & Baruzzi, G. (2020). Influence of cold storage on pear physico-chemical traits and antioxidant systems in relation to superficial scald development. *Foods, 9*(9), 1175.

Carvajal, F., Palma, F., Jim enez-Munoz, R., Jamilena, M., Pulido, A., & Garrido, D. (2017). Unravelling the role of abscisic acid in chilling tolerance of zucchini during postharvest cold storage. *Postharvest Biology and Technology, 133*, 26–35.

Chen, S., Chen, M., Li, Y., Huang, X., Niu, D., Rashid, A., & Wang, K. (2021). Adjustments of both phospholipids and sphingolipids contribute to cold tolerance in stony hard peach fruit by continuous ethylene. *Postharvest Biology and Technology, 171*, 111332.

Chigwaya, K. (2020). *Browning and watercore disorders in'Fuji'apples explored by means of X-ray computed tomography (CT)* (Doctoral dissertation, Stellenbosch University).

D'Aquino, S., Strano, S. M., Gentile, A., & Palma, A. (2022). Decay incidence and quality changes of film packaged 'Simeto' mandarins treated with sodium bicarbonate. *Horticulturae, 8*, 354. https://doi.org/10.3390/horticulturae8050354

Dias, C., Amaro, A. L., C Salvador, Â. C., Silvestre, A. J. D., Rocha, S. M., Isidoro, N., & Pintado, M. (2020). Strategies to preserve postharvest quality of horticultural crops and superficial scald control: From diphenylamine antioxidant usage to more recent approaches. *Antioxidants, 9*(4), 356.

Drogoudi, P., Pantelidis, G. E., & Vekiari, S. A. (2021). Physiological disorders and fruit quality attributes in pomegranate: Effects of meteorological parameters, canopy position and acetylsalicylic acid foliar sprays. *Frontiers in Plant Science, 12*, 645547.

Eaks, I. L. (1956). Effect of modified atmosphere on cucumbers at chilling and non-chilling temperatures. *Proceedings of the American Society for Horticultural Science, 67*, 473–478.

Elbagoury, M. M., Turoop, L., Runo, S., & Sila, D. N. (2021). Regulatory influences of methyl jasmonate and calcium chloride on chilling injury of banana fruit during cold storage and ripening. *Food Science and Nutrition, 9*(2), 929–942.

Fallik, E., & Ilic, Z. (2019). Positive and negative effects of heat treatment on the incidence of physiological disorders in fresh produce. In *Postharvest physiological disorders in fruits and vegetables* (pp. 111–122). CRC Press.

Fawole, O. A., & Opara, U. L. (2013). Effects of storage temperature and duration on physiological responses of pomegranate fruit. *Industrial Crops & Products, 47*, 300–309.

Freitas, S. T. D., & Pareek, S. (2019). Postharvest physiological disorders in fruit and vegetables. In *Postharvest physiological disorders in fruits and vegetables* (pp. 3–14). CRC Press.

García, A., Aguado, E., Cebrián, G., Iglesias, J., Romero, J., Martínez, C., ... Jamilena, M. (2020). Effect of ethylene-insensitive mutation etr2b on postharvest chilling injury in zucchini fruit. *Agriculture, 10*(11), 532.

Ge, W., Zhao, Y., Kong, X., Sun, H., Luo, M., Yao, M., & Ji, S. (2020). Combining salicyli acid and trisodium phosphate alleviates chilling injury in bell pepper (*Capsicum annuum* L.) through enhancing fatty-acid desaturation efficiency and water retention. *Food Chemistry, 327*, 127057.

Habibi, F., Ramezanian, A., Guillén, F., Martínez-Romero, D., Serrano, M., & Valero, D. (2020). Susceptibility of blood orange cultivars to chilling injury based on antioxidant system and physiological and biochemical responses at different storage temperatures. *Foods, 9*(11), 1609.

Hu, S., Wang, T., Shao, Z., Meng, F., Chen, H., Wang, Q., ... Liu, L. (2022). Brassinosteroid biosynthetic gene SlCYP90B3 alleviates chilling injury of tomato (*Solanum lycopersicum*) fruits during cold storage. *Antioxidants, 11*(1), 115.

Jing, Y., Ma, X., Jin, P., & Zhu, X. (2018). Effects of harvest maturity on chilling injury and storage quality of apricots. *Journal of Food Quality, 2018*(1), 1–7.

Kader, A. A. (2002). Modified atmospheres during transport and storage. In A. A. Kader (Ed.), *Postharvest technology of horticultural crops* (pp. 135–144). University of California, Davis.

Kader, A. A., & Morris, L. L. (1975). Amelioration of chilling injury symptoms on tomato fruits. *Horticultural Science, 10*, 324.

Kasso, M., & Bekele, K. (2018). Post-harvest loss and quality deterioration of horticultural crops in Dire Dawa Region, Ethiopia. *Journal of the Saudi Society of Agricultural Sciences, 17*, 88–96.

Kawhena, T. G., Fawole, O. A., & Opara, U. L. (2021). Application of dynamic controlled atmosphere technologies to reduce incidence of physiological disorders and maintain quality of 'Granny Smith' apples. *Agriculture, 11,* 491. https://doi.org/10.3390/ agriculture11060491

Khaliq, G., Ali, S., Gapper, N., & Nicola, S. (2023). Recent advances and approaches in the application of elicitors to enhance resistance mechanisms in fresh produce. *Frontiers in Sustainable Food Systems, 6,* 1061079.

Khan, L., Khaliq, G., Ullah, E., Abdul-Rahaman, A., El-Mogy, M. M., Kesba, H. H., Ismail, A. M., & El-Beltagi, H. S. (2025). Guggul gum and clove essential oil synergistically reduced anthracnose disease and modulated physicochemical changes in tomato during storage. *Journal of Stored Products Research, 114,* 102713.

Klieber, A., Bagnato, N., Barrett, R., & Sedgley, M. (2002). Effect of post-ripening nitrogen atmosphere storage on banana shelf life, visual appearance and aroma. *Postharvest Biology and Technology*, *25*, 15–24.

Kong, X., Ge, W., Wei, B., Zhou, Q., Zhou, X., & Zhao, Y. (2020). Melatonin ameliorates chilling injury in green bell peppers during storage by regulating membrane lipid metabolism and antioxidant capacity. *Postharvest Biology and Technology, 170*, 111315.

Lal, S., Khaliq, G., Mastoi, A. H., Irshad, A., Jaffar, M. A., Khan, L., & Muhammad, N. (2025). Effect of *Fagonia indica* and garlic extract on *Alternaria alternata* and post-harvest quality of tomato fruit during storage. *Journal of Crop Health, 77*(6), 1–13.

Larrigaudière, C., & Giné-Bordonaba, J. (2019). Oxidative stress and physiological disorders. In *Postharvest physiological disorders in fruits and vegetables* (pp. 30–51). CRC Press.

Lee, J., Leisso, R., Rudell, D. R., & Watkins, C. B. (2022). 1-Methylcyclopropene differentially regulates metabolic responses in the stem-end and calyx-end flesh tissues of 'Empire'apple during long-term controlled atmosphere storage. *Postharvest Biology and Technology, 192*, 112018.

Lee, J., Mattheis, J. P., & Rudell, D. R. (2019). High storage humidity affects fruit quality attributes and incidence of fruit cracking in cold-stored 'Royal Gala'apples. *Horticultural Science, 54*(1), 149–154.

Li, L., Kitazawa, H., Zhang, R., Wang, X., Zhang, L., Yu, S., & Li, Y. (2019). New insights into the chilling injury of postharvest white mushroom (*Agaricus bisporus*) related to mitochondria and electron transport pathway under high O_2/CO_2 controlled atmospheres. *Postharvest Biology and Technology, 152*, 45–53.

Li, X., Yang, H., & Lu, G. (2018). Low-temperature conditioning combined with cold storage inducing rapid sweetening of sweetpotato tuberous roots (*Ipomoea batatas* (L.) Lam) while inhibiting chilling injury. *Postharvest Biology and Technology, 142*, 1–9.

Li, J., Yao, T., Xu, Y., Cai, Q., & Wang, Y. (2022). Elevated CO_2 exposure induces core browning in Yali pears by inhibiting the electron transport chain. *Food Chemistry, 378*, 132101.

Liang, C., Jiang, L., Jin, M., Li, H., & Rao, J. (2017). Effects of oxalic acid treatment on chilling injury antioxidant capacity and energy status in harvested kiwifruits under low temperature stress. *Acta Horticulturae Sinica, 44*(2), 279–287.

Liang, S. M., Kuang, J. F., Ji, S. J., Chen, Q. F., Deng, W., Min, T., & Lu, W. J. (2020). The membrane lipid metabolism in horticultural products suffering chilling injury. *Food Quality and Safety, 4*(1), 9–14.

Lipton, W. J. (1990). Cauliflower. In P. D. Lister, G. D. Blanpied & R. K. Prange (Eds.), *Controlled atmosphere disorders of commercial fruits and vegetables* (pp. 31–32). Agriculture and Agri-Food Canada, Kentville, Canada.

Lipton, W. J., & Mackey, B. E. (1987). Physiological and quality responses of Brussel sprouts to storage in controlled atmospheres. *Journal of the American Society for Horticultural Science, 112*, 491–496.

Lufu, R., Ambaw, A., & Opara, U. L. (2021). The influence of internal packaging (Liners) on moisture dynamics and physical and physiological quality of pomegranate fruit during cold storage. *Foods, 10*, 1388. https://doi.org/10.3390/foods10061388

Luo, M., Zhou, X., Sun, H., Zhou, Q., Wei, B., & Cheng, S. (2020). Glycine betaine treatment alleviates loss of aroma-related esters in cold-stored 'Nanguo'pears by regulating the lipoxygenase pathway. *Food Chemistry, 316*, 126335.

Lurie, S., & Watkins, C. B. (2012). Superficial scald, its etiology and control. *Postharvest Biology and Technology, 65*, 44–60.

Łysiak, G. P., Rutkowski, K., & Walkowiak-Tomczak, D. (2021). Effect of storage conditions on storability and antioxidant potential of pears cv. 'conference'. *Agriculture, 11*(6), 545. https://doi.org/10.3390/agriculture11060545

Ma, M., Zhu, Z., Cheng, S., Zhou, Q., Zhou, X., Kong, X., & Ji, S. (2020). Methyl jasmonate alleviates chilling injury by regulating membrane lipid composition in green bell pepper. *Scientia Horticulturae, 266*, 109308.

Martínez-Romero, D., Bailén, G., Serrano, M., Guillén, F., Juan M., Zapata, P., Castillo, S., & Valero, D. (2007). Tools to maintain postharvest fruit and vegetable quality through the inhibition of ethylene action: A review. *Critical Reviews in Food Science and Nutrition, 47*(6), 543–560

Mditshwa, A., Fawole, O. A., Vries, F., van der Merwe, K., Crouch, E., & Opara, U. L. (2016). Classification of 'Granny Smith' apples with different levels of superficial scald severity based on targeted metabolites and discriminant analysis. *Journal of Applied Botany and Food Quality, 89*, 49–55.

Mencarelli, F. (1987). Effect of high CO2 atmospheres on stored zucchini squash. *Journal of American Society for Horticultural Science, 112*, 985–988.

Park, M. H., Sangwanangkul, P., & Choi, J. W. (2018). Reduced chilling injury and delayed fruit ripening in tomatoes with modified atmosphere and humidity packaging. *Scientia Horticulturae, 231*, 66–72.

Pesis, E., Feygenberg, O., Sabban-Amin, R., Ebeler, S. E., Mitcham, E. J., & Ben-Arie, R. (2014). Low oxygen pre-storage treatment is effective in reducing chilling injuries of deciduous fruit. *Journal of Postharvest Technology and Innovation, 4*(1), 23–32.

Purvis, A. C., & Cubbedge, R. H. (1990). In P. D. Lister, G. D. Blanpied & R. K. Prange (Eds.), *Controlled atmosphere disorders of commercial fruits and vegetables* (pp. 36–38). Agriculture and Agri-Food Canada, Kentville, Canada.

Riaño, C., Ribba, T., Juan I., Marchant, J. I., O'Brien, J. A., Contreras, C., & Zoffoli, J. P. (2022). Ultra-low oxygen and preconditioning storage regulate ethylene synthesis to prevent corky disorders in 'Fuji' apple. *Frontiers in Plant Science, 13*, 910139. DOI: 10.3389/fpls.2022.910139

Robitaille, H. A., & Badenhop, A. F. (1981). Mushroom response to postharvest hyperbaric storage. *Journal of Food Science, 46*, 249–253.

Ruan, J., Li, M., Jin, H., Sun, L., Zhu, Y., & Xu, M. (2015). UV-B irradiation alleviates the deterioration of cold-stored mangoes by enhancing endogenous nitric oxide levels. *Food Chemistry, 169*, 417–423.

Schotsman, W. C., DeLong, J. M., Larrigaudiere, C., & Prange, R. K. (2009). Effects on physiological Disorders. In E. M. Yahia (Ed.), *Modified and controlled atmospheres for the storage, transportation, and packaging of horticultural commodities*. CRC Press.

Seo, H. J., Wang, Y. S., Lwin, H. P., Choi, J. H., Chun, J. P., Roan, S. F., & Lee, J. (2019). Early season 'Wonhwang' pear fruit quality following international transport and storage is negatively impacted by fruitlet stage gibberellic acid 4+7 (GA 4+7) application but improved by postharvest 1-methylcyclopropene (1-MCP). *Scientia Horticulturae, 256*, 108549.

Steffens, C. A., Santana, G. R. O., do Amarante, C. V. T., Antonovviski, J. L., Miqueloto, T., Anami, J. M., & Fenili, C. L. (2022). Treatment with nitric oxide in controlled atmosphere storage to preserve the quality of 'Laetitia'plums. *LWT, 158*, 113033.

Strano, M. C., Altieri, G., Allegra, M., Di Renzo, G. C., Paterna, G., Matera, A., & Genovese, F. (2022). Postharvest technologies of fresh citrus fruit: Advances and recent developments for the loss reduction during handling and storage. *Horticulturae, 8*(7), 612.

Strano, M. C., Di Silvestro, S., Allegra, M., Russo, G., & Caruso, M. (2021). Effect of cold storage on the postharvest quality of different Tarocco sweet orange clonal selections. *Scientia Horticulturae, 285*, 110167.

Tarmizi, N. H. A., & Dolhaji, N. H. (2022). A review on chilling injury and antioxidant metabolism of pineapple (*Ananas comosus*). *Food Research*, *6*(1), 13–24.

Techavuthiporn, C., & Boonyaritthongchai, P. (2016). Effect of prestorage short-term anoxia treatment and modified atmosphere packaging on the physical and chemical changes of green asparagus *Postharvest Biology and Technology, 117*, 64–70.

Thewes, F. R., Thewes, F. R., Both, V., Schultz, E. E., Berghetti, M. R. P., Ludwig, V., Brackmann, A. (2021). Static × dynamic controlled atmosphere: Impacts of aerobic and anaerobic metabolism on physiological disorders and overall quality of 'Royal Gala' apples. *Food Science and Technology, 141*, 110922. https://doi.org/10.1016/j.lwt.2021.110922

Thompson, A. K. (1998). In *Controlled atmosphere storage of fruits and vegetables*. CAB International.

Thompson, A. K. (2016). Springer briefs in food, health and nutrition. In *Fruit and vegetable storage, hypobaric, hyperbaric and controlled atmosphere*.

Vazquez-Hernandez, M., Blanch, M., Sanchez-Ballesta, M. T., Merodio, C., & Escribano, M. I. (2020). High CO_2 alleviates cell ultrastructure damage in Autumn Royal table grapes by modulating fatty acid composition and membrane and cell oxidative status during long-term cold storage. *Postharvest Biology and Technology, 160*, 111037.

Venkatachalam, K. (2018). Exogenous nitric oxide treatment impacts antioxidant response and alleviates chilling injuries in longkong pericarp. *Scientia Horticulturae, 237*, 311–317.

Wang, Z., & Dilley, D. R. (2000). Hypobaric storage removes scald-related volatiles during the low-temperature induction of superficial scald of apples. *Postharvest Biology and Technology, 18*(3), 191–199.

Wang, Y., Jiang, H., Mao, Z., Liu, W., Jiang, S., Xu, H., ... Chen, X. (2021). Ethylene increases the cold tolerance of apple via the MdERF1B–MdCIbHLH1 regulatory module. *The Plant Journal*, *106*(2), 379–393.

Wang, C. Y., & Wallace, H. A. (2004). Chilling and freezing injury. In *The commercial storage of fruits, vegetables, and florist and nursery stocks* (p. 66). USDA Agriculture.

Wang. C. Y., Watada, A. E., & Anderson, R. E. (1990). Peaches and nectarines. In P. D. Lister, G. D. Blanpied & P. K. Prange (Eds.), *Controlled atmosphere disorders of commercial fruits and vegetables* (pp. 47–49). Agriculture and Agri-Food Canada.

Watkins, C. B. (2016). Postharvest physiological disorders of fresh crops. In *Encyclopedia of applied plant science* (pp. 315–322).

Weber, A., Neuwald, D. A., Kittemann, D., Thewes, F. R., Both, V., & Brackmann, A. (2020). Influence of respiratory quotient dynamic controlled atmosphere (DCA – RQ) and ethanol application on softening of Braeburn apples. *Food Chemistry*, *303*, 125346. https://doi.org/10.1016/j.foodchem.2019.125346

Wills, R., & Golding, J. (2016). In *Postharvest: An introduction to the physiology and handling of fruit and vegetables* (Sixth Ed.). CAB International.

Zhai, R., Liu, J., Liu, F., Zhao, Y., Liu, L., Fang, C., & Xu, L. (2018). Melatonin limited ethylene production, softening and reduced physiology disorder in pear (*Pyrus communis* L.) fruit during senescence. *Postharvest Biology and Technology, 139*, 38–46.

Zhang, X., Ma, M., Ye, B., Liu, L., & Ji, S. (2021). Calcium ion improves cold resistance of green peppers (*Capsicum annuum* L.) by regulating the activity of protective enzymes and membrane lipid composition. *Scientia Horticulturae, 277*, 109789.

Zhang, W., Zhao, H., Jiang, H., Xu, Y., Cao, J., & Jiang, W. (2020). Multiple 1-MCP treatment more effectively alleviated postharvest nectarine chilling injury than conventional one-time 1-MCP treatment by regulating ROS and energy metabolism. *Food Chemistry, 330*, 127256.

Zhao, H., Liu, B., Zhang, W., Cao, J., & Jiang, W. (2019). Enhancement of quality and antioxidant metabolism of sweet cherry fruit by near- freezing temperature storage. *Postharvest Biology and Technology, 147*, 113–122.

Zhou, H., Tian, M., Huang, W., Luo, S., Hu, H., Zhang, Y., & Li, P. (2020). Physiological and transcriptomic analysis of 'Whangkeumbae' pear core browning during low-temperature storage. *Gene Expression Patterns, 36*, 119113.

Zhu, Y., Wang, K., Wu, C., Zhao, Y., Yin, X., Zhang, B., ... Xu, C. (2019). Effect of ethylene on cell wall and lipid metabolism during alleviation of postharvest chilling injury in peach. *Cells, 8*(12), 1612.

Zoffoli, J. P., Sanguedolce, V., Naranjo, P., & Contreras, C. (2018). Prestorage temperature and ultra-low oxygen treatments affect incidence of physiological disorders in organic 'Granny Smith' apples. *HortScience, 53*(10), 1475–1481.

Zou, Y., Zhang, L., Rao, S., Zhu, X., Ye, L., Chen, W., & Li, X. (2014). The relationship between the expression of ethylene-related genes and papaya fruit ripening disorder caused by chilling injury. *PLoS One, 9*(12), e116002.

CHAPTER 6

Temperature-Related Physiological Disorders of Fruits and Vegetables

JAHAN ZAIB ASHRAF[1], HAFIZ NOOR UL AIN[2], MAIRA ANAM[3], and AYSHA SALEEM[1*]

[1]*Department of Agriculture, Food, Natural Resources and Engineering (DAFNE), University of Foggia, Italy*

[2]*Department of Food Science and Technology, University of Central Punjab, Lahore, Pakistan*

[3]*Department for Innovation in Biological, Agrofood and Forest Systems, University of Tuscia, Viterbo, Italy*

*Corresponding author

ABSTRACT

The wide variety of physiological disorders has several different root causes, including (but not limited to) environmental factors; micronutrient deficiencies; cultural practices such as poor soil conditions, watering training, and pruning practices of different commodities; and environmental stress resulting from unpredictable and uncontrollable temperature, either high or low, during pre- and postharvest life. Breakdown and necrosis of tissues, pitting and water drenching, loss of taste and smell, sensitivity to degradation, and ultimately tissue loss are all indications of chilling injury (CI) disorders. There is an inverse relationship between the temperature and the intensity of the symptoms in potentially dangerous conditions. Heat can ruin a crop at any time, before or after harvest. If the temperature was raised by 10°C, the rate of decline would double or even quadruple. Visual inspection and measurement of physical temperature,

Physiological Disorders of Fruits and Vegetables: Causes and Remedies.
Ghulam Khaliq, Sajid Ali, and Mohammed Wasim Siddiqui (Eds.)

DOI: 10.1201/9781998511334-6

cryo-scanning electron microscopy, ocular inspection, electrolyte leakage, specimen holder (SEM), and UV or chlorophyll fluorescence are all viable methods for diagnosing physiological disorders without causing any damage. Correct pre- and postharvest handling, rather than postharvest control therapies, is necessary to avoid physiological complications.

6.1 INTRODUCTION

Different types of horticultural crops that include fruits (tropical, sub-tropical, and temperate) and vegetables (fruit vegetables, roots, tuber, bulb, leafy, and minor vegetable crops) are considered disease-preventing foods as they are loaded with thousands of functional compounds. They are the cheapest source of natural protective foods contributing vitamins, minerals, proteins, and calories, and are essential parts of a balanced diet. Total fruit production of the world has been estimated at more than 865 million metric tons, while total annual vegetable production has been estimated to be around 1090 million metric tons (FAO, 2018) showcasing the highly important segment of the agriculture industry with ever-increasing industry size globally. Fresh fruits and vegetables are not only rich in nutrients but also play a vital role in economic development.

Fruit and vegetable losses are between 40 and 50% worldwide, with 54% occurring during the manufacturing, postharvest processing, and storage stages (Santos et al., 2019; Rutta, 2024). Every year, fruits lose 16–36% of their value due to technical, microbiological, and physiological factors. After harvest, there are several phases, such as processing, storing, packaging, or transport, where matters can go wrong (Conrad et al., 2018).

More than 20% of perishable items such as vegetables and fruits are lost due to insufficient refrigeration equipment or lack of access to energy. Temperature extremes can cause physiological disorders such as freezing injury, chilling injury, heat damage, and sunburn. Chilling injury occurs when temperatures are held above freezing but below 49–59°F, resulting in browning and/or the product seeming water-drenched. Heat harm is caused by direct sunlight or high heat and manifests as bleaching, scorching, or product softness (Zhang et al., 2021).

Removing a crop from the field immediately after harvest, avoiding direct sunlight, and managing chilling and holding will help to ensure a high-quality product with low losses. Being highly perishable, both quantity and quality of fresh produce are lost due to various abiotic (such as temperature, humidity, light, moisture, storage, and packaging) and biotic factors during pre- and postharvest periods making them unsuitable for the market (Madani et al., 2019). Physiological disorders have a major role in hammering the yield and quality of fresh vegetables. Physiological or abiotic disorders are different from other disorders since they are produced by nonliving, abiotic circumstances and affect normal growth (Ali et al., 2024; Khan et al., 2025; Lal et al., 2025). They are externally caused by physical and chemical plant alterations. Some physiological disorders may start before, continue after, or arise after harvest. Low or elevated temperature is a crucial abiotic stressor that affects vegetable development, development, and quality, leading to economic and nutritional losses (Duart-Sierra et al., 2020).

If taken care of by this factor, the shelf life and quality of fresh produce can be maintained and enhanced. Care is needed in maintaining both pre- and postharvest factors for overall improved quality, including the nutritional, physiochemical, and biological processes. Some important temperature-related physiological disorders, their symptoms, and causes are described in this manuscript. This chapter aims to overlay the way for upcoming exploration to conquer the trials of temperature and development of crop administration in these circumstances.

6.2 CAUSES OF POSTHARVEST PHYSIOLOGICAL DISORDERS

Any deformity in the edible or economically valuable part of fruits or vegetables that affects yield or quality is considered a physiological disorder. These disorders may appear as external or internal blemishes. They result from improper environmental or cultural conditions during growth, harvest, or postharvest handling. Such conditions disturb the normal physiological processes of the produce. Physiological disorders are not caused by fungi, bacteria, viruses, or insect pests. Most of the biological disorders are not reversible once they have happened (Kathayat and Rawat, 2019). The major causes of these physiological disorders are diverse and include (i) environmental factors, (ii) micronutrient deficiency, (iii) cultural practices such as improper soil conditions, watering training, and pruning practices of different commodities, and

(iv) environmental stress resulting from unpredictable and uncontrollable temperature, either high or low, during pre/postharvest life (Zheng et al., 2020).

6.2.1 LOW TEMPERATURE

Temperature regulates plant growth and development. Temperature regulation is crucial for handling perishable horticulture products after harvest. Different horticulture products respond differently to temperatures, depending on their origin (tropical, subtropical, or temperate). Temperature deviations can cause stress and harm.

6.2.1.1 FREEZING INJURY

Freezing stress is another serious hazard to agricultural production. Growing or storing crops below freezing before or after harvest might cause tissue collapse. Freezing injury is a key abiotic pressure factor that reduces plant development, and postharvesting (Dalvi-Isfahan et al., 2019). The temperature at which crops freeze (below 0°C) depends on tissue and cell solute concentrations (e.g., sugars). Real water freezes at 0°C and lettuce, which is mainly water, freezes at 0.2°C (Figure 6.1) (Yahia et al., 2019).

FIGURE 6.1 Carrots and lettuce, both fresh (on the left) and frozen (right), displaying the dehydration and water-soaking symptoms that are characteristic of freezing injury.

Source: Post harvest management of Vegetables.net (2020).

Freezing damage occurs when ice crystals form in the soft tissues of fruits or vegetables at −1.5°C or below, either in the field or during storage. For example, a water-soaked patch at the proximal end of a tuber changes color and loses market value (Li et al., 2021). Ice crystals can breach cell barriers. When the produce thaws, the cells break down, causing a water-soaked look and damage to structural veracity. Freezing frequently causes tissue breakdown and product loss. Freezing disrupts metabolic and hormonal processes in fruits and vegetables. When a product freezes slowly, bigger ice crystals accumulate, causing more harm.

Freezing or cold stress can disrupt growth by inhibiting a plant's full genetic potential, which alters metabolic procedures, enzyme action, and membrane flexibility, resulting in physical and metabolic transport problems. Cold and freezing tension also cause cold-stimulated osmotic stress, when plants dehydrate owing to reduced water consumption. Fast-blast freezers minimize the sizing of ice crystals and protect the quality of frozen fruits and vegetables. Plants have evolved cold response systems over time. Cold adaptation allows plants to tolerate freezing temperatures (Muthuramalingam et al., 2022). Some vegetables, such as cabbage, can withstand freezing if defrosted gently and briefly. Always keep fresh horticulture commodities at higher freezing temperatures to avoid damage. Chilling and freezing injury temperatures of some common fruits and vegetables are listed in Tables 6.1 and 6.2, respectively.

TABLE 6.1 Chilling and Freezing Injury Temperatures of Some Common Fruits.

Fruit Name	Normal Storage Temperature (°C)	Chilling Injury Temperature (°C)	Freezing Injury Temperature (°C)
Plums	4	3	−1.7°C or lower
Banana	11.5–13	Below 13	Below 12.5°C
Honey dew melons	7–10	10	5
Guava	4.5	Below 8°C for few days	2
Grapefruit	10	10–15	−1.5
Mango	10–13	5–12	Below 10
Apple	2–3	2–3	2.2–1.7
Pomegranate	4.5	2	−1.2
Watermelon	4.5	10	−0.4
Lychee	3	4–9	0–2

TABLE 6.2 Chilling and Freezing Injury Temperatures of Some Common Vegetables.

Vegetable Name	Normal Storage Temperature (°C)	Chilling Injury Temperature (°C)	Freezing Injury Temperature (°C)
Cucumber	7	7	0.9–0.8
Eggplant	7	7	5.5–6
Ginger	7	5	4–5
Lemon	11–13	10	7
Potato	3	0–1	1.8–1.7
Okra	7	3	0–1
Pumpkin	10	5	2
Ripe tomato	7–10	2–3	−1.1 to −0.6
Lettuce	0–2	Below 0	0.6–0.3

6.2.1.2 CHILLING INJURY

Because it occurs between 0 and 15°C above freezing, chilling injury differs from freezing injury. Tissue collapse and necrosis, pitting, water soaking, flavor and scent loss, sensitivity to deterioration, and ultimately, tissue death are all signs of chilling injury (CI) disorders and their symptoms. The most popular are capsicums (Figure 6.2), peaches (Figure 6.3), and cucurbits. It is crucial to notice that each has a specific temperature limit at which it can be properly held. Depending on their maturity level and intrinsic variances, citrus fruits, for example, have varying storage behaviors (Strano et al., 2022). The storage temperatures indicated, however, are those considered most favorable for most of them. Grapefruit stored between 32 and 40°F for 2–3 months could experience pitting, watery disintegration, scald, and browning of the oil glands. Pits may form anywhere on the surface and often are numerous. Pits generated on fruits stored at 32°F are smaller than those on fruits stored at 36 and 40°F. During chilling injury, changes take place at the cell level in fruit skins and forms that ultimately impact covering permeability. Consequent to these alterations, there is electrolyte leakage, loss in metabolic energy, and cellular breakdown (Zainalabidin et al., 2019; Ali et al., 2023). Oxidative stress at the tissue level leads to chilling injury and thermal treatments alter the antioxidant systems of the mandarin fruit which in turn reduces the formation of spontaneous oxygen species (Wang et al., 2020). Watery breakdown may develop in citrus fruits stored at low temperatures for 2 to 3 months. Grapefruit picked late in the season is more sensitive than those collected early. Affected fruits are mushy, spongy, and water-soaked in both peel and meat and look as if they had been frozen while acquiring a fermented stench when they are held at room temperature. Chilling injury is localized mostly to the peel, in which some of the surface cells are killed.

FIGURE 6.2 Capsicums that have matured and turned red are more resistant to damage caused by frost than green fruit. The picture depicts how the quality of the capsicum changes depending on whether it is stored for 3 weeks at a temperature of 0, 2, 4, or 6°C. Green fruit kept at temperatures of 0 and 2°C exhibits light pitting, an early sign of chilling damage.

Source: Post harvst management of Vegetables.net (2020).

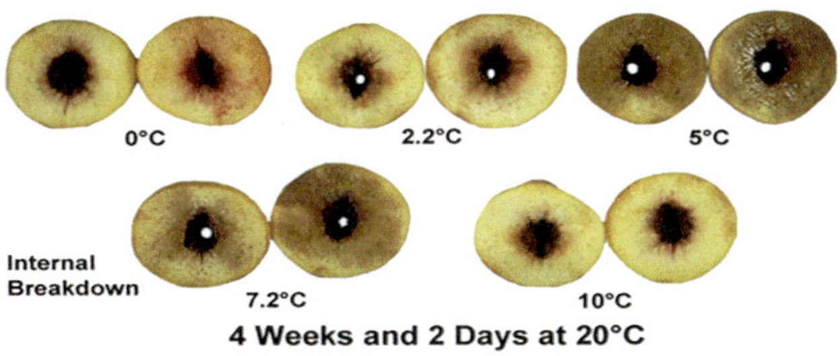

FIGURE 6.3 Chilling injury in peaches.

Source: Reprinted from Crisosto et al. (1996).

Chilling injury may not be immediately visible on mature fruits until they are stored at low temperatures for extended periods. The characteristic dull appearance soon appears, however, when frozen ripe fruits are shifted to higher temperatures (Zhao et al., 2019). The cold chain is the highly significant postharvest method for preserving horticulture items; hence, CI is of good monetary significance. Many chilling-susceptible goods cannot be maintained at very minimal temperatures for lengthy intervals of time, resulting in significant losses. CI may be caused by membrane lipids shifting from liquid to solid form. Table 6.3 shows chilling damage signs in common fruits and vegetables. Recurrent warming up, elevated CO_2 atmospheres if allowed by the tissue, elevated relative humidity, high-temperature treatments, and treatments with specific medicines, such as the fungicide thiabendazole, help alleviate CI harm (Valenzuela et al., 2017).

TABLE 6.3 Chilling Injury Symptoms in Some Common Vegetables and Fruits.

Commodity	Symptoms of Chilling Injury
Apple	Inner browning, brown middle, damp cultivars collapse, delicate scald, flesh discoloration, and mealiness
Cranberries	Rubbery surface, red skin
Cucumber	Pitting, water-soaked stains, rotting
Lemons	Pitting, membranous staining, red blotch
Tomato	Water soaking and softening, decay
Potato	Mahogany browning, sweetening
Oranges	Pitting, brown stain
Mango	Grayish scald like staining of skin, rough maturation
Sweet potatoes	Deterioration, pitting, inner staining, hardcore when cooked
Pineapple	Dull green when mature, internal intense browning, tissue collapse: particularly subepidermal tissue

6.2.1.2.1 Factors Influencing the Onset of Chilling Injury Symptoms

1. **Temperature:** In risky temperatures, the severity of the symptoms increases as temperature decreases.
2. **Time:** Below hazardous temperatures, the lengthier the duration of experience, the more severe the signs.
3. **Cultivars:** Sensitivity to chilling injury varies widely among cultivars of the same crop. Some cultivars are highly susceptible and develop symptoms even at mild low temperatures, while others show greater tolerance (Loayza et al., 2021).
4. **CI accumulates:** The accumulative exposures to dangerously chilling temperatures before harvest are added to the exposures after harvest.

5. Symptoms might not manifest up until produce is returned to room temperature.

6.2.1.2.2 Factors Influencing Chilling Injury

Internal (About the Commodity)

1. **Genetics:** Several varieties, and sometimes even distinct cultivars of a similar genus, can vary in their sensitivity to chilling (Lurie, 2021).
2. **Phase of development:** With maturing and aging, chilling sensitivity decreases.

External (Regarding the Atmosphere Close to the Produce)

1. **Acclimation:** Contact to low, but reliable, temperatures for brief intervals.
2. **Exposure to high temperatures before exposure to low temperatures:** For instance, 2 days surrounded by the air at 38°C or 10 mins in water at 53°C can enhance the endurance of certain berries to CI.
3. **Alternating warming up:** It is believed that this may also stimulate the purification of toxic complexes that end in the damage, thereby increasing resistance to CI.
4. **Humidity relative to dew point through storage:** This reduces water damage as well as the occurrence of pitting.
5. **Modified (MA) and controlled (CA) environments:** Various products, such as mangoes, avocados, and grapefruit, are more resistant to CI when exposed to MA/CA, particularly with high concentrations of CO_2 (Dogan, 2022).
6. **Some fungicides:** Thiabendazole applied to grapefruit has been shown to alleviate CI.

6.2.2 HIGH TEMPERATURE

High field temperature is among the foremost causes liable for physiological and biochemical failure of the produce. Increased respiration and changes in the reserve sugars have been observed during the high-temperature conditions.

6.2.2.1 HEAT INJURY

Before or after harvest, heat damage can occur. When horticultural commodities are exposed to high temperatures, such as immediate daylight or during (nonrefrigerated) storage or transportation, their temperature can rise substantially, causing qualitative and quantitative losses. For every 10°C increase in temperature, the rate of deterioration may double or triple (Sharma and Manjeet, 2020). Temperature abuses during postharvest handling occur at several phases of the postharvest chain, including after yield, before refrigeration, before packaging, before shipment, before selling, and so forth.

Broken cold chains can cause excessive temperatures disruption, and injury. Above-optimal temperatures increase metabolic activity, limit fruit life, and cause other qualitative and quantifiable failures. Heat damage causes bleaching, scalding, irregular ripening, too much softening, and desiccation. Numerous fruits will not ripen at 35–40°C because ethylene synthesis is suppressed. Heat kills tissue. Heat stress alters RNA species, proteins, cytoskeleton structures, membranes, and cellular enzymatic processes, causing a metabolic imbalance. When metabolite flow is interrupted, reactive oxygen species (ROS) might accumulate.

Plants' transcriptome, metabolome, and lipidome change during heat pressure. These alterations produce a new-found stable-state equilibrium of metabolic activities that allow the bacterium to perform at the new temperatures, which might, likewise, remain advantageous in reacting toward cold temperatures. Many commodities use elevated-temperature treatments (in the atmosphere or water) to reduce pathogen inoculum, insect pests, and chilling injury symptoms. Treatment timeframes and temperatures vary from days in 35–40°C air to fewer than a minute in 63°C water. Certain commercially approved pest control techniques include vapor heat or hot water soaking of the fruit up until the core temperature achieves a predetermined degree. Hot water remedies at 46.1°C for 65–110 min kill fruit flies on mangoes. Hot water dips at 48–56°C are utilized to reduce degradation. Very short (10–25 s) HWB treatments at 63°C have also been used.

Several of these mechanisms involve inhibiting ethylene synthesis and additional ripening and senescence-linked activities, inducing defensive compounds counter to pathogen damage, and developing low-temperature stress resistance (Bulgari et al., 2019). High temperatures are usually associated with lower relative humidity (high vapor pressure deficit), which can increase the transpiration rate of leaves and fruits. Xylem sap containing essential nutrients flows into leaves and fruit in response to the low water potential developed in these organs mainly in response to transpiration. Therefore, increasing temperature and vapor pressure deficit can improve nutrient uptake in leaves and fruit. However, extremely high temperature and vapor pressure deficits can increase leaf transpiration at higher rates than fruit transpiration, in some cases resulting in xylem sap backflow from fruit to plant. In this scenario, high temperatures leading to high vapor pressure deficit can have an undesirable effect on the accretion of essential nutrients such as calcium and boron in the fruit, increasing fruit susceptibility to physiological disorders. Oscillations in water availability to plants have also been associated with fruit cracking, sunburn, and other disorders (Bustamante et al., 2018). In addition, high environmental temperature and drought can lead to stress conditions that disturb normal cellular homeostasis, reducing growth and development, and in severe cases, resulting in cell death.

High-temperature stress causes plants to accumulate stress- and thermotolerance-related heat shock proteins. Many heat shock proteins get a chaperone task that stabilizes nascent proteins to guarantee proper folding or refolds high-temperature-damaged proteins, which is critical for protein conformation and function in the cell. Different genotypes may regulate heat shock proteins in ways that affect susceptibility to heat-related diseases (Kumar et al., 2018).

6.2.2.1.1 Different Cooling Methods to Overcome This Issue

1. Pack in light-colored containers and make regular journeys stuck between the fields and the colder to avoid fleeting field storing.
2. If containers are left out in the sun, make sure they are covered.
3. Temporary field storage might be done in a shaded area (Brizzolara et al., 2020).
4. Keep in mind that the shadow cast by a tree change throughout the day.
5. Short-distance deliveries should be made using covered trucks.
6. Refrigerant-hauling vehicles are required over long distances.
7. As soon as the product has arrived at the cooling location, it should be chilled as shown in Tables 6.4 and 6.5 (Rajapaksha et al., 2021).

6.3 MECHANISM OF PHYSIOLOGICAL DISORDER DEVELOPMENT

Disorder development initiates with a phase-transition hypothesis. When the rate of mitochondrial respiration suddenly drops, it signals that a crucial enzyme in respiratory metabolism's activation energy has suddenly changed. In chilling-sensitive plants, the membranes transform from a liquid to a crystallized condition at threshold temperatures.

Threshold temperatures at which sensitive plant membranes transition closely align with the point where mitochondrial respiration slows sharply on an Arrhenius plot, suggesting a change in the activation energy of a

TABLE 6.4 Suggested Cooling Methods for Several Horticultural Commodities.

Product	Range of Operations		
	Larger	Smaller	Remarks
	Tree Fruits		
Stone crops	FA, HC	FA	Apricots cannot be HC
Pome fruits	FA, R, HC	R	–
Citrus	R, FA	R	–
Sub-tropical	HC, R, FA	FA	–
Tropical			
Grapes	FA	FA	Requires fast refrigeration services compliant to SO_2 sterilization
Root Vegetables			
Irish potatoes	R, w/evap coolers	–	Along with evap chillers, facilities must be modified to curing
Sweet potatoes	HC	R	–
With tops	FA, PI, and HC	HC, FA	Carrots may be VC
	Leafy vegetables		
Kale	VC, R, WVC	FA	–
Lettuce	VC	FA	–
Cabbage	FA, VC	FA	–
Mushrooms	FA, VC	FA	–
Ornamentals			
Plants in pots	R	R	–
Cut flower	FA, R	FA	When packed, FA is utilized
	Fruit-form vegetables		
Summer squashes	R, FA, FA-EC	FA, FA-EC	–
Winter squashes	R	R	–
Cucumbers, eggplant	R, FA, FA-EC	FA, FA-EC	Fruit-type vegetables are chilling sensitive at varying temperatures
	Pod-vegetables		
Peas	FA, PI, VC	FA, PI	–
Beans	HC, FA	FA	–

FA, forced air cooling; FA-EC, forced-air evaporative cooling; HC, hydro-cooling; PI, package-icing; R, room-cooling; VC, vacuum-cooling; WVC, water spray vacuum cooling.

TABLE 6.5 Commonly Applied Chilling Techniques.

Room Chilling	Cooling Bays	Forced Air Chilling
Tunnel typed pressurized air chilling	Cold wall	Serpentine cooling
Forced air evaporative cooling	Container venting	Refrigeration in transport
Hydro-cooling	Package cool	Vacuum cooling
Pre-packaging cooling	–	–

critical enzyme in respiratory metabolism. Threshold temperatures are close to temperatures at which membranes of chilling-sensitive plants change from a liquid to a crystalline state.

The temperature point at which no visible damage occurs regardless of storage duration (3–15 days) typically falls within the range of −0.5°C to 16–20°C, although this varies by commodity (e.g., pineapple is very sensitive). The pattern of disorder development may be linear or nonlinear depending on temperature, as seen in some peach varieties. After peroxidation of membrane lipids leading to cell death, antioxidant defenses and enzyme repair systems may not be able to mobilize in sensitive plants. The cell death is followed by fast inhibition of cytoplasmic streaming (in min). The inability to transport substrates, metabolites, and control molecules causes possible metabolic imbalance and accumulation of hazardous respiratory intermediates. Enzyme temperature sensitivity happens due to phosphofructokinase, one of the most important regulating enzymes which is temperature sensitive. They could also be responsible for the symptoms of chilling injury if they malfunction (Selvarajan et al., 2018).

6.4 INSPECTION AND QUANTIFICATION OF VISIBLE PHYSIOLOGICAL TEMPERATURE-RELATED DISORDERS

Inspection and quantification of visible physiological temperature-related disorders involve the visual assessment of fruits and vegetables for symptoms such as sunscald, chilling injury, heat damage, or freeze injury. These disorders are characterized by discoloration, surface pitting, water-soaked areas, or tissue collapse. Standardized scoring systems, imaging tools, or digital analysis software are often used to quantify the extent and severity of the damage, enabling objective evaluation for research or quality control purposes.

6.4.1 CRYO-SCANNING ELECTRON MICROSCOPY, SCANNING ELECTRON MICROSCOPY (DESTRUCTIVE)

As opposed to traditional specimen holder, CryoSEM allows the study of freezing biologic specimens and hydrous polymer samples in their normal physiological circumstances. Using a CT-100C cryo-transfer device connected to a scanning electron microscope (as shown in Figure 6.4), slush N2 was used to freeze the sample and adhere it to the SEM. Fractured and moved from the cryo storage, water condensing on its surface was then heated to a regulated 90°C temperature in the microscope sample stage. In the following steps, the sample was moved back to the cryo storage and coated with gold using the sputtering method. An acceleration voltage of 15 KeV was then used to examine the material once more in the microscope. The lens breaks because of damage to the material caused by freezing. These stress spots, on the other hand, provided excellent opportunities to study the polymer's structure and bulk from the inside out.

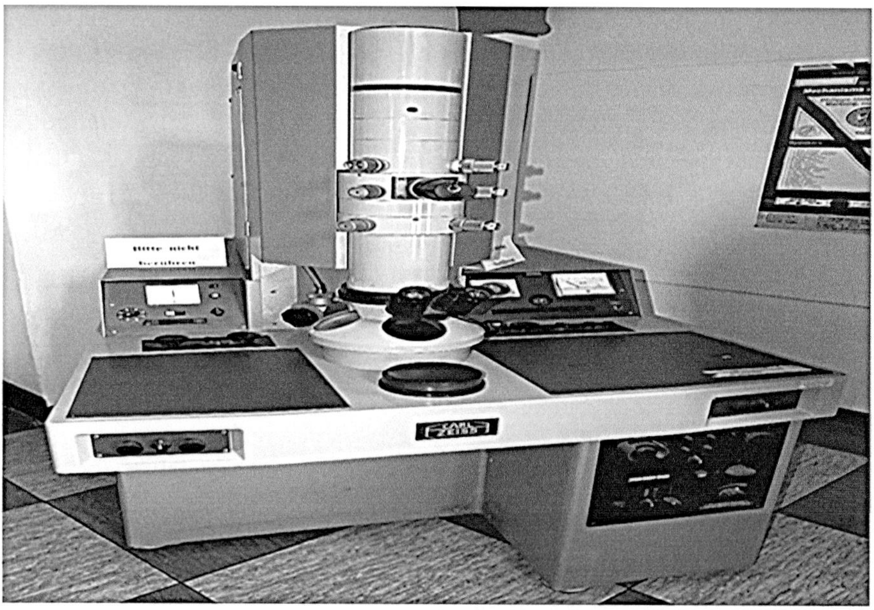

FIGURE 6.4 Cryo scanning electron microscopy instrument.

Source: Photo of Hydro from Wikipedia. https://creativecommons.org/licenses/by-sa/4.0/deed.fr

6.4.2 CHLOROPHYLL FLUORESCENCE OR UV LIGHT (NONDESTRUCTIVE)

Pulse amplitude modulation measures of chlorophyll fluorescence have been used to quantify the amount of CI in multiple tissues. Increases in the no-photochemical-reducing coefficient (qNP) correlate favorably with dangerous chilling temperatures, whereas variations in the photochemical-quenching coefficient (qP) do not. Using a chlorophyll fluorimeter, it is simple to quantify variations in qNP. As a result of a phase change in the triglycerides of the thylakoid tissues, storage at dangerously minimal temperatures has been shown to cause loss to photosystem II (Cakmak, 2019). On the thylakoid membranes of apples and cucumbers, it had been determined that photosystem II damage is a useful indicator of chilling-induced damage. The chlorophyll fluorescence approach has also been used to assess the formation of CI during the storage capacity of bananas, mangoes, and green peppers at minimal temperatures (Figure 6.5). As a nondestructive method, this technique can offer a quick, precise, and useful estimate of the chilling harm to fruits and vegetables. It might identify CI even before the onset of visible signs and in fruits that are believed to possess inadequate chlorophylls, such as ripe apples.

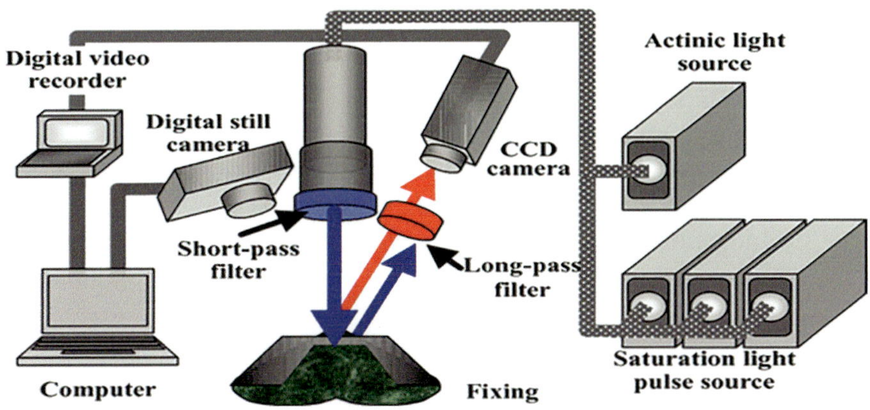

FIGURE 6.5 Chlorophyll fluorescence instrument.

Source: Reprinted from Takayama et al. (2003).

6.4.3 ELECTROLYTE LEAKAGE (INDIRECT)

Extended introductions of vulnerable fruits and vegetables to improper cooling temperatures might cause cell membrane collapse and ion release. Ion leakage measurements are often used to evaluate CI in low-temperature-stressed plant tissues. Ion leakage can be generated by two processes: (i) phase shift in membrane fatty acid and (ii) oxidative stress. Lipid layers change from a liquid-crystalline to a gel stage, causing phase separation and increased permeability. Alternatively, oxidative stress is induced by enzyme failure and cellular calcium redistribution, which leads to a buildup of ROS that oxidize cell coatings and encourage membrane bilayer rupture.

6.4.4 VISUAL INSPECTION WITH FOUR TO FIVE DEGREES OF DISORDER PREVIOUSLY QUANTIFIED

The CI can be assessed on an individual basis by the use of a visual examination of the harm to the skin tissue. It is common to practice evaluating CI based on the presence of injury symptoms such as pitting, water-drenched regions, and rotting. This methodology is utilized rather frequently in the process of doing repetitive evaluations of produce injury. The visual assessment of harm for the various concentrations of indications can be rated using a particular scale, such as the one that goes as follows: 0 = No damage (ND).

1. Indicates that there is very little damage (VLD, less than 5% of the area impacted).
2. Indicates a moderate amount of damage (LD, 6–10% of the area concerned).
3. Indicates that there has been moderate damage (MD, 11–25% of the area impacted).
4. Indicates severe damage (standard deviation: between 26 and 50% of the surface is affected).
5. Indicates very severe damage (VSD, more than half of the surface has been impacted).

The following formula is then used to determine the CI index for a given batch of fruits or vegetables: The formula for calculating the CI index is as follows: CI index = (1NDFVLD /TDF) + (2NDFLD /TDF) + (3NDMD /TDF) + (4NDFSD /TDF) + (5NDFVSD /TDF)

- The formula for calculating NDFVLD is as follows: (number of fruits displaying VLD by pitting) plus (number of fruits showing VLD by water-soaked areas) plus (quantity of fruits showing VLD by decay) plus (number of fruits displaying VLD of another kind (s), if any).
- The formula for calculating NDFLD is as follows: (quantity of fruits exhibiting LD by pitting) plus (number of fruits revealing LD by water-drenched regions) plus (number of fruits showing LD by decay) plus (sum of fruits displaying LD in another type (s), if any).
- NDFMD is equal to the following: (number of fruits displaying MD by pitting) plus (total of fruits showing MD by water-soaked areas) plus (number of fruits showing MD by decay) plus (quantity of fruits showing MD of another type (s) of symptom(s), if any).
- Number of fruits showing signs of disturbance (NDFSD) = (number of fruits displaying symptoms of disturbance due to pitting) + (number of fruits showing signs of disturbance due to water-soaked areas) + (quantity of fruits displaying signs of disturbance due to decay) + (quantity of fruits showing signs of disturbance due to further type(s).
- NDFVSD equals the number of fruits that exhibit VSD due to pitting, the figure of fruits that exhibit VSD due to water-soaked areas, the number of fruits that exhibit VSD due to decay, and the number of fruits that exhibit VSD due to any other type of symptom(s), if any.

Where,

TDF is the total of losses collected from all the concentrations of damages of all the kinds of symptoms:

$$TDF = NDF_{VLD} + NDF_{LD} + ND_{MD} + NDF_{SD} + NDF_{VSD}$$

6.5 PREVENTIVE MEASUREMENTS AND RECOMMENDATIONS

1. Preharvest sensitivity should be evaluated to avoid unnecessary storage of fruits with remarkable sensitivity (i.e., commodities without alternating day/night temperature or subjected to abnormally high temperatures during development).
2. Selection of commodities/cultivars. Start a breeding program to select new cultivars with reduced sensitivity to disorders.

3. Avoid storing commodities beyond the latency period to chilling injury.
4. Try stepwise temperature cooling for sensitive commodities.
5. Strict temperature and RH management minimize physiological disorders.
6. Evaluation of coadjutants (pretreatments, controlled or modified atmospheres, authorized antioxidants, coatings, or low residue chemicals).
7. Optimize storage time to the minimum.
8. Monitored sensitivity to disorders by using biomarkers alone or in combination (ethylene and other volatiles, and so forth).
9. Use strict control of respiration above the O_2 extinction point to avoid scalding but also fermentation in DCA systems.

6.6 SUCCESS AND PERSPECTIVES

6.6.1 *IDENTIFICATION OF QUANTITATIVE TRAIT LOCUS (QTLS) RESPONSIBLE FOR SENSITIVITY TO PHYSIOLOGICAL DISORDERS (PEACH, APPLE, MELONS)*

Melon cultivars can ripen in either a climacteric or nonclimacteric manner, depending on the variety. Climacteric ripening and fruit softening were explored with the help of a set of near-isogenic lines (NILs) that were produced from the nonclimacteric melon parental line-ups PI 161375 (SC) and "Piel de Sapo" (PS). The identification of eth3.5, a quantitative trait locus (QTL) associated with critical maturation, was achieved through a cross between SC and PS, two non-climacteric melon varieties. In some cases, the SC allele at V10.2 was found to improve flesh firmness (Pereira et al., 2020).

6.6.2 *BREEDING PROGRAMS TO DEVELOP CULTIVARS WITH REDUCED SENSITIVITY TO DISORDERS ARE IN THE MARKET (APPLE) OR UNDERWAY (PEACH, AND SO FORTH)*

Modern breeding programs are driven by what people want, with a lot of focus on things such as color, taste, texture, and, health (Massey et al., 2018). The consumer expects that the food they buy will keep for a certain amount of time but maybe does not put enough weight on this when making a purchase decision. This is shown by how much food is wasted in the home.

There are new varieties of apples or stone fruits that do not store well, and they are often very good. Even though there was hope at first, the genetics of many diseases are still unknown. In the breeding programs, both traditional breeding and new gene-editing technology can be used to make successful cultivars with new colors and tastes, but it is harder to make cultivars that last longer and are stronger when stored. This problem is even worse for many tropical crops, where long-term breeding programs are not as thorough because they do not have enough resources or research. Changing the order of breeding priorities could have a much bigger effect on reducing food loss and waste than just using postharvest technology (Shipman et al., 2021).

6.6.3 BIOMARKERS OF DISORDERS HAVE BEEN IDENTIFIED (METABOLITES IN APPLE)

Even though biomarkers cannot replace traditional ways of analyzing nutrition, finding strong biomarkers will make food incidence surveys and other traditional techniques stronger and more reliable, and may one day be able to replace them. Phloretin, epicatechin, and procyanidin B2 are some of the short-term biomarkers that can be derived from apple eating. They are biomarkers for apple eating, but they can also differentiate between low and high apple consumption since they can detect how many apples were consumed. A diversified diet does not affect any of the three biomarkers, which means that they may all be used to determine how many apples a person consumes. It is recommended that spot urine experiments be taken over 1–3 months to figure out how much an individual eats apple to get around the fact that the compounds found are only short-duration biomarkers and are typically no longer noticeable 24 h after eating an apple. This is because the compounds found are only found in apples (Ulaszewska et al., 2018).

6.6.4 POTENTIAL APPLICATIONS OF GENE EDITION ON TARGET GENES COMING FROM RESEARCH USING ACCESSIONS GERMPLASM OR WILD SPECIES, AND NEW OMIC TOOLS

Over the last few decades, numerous omics technologies have developed (Habibi et al., 2021). Such omics-based methodologies have proven to remain beneficial for analyzing the molecular and genetic underpinning of crop expansion by looking at DNA, transcript levels, proteins, metabolites, and mineral nutrients as opposed to the background of environmental and physiological stress responses. Genomes, epigenomics, transcriptomics, proteomics, metabolomics, and phenomics are all incorporated into the panoramic program, which then allows for the construction of models that predict agronomically significant features for precision breeding (Yang et al., 2021). Integrating structures biology with big omics datasets has also improved the understanding of agricultural molecular regulator groups. These omics techniques have been used in wheat (*Triticum aestivum L.),* soybean *(Glycine max)*, tomato *(Solanum Lycopersicum),* maize *(Zea mays* L.), and rice *(Oryza sativa L.).*

6.7 CONCLUSION

Fruit and vegetables are an essential part of a healthy diet. However, these commodities commonly suffer from a wide range of physiological disorders affecting the overall quality and resulting in considerable loss. Physiological disorders include all fruit alterations and quality defects that are not attributable to pathogens. These noninfectious disorders are sometimes difficult to identify. Further, once symptoms are recognized, corrective measures often are not economically viable. Therefore, focus on prevention. Several steps may be taken during pre as well as postharvest operations to minimize the occurrence of abiotic disorders. These include proper time of sowing, management of uniform soil moisture, supply of balanced nutrients, harvesting at the proper stage, and accurate post-harvest handling and storage conditions. Physiological disorders in fresh horticultural produce can lead to significant qualitative and quantitative losses, particularly chilling injury.

KEYWORDS

- **fresh-cut**
- **moisture**
- **physiological disorder**
- **vegetable**
- **temperature**

REFERENCES

Ali, S., Anjum, M. A., Ullah, S., Nawaz, A., Ejaz, S., & Khaliq, G. (2023). Salicylic acid mediated postharvest chilling and disease stress tolerance in horticultural crops. In Plant stress mitigators (pp. 69–85). Academic Press.

Ali, S., Khaliq, G., Iftikhar, A., Mir, S. A., & Iqbal, S. (2024). Nanoemulsions-based edible coatings for postharvest quality preservation of fruits and vegetables. In *Sustainable postharvest technologies for fruits and vegetables,* (pp. 213–226). CRC Press.

Bhatia, K., & Asrey, R. (2019). Minimal processing of pomegranates (*Punica granatum L.)*: A review on processing, quality and shelf life. *Journal of Food Processing and Preservation, 43*, 14281.

Brizzolara, S., Manganaris, G. A., Fotopoulos, V., Watkins, C. B., & Tonutti, P. (2020). Primary metabolism in fresh fruits during storage. *Frontiers in Plant Science, 11*, 80.

Bulgari, R., Franzoni, G., &Ferrante, A. (2019). Biostimulants application in horticultural crops under abiotic stress conditions. *Agronomy, 9*, 306.

Bustamante, C. A., Brotman, Y., Monti, L. L., Gabilondo, J., Budde, C. O., & Lara, M. V. (2018). Differential lipidome remodeling during postharvest of peach varieties with different susceptibility to chilling injury. *Physiologia Plantarum, 163*, 2–17.

Cakmak, H. (2019). Assessment of fresh fruit and vegetable quality with non-destructive methods. In *Food quality and shelf life* (Vol. 23, pp. 303–331).

Conrad, Z., Niles, M. T., Neher, D. A., Roy, E. D., Tichenor, N. E., & Jahns, L. (2018). Relationship between food waste, diet quality, and environmental sustainability. *PloS One, 13*, 405–425.

Crisosto, C. H., Mitcham, E. J., & Kader, A. A. (1996). Peach and nectarine recommendations for maintaining postharvest quality. *Postharvest Technology Center, University of California: Davis, CA, USA*.

Dalvi-Isfahan, M., Jha, P. K., Tavakoli, J., Daraei-Garmakhany, A., Xanthakis, E., & Le-Bail, A. (2019). Review on identification, underlying mechanisms and evaluation of freezing damage. *Journal of Food Engineering, 255*, 50–60.

Dogan, A. (2022). Effects of different oxygen levels with high-carbon dioxide atmosphere on postharvest quality of fresh fig under palliflex storage systems. *Horticulture, 8*, 353.

Duarte-Sierra, A., Tiznado-Hernandez, M. E., Jha, D. K., Janmeja, N., & Arul, J. (2020). Abiotic stress hormesis: An approach to maintain quality, extend storability, and enhance phytochemicals on fresh produce during postharvest. *Comparehensive Reviews in Food Science and Food Safety, 19*, 3659–3682.

FAO. (2018). The future of food and agriculture – Trends and challenges. Food and Agriculture Organization of the United Nations Rome (pp. 16–180).

Habibi, F., Serrano, M., Zacarias, L., Valero, D., & Guillen, F. (2021). Postharvest application of 24-epibrassinolide reduces chilling injury symptoms and enhances bioactive compounds content and antioxidant activity of blood orange. *Frontiers in Plant Science, 12*, 145.

Kathayat, K., & Rawat, M. (2022). Physiological disorders in vegetable crops. *Advances in Horticultural Management and Value Addition, 16*, 307.

Khaliq, G., Ali, S., Gapper, N., & Nicola, S. (2023). Recent advances and approaches in the application of elicitors to enhance resistance mechanisms in fresh produce. *Frontiers in Sustainable Food Systems, 6,* 1061079.

Khan, L., Khaliq, G., Ullah, E., Abdul-Rahaman, A., El-Mogy, M. M., Kesba, H. H., Ismail, A. M., & El-Beltagi, H. S. (2025). Guggul gum and clove essential oil synergistically reduced anthracnose disease and modulated physicochemical changes in tomato during storage. *Journal of Stored Products Research, 114,* 102713.

Kumar, R. R., Goswami, S., Singh, K., Dubey, K., Rai, G. K., Singh, B., & Praveen, S. (2018). Characterization of novel heat-responsive transcription factor (TaHSFA6e) gene involved in the regulation of heat shock proteins (HSPs)—A key member of heat stress-tolerance network of wheat. *Journal of biotechnology,* 279, 1–12.

Lal, S., Khaliq, G., Mastoi, A. H., Irshad, A., Jaffar, M. A., Khan, L., & Muhammad, N. (2025). Effect of *Fagonia indica* and garlic extract on *Alternaria alternata* and post-harvest quality of tomato fruit during storage. *Journal of Crop Health, 77*(6), 1–13.

Li, H., Luo, W., Ji, R., Xu, Y., Xu, G., Qiu, S., & Tang, H. (2021). A comparative proteomic study of cold responses in potato leaves. *Heliyon, 7*, 6002–6018.

Loayza, F. E., Brecht, J. K., Simonne, A. H., Plotto, A., Baldwin, E. A., Bai, J., & Lon-Kan, E. (2021). A brief hot-water treatment alleviates chilling injury symptoms in fresh tomatoes. *Journal of the Science of Food and Agriculture, 101*, 54–64.

Lurie, S. (2021). Genomic and transcriptomic studies on chilling injury in peach and nectarine. *Postharvest Biology and Technology, 174*, 111444.

Madani, B., Mirshekari, A., & Imahori, Y. (2019). Physiological responses to stress. In *Postharvest physiology and biochemistry of fruits and vegetables* (Vol. 1, pp. 405–423). Woodhead Publishing.

Massey, L. M., Hettiarachchy, N. S., Horax, R., Rayaprolu, S. J., Kumar-Phillips, G., Martin, E. M., & Ricke, S. C. (2018). Efficacy of organic acid electrostatic spray for decontaminating Salmonella on cantaloupe cubes and cherry tomatoes. *Journal of Food Processing and Preservation, 42*, 13748.

Muthuramalingam, P., Shin, H., Adarshan, S., Jeyasri, R., Priya, A., Chen, J. T., & Ramesh, M. (2022). Molecular insights into freezing stress in peach based on multi-omics and biotechnology: An overview. *Plants, 11*, 812.

Pereira, L., Santo Domingo, M., & Ruggieri, V. (2020). Genetic dissection of climacteric fruit ripening in a melon population segregating for ripening behavior. *Horticulture Research, 7*, 187.

Postharvest Management of Vegetables. (2020) Resources for Australian supply chains; 1; pp. 2–3.

Rajapaksha, L., Gunathilake, D. C., Pathirana, S. M., & Fernando, T. (2021). Reducing post-harvest losses in fruits and vegetables for ensuring food security—Case of Sri Lanka. *MOJ Food Processing and Technology, 9*, 7–16.

Rutta, E. W. (2024). Postharvest food loss reduction and agriculture policy framework in Tanzania: status and way forward. Agriculture & Food Security, 13(1), 36.

Santos, S. F., Cardoso, R. D. C. V., Borges, I. M. P., Almeida, E. A. C., Andrade, E. S., Dos Santos, S. F., & Ramos, L. D. C. (2019). Post-harvest losses of fruits and vegetables in supply centers in Salvador, Brazil: Analysis of determinants, volumes and reduction strategies. *Waste Management, 101*, 161–170.

Selvarajan, D., Mohan, C., Dhandapani, V., Nerkar, G., Jayanarayanan, A. N., Mohanan, V. M., & Chinnaswamy, A. (2018). Differential gene expression profiling through transcriptome approach of Saccharum spontaneum L. under low-temperature stress reveals genes potentially involved in cold acclimation. *Biotechnology, 8*, 1–18.

Sharma, S., & Manjeet. (2020). Heat Stress Effects in Fruit Crops. Agricultural Reviews, *41*, 73–78.

Shipman, E. N., Yu, J., Zhou, J., Albornoz, K., & Beckles, D. M. (2021). Can gene editing reduce postharvest waste and loss of fruit, vegetables, and ornamentals? *Horticulture Research, 8*, 1–15.

Strano, M. C., Altieri, G., Allegra, M., Di Renzo, G. C., Paterna, G., Matera, A., & Genovese, F. (2022). Postharvest technologies of fresh citrus fruit: Advances and recent developments for the loss reduction during handling and storage. *Horticulture, 8*, 612.

Takayama, K., Konishi, A., & Omasa, K. (2003). Diagnosis of invisible photosynthetic injury caused by a herbicide (Basta) with chlorophyll fluorescence imaging system. *Agricultural Engineering International: CIGR Journal.*

Ulaszewska, M., Vazquez-Manjarrez, N., & Garcia-Aloy, M. (2018). Food intake biomarkers for apple, pear, and stone fruit. *Genes and Nutrition, 13*, 29.

Valenzuela, J. L., Manzano, S., Palma, F., Carvajal, F., Garrido, D., & Jamilena, M. (2017). Oxidative stress associated with chilling injury in immature fruit: postharvest technological and biotechnological solutions. *International Journal of Molecular Sciences, 18*, 1467.

Wang, Z., Kannemeyer, R., McGlone, A., & Burdon, J. (2020). Potential of Vis-NIR spectroscopy for detection of chilling injury in kiwifruit. *Postharvest Biology and Technology, 164*, 111160–111175.

Wikipedia. (2015). Old Electron microscope in the Institute for Cell Biology in Marburg.

Yahia, E. M., Carrillo-Lopez, A., & Sanudo, A. (2019). Physiological disorders and their control. In E. M. Yahia (Ed), *Postharvest technology of perishable horticultural commodities* (Vol. 1, pp. 499–500).; Elsevier Publishers.

Yang, Y., Saand, M. A., Huang, L., Abdelaal, W. B., Zhang, J., Wu, Y., Li, J., Sirohi, M. H., & Wang, F. (2021). Applications of multi-omics technologies for crop improvement. *Frontiers in Plant Science, 12*, 563953–563974.

Zainalabidin, F., Sagarin, M., Azmi, W., & Ghazali, A. (2019). Optimum postharvest handling-effect of temperature on Quality and shelf life of tropical fruits and vegetables. *Journl of Tropical Resources and Sustainable Science, 7*, 23–30.

Zhang, W., Jiang, H., Cao, J., & Jiang, W. (2021). Advances in biochemical mechanisms and control technologies to treat chilling injury in postharvest fruits and vegetables. *Trends in Food Science and Technology, 113*, 355–365.

Zhao, H. D., Jiao, W. X., Cui, K. B., Fan, X. G., Shu, C., & Zhang, W. L. (2019). Near-freezing temperature storage enhances chilling tolerance in nectarine fruit through its regulation of soluble sugars and energy metabolism. *Food Chemistry, 289*, 426–435.

Zheng, Y., Ma, Y., Liu, W., & Qiu, F. (2020). Plant nutrition and physiological disorders in fruit crops. In *Frute Crops* (Vol. 18, pp. 47–58).

CHAPTER 7

Low-Temperature-Related Physiological Disorders

YOSHIHIRO IMAHORI

Graduate School of Life and Environmental Sciences, Osaka Prefecture University, Sakai, Osaka, Japan

ABSTRACT

Low temperature is the most effective postharvest technology for keeping the quality and prolonging the shelf life of horticultural products. However, when commodities are stored at or exposed to temperature below a certain threshold temperature, low-temperature physiological disorder takes place in them. There are two types of physiological disorders in low-temperature storage. Generally, freezing injury is caused in lower temperatures below the freezing temperature of water. The main symptom of injury by freezing occurs in the plasma membrane. Common symptoms of freezing injury include the water-soaking appearance, internal discoloration, and destruction of structural integrity. Cell membranes have an important role in the survival of plant cells under freezing temperature. Chilling injury (CI) is a typical physiological disorder of horticultural products that occurs at low temperature above freezing point of water. This disorder is caused in several horticultural products, subtropical and tropical crops, when they are exposed to or stored at lower temperatures than specific threshold temperatures. In chilling-sensitive crops several physiological and biochemical changes, cellular dysfunctions, and metabolic process alterations occur in response to chilling stress. The symptoms of CI appear in all development stages in crops, but its incident intensity is dependent

Physiological Disorders of Fruits and Vegetables: Causes and Remedies.
Ghulam Khaliq, Sajid Ali, and Mohammed Wasim Siddiqui (Eds.)

DOI: 10.1201/9781998511334-7

on the storing temperature and duration, plant organ, and the developmental stage. Many approaches are developed on preventive methods of CI during low-temperature storage in horticultural products. These technologies involve physical and chemical treatment methods by modifying the storage environment or treating the products.

7.1 INTRODUCTION

Horticultural products contain important nutrition of carbohydrates, protein, minerals, vitamins, and dietary fiber and are not only utilized for human diet but also benefit human health. They also contain high water content and are highly susceptible to dehydration and mechanical damage. As horticultural crops are living organisms, they are subject to continuous physiological and biochemical changes. Some changes are desirable, but most are not (Imahori, 2012, 2014; Madani et al., 2019). Thus, since horticultural commodities are fresh products, they are subject to widespread losses through microbial effect, physical damage, and senescence after harvest, and are perishable. The losses in quality and quantity after harvest significantly affect the relationship between harvest and consumption of horticultural products. Although the irreversible changes are unstoppable, they can be slowed down within certain thresholds (Imahori, 2014; Madani et al., 2019). Therefore, the maintenance of quality and improvement of shelf life are important for horticultural products. Prolonged shelf life of the commodities is necessary for reducing the loss of economical values (Imahori, 2014; Madani et al., 2019).

Low-temperature storage is usually the effective technique for maintaining the quality of horticultural products and prolonging their shelf life after harvest and is the greatest extensively utilized postharvest technology (Endo et al., 2019a, b; Imahori et al., 2021; Kantakhoo et al., 2022). Reducing storage temperature substantially can decline many metabolic process rates that introduce deterioration, senescence, and quality loss of horticultural products after harvest (Imahori et al., 2008, 2014). The low temperatures can effectively promote the decrease in reaction velocity in many metabolic processes that change rate of ripening, delay the development of fruit color, and reduce the fruit softening (Khan et al., 2025; Lal et al., 2025). Therefore, low temperature is generally the most effective implementation for prolonging storage life of products (Imahori, 2012,

2021; Kantakhoo et al., 2022). Storage temperatures are determined to maintain the postharvest quality properties such as appearance, texture, flavor, and nutritional in horticultural products (Hodges et al., 2004; Tao et al., 2024).

Storage temperatures that do not cause physiological disorders are usually recommended in temperature as low as possible but slightly above the freezing point. However, while the storage by low temperature is generally the most effective preservative means, there are crop-specific temperature thresholds (Hodges et al., 2004). The divergence from temperature thresholds results in a source of stress and physiological disorder. When horticultural products are stored at or exposed to temperature below a certain threshold temperature, low-temperature physiological disorder takes place in horticultural products (Imahori et al., 2008, 2021). Two types of physiological disorders in low-temperature storage exist. Freezing injury can occur if storage temperature was below the freezing point of horticultural products, while chilling injury (CI) can occur at temperature above freezing point (Rai et al., 2022). Approaches to control or modulate physiological disorder in horticultural products on low-temperature storage are significant for maintaining the quality after postharvest and prolonging their shelf life. Understanding and control of low-temperature physiological disorder need the detailed information on physiological and biochemical characteristics of physiological disorder. This chapter reviews the knowledge in low-temperature physiological disorder in horticultural products.

7.2 FREEZING INJURY

The nonfreezing range of subzero temperature can maintain the quality of horticultural products in living state without the destruction of their tissues. This temperature zone can decrease the physiological and biochemical reactions and minimize the growth of pathogenic microorganisms in horticultural products (Liu et al., 2020). Because the storage temperature is only slightly above the freezing point, freezing injury occurs in horticultural products when products are kept below the freezing temperature (Yahia et al., 2019). For example, the freezing temperature of commodities is −2.2 to −1.7°C in apples, −0.9 to −0.8°C in cucumber, and −0.6 to −0.3°C in lettuce, respectively (Yahia et al., 2019). The susceptibility to freezing injury differs between horticultural products. The susceptibility in freezing stress is influenced by the variation in freezing sensitivity in plant tissues (Madani et al., 2019). The freezing temperature constrains the growth and development in plants and distribution in plant species. Their cultivars, locations, and growing conditions affect the freezing temperature. Some plants can resist freezing stress with no apparent damage, while other plants are damaged with even a slight freezing temperature. Generally, freezing injury occurs with lower temperatures below the freezing point of water (Madani et al., 2019). Common symptoms of freezing injury include the water-soaking appearance, internal discoloration, and destruction of structural integrity. As wilting or softening the deterioration is observed in plant tissues. The tissues of horticultural products lose rigidity and become mushy upon thawing (Yahia et al., 2019). When products freeze, the ice crystals form into and between the plant cells, leading to the rupture of cell membranes, and being lost cellular integrity, and finally inducing the death of plant tissues (Madani et al., 2019). Freezing injury becomes more severe by forming larger ice crystals when it freezes slowly (Yahia et al., 2019). The main symptom of injury by freezing occurs in the plasma membrane. Therefore, cell membranes have an important role in the survival ability in plant cell to tolerate or be influenced with freezing (Madani et al., 2019).

Plants have the strategy that survives freeze either avoiding the ice crystallization by supercooling of water within their tissues (freezing avoidance) or tolerating the crystallization of ice into the apoplast (freezing tolerance). Freezing avoidance implies supercooling and circumvent of the ice crystallization into the apoplast. Pure water can be supercooled to a certain freezing temperature below 0°C without the ice nucleation (Galindo et al., 2007; Madani et al., 2019). As a strategy some particular cells and organs in plants utilize supercooling. The common survival strategy in plants is freezing tolerance (freezing accumulation) and achieved by several changes in physiological response through the regulation of gene expression. This tolerance property is induced by freezing acclimation, but it is quickly absent upon returning to a warmish nonacclimating temperature. The freezing tolerance in plants is acquired by the accumulation of osmotic regulation substances such as sugars, organic acid, proline, and glycinebetaine (Galindo et al., 2007). The high amounts of sugars in freezing acclimated plant function performance as osmotic regulation, while less amount sugars play a role as cryoprotective mechanism or signaling molecules introduced due to freezing stress (Mantri et al., 2012).

7.3 CHILLING STRESS METABOLISM

Changes in the external environmental circumstances possibility cause various stresses to plants (Capanoglu, 2010). Stress is usually defined as an extensive environmental factor that has potential influences on organisms (Madani et al., 2019). The environmental factors may interrupt, restrict, and accelerate the metabolism in plants. Various stresses induce a series of morphological, physiological, biochemical, and molecular changes that harmfully influence the growth and productivity of plants. Consequently, the losses of quality after harvest in horticultural products directly or indirectly are attributed to a combination of stress and stress-initiated senescence (Lester, 2003; Madani et al., 2019). Exposing to unfavorable environmental conditions, plants cause some amount of stress and express a part of the plant potential genes. Plants can also acclimate to disadvantageous circumstances by genetically decided stress tolerance (Imahori, 2012; Madani et al., 2019). These plants can respond to quicker or stronger stimulus in defense responses inducing increased resistance to stress and obtain resistance to stresses (Capanoglu, 2010; Madani et al., 2019). Thus, the study on the influences of chilling stress in metabolism and performance of plants is a main focus topics on plant research, particularly on postharvest physiology.

7.3.1 CHILLING STRESS

Chilling stress is major abiotic stress that delays plant's growth and development and restricts geographical distribution of many crop species and decreases agricultural productivity (Imahori, 2011, 2014). Chilling temperature can usually decrease the velocity of metabolic process which relates to senescence, deterioration, and quality loss of horticultural products (Imahori, 2014). The metabolism condition and the nutritional composition of horticultural products at the time that is exposed in chilling temperature possibility affect their sensitivity to the chilling stress (Wang, 2010; Imahori, 2014). At low temperatures, the plant tissues become weak because they cannot act in normal metabolism condition. In chilling-sensitive plants various changes in physiological and biochemical processes and cellular structure are caused in response to chilling stress (Wang, 2010). In addition, chilling stress detracts the cellular energy state and also alters the integrity of membrane. Chilling-sensitive plants typically indicate structural injuries and may be disadvantaged by the dysfunction of metabolism subjected to chilling temperature (Mahajan and Tuteja, 2005).

The metabolic condition in plant and the chemical composition that exists in plant tissues can influence the sensitivity to the chilling stress when exposed at chilling temperatures (Wang, 2010; Imahori, 2014). Chilling-resistant plants generally have a higher ratio of unsaturated fatty acid and, thus indicate a lower transition temperature in the membrane lipids (Mahajan and Tuteja, 2005; Imahori, 2014). In contrast, chilling-sensitive plants have a higher ratio of saturated fatty acid and, consequently, indicate a higher transition temperature. Chilling-tolerant plants are thus liable to contain a higher unsaturation degree of fatty acids than chilling-sensitive plants in membrane lipids (Wang, 2010; Singh and Singh, 2013). Also, chilling stress changes the equilibrium between the reactive oxygen species (ROS) accumulation and defense machinery when oxidatively initiated stress is observed (Hodges et al., 2004; Imahori et al., 2008, 2011, 2014).

7.3.2 EFFECT OF CHILLING STRESS ON PHYSIOLOGICAL CHANGES

Low temperature changes the respiration rate in horticultural products. Generally, crops have a low respiration rate, when low temperature is at 0°C. Respiration rate varies in some cases. It tends to be more stable and lower when storage temperatures vary between 0 and 5°C (Rodrigues et al., 2020). On the other hand, the increase in respiration rate occurs by the dysfunction of metabolism and increase in partly oxidized intermediates during low temperature or after chilling exposure. The decrease in respiration and increase in phosphates consumption during chilling stress results in the reduction in ATP levels (Lukatkin et al., 2012).

Subcellar organelles, mitochondria and chloroplasts are connected with the energy production that sustains the metabolism of an alive cell in plants. Low temperature decreases mitochondrial respiration, together with the changed function of cellular membrane bound ATPase system, which consequently causes a severe suppression in ATP production (Liang et al., 2020). Chilling stress directly affects the physiological functions of chloroplast. The lipid membrane state in chloroplast alters by chilling temperature. Photosynthesis rate is depressed due to rigidification of thylakoid membranes and inactivation of the related enzymes, and excess ROS produces in chloroplasts. Excess ROS that is produced by chilling stress destructs the lipid membrane in chloroplasts (Liu et al., 2018b).

Chilling stress affects cell membrane and changes its lipid composition; as a result, it deteriorates the function of mitochondria and chloroplasts (Liang et al., 2020). The physical characteristics of cellular membranes are influenced by chilling stress and are related to chilling stress-provoked quality losses in horticultural products. The defects in cell membrane are caused by the decrease in the rate of unsaturated fatty acids and the decrease in permeability in cellular membrane (Imahori et al., 2008, 2012, 2014). The decrease in the rate of unsaturated fatty acids against saturated fatty acids is associated with an increase in the peroxidation of lipid during low temperature (Imahori et al., 2008, 2014). As the peroxidation of lipid in cellular membrane is the first phenomenon in the manifestation of chilling stress, chilling stress results to induce lipid degradation in chilling-sensitive horticultural products and enhances lipid peroxidation in the cell membranes (Imahori et al., 2008, 2021; Endo et al., 2019a; Kantakhoo et al., 2022). Oxidative damage that is occurred by oxidative stress is reflected to be a reaction of sensitive plant tissues to cold stress. ROS that is produced in response to chilling stress causes damage to cellular membrane in horticultural products. Thus, oxidative stress can be associated with the appearance of physiological disorder that occurs in horticultural products below low temperature (Imahori, 2014; Endo et al., 2019a; Imahori et al., 2021; Kantakhoo et al., 2022).

7.4 CHILLING INJURY

When horticultural products are stored or exposed under temperature below a certain threshold temperature, these crops cannot conduct their normal metabolic activities, and then physiological and biochemical alterations occur. The changes in multiple processes in metabolism and the visible symptoms occur in chilling-sensitive species, and cell death finally causes, if these crops are subjected to the unfavorable low temperature for a longer duration. This physiological disorder is called CI and occurs in several horticultural products (Endo et al., 2019a, b; Imahori et al., 2021; Kantakhoo et al., 2022). CI is a typical physiological disorder that occurs in several horticultural products, when they are stored or exposed at low temperatures below certain threshold temperatures. CI occurrence is the property for the specific species and it occurs in several horticultural products that especially originate in the tropics and subtropics (Imahori et al., 2008, 2021; Kantakhoo et al., 2022). Usually, this physiological disorder arises at temperatures over 0°C, about 10–12°C for tropical crops, and about 8°C for subtropical crops (Sevillano et al., 2009; Zhang et al., 2021a).

From the sensitivity of low temperature on horticultural products, the crops can be classified into three categories. One is chilling-tolerant crops. This group is resistant to low temperature and does not cause CI. The lower the temperature, the longer the storage duration, as long as the storage temperature is not below the freezing point. Another is the group of chilling-sensitive crops, or slightly chilling-sensitive crops. Each crop is sensitive to low temperature and causes CI below storage temperature down to a certain point. This temperature is called critical chilling temperature. The critical chilling temperature in slightly chilling crops is usually 10–13°C. Most tropical and subtropical commodities are a member of slightly chilling crops. The critical chilling temperature in chilling-sensitive crops is lower than that in slightly chilling-sensitive products. The critical chilling temperature in chilling-sensitive crops is usually 5–8°C (Wang, 1994; Zhang et al., 2021a). However, CI does not appear below low temperature in some crops but there is a case in which CI symptom occurs when it is returned to ambient temperature (Zhang et al., 2021a). In general, the lower storage temperature below the critical chilling temperature causes more severe CI in chilling-sensitive horticultural products. However, mature green mume fruit showed a more severe symptom of CI at 6°C storage than 1°C storage (Imahori et al., 2008). Similarly, CI incidence in papaya fruits at 1°C storage was lower than that at 6°C storage for the reason of higher sucrose synthesis in fruits stored at 1°C (Zhu et al., 2022). These phenomena are known as abnormal CI behavior (Zhu et al., 2022).

Apart from low temperature, relative humidity in storage environment affects the occurrence and development of CI. Storage under low relative humidity generally causes more severe CI than that below highly relative humidity during low temperature. Light in storage environment also can raise CI occurrence during low-temperature storage (Wang, 2010). The occurrence of CI can be seen in all growing and developmental stages of plants, and its incidence is influenced by the storage temperature and exposing duration (Sevillano et al., 2009; Zhang et al., 2021).

Preharvest conditions influence the occurrence of CI. Exposing fruits on the tree to high temperature at harvest or close to harvest induces resistance to CI during low-temperature storage. Kiwifruit that acclimates in cold temperature on the tree before harvest increases the tolerance to CI. "Delicious" apples are affected by low temperatures, rainfall, and sunshine before harvest, and reduce susceptibility of CI symptom called superficial scald, while causing higher susceptibility to scald in high-temperature season (Sevillano et al., 2009).

7.4.1 SYMPTOMS OF CHILLING INJURY

The symptoms of CI appear regardless of developmental stages in crops, but their level depends on the exposing temperature and period, plant organ, and the development stage. The symptoms alter as the species and even of variety and also depend on the environmental situations such as soil, climate, watering, where horticultural crops have been grown (Sevillano et al., 2009; Zhang et al., 2021). In addition, CI development varies on species, cultivars, maturity, tissue types, and environmental factors (Imahori, 2014). The visual symptoms in horticultural products include surface pitting, internal breakdown, water-soaked appearance, discoloration, off-flavor, failure to ripen, and decay (Wang, 2010). Figure 7.1 exhibits the symptom of CI in banana and eggplant. The symptom of CI is specific to horticultural crops (Wang, 1994). Pitting is circular or irregular-shaped pits that occur on the surface, and is also the most popular CI (Wang, 1994). Surface pitting is observed on grapefruit, mume, lemon, orange, cucumber, eggplant, okra, melon, sweet pepper, tomato, and watermelon (Table 7.1). This symptom is the most general symptom in tropic and subtropic products (Imahori, 2014). Failure to ripen is also a general symptom and occurs in tropic and subtropic originated fruits such as avocado, banana, mango, melon, and tomato (Yamawaki, 2021) (Table 7.1).

FIGURE 7.1 Chilling injury on banana and eggplant.

TABLE 7.1 Temperature and Symptoms of Chilling Injury in Some Fruits and Vegetables (Yamawaki, 2021).

Commodity	Temperature of Chilling Injury (°C)	Symptoms of Chilling Injury
	Fruits	
Apple	0–3.5	Flesh browning, soft scald
Avocado	5–10	Failure to ripen normally, flesh browning, off flavor
Banana	12–14.5	Peel browning, off flavor

TABLE 7.1 *(Continued)*

Commodity	Temperature of Chilling Injury (°C)	Symptoms of Chilling Injury
Grapefruit	8–10	Pitting, off flavor
Mume mature green	5–6	Pitting, flesh browning
Lemon ripe	0–4	Pitting, browning
mature green	11–14.5	Off flavor
Olive	6–7	Flesh browning
Orange	2–7	Pitting
Papaya	7–8.5	Pitting, off flavor
Peach	2–5	Flesh browning, wooly texture
Pineapple	4–7	Internal browning
Mango	7–10	Water soaking, failure to ripen
	Vegetables	
Cucumber	7–8	Pitting
Eggplant	7–8	Pitting, surface scald
Okra	6–7	Pitting, water soaking
Melon cantaloupe	2–4	Pitting, failure to ripe, off flavor
Honey dew	7–10	Pitting, failure to ripe, off flavor
Sweet pepper	6–8	Pitting, sheet pitting, calyx, and seed browning
Sweet potato	9–10	Internal browning, pitting, hardcore
Pumpkin	7–10	Internal browning, pitting
Tomato ripe	7–9	Discoloration, off flavor
mature green	12–13	Pitting, failure to ripe
Watermelon	4–5	Pitting, off flavor

The symptoms are similar in all plants at a microscopic level and contain expansion and disorganization of mitochondria and chloroplasts, decrease in the number and scale in starch granules, increase in lipid droplets within the chloroplasts and condensation in nuclear chromatin. Additionally, in chloroplasts, thylakoid dilation and grana unstacking occur (Sevillano et al., 2009). On the other hand, at macroscopic level CI symptoms are distinguished into two classes. One is metabolic and developmental CI symptoms like failure to ripen and deficient flavor and aroma, related to qualitative nature. Another is physiological CI symptoms like surface pitting, discolorations, water soaking, woolly in the pulp, internal break-down, low tolerance to mechanical injury, and susceptibility to fungal and microbial infection (Sevillano et al., 2009; Aghdam and Bodbodak, 2014). However, the decay after chilling storage is one of common phenomena after CI and occurs due to lower tolerance to microbial attack than as a direct symptom of CI. The critical temperature and appearance in CI symptom are different among horticultural products. The main apparent symptoms are surface water-soaked in cucumber and kiwifruit, surface depression and browning in banana, orange, eggplant, green pepper, and zucchini, woolly texture such as peach and nectarine, failure to ripen such as tomato and guava, and seed and calyx browning in sweet pepper.

7.4.2 MECHANISM OF CHILLING INJURY

The event initiating CI occurs when storage temperature decreases below the threshold temperature particular for crops and the cells in crops sense the lower temperature during low-temperature storage (Wills and Golding, 2016; Rai et al., 2022). This primary response leads a cascade of secondary responses as visible CI symptoms (Rai et al., 2022). The reactions in chilling-induced events are distinguished into two categories. The primary phenomenon is transition on membrane lipids, changes in kinetic property or substrate specificity of regulatory enzyme, structure alteration in cytoskeleton, and rise of Ca ion concentration in cytosol (Wang, 1994). These responses are more or less instantaneous, reversible during chilling exposing period, and not still conclusive and pinpoint response (Wills and Golding, 2016). This primary event leads to secondary response that contains physiological and biochemical changes. The secondary event is multitudinous response to low temperature, is irreversible, and induces necrosis and various CI symptoms that are visible (Wang, 1994; Sevillano et al., 2009; Wills and Golding, 2016).

7.4.2.1 ALTERATIONS OF CELL MEMBRANE

Cellular membrane performs as a barrier protecting the plant cell against various damage and destruction and maintains a certain function in cellular biochemical processes within the cell (Liang et al., 2020). The cellular membrane is the first cell structure influenced by chilling stress (Aghdam and Bodbodak, 2014). Low temperature alters the fluidity of membrane, lipid composition, and lipid–protein interactions in membrane, changes membrane microdomains, and relays the cues to downstream signaling pathways (Sossi et al., 2022). The membrane fluidity indicates the double bond index and the acyl chain length. The saturation degree in fatty acids and length in the acyl chains affect the physicochemical properties and play significant roles in membrane structure and function (Sossi et al., 2022). Once cell membranes are under the influence of chilling stress, intracellular reactions are affected, and as a result the membrane destruction and cell death occur (Liang et al., 2020). The physical characteristics of cell membranes are influenced by chilling stress and are related to chilling-induced quality defects. CI is associated with cell membrane deterioration, which results in alterations of permeability in cell membrane. These changes would lead to CI symptom development (Imahori, 2014).

The primary response for CI development was proposed at molecular level (Aghdam and Bodbodak, 2014). Physical-phase transition in cellular membranes from a flexible liquid crystalline to a solid gel structure occurs and results in the reduction in cellular membrane fluidity, and consequently the loss in cellular membrane permeability increases, when subjected to low temperature (Lyons, 1973). The more unsaturated fatty acids include cellular membrane, the lower transition temperature, and the higher fluidity they contain (Ge et al., 2019). The occurrence of CI is caused by the decline in the degree of unsaturated fatty acids and reduction in the membrane permeability of cell membrane (Imahori et al., 2008, 2011, 2014; Kantakhoo et al., 2022). The increase in degree in fatty acids saturation, degeneration of phospholipids and galactolipids, and enhancement of the proportion in sterol to phospholipids induces decrease in fluidity and function in cellular membrane (Aghdam and Bodbodak, 2014). The reduction in the rate of unsaturated to saturated fatty acids shows an increase in the peroxidation of lipid related to CI development during low-temperature storage (Imahori et al., 2008, 2014).

The peroxidation of lipid in cellular membrane is the responses in CI development by low-temperature stress and one of the initial phenomena in the appearance of CI (Endo et al., 2019a, b; Kantakhoo and Imahori, 2021; Kantakhoo et al., 2022). The degree of lipid peroxidation depends on the strength of chilling stress and affects the extent of CI development (Imahori, 2014). Chilling stress has been shown to induce lipid degradation in chilling-sensitive horticultural products, and consequently may indicate that CI increases lipid peroxidation in cellular membranes (Imahori et al., 2008, 2021; Endo et al., 2019a; Kantakhoo and Imahori, 2021).

The lipid peroxidation in cell membrane can be assessed by malondialdehyde (MDA) content that is a secondary end substance caused by the oxidation of polyunsaturated fatty acid (Imahori et al., 2008). The increase in MDA content relates to a clear deterioration of integrity in cell membrane and promotion of peroxidation in membrane lipid on horticultural products (Imahori, 2014, 2016). Being exposed to cold temperature changes the membrane structure due to peroxidation of lipid in membrane. Due to lipid peroxidation the polyunsaturated fatty acids degrade and at the same alter MDA, consequently causing membrane destruction and cell death. Thus, alterations in MDA content in plant tissues can become a significant indicator of integrity degree in the structure of cell membranes due to chilling temperature (Endo et al., 2019a, b; Kantakhoo and Imahori, 2021; Kantakhoo et al., 2022). In addition, at low temperature for a very long duration membrane destruction occurs (Aghdam and Bodbodak, 2014). Similar to MDA the electrolyte leakage indicates the structural integrity of cell membranes (Endo et al., 2019b; Kantakhoo and Imahori, 2021; Kantakhoo et al., 2022).

The function and integrity in cell membrane is associated with the enzyme activity related to lipid metabolism of membrane (Ma et al., 2020). Fatty acid desaturase (FAD) promotes the degree of fatty acids unsaturation and membrane fluidity in the lipid bilayer constructing cellular membrane (Aghdam and Bodbodak, 2014; Ge et al., 2019). This enzyme group consists in acyl carrier protein desaturases, acyl-lipid desaturases, and acyl-CoA desaturases (Ge et al., 2019). It is considered that chilling tolerance in plants is associated with the improvement in membrane fluidity with the increase in unsaturation degree in fatty acid by the enhancement of FAD activity (Aghdam and Bodbodak, 2014). Under chilling temperature, phospholipase D (PLD) and lipoxygenase (LOX) in horticultural products promote the degradation of unsaturated fatty acid, and consequently decrease the rate of unsaturated/saturated fatty acid and induce lipid peroxidation, and reduce membrane integrity (Aghdam and Bodbodak, 2014; Madani et al., 2019; Zhang et al., 2021a). PLD hydrolyzes structural membrane phospholipid to phosphatidic acid. The hydrolyzation product reacts to signal transduction function as the second messenger (Ma et al., 2020).

7.4.2.2 IMBALANCE OF ENERGY METABOLISM

Mitochondria and chloroplasts are associated with energy production in plants. Destruction of membrane by CI affects the function of mitochondria and chloroplasts subjected to cold stress. Low temperature decreases the respiration metabolism in mitochondria and destructs the ATPase system bound with membrane, resulting in sharp suppression in ATP production (Liang et al., 2020). Photosynthesis rate is also suppressed by low temperature and causes the destruction of thylakoid membranes by CI (Liang et al., 2020). Therefore, the occurrence of CI associates with energy levels that are necessary for the function and viability in plant cell (Shan et al., 2022a, b). If these energy manufacturing systems are damaged by CI, the ATP level in plant cell is lower and the deficiency in energy occurs (Liang et al., 2020).

Chilling stress induces physical-phase transition of cell membranes, leading to more severe electrolyte leakage that suppresses ATPase activity. Insufficient intracellular ATP supply is caused by higher PLD and LOX activities with higher ROS accumulation. Higher PLD and LOX activities cause higher cytosolic Ca^{2+} along with the reduction in Ca^{2+}-ATPase activity. Their enzymes also accumulate at higher ROS level, affecting membrane integrity along with MDA accumulation and membrane peroxidation (Aghdam et al., 2018). ATP can be divided into intracellular ATP and extracellular ATP. Intracellular ATP in plants maintains normal physiological conditions and causes important reactions such as respiratory metabolism (Shan et al., 2022a, b). Intracellular level of ATP supply within cell can be indicated by energy charge that is presented as [ATP + 0.5ADP]/[ATP + ADP + AMP]. CI such as internal flesh browning and external pitting in blueberry is caused by the lack of intracellular ATP supply, lower ATP and ADP levels, and higher AMP level leading to lower energy charge, because H^+-ATPase, Ca^{2+}-ATPase, cytochrome c oxidase, and succinate dehydrogenase activities are suppressed by low temperature (Aghdam et al., 2018). Similarly, the development of banana CI relates to lower intracellular ATP supply, lower ATP and ADP levels, and higher AMP level leading to lower energy charge, and higher PLD and LOX activities (Aghdam et al., 2018).

Since the maintenance of cellular energy is important for metabolism, the preservation and use of cellular energy can retain intracellular ATP supply and the continual supply of intracellular ATP can suppress the occurrence of CI (Shan et al., 2022a, b). But the deficiency of intracellular ATP and limitation of energy utilization induces membrane breakdown and lipid peroxidation and causes CI in horticultural products stored at low temperature by changing a series of biochemical and physiological reactions (Shan et al., 2022a, b). Intracellular ATP alters extracellular ATP by being released into the extracellular matrix such as apoplast from cytoplasm with the action of anion channels, gap junctions, ATP-binding cassette transporters, or vesicle exocytosis in reaction to stresses (Aghdam et al., 2018; Shan et al., 2022a, b). Extracellular ATP plays a role as a signaling molecule to mediate physiological metabolism in reaction against various stresses, acts as a damage-associated molecular pattern molecule to trigger downstream signaling pathway, alters gene expression, and relates to physiological reactions such as apoptosis and stress response (Aghdam et al., 2018). The level of extracellular ATP depends on environmental factors such as chilling stress. Chilling stress induces the increase in extracellular ATP level related to the occurrence of CI and leads to an imbalance of energy metabolism. As extracellular ATP is a strong charge molecule, it cannot independently diffuse across the plasma membrane to enter to the cytoplasm, as result damaging intracellular mitochondrial and chloroplastic ATP production system and exerting as a molecular switch for regulating intracellular ATP production (Aghdam et al., 2018). Chilling-induced deficiency of intracellular ATP and increase in extracellular ATP by dysfunction of membrane causes CI occurrence (Shan et al., 2022a, b). The homeostasis of extracellular ATP/intracellular ATP ratio indicates the release and transport of ATP inside cells and is regulated by internal and external factors. Extracellular ATP/intracellular ATP homeostasis induces cascade reactions of biochemical and physiological processes in CI occurrence (Shan et al., 2022a, b).

7.4.2.3 OXIDATIVE STRESS

Oxidative stress is generated when the homeostasis in plant cell becomes imbalanced by abiotic or biotic stress, results in damaging cell components (Madani et al., 2019). Exposure to low temperature increases oxidative stress and causes oxidative damage in horticultural products. Oxidative damage caused by oxidative stress is considered to be a preliminary reaction to low temperature in chilling-sensitive tissues (Imahori et al., 2008, 2021; Endo et al., 2019a, b; Kantakhoo and Imahori, 2021; Kantakhoo et al., 2022). Oxidative stress induces a disproportionate increase in ROS production in association with the appearance of CI in horticultural products. Excessive production of ROS in reaction to chilling stress causes the destruction in cell membrane of horticultural products (Imahori, 2014, 2021; Endo et al., 2019a, b; Kantakhoo and Imahori, 2021; Kantakhoo et al., 2022). The ROS are divided into free radical form and nonradical (molecular) form. Free radical form consists of superoxide radical (O_2^-), hydroxyl radical (·OH), perhydroxy radical ($HO_2^·$), and alkoxy radicals (RO·), while nonradical (molecular) form contains hydrogen peroxide (H_2O_2) and singlet oxygen ($^1O_2$) (Gill and Tuteja, 2010; Mittler, 2016). ROS contain activated atoms of oxygen and stimulate oxidative stress by oxidating the compounds (Madani et al., 2019). The production of ROS is usually maintained in all organisms at low level during normal life activity. But if the balance between the generation and detoxifying of ROS is disturbed by severe stresses and the ROS are produced at higher levels, these ROS accumulate, resulting in the loss of the function in cell membrane of plant (Gill and Tuteja, 2010; Ali et al., 2019). The ROS accumulation causes oxidative damage to cell membrane by the loss of cell membrane integrity that occurs with the peroxidation of lipid, the oxidation of protein, the inactivation of enzymatic activity, and destruction of DNA and RNA, and finally leads to cell death (Sevillano et al., 2009; Aghdam and Bodboadak, 2014; Madani et al., 2019).

Lipid peroxidation of cell membrane is considered in one phenomenon of the responses in CI appearance by cold stress (Hodges et al., 2004; Imahori, 2014, 2021; Endo et al., 2019a, b; Kantakhoo and Imahori, 2021; Kantakhoo et al., 2022). The equilibrium between the production and scavenging of ROS in plant cell is significant to the survival of plant under chilling stress. Low temperature changes the balance between the formation of ROS and defense mechanisms against ROS (Hodges et al., 2004; Imahori, 2014, 2021; Endo et al., 2019a, b; Kantakhoo and Imahori, 2021; Kantakhoo et al., 2022). If the imbalance by chilling stress becomes serious, plant cells will stop the functional property and eventually lose the integrity of plant cell. Accumulation of ROS may cause oxidative destruction to lipids in cell membrane, consequently producing toxic substances such as MDA, which is a final secondary metabolite in the oxidation of polyunsaturated fatty acid that evaluates the levels of membrane damage and is generally an indicator of oxidative stress (Endo et al., 2019a, b; Imahori et al., 2021; Kantakhoo and Imahori, 2021; Kantakhoo et al., 2022).

Scavenging system of ROS is important for maintaining the equilibrium between the formation and detoxifying of ROS. These systems consist of enzymatic and nonenzymatic antioxidative mechanisms. The former consists of antioxidant enzymes that involve superoxide dismutase, catalase (CAT), ascorbate (AsA) peroxidase, monodehydroascorbate reductase, dehydroascorbate reductase, and glutathione (GSH) reductase, whereas the latter system involves AsA and GSH (Imahori et al., 2016). In addition to nonenzymatically reacting between ROS and antioxidants, reducing substances compose an AsA-GSH cycle. This cycle plays a significant role to detoxify hydrogen peroxide in the plant (Endo et al., 2019a, b; Imahori et al., 2021; Kantakhoo et al., 2022). There is also another defense system that scavenges oxidative stress. Enhancement of alternative oxidase (AOX) gene expression and activation of AOX enzyme can accelerate ROS avoidance. The AOX pathway branches from the respiratory electron transport chain that contains mitochondria and is the final step that catalyzes ubiquinol oxidation. AOX diverts the excess electrons that are produced from the respiratory electron transport chain and utilizes them for catalyzing oxygen to water. If excessive reduction takes place in this respiratory electron transport chain, AOX mediates to inhibit the extreme reduction in ubiquinol and acts to avoid ROS accumulation (Aghdam, 2013; Aghdam and Bodbodak, 2013; Aghdam and Bodbodak, 2014).

7.5 ALLEVIATION OF CHILLING INJURY

Generally, low-temperature storage is the effectual postharvest technology that maintains the quality and prolongs the shelf life of horticultural products during storage and is most widely used in the world. Cold temperature decreases the metabolic velocity of horticultural products (Endo et al., 2019a, b; Imahori et al., 2021; Kantakhoo and Imahori, 2021; Kantakhoo et al., 2022). However, although low temperature can generally preserve the quality of horticultural products, depending on the cold tolerance of horticultural products and period of storage, chilling stress often causes severe CI, when being exposed to too low storage temperature below suitable storage temperature (Endo et al., 2019a, b; Imahori et al., 2021; Kantakhoo and Imahori, 2021; Kantakhoo et al., 2022). It has been shown that many approaches are increasing preventive methods of CI during low-temperature storage in horticultural products. These technologies involve physical treatment and chemical treatment methods by modifying the storage environment or treating the products (Yahia et al., 2019; Zhang et al., 2021a) (Table 7.2).

TABLE 7.2 Postharvest Treatments and Their Main Effects for Alleviation of Chilling Injury in Horticultural Products.

Treatment	Commodity	Storage	Main Effects	Reference
		Hot water treatment		
45°C for 15 min	Sweet pepper	6°C for 13 days	Increasing antioxidant enzymes	Endo et al. (2019a)
			Decreasing H_2O_2 and MDA contents	
45°C for 5 min	Mume	6°C for 4 weeks	Increasing antioxidant enzymes	Endo et al. (2019b)
			Decreasing H_2O_2 and MDA contents	
52°C for 5 min	Tomato	5°C for 2 weeks	Increasing lycopene, ascorbate, and phenolics content	Loayza et al. (2020)
45°C for 15 min	Zucchini	4°C for 15 days	Enhancing arginine pathways	Bokhary et al. (2020)
			Increasing polyamine and proline contents	
45°C for 10 min	Eggplant	1°C for 15 days	Enhancing phenolic metabolism	Kantakhoo et al. (2022)
			Decreasing H_2O_2 and MDA contents	
		Hot air treatment		
45°C for 3 h	Loquat	5°C for 35 days	Enhancing membrane integrity	Shao et al. (2013)
			Increasing antioxidant enzymes	
40°C for 4 h	Peach	1°C for 35 days	Regulating metabolism of sugar and organic acid	Zhou et al. (2021)
		Low-temperature conditioning treatment		
15°C for 10 days	Pomegranate	1°C for 2 weeks	Mitigating the changes in transcript associated with cold storage	Kashash et al. (2018)
			Modification of specific transcription factor, signaling, and stress-related genes	
13°C for 2 days	Eggplant	4°C for 9 days	Increasing activity and gene expressions of PPO and CAT	Shi et al. (2018)
			Decreasing MDA content	

TABLE 7.2 *(Continued)*

Treatment	Commodity	Storage	Main Effects	Reference
12°C for 5 days	Sweet potato	4°C for 35 days	Increasing activity and gene expression of AOX	Zhang et al. (2021c)
10°C for 2 days	Eggplant	4°C for 17 days	Improving phenolic antioxidants	Darré et al. (2022)
Intermittent warming treatment				
20°C for 1 day every 20 days	Pear	0°C for 120 days	Decreasing enzyme activity and gene expression of PLD	Wang et al. (2018)
			Decreasing MDA content	
20°C for 1 day on the 15th day	Pomegranate	2°C for 70 days	Increasing antioxidant enzyme activity	Taghipour et al. (2021)
Controlled atmosphere and modified atmosphere				
O_2 (14.9–16.7%) and CO_2 (4.2–7.3%)	Tomato	4°C for 14 days	Maintenance of relative humidity	Park et al. (2018)
			Delay of ripening and color development	
CO_2 (5%)	Guava	10°C for 3 weeks	Enhancing membrane integrity	Alba-Jiménez et al. (2018)
Ultraviolet treatment				
UV-B at 5 kJ m^{-2} for 4 h at 30°C	Mango	6°C for 10 days	Enhancing antioxidant system	Jiang et al. (2015)
UV-B at 5 kJ m^{-2} for 4 h at 29°C	Mango	6°C for 10 days	Enhancing endogenous nitric oxide production	Ruan et al. (2015)
UV-C at 0.03 kJ m^{-2}	Banana	8°C for 21 days	Maintaining membrane integrity	Pongprasert et al. (2011)
			Decreasing PPO activity	
UV-C at 1.5 kJ m^{-2}	Peach	1°C for 35 days	Enhancing emissions of aroma volatiles	Zhou et al. (2019)
Polyamine treatment				
1 and 2 mM putrescine	Blood orange	5°C for 60 days	Increasing total phenolic and ascorbate contents	Habibi and Ramezanian (2017)
			Enhancing antioxidant activity	
2 mM putrescine	Okra	4°C for 16 days	Increasing total phenolic compounds content	Phornvillay et al. (2019)
0.5 mM putrescine plus 0.5 mM spermidine	Jujube	−2°C for 35 days	Decreasing MDA accumulation	Zhang et al. (2020b)
0.5 mM putrescine plus 0.5 mM spermidine	Lemon	−2°C for 45 days	Enhancing POD and PPO activities	Kowaleguet et al. (2022)
Jasmonate treatment				

TABLE 7.2 *(Continued)*

Treatment	Commodity	Storage	Main Effects	Reference
50 μM methyl jasmonate	Blood orange	3°C for 150 days	Enhancing total antioxidant activity	Habibi et al. (2020)
			Increasing ascorbate content	
5 mM methyl jasmonate	Pomegranate	2°C for 90 days	Maintaining membrane integrity	García-Pastor et al. (2020)
			Enhancing antioxidant activity	
50 μM methyl jasmonate	Pepper	2°C for 25 days	Enhancing antioxidant activity	Seo et al. (2020)
			Reducing seed browning	
30 μM jasmonate	Peach	4°C for 35 days	Regulating ethylene and sugar metabolism	Zhao et al. (2021a)
		Salicylates treatment		
0.1, 0.5, and 1.0 mM acetyl salicylate	Pomegranate	2°C for 84 days	Enhancing antioxidant activity	Sayyari et al. (2011)
			Maintaining total phenolics contents	
0.05 mM methyl salicylate plus hot water (45°C for 15 min)	Sweet pepper	5°C for 20 days	Enhancing antioxidant enzyme activities	Rehman et al. (2021)
		Combination treatment of jasmonate and salicylate		
10 μM methyl jasmonate plus 2 mM methyl salicylate	Lemon	4.5 °C for 28 days	Increasing antioxidant enzyme activities	Siboza et al. (2017)
			Enhancing heat shock protein accumulation	
100 μM methyl jasmonate plus 100 μM methyl salicylate	Avocado	2°C for 21 days	Maintaining membrane integrity	Glowacz et al. (2017)
			Decreasing lipoxygenase gene expression and enzyme activity	
		Nitric oxide treatment		
60 μL L^{-1} nitric oxide for 3 h at 25°C	Hami melon	1 ± 0.5°C for 49 days	Enhancing antioxidant enzyme activities	Zhang et al. (2017)
25 μM sodium nitroprusside	Zucchini	4°C for 14 days	Enhancing antioxidant enzyme activities	Jiménez-Muñoz et al. (2021)

TABLE 7.2 *(Continued)*

Treatment	Commodity	Storage	Main Effects	Reference
			Maintaining cellular redox homeostasis	
10 μM sodium nitroprusside	Peach	0°C for 35 days	Regulating cell wall and lipid metabolism	Zhao et al. (2021b)
		Hydrogen sulfide treatment		
3 mM hydrogen sulfide	Eggplant	7C for 21 days	Enhancing ROS scavenging system	Barzegar et al. (2021)
3 mM hydrogen sulfide	Persimmon	2C for 45 days	Accumulating phenols and flavonoids	Niazi et al. (2021)
10 μM hydrogen sulfide	Peach	1°C for 49 days	Maintaining antioxidants	Wang et al. (2022)
			Suppressing cell-wall degradation	
		Melatonin treatment		
100 μM melatonin	Peach	1°C for 28 days	Decreasing MDA content and LOX activity	Gao et al. (2016)
			Maintaining higher levels linolenic acid and linoleic acid	
100 μM melatonin	Pomegranate	4°C for 120 days	Enhancing antioxidant enzyme activities	Jannatizadeh (2019)
100 μM melatonin	Tomato	4°C for 28 days	Enhancing arginine pathways	Aghdam et al. (2019)
100 μM melatonin	Bell pepper	4°C for 20 days	Decreasing membrane lipid peroxidation	Kong et al. (2020)
			Activating antioxidant defense system	
100 μM melatonin	Kiwifruit	0°C for 100 days	Inhibiting activity of lignin metabolism enzymes	Jiao et al. (2022)
			Increasing antioxidant enzyme and antioxidant substance	
		Ethylene treatment		
100 μL L^{-1} ethylene for 24 h	Pear	0°C for 30 days	Enhancing antioxidant activity	Wei et al. (2019)
			Promoting proline accumulation	
Continuous 14 μL L^{-1} ethylene	Peach	0°C for 59 days	Maintaining membrane integrity	Chen et al. (2021)
		1-methylcyclopropene treatment		

TABLE 7.2 *(Continued)*

Treatment	Commodity	Storage	Main Effects	Reference
5 μL L^{-1} 1-methylcyclopropene plus intermittent warming (two cycles of 1 day by exposing to 20°C)	Peach	2°C for 14 days	Enhancing antioxidant activity	Liu et al. (2018a)
multi-high dose 1-methylclopropene (1μL L^{-1})	Nectarine	0°C for 35 days	Enhancing energy metabolism	Zhang et al. (2020c)
		γ-aminobutyric acid treatment		
1 mM γ-aminobutyric acid	Zucchini	4°C for 14 days	Enhancing energy metabolism	Palma et al. (2019)
		Oxalic acid treatment		
5 mM oxalic acid for 10 min	Kiwifruit	0°C for 90 days	Enhancing antioxidant activity	Liang et al. (2017)
			Enhancing energy metabolism	
5 mM oxalic acid for 10 min plus 1 μL L^{-1} 1-methylclopropene	Sweet persimmon	1°C for 70 days	Enhancing antioxidant activity	Li et al. (2018)
		Calcium ion treatment		
1% $CaCl_2$	Loquat	1°C for 35 days	Enhancing energy metabolism	Li et al. (2020)
			Accumulation of endogenous osmotic substances	
6% $CaCl_2$	Green pepper	4 °C for 20 days	Delaying degradation of membrane lipid	Zhang et al. (2021b)
			Maintaining cell membrane stability	

AOX, alternative oxidase; ATP, adenosine-5`-triphosphate; CAT, catalase; CBF, C-repeat binding factor; MDA, malondialdehyde; PLD, phospholipase D; PO, polyphenol oxidase; POD, peroxidase; ROS, reactive oxygen species.

7.5.1 HEAT TREATMENT

Postharvest heat treatments in horticultural products are utilized for insect control, disease control, adjustment of maturity, modification of reactions to stresses such as chilling stress, and reduction of quality loss during low-temperature storage (Paull and Chen, 2000; Lurie and Pedreschi, 2014). These heat conditioning treatments cause an adaptable response of chilling stress against horticultural products, and reduce CI incidence and development, and alleviate external damage during low-temperature storage (Fallik, 2004; Kantakhoo and Imahori, 2021; Kantakhoo et al., 2022). Horticultural products are treated by heat medium like hot water, vapor heat, or hot air in previous cold storage (Fallik, 2004; Endo et al., 2019b; Kantakhoo et al., 2022). The appropriate treatment temperature and exposure duration dependent on the type, mature stage, size, and conditions for cultivation of commodities are empirically determined every crops. Treatment conditions widely vary at treatment temperature from 30 to 40°C for treatment times ranging from several hours to days in hot air treatment, and at 50–56°C for several minutes in hot water treatment, and additionally temperature of up to 63°C for shorter times (10–25 s) in hot water rinsing and brushing (Lurie and Pedreschi, 2014). Hot water is more chosen for heat medium because of a more effective heat transferred medium than hot air (Endo et al., 2019a, b; Kantakhoo and Imahori, 2021; Kantakhoo et al., 2022). Mature green mume fruit treated with hot water at 45°C for 5 min significantly increased tolerance of fruit to CI at 6°C storage for 4 weeks (Endo et al., 2019b). Brief hot water treatment at 52°C for 5 min considerably alleviated symptoms of CI in tomato fruit during cold storage at 5°C for 2 weeks (Loayza et al., 2020). Zucchini squashes treated with hot water treatment at 45°C for 15 min increased chilling tolerance and decreased CI development at 4°C storage for 15 days (Bokhary et al., 2020). Hot water treatment at 45°C for 10 min deceased the CI development in treated eggplant fruit during cold storage at 1°C storage for 15 days (Kantakhoo et al., 2022). On the other hand, postharvest hot air treatment at 40°C for 4 h alleviated CI in peach fruit at 1°C storage for 35 days (Zhou et al., 2021) (Table 7.2).

Heat stress caused by heat treatment influences stabilities of proteins, cellular membranes, nucleic acid and cytoskeleton, and changes reactive functions of enzymes and raises the metabolic imbalance in horticultural products (Lurie and Pedreschi, 2014; Madani et al., 2019). Thermal tolerance is related to heat shock proteins (HSPs), and its induction depends on protein production (Imahori, 2014). The heat stress response is observed in most organisms as initiation or increase of HSPs biosynthesis and is predominantly regulated on the transcriptional levels (Ferguson et al., 2000; Wang et al., 2003). These HSPs prevent irreversible denaturation and destruction of proteins that would be injurious to the cellular integrity, conferring heat tolerance to organisms (Ferguson et al., 2000; Paull and Chen, 2000; Imahori, 2014). The HSPs are stress-responsive protein that has the molecular weights of range between 15 and 115 kDa and classified with five families consisting in HSP60s (chaperonins), HSP70s, HSP90s, HSP100s, and small HSPs that have low molecular weight (Aghdam et al., 2013; Aghdam and Bodbodak, 2014).

Heat treatment may encourage oxidative stress in horticultural products. Oxidative stress arises when ROS in horticultural products is excessively produced more than the capacity maintaining cellular redox homeostasis or scavenging ROS. Consequently, ROS damages the membrane and protein of cell in horticultural products (Paull and Chen, 2000; Hodeges et al., 2004; Vicente et al., 2006). Reduction in antioxidant levels in plant tissues subjected to stress results in higher ROS levels that may cause cellular injury such as CI. However, moderate heat stress may initiate cellular signaling pathways and response in plants, as a result to contribute to the promotion of ROS scavengers (Wang et al., 2003). Protective function contrary to oxidative stress is a significant element in deciding the survival of horticultural products against heat stress. Horticultural products have developed both enzymatic and nonenzymatic antioxidants to detoxify ROS with object to protect them from oxidative stress (Imahori, 2014). Consequently, the nondamage heat treatment causes a moderate stress in the horticultural products that are activated antioxidant responses. Moderate heat treatment can reduce the effects of oxidative stress in some horticultural products before low-temperature storage (Lurie, 1998; Hodges et al., 2004).

7.5.2 LOW-TEMPERATURE CONDITIONING TREATMENT

Low-temperature conditioning treatment is an alternative storage technology and consists in the temperature differences between conditioning and cold storage, and the period of the treatment (Imahori, 2012; Zhang et al., 2021a). This is the treatment for increasing tolerance to cold storage temperatures in horticultural products and involves holding chilling-sensitive products at temperatures just above a critical temperature that CI occurs, to provoke tolerance to normally injurious cold temperatures (Imahori, 2012; Zhang et al., 2021a).

Low-temperature conditioning treatments (12°C for 5 days) effectively alleviated CI of sweetpotato roots during storage for 35 days at 4°C (Zhang et al., 2021c). American eggplant treated by low-temperature conditioning (10°C for 2 days) alleviated CI and improved phenolic antioxidants during cold storage at 4°C for 17 days (Darré et al., 2022) (Table 7.2).

7.5.3 INTERMITTENT WARMING TREATMENT

Intermittent warming treatment is made up of the periodic transfer of horticultural products from low-temperature storage to short periods of warming temperature storage. This treatment enhances the cold resistance and is effective for alleviating CI in horticultural products (Sevillano et al., 2009; Biswas et al., 2016). For applying intermittent warming, it is important to find the optimal conditions such as temperature and duration. The optimal conditions change with products and their cultivars (Sevillano et al., 2009; Biswas et al., 2016). Intermittent warming treatment must apply before CI becomes irreversible and when CI development is still reversible (Sevillano et al., 2009). In pear fruit, intermittent warming treatment by exposing to 20°C for 1 day every 20 days under cold storage at 0°C for 120 days and shelf-life duration at 20°C for 15 days alleviated CI of peel browning and showed higher firmness by increasing oxidative phosphorylation, and decreasing enzyme activity and gene expression of PLD (Wang et al., 2018). Pomegranate fruit stored at 2°C for 70 days decreased CI damage and enhanced antioxidant enzymes activity by intermittent warming treatment at 20°C for 1 day on the 15th day (Taghipour et al., 2021) (Table 7.2).

7.5.4 CONTROLLED ATMOSPHERE STORAGE AND MODIFIED ATMOSPHERE TREATMENT

Controlled atmosphere (CA) and modified atmosphere (MA) using decreased O_2 gas and/or increased CO_2 gas are well known for maintaining the quality and prolonging shelf life in horticultural crops. The compositional CA treatment gas is usually monitored and differences from the set points are corrected. MA treatment differs in that they are not dynamically adjusted, and the gas composition of MA results from a balance between gas consumption and/or production in horticultural crops themself and gaseous diffusion through a permeable membrane. The advantageous effects of CA and MA treatment indicate deletion of ripening, reduction in physiological and pathological disorders such as CI, and the possibility in disinfection of horticultural products (Imahori, 2012). CA condition with elevated CO_2 (5%) decreases CI levels and increases shelf life by maintenance of membrane functionality for guava fruit (Alba-Jiménez et al., 2018). The MA condition with reduced O_2 (14.9–16.7%) and elevated CO_2 (4.2–7.3%) for tomato fruit prevents the CI symptoms and maintains quality during storage for 14 days at 4°C (Park et al., 2018) (Table 7.2). The hypoxia pretreatment with 3% O_2 at room temperature for 24 h decreased the symptoms of CI in avocado fruit during storage at 2°C for 3 weeks. The pretreated fruit indicated lower respiration rate, ethylene level, and ion leakage than control. Pretreatment with 40% CO_2 for 3, 6, or 9 days decreased CI symptoms of lemon fruit after 30, 60, and 90 days of storage at 0°C (Sevillano et al., 2009).

Packaging of horticultural products with plastic films generates reduced O_2 and elevated CO_2 inside the atmospheres of packaging by their respiration and the gaseous diffusion through the package films, and results in MA conditions. The MA packaging with plastic films reduces levels of respiration and ethylene, and delays the maturation and senescence of horticultural products, and maintains high relative humidity around the horticultural products, and consequently prevents water loss of the horticultural products (Sevillano et al., 2009; Wang, 2010). The maintenance of highly relative humidity prevents the CI development during low-temperature storage in horticultural products such as zucchini (Zuo et al., 2021) and tomato (Park et al., 2018). Prevention of water loss from the plant tissues relates to inhibition of breakdown in epidermal and underlying cells and promotes the reduction in pitting formation during cold storage in horticultural products (Wang, 2010).

7.5.5 ULTRAVIOLET TREATMENT

Ultraviolet (UV) light induces various biological stress and protection mechanisms to plants. It is divided into long: UV-A (315–400 nm), medium: UV-B (280–315 nm), and short-wave UV-C (100–280 nm) according to wavelength and indicates the different effects among these ranges (Imahori, 2012; Madani et al., 2019). UV irradiation induces delay of growth in plants and changes metabolic processes like photosynthesis and respiration. UV light affects the plant physiological processes and induces deleterious effects such as reduction in protein synthesis, impairment of chloroplast, and destruction of DNA (Imahori, 2012; Madani et al., 2019). Damages by UV irradiation access to accelerate ROS production, inducing in oxidative stress. However, plants have various antioxidative strategies scavenging ROS and can enhance the tolerance to UV as increasing antioxidant defense (Wang et al., 2003). Plant secondary metabolism such as phenolic metabolism is also promoted along with the enzyme antioxidant system in plant (González-Aguilar et al., 2010). In addition, as low doses of UV irradiation can inspire the vital processes in plant and cause the positive alterations in homeostasis of plant, UV irradiation has been utilized for maintaining the quality and prolonging the shelf-life in horticultural products (Imahori, 2012; Madani et al., 2019).

Short irradiation of UV-B at 5 kJ m^{-2} to mango fruit for 4 h at 30°C reduced oxidative stress and alleviated CI in fruit by enhancing the antioxidant systems during cold storage at 6°C (Jiang et al., 2015). Similarly, UV-B irradiation at 5 kJ m^{-2} to mango fruit for 4 h at 29°C also enhanced chilling tolerance and reduced CI occurrence by enhancing endogenous nitric oxide production in fruit during storage at 6°C (Ruan et al., 2015). The exposure of peach fruits with short irradiation of UV-C alleviated CI during low-temperature storage at 5°C plus shelf life for 7days, and increased polyamine levels. As a result, UV-C treatment enhanced the tolerance of fruit tissue to quality degradation and CI (González-Aguilar et al., 2004). The UV-C irradiation with 1.5 kJ m^{-2} to peach fruit reduced CI, and increased emissions of aroma volatiles such as esters and lactones in peach during cold storage at 1°C (Zhou et al., 2019). UV irradiation enhanced antioxidant system of horticultural products and activated their defense system of oxidative stress, and as a result reduced the secondary damage of oxidative stress during cold storage (Zhang and Jiang, 2019) (Table 7.2).

7.5.6 CHEMICAL TREATMENTS

Many attempts have been made for conferring CI alleviation in horticultural products during low-temperature storage by chemical treatments, for example polyamine treatment, jasmonate treatment, salicylate treatment, nitric oxide treatment, and melatonin treatment. Chemical substances involve plant hormones and natural elicitors.

7.5.6.1 POLYAMINE TREATMENT

Polyamines (PAs) are plant growth regulators, and low molecular weight, and present in the form of aliphatic amines in plant. The three common forms in PAs consist of putrescine (butan-1,4-diamine, diamine), spermidine

(*N*-(3-aminopropyl) butane-1,4-diamine, triamine), and spermine (*NN'*-bis-(3-aminopropyl) butane-1,4-diamine, tetramine) (Phornvillay et al., 2019; Zhang et al., 2021a). PAs ubiquitously exist in all organisms containing plants. In plants, they play an important role in various physiological and metabolic processes, and function related with growth and development in plant such as division and development of cell, dormancy breaking, fruit set and growth, and fruit ripening and senescence (Valero et al., 2002; Groppa and Benavides, 2008). In addition, PAs respond to abiotic stress, and induce defense mechanisms to various stresses, as a result decreasing the damage to plant cells affected by abiotic stress (Sevillano et al., 2009; Zhang et al., 2021a). PAs interact with DNA and RNA, phospholipids and proteins due to their polycationic nature charged at physiological pH. The binding of PAs with negatively charged membrane phospholipid components or anionic sites on cellular membranes stabiles cellular membranes under stress situation, thus delaying the destruction of cellular membranes (Valero et al., 2002; Sevillano et al., 2009). Second, PAs have antioxidant capacity, thus protecting cell membranes against lipid peroxidation by cold stress (Sevillano et al., 2009; Gupta et al., 2013). Postharvest application of PAs could alleviate CI development during cold storage in horticultural products (Zhang et al., 2021a). Okra pods were dipped into 2 mM putrescine and stored at low temperature (4°C); thus, putrescine treatment increased total phenolic compounds, total antioxidant activity, and antioxidant enzymes (CAT and SOD), reduced CI such as seed browning during storage (Phornvillay et al., 2019). Treatments with 0.5 mM putrescine plus 0.5 mM spermidine maintained the quality of jujube fruit by decreasing incidence in CI and MDA accumulation during cold storage at −2°C for 35days (Zhang et al., 2020b). Lemon fruit treated 1 mM putrescine, 1 mM sperimidine, or 0.5 mM putrescine plus 0.5 mM sperimidine reduced CI incidence and maintained the quality by higher peroxidase (POD) and PPO activities during low-temperature storage (−2°C, 45 days) (Kowaleguet et al., 2022) (Table 7.2).

7.5.6.2 JASMONATE TREATMENT

Jasmonates are one of plant volatiles and significant plant hormones. They act as protection system in plant development process as signaling molecular in plant cellular reactions and modulation of other plant hormone synthesis (Wang et al., 2021; Zhang et al., 2021a, Hasan et al., 2025). They are synthesized in the linolenic acid pathway on chloroplast, peroxisome, and cytoplasm (Seo et al., 2020; Hasan et al., 2025). Methyl jasmonate (MeJA) is the derivative of jasmonates and synthesized in the octadecanoid pathway from lineolate (Wang et al., 2021).

Alleviation of CI was attributed to antioxidant capacity and membrane integrity promoted by MeJA (Wang et al., 2021). Blood orange fruit with 50 μmol L^{-1} MeJA treatment maintained fruit quality and indicated higher total antioxidant activity and ascorbate content than control fruit during low-temperature storage at 3°C for 150 days (Habibi et al., 2020). Preharvest 5 mM MeJA treatments alleviated CI and enhanced the cold storability of pomegranate fruit by maintaining membrane stability and level of bioactive compounds related to antioxidant activity (García-Pastor et al., 2020). White-fleshed peach fruit immersed with 30 μmol L^{-1} JA treatment increased chilling tolerance during cold storage at 4°C for 35 days by regulating ethylene and sugar metabolism (Zhao et al., 2021a) (Table 7.2).

7.5.6.3 SALICYLATES TREATMENT

Salicylates are phenolic molecule and significant plant hormones and are synthesized in the shikimic acid pathway on the chloroplast (Rasouli et al., 2019; Seo et al., 2020). Their derivatives are acetyl salicylate and methyl salicylate (MeSA) (Aghdam and Bodbodak, 2013). As a signaling molecule, salicylates play a significant function in controlling many physiological processes such as systemic resistance, and they induce the defense compounds like pathogenesis-related proteins and polyphenols (Rasouli et al., 2019; Ali et al., 2023)

Pomegranates treated with acetyl salicylate reduced CI and maintained fruit quality by enhancing antioxidant capacity during cold storage for 84 days at 2°C plus a subsequent period of 4 days at 20°C (Sayyari et al., 2011). The combination of 2 mM salicyliate and 30% *Aloe vera* gel reduced CI and decay and maintained the quality of orange fruit during storage for 80 days at 4°C (Rasouli et al., 2019). In pepper fruit treated by MeJA or MeSA, 50 μM MeJA reduced seed browning and CI symptom during storage at 2°C for 25 days, and 250 μM MeSA induced chilling sensitivity in fruit (Seo et al., 2020) (Table 7.2).

7.5.6.4 NITRIC OXIDE TREATMENT

Nitric oxide is a stress signaling compound and plays an important role as several defense reactions to both biotic and abiotic stresses, and responds as an antioxidant for ROS scavenging to prevent oxidative stress. It also acts as plant growth signal for plant physiological functions (Zhang et al., 2020a; Jiménez-Muñoz et al., 2021). Fumigation of sweet orange fruit with 10 μL L^{-1} nitric oxide gas reduced CI and maintained fruit quality during storage at 4°C for 90 days (Rehman et al., 2019). Pretreatment in Hami melon fruit with nitric oxide at 60 μL L^{-1} for 3 h at 25°C delayed CI incidence and alleviated CI during storage at 1 ± 0.5°C for 7 days by increasing antioxidant enzyme activities and upregulating CmCBF1 and CmCBF3 genes expression (Zhang et al., 2017). Treatment of 25 μM sodium nitroprusside, nitric oxide donor, decreased weight loss and CI incident and improved the quality in zucchini fruit during storage at 4°C by higher antioxidant enzyme activity, and mentation of cellular redox homeostasis (Jiménez-Muñoz et al., 2021). Peach fruit treated with 10 μM sodium nitroprusside alleviated CI during cold storage at 0°C for 35 days by regulating cell wall and lipid metabolism (Zhao et al., 2021b) (Table 7.2).

7.5.6.5 HYDROGEN SULFIDE TREATMENT

Hydrogen sulfide is a flammable and colorless gas like nitric oxide and an important gaseous transmitter for plant physiology (Ali et al., 2019; Ali et al., 2022). Its gas is also an imperative signaling molecule that plays significant roles in physiological actions in plant (Ali et al., 2022). Treatment with 3 mM hydrogen sulfide significantly decreased CI development in eggplant fruits during storage at 7°C for 21 days by the enhancement of ROS scavenging system (Barzegar et al., 2021). Persimmon fruit exposed to 3 mM hydrogen sulfide alleviated CI and maintained fruit quality during storage at 2°C for 45 days by higher accumulation of phenols and flavonoids (Niazi et al., 2021). Hydrogen sulfide treatment (10 μM) mitigated CI of refrigerated peach fruits at 1°C for 49 days by maintaining antioxidants and suppressing cell-wall degradation (Wang et al., 2022) (Table 7.2).

7.5.6.6 MELATONIN TREATMENT

Melatonin is N-acetyl-5-methoxytryptamine, a kind of indoleamine, and is an indole derivative of tryptophan, and is well known as an animal hormone causing many physiological processes (Kong et al., 2020; Jayarajan and Sharma, 2021). Melatonin is also found in plants, and widely contained in various plant organs (Zhang et al., 2020a). In addition to several plant physiological actions as darkness signaling and plant growth regulator, melatonin has an important role in antioxidant activity as the protection to abiotic stress like oxidative stress (Gao et al., 2016; Reiter et al., 2015). Treatment with 100 μM melatonin alleviated CI in pomegranate fruit during storage at 4°C for 120 days by decreasing electrolyte leakage and reducing the MDA and H_2O_2 accumulation and enhancing the activities of ROS scavenging enzymes (Jannatizadeh, 2019). Green bell pepper treated with melatonin (100 μM) alleviated CI during cold storage at 4°C for 20 days by decreasing the peroxidation of membrane lipid and enhancing the activation of antioxidant defense system (Kong et al., 2020). Melatonin treatment (100 μM) reduced CI development and inhibited the lignin metabolism enzyme activities and the expression of structural genes expression while enhancing the antioxidant enzyme activities and the antioxidant substance level during cold storage of kiwifruit fruit at 0°C for 100 days (Jiao et al., 2022) (Table 7.2).

7.5.6.7 ETHYLENE TREATMENT AND 1-METHYLCYCLOPROPENE TREATMENT

Ethylene plays a significant role in the maturation and senescence in horticultural products. Ethylene synthesis is essential for fruit ripening of climacteric fruit (Wei et al., 2019; Chen et al., 2021). Ethylene treatment exposing at 100 μL L^{-1} for 24 h alleviated CI development without significantly affecting fruit quality and enhanced proline accumulation and antioxidant activity in pear fruit during cold storage at 0°C for 30 days (Wei et al., 2019) (Table 7.2).

1-methylcyclopropene (1-MCP), an ethylene inhibitor, has been used to prolong the shelf life and reduce CI in horticultural products (Zhang et al., 2020c). Multiple 1-MCP treatment enhances cold tolerance and alleviates CI in nectarine fruits during cold storage at 0°C for 35 days (Zhang et al., 2020c) (Table 7.2).

7.5.6.8 OTHER CHEMICAL TREATMENTS

γ-Aminobutyric acid (GABA) is classified in a non-proteinogenic amino acid and acts in growth and development of plant and accumulates under various abiotic stresses like temperature, dehydration, hypoxia, and mechanical injury (Palma et al., 2019). Treatment with 1 mM GABA alleviated CI development, weight loss, cell death, and electrolyte leakage and improved fruit quality of zucchini fruit during storage at 4°C for 14 days (Palma et al., 2019). During cold storage, GABA functioned as inhibiting MDA during lipid oxidation by oxidative stress, maintaining the integrity of membrane, and enhancing antioxidant and GABA-shunt pathway (Palma et al., 2019) (Table 7.2).

Oxalic acid is dibasic acid and has a strong chelating ability and acts as a positive abiotic elicitor (Li et al., 2018; Zhang et al., 2021a). Combined treatment of 5 mM oxalic acid and 1 $\mu L\ L^{-1}$ 1-MCP reduced CI incidence and inhibited chilling-induced membrane destruction in sweet persimmon fruit during storage at 1 ± 0.5°C for 70 days by suppressing phenolics oxidation (Li et al., 2018) (Table 7.2).

Calcium ion plays a significant role for growth and development in plant as a signal substance and connects with calmodulin for the regulation of plant physiological and biochemical reaction (Ranty et al., 2016). Free calcium ion in cytoplasm increases by abiotic stress and combines with calmodulin. Consequently, calcium ion causes plant physiological and biochemical processes and changes their metabolisms (Li et al., 2020). Immersion treatment of 1% $CaCl_2$ solution enhanced chilling resistance in loquat fruit during cold storage at 1°C for 35 days due to the high energy level and increase in endogenous osmotic substances like proline, GABA, and PA (Li et al., 2020). Green pepper fruits treated with 6% $CaCl_2$ solution alleviated CI, delayed the membrane lipid degradation, and maintained cell membrane stability by decreasing the expression of phospholipase C gene and diacylglycerol kinase gene (Zhang et al., 2021b) (Table 7.2).

7.5.7 ALLEVIATION MECHANISMS OF CHILLING INJURY

Postharvest treatments for the mitigation of CI that causes a string of structural, physiological, biochemical, and molecular alterations benefit horticultural products under low temperature storage and lead to phytochemical accumulation by promoting the increase or decrease in enzyme activity and gene expression in key enzymes of metabolic pathways (Imahori, 2012, 2014). Exposing a moderate stress to horticultural products not only induces the tolerance to same kind of more severe stress, but also can progress resistance against other stresses. Such cross-protection has been indicated for different types of stress like chilling tolerance induced by postharvest treatment (Imahori, 2012, 2016; Madani et al., 2019).

Postharvest treatments maintain the structure in membrane and increase the integrity of membrane. Below low temperature, the enhancement of PLD and LOX activities in horticultural products leads to the degradation in unsaturated fatty acid and decrease in unsaturated fatty acid/saturated fatty acid ratios and induces lipid peroxidation, and reduces membrane integrity (Aghdam and Bodbodak, 2014; Madani et al., 2019; Zhang et al., 2021a). However, heat treatments such as hot water or hot air can alleviate CI and maintain fruit quality in products during cold storage by decreasing the activities of PLD and LOX and enhancing the unsaturated fatty acid/saturated fatty acid ratios (Aghdam and Bodbodak, 2014; Zhang et al., 2021a). Melatonin treatment also alleviated CI, decreased MDA level and LOX activity, maintained higher ratio of unsaturated fatty acid/saturated fatty acid, and improved phenolic acid accumulation during low-temperature storage of horticultural products (Zhang et al., 2020a). HSPs are a stress-responsive protein and accumulate quickly in response to heat shock such as high temperature in plants. Synthesis and accumulation of HSPs are proportional to the temperature and duration of heat stress. The HSPs accumulation not only relates to defense for the stress but also against consequent stressful conditions. The postharvest stress can influence the ability of protein biosynthesis in horticultural crops, and as a result, affect in greater or lesser synthesizing the proteins present and new protein, HSPs (Imahori, 2014). HSPs protect plant cells against stress by the function of molecular chaperones, and help in folding, assembly, and transport of protein, and targeting in damaging proteins for proteolysis. Thus, they can also support protection against the plant cells in chilling stress conditions. Postharvest treatments like hot water treatment and MeJA and MeSA treatments induce the HSPs gene expression and accumulate HSPs protein and prevent or alleviate their CI during storage at low temperature (Aghdam et al., 2013; Aghdam and Bodbodak, 2014; Si et al., 2022).

Oxidative damage is considered to be an initial reaction in chilling-sensitive plants to chilling stress. Chilling stress may change the balance between production and defense mechanisms in ROS such that oxidatively promoted CI is detected (Imahori, 2014, 2021; Endo et al., 2019a, b; Kantakhoo and Imahori, 2021; Kantakhoo et al., 2022). The equilibrium between ROSs production and their detoxification is critical to plant survival under cold storage. Promotion of some scavenging enzymes under cold storage is the result of an introduction of these enzymes through ROS production such as $O_2^{\cdot-}$ and H_2O_2 (Imahori, 2014, 2021; Endo et al., 2019a, b; Kantakhoo and Imahori, 2021; Kantakhoo et al., 2022). Various postharvest treatments such as hot water treatment, 1-MCP treatment, and MeJA treatment induce antioxidant system; cold prevents or alleviates their CI during low-temperature storage. The antioxidant activity involves in the neutralization of free radicals, the increase in enzymatic and nonenzymatic antioxidative activities, the protection of intercellular antioxidant enzymes, and the inhibition of free radical production (Imahori, 2014, 2021; Endo et al., 2019a, b; Kantakhoo and Imahori, 2021; Kantakhoo et al., 2022).

Arginine is one of amino acids and is used not only a nitrogen supply for protein synthesis, but also a precursor for signal molecules such as PA, nitric oxide, and proline. PA or proline has a significant role as CI alleviation of horticultural products under low-temperature storage (Aghdam and Bodbodak, 2013, 2014; Zhang et al., 2021a). Hot water treatment alleviates CI under cold storage by enhancing arginine pathway and PA and proline accumulations, consequently enhancing the cold tolerance (Bokhary et al., 2020). Sugar metabolism has an important role in many plant metabolisms and provides substances for various metabolic processes. In addition, sugar metabolism relates with the synthesis of antioxidant substances (Aghdam and Bodbodak, 2014; Zhang et al., 2021a). C-repeat binding factor (CBF), transcription factor, has a critical role in response against chilling stress in plants. As plants are exposed to cold stress, CBF identifies and binds to the C-repeat/dehydration-responsive DNA regulatory elements and activates chilling-responsive genes expression (Aghdam and Bodbodak, 2013). Melatonin treatment alleviates CI during low-temperature storage by upregulating *SIZAT2/6/12* increasing to trigger CBF1 gene expression (Aghdam et al., 2019). Additionally, MeJA treatment, nitric oxide treatment, and low-temperature conditioning induce the gene expression of CBF and prevent or alleviate their CI during low-temperature storage (Zhang et al., 2021; Jiao, 2021; Fu et al., 2022).

7.6 CONCLUSION

Fresh horticultural products are perishable commodities with actively metabolic actions, as they are living organisms subject to continuous changes also after being harvested, resulting in a shorter shelf life by reason of rapid deterioration of quality after harvest when stored under ambient conditions. Therefore, keeping or improving quality and shelf life of horticultural products is required during postharvest. Maintaining the quality of products, low-temperature storage is generally the more enabled technique and can considerably reduce many metabolic process rates. It is also the most extensively utilized postharvest technology and can efficiently prolong the storability of horticultural products. However, low temperature causes physiological disorder, namely chilling and freezing injury, in products when exposing to or storing at temperature under a specific threshold temperature. Depending on the tolerance of low temperature in products and storage period, a too lower storage temperature can cause physiological disorder, especially in subtropical and tropical origin crops. Therefore, physiological disorder incidence restricts the utilization of low-temperature storage. The influence of physiological disorder may cause the serious economic losses on agricultural and food industries. Considering the physiological, biochemical, and molecular mechanisms related to the development of this physiological disorder can permit the strategies to avoid or delay the incidence of physiological disorder. This chapter describes the mechanisms of appearance and symptom in low-temperature physiology disorder which cold stress exerts. The development of physiology disorder contributes to differences of sensitivity of tolerance to chilling stress, and depends on maturity stage, plant tissues, and environmental conditions. At the same time, several strategies alleviating CI are shown in this chapter. The effectiveness of these technologies in mitigating CI varies with products, and is affected by the stage of maturity, physiological and environmental conditions. Additionally, these treatment conditions are specific for species or cultivar. Further study is necessary for a proper comprehension of the physiological, biochemical, and molecular machinery of low-temperature physiological disorder. This information will be important in the improvement of physiological disorder alleviated treatments to avoid low-temperature-induced postharvest losses of horticultural products for practical commercial use.

KEYWORDS

- **cell membrane**
- **chemical treatment**
- **chilling injury**
- **chilling stress**
- **energy metabolism**
- **freezing injury**
- **heat treatment**
- **oxidative stress**
- **physiological disorder**

REFERENCE

Aghdam, M. S. (2013). Role of alternative oxidase in postharvest stress of fruit and vegetables: Chilling injury. *African Journal of Biochemistry, 12*, 7009–7016.

Aghdam, M. S., & Bodbodak, S. (2013). Physiological and biochemical mechanisms regulating chilling tolerance in fruits and vegetables under postharvest salicylates and jasmonates treatments. *Scientia Horticulturae, 156*, 73–85.

Aghdam, M. S., & Bodbodak, S. (2014). Postharvest heat treatment for mitigation of chilling injury in fruits and vegetables. *Food Bioprocess Technology, 7*, 37–53.

Aghdam, M. S., Jannatizadeh, A., Luo, Z., & Paliyath, G. (2018). Ensuring sufficient intracellular ATP supplying and friendly extracellular ATP signaling attenuates stresses, delays senescence and maintains quality in horticultural crops during postharvest life. *Trends in Food Science and Technology, 76*, 67–81.

Aghdam, M. S., Luo, Z., Jannatizadeh, A., Sheikh-Assadi, M., Sharafi, Y., Farmani, B., Fard, J. R., & Razavi, F. (2019). Employing exogenous melatonin applying confers chilling tolerance in tomato fruits by upregulating *ZAT2/6/12* giving rise to promoting endogenous polyamines, proline, and nitric oxide accumulation by triggering arginine pathway activity. *Food Chemistry, 275*, 549–556.

Aghdam, M. S., Sevillano, L., Flores, F. B., & Bodbodak, S. (2013). Heat shock proteins as biochemical markers for postharvest chilling stress in fruits and vegetables. *Scientia Horticulturae, 160*, 54–64.

Alba-Jiménez, J. E., Benito-Bautista, P., Nava, G. M., Rivera-Pastrana, D. M., Vázquez-Barrios, M. E., & Mercado-Silva, E. M. (2018). Chilling injury is associated with changes in microsomal membrane lipids in guava fruit (*Psidium guajava* L.) and the use of controlled atmospheres reduce these effects. *Scientia Horticulturae, 240*, 94–101.

Ali, S., Anjum, M. A., Ullah, S., Nawaz, A., Ejaz, S., & Khaliq, G. (2023). Salicylic acid mediated postharvest chilling and disease stress tolerance in horticultural crops. In Plant stress mitigators (pp. 69-85). Academic Press.

Ali, S., Nawaz, A., Ejaz, S., Haider, S. T-A., Alam, M. W., & Javed, H. U. (2019). Effects of hydrogen sulfide on postharvest physiology of fruits and vegetables: An overview. *Scientia Horticulturae, 243*, 290–299.

Ali, S., Nawaz, A., Naz, S., Ejaz, S., Maqbool, M., Siddiqui, M. H., Kalaji, H. M., Wróbel, J., Telesiński, A., & Auriga, A. (2022). Hydrogen sulfide mitigates chilling injury of postharvest banana fruits by regulating γ-aminobutyric acid shunt pathway and ascorbate-glutathione cycle. *Frontiers in Plant Science, 13*, 941246.

Barzegar, T., Najafi, R., Razavi, F., & Ghahremani, Z. (2021). Hydrogen sulfide and phenylalanine alleviate chilling injury in eggplant fruits during cold storage by enhancing antioxidant activities and membrane stability. *Journal of Food Processing and Preservation, 45*, e15933.

Biswas, P., East, A. R., Hewett, E. W., & Heyes, J. A. (2016). Intermittent warming in alleviating chilling injury-a potential technique with commercial constraint. *Food and Bioprocess Technology, 9*, 1–15.

Bokhary, S. U. F., Wang, L., Zheng, Y., & Jin, P. (2020). Pre-storage hot water treatment enhances chilling tolerance of zucchini (*Cucurbita pepo* L.) squash by regulating arginine metabolism. *Postharvest Biology and Technology, 166*, 111229.

Capanoglu, E. (2010). The potential of priming in food production. *Trends in Food Science and Technology, 21*, 399–407.

Chen, S., Chen, M., Li, Y., Huang, X., Niu, D., & Rashid, A. (2021). Adjustments of both phospholipids and sphingolipids contribute to cold tolerance in stony hard peach fruit by continuous ethylene. *Postharvest Biology and Technology, 171*, 111332.

Darré, M., Valerga, L., Zaro, M. J., Lemoine, M. L., Concellón, A., & Vicente, A. R. (2022). Low temperature conditioning improves American eggplant (*Solanum melongena* L.) storage compatibility. The *Journal of Horticultural Science and Biotechology, 97*, 773–784.

Endo, H., Miyazaki, K., Ose, K., & Imahori, Y. (2019a). Hot water treatment to alleviate chilling injury and enhance ascorbate-glutathione cycle in sweet pepper fruit during postharvest cold storage. *Scientia Horticulturae, 257*, 108715.

Endo, H., Ose, K., Bai, J., & Imahori, Y. (2019b). Effect of hot water treatment on chilling injury incidence and antioxidative responses of mature green mume (*Prunus mume*) fruit during low temperature storage. *Scientia Horticulturae, 246*, 550–556.

Fallik, E. (2004). Prestorage hot water treatments (immersion, rinsing and brushing). *Postharvest Biology and Technology, 32*, 125–134.

Ferguson, I. B., Ben-Yehoshua, S., Mitcham, E. J., & McDonald, R. E. (2000). Postharvest heat treatment: Introduction and workshop summary. *Postharvest Biology and Technology, 21*, 1–6.

Fu, A., Zheng, Y., Lv, Y., Watkins, C. B., Bai, C., Ma, L.. Yuan, S., Jia, L., Gao, L., Wang, Q., Mu, J., & Zuo, J. (2022). Multi-omics analysis reveals specific modifications associated with reduced chilling injury in bell pepper fruit by methyl jamonate. *Postharvest Biology and Technology, 185*, 111799.

Galindo, F. G., Sjöholm, I., Rasmusson, A. G., Widell, S., & Kaack, K. (2007). Plant stress physiology: Opportunities and challenges for the food industry. *Critical Reviews in Food Science and Nutrition, 47*, 749–763.

Gao, H., Lu, Z., Yang, Y., Wang, D., Yang, T., Cao, M. & Cao, W. (2018). Melatonin treatment reduces chilling injury in peach fruit through its regulation of membrane fatty acid contents and phenolic metabolism. *Food Chemistry, 245*, 659–666.

Gao, H., Zhang, Z. K., Chai, H. K., Cheng, N., Yang, Y., & Wang, D. N. (2016). Melatonin treatment delays postharvest senescence and regulates reactive oxygen species metabolism in peach fruit. *Postharvest Biology and Technology, 118*, 103–110.

García-Pastor, M. E., Serrano, M., Guillén, F., Zapata, P. J., & Valero, D. (2020). Preharvest or a combination of preharvest and postharvest treatments with methyl jasmonate reduced chilling injury, by maintaining higher unsaturated fatty acids, and increased aril colour and phenolics content in pomegranate. *Postharvest Biology and Technology, 167*, 111226.

Ge, W., Kong, X., Zhao, Y., Wei, B., Zhou, Q., & Ji, S. (2019). Insights into the metabolism of membrane lipid fatty acids associated with chilling injury in post-harvest bell peppers. *Food Chemistry, 295*, 26–35.

Gill, S. S., & Tuteja, N. (2010). Reactive oxygen species and antioxidant machinery in abiotic stress tolerance in crop plants. *Plant Physiology and Biochemistry, 48*, 909–930.

Glowacz, M., Bill, M., Tinyane, P. P., & Sivakumar, D. (2017). Maintaining postharvest quality of cold stored 'Hass' avocados by altering the fatty acids content and composition with the use of natural volatile compounds – methyl jasmonate and methyl salicylate. *Journal of the Science of Food and Agriculture, 97*, 5186–5193.

Gonzalez-Aguilar, G., Wang, C. Y., & Buta, G. J. (2004). UV-C irradiation reduces breakdown and chilling injury of peaches during cold storage. *Journal of the Science of Food and Agriculture, 84*, 415–422.

Gonzalez-Aguilar, G. A., Villa-Rodriguez, J. A., Ayala-Zavala, J. F., & Yahia, E. M. (2010). Improvement of the antioxidant status of tropical fruits as a secondary response to some postharvest treatment. *Trends in Food Science and Technology, 21*, 475–482.

Groppa, M. D., & Benavides, M. P. (2008). Polyamines and abiotic stress: Recent advances. *Amino Acids, 34*, 35–45.

Gupta, K., Dey, A., & Gupta, B. (2013). Plant polyamines in abiotic stress responses. *Acta Physiologiae Plantarum, 35*, 2015–2036.

Habibi, F., & Ramezanian, A. (2017). Vacuum infiltration of putrescine enhances bioactive compounds and maintains quality of blood orange during cold storage. *Food Chemistry, 227*, 1–8.

Habibi, F., Ramezanian, A., Guillén, F., Serrano, M., & Valero, D. (2020). Blood oranges maintain bioactive compounds and nutritional quality by postharvest treatments with γ-aminobutyric acid, methyl jasmonate or methyl salicylate during cold storage. *Food Chemistry, 306*, 125634.

Hasan, M. U., Singh, Z., Shah, H. M. S., Kaur, J., Woodward, A., Afrifa-Yamoah, E., & Vithana, M. D. K. (2025). Preharvest methyl jasmonate application regulates ripening, colour development and improves phytochemical quality of fruits: A review. Scientia Horticulturae, 339, 113909.

Hodges, D. M., Lester, G. E., Munro, K. I. D., & Toivonen, P. M. A. (2004). Oxidative stress: Importance for postharvest quality. *HortScience, 39*, 924–929.

Imahori, Y. (2011). Postharvest oxidative stress in fruits and vegetables. In P. Ahmad & S. Umar (Eds.), *Oxidative stress: Role of antioxidants in plants* (pp. 251–268). Studium Press Pvt. Ltd, New Delhi.

Imahori, Y. (2012). Postharvest stress treatments in fruits and vegetables. In P. Ahmad & M. N. V. Prasad (Eds.), *Abiotic stress responses in plants, metabolism, productivity and sustainability* (pp. 347–358). Springer, Science1Business Media, New York.

Imahori, Y. (2014). Role of ascorbate peroxidase in postharvest treatment of horticultural crops. In A. Ahmad (Ed.), *Oxidative damage to plants: Antioxidant networks and signaling* (425–451). Elsevier.

Imahori, Y., Bai, J., & Baldwin, E. (2016). Antioxidative responses of ripe tomato fruit to postharvest chilling and heating treatments. *Scientia Horticulturae, 198*, 398–406.

Imahori, Y., Bai, J., Ford, B. L., & Baldwin, E. A. (2021). Effect of storage temperature on chilling injury and activity of antioxidant enzymes in carambola "Arkin" fruit. *Journal of Food Processing and Preservation, 45*, e15178.

Imahori, Y., Takemura, M., & Bai, J. (2008). Chilling-induced oxidative stress and antioxidant responses in mume (*Prunus mume*) fruit during low temperature storage. *Postharvest Biology and Technology, 49*, 54–60.

Jannatizadeh, A. (2019). Exogenous melatonin applying confers chilling tolerance in pomegranate fruit during cold storage. *Scientia Horticulturae, 246*, 544–549.

Jayarajan, S., & Sharma, R. R. (2021). Melatonin: A blooming biomolecule for postharvest management of perishable fruits and vegetables. *Trends in Food Science and Technology, 116*, 318–328.

Jiang, Z., Zheng, Y., Qiu, R., Yang, Y., Xu, M., Ye, Y., & Xu, M. (2015). Short UV-B exposure stimulated enzymatic and nonenzymatic antioxidants and reduced oxidative stress of cold-stored mangoes. *Journal of Agricultural and Food Chemistry, 63*, 10965–10972.

Jiao, C. (2021). IP3 mediates NO-enhanced chilling tolerance in postharvest kiwifruit. *Postharvest Biology and Technology, 176*, 111463.

Jiao, J., Jin, M., Liu, H., Suo, J., Yin, X., Zhu, Q., & Rao, J. (2022). Application of melatonin in kiwifruit (*Actinidia chinensis*) alleviated chilling injury during cold storage. *Scientia Horticulturae, 296*, 110876.

Jiménez-Muñoz, R., Palma, F., Carvajal, F., Castro-Cegrí, A., Pulido, A., Jamilena, M., Romero-Puertas, C., & Garrido, D. (2021). Pre-storage nitric oxide treatment enhances chilling tolerance of zucchini fruit (*Cucurbita pepo* L.) by *S*-nitrosylation of proteins and modulation of the antioxidant response. *Postharvest Biology and Technology, 171*, 111345.

Kantakhoo, J., & Imahori, Y. (2021). Antioxidative responses to pre-storage hot water treatment of red sweet pepper (*Capsicum annuum* L.) fruit during cold storage. *Foods, 10*, 3031.

Kantakhoo, J., Ose, K., & Imahori, Y. (2022). Effects of hot water treatment to alleviate chilling injury and enhance phenolic metabolism in eggplant fruit during low temperature storage. *Scientia Horticulturae, 304*, 111325.

Kashash, Y., Doron-Faigenboim, A., Holland, D., & Porat, R. (2018). Effects of low-temperature conditioning and cold storage on development of chilling injuries and the transcriptome of 'Wonderful' pomegranate fruit. *International Journal of Food Science and Technology, 53*, 2064–2976.

Khan, L., Khaliq, G., Ullah, E., Abdul-Rahaman, A., El-Mogy, M. M., Kesba, H. H., Ismail, A. M., & El-Beltagi, H. S. (2025). Guggul gum and clove essential oil synergistically reduced anthracnose disease and modulated physicochemical changes in tomato during storage. *Journal of Stored Products Research, 114,* 102713.

Kong, X., Ge, W., Wei, B., Zhou, G., Zhou, X., Zhao, Y., & Ji, S. (2020). Melatonin ameliorates chilling injury in green bell peppers during storage by regulating membrane lipid metabolism and antioxidant capacity. *Postharvest Biology and Technology, 170,* 111315.

Kowaleguet, M. G., Chen, F. Y., Shi, W. L., Wu, Z. B., Wang, L. Y., Ban, Z. J., Liu, L. L., Wang, L. J., & Wu, Y. F. (2022). Exogenous polyamines alleviate chilling injury of Citrus limon fruit. *International Food Research Journal, 29,* 698–706.

Lal, S., Khaliq, G., Mastoi, A. H., Irshad, A., Jaffar, M. A., Khan, L., & Muhammad, N. (2025). Effect of *Fagonia indica* and garlic extract on *Alternaria alternata* and post-harvest quality of tomato fruit during storage. *Journal of Crop Health, 77*(6), 1–13.

Lester, G. E. (2003). Oxidative stress affecting fruit senescence. In D. M. Hodges (Ed.), *Postharvest oxidative stress in horticultural crops* (pp. 113–129). Haworth Press Inc.

Li, J., Han, Y., Hu, M., Jin, M., & Rao, J. (2018). Oxalic acid and 1-methylcyclopropene alleviate chilling injury of 'Youhou' sweet persimmon during cold storage. *Postharvest Biology and Technology, 137*, 134–141.

Li, Z., Wang, L., Xie, B., Hu, S., Zheng, Y., & Jin, P. (2020). Effects of exogenous calcium and calcium chelant on cold tolerance of postharvest loquat fruit. *Scientia Horticulturae, 269*, 109391.

Liang, C., Jiang, L., Jin, M., Li, H., & Rao, J. (2017). Effects of oxalic acid treatment on chilling injury, antioxidant capacity and energy status in harvested kiwifruits under low temperature stress. *Acta Horticulturae Sinica, 44*, 279–287.

Liang, S.-M., Kuang, J.-F., Ji, S.-J., Chen, Q.-F., Deng, W., Min, T., Shan, W., Chen, J.-Y., & Lu, W.-J. (2020). The membrane lipid metabolism in horticultural products suffering chilling injury. *Food Quality and Safety, 4*, 9–14.

Liu, D.-K., Xu, C.-C., Guo, C.-X., & Zhang, X.-X. (2020). Sub-zero temperature preservation of fruits and vegetables: A review. *Journal of Food Engineering, 225*, 109881.

Liu H., Jiang, W., Cao, J. & Ma, L. (2018a). A combination of 1-methylcyclopropene treatment and intermittent warming alleviates chilling injury and affects phenolics and antioxidant activity of peach fruit during storage. *Scientia Horticulturae*, *229*, 175–181.

Liu, X., Zhou, Y., Xiao, J., & Bao, F. (2018b). Effects of chilling on the structure, function and development of chloroplasts. *Frontiers Plant Science, 9*, 1715.

Loayza, F. E., Brecht, J., Simonne, A. H., Plotto, A., Baldwin, E. A., Bai. J., & Lon-Kan, E. (2021). A brief hot-water treatment alleviates chilling injury symptoms in fresh tomatoes. *Journal of the Science of Food and Agriculture, 101*, 54–64.

Lukatkin, A. S., Brazaitytė, A., Bobinas, Č., & Duchovskis, P. (2012). Chilling injury in chilling-sensitive plants: A review. *Žemdirbystė=Agriculture, 99*, 111–124.

Lurie, S. (1998). Postharvest heat treatments. *Postharvest Biology and Technology, 14*, 257–269.

Lurie, S., & Pedreschi, R. (2014). Fundamental aspects of postharvest heat treatments. *Horticulture Research, 30*, 14030.

Lyons, J. M. (1973). Chilling injury in plants. *Annual Review of Plant Physiology, 24*, 445–466.

Ma, M., Zhu, Z., Cheng, S., Zhou, Q., Zhou, X., Kong, X., Hu, M., Yin, X., Wei, B., & Ji, S. (2020). Methyl jasmonate alleviates chilling injury by regulating membrane lipid composition in green bell pepper. *Scientia Horticulturae, 266*, 109308.

Madani, B., Mirshekari, A., & Imahori, Y. (2019). Physiological responses to stress. In E. M. Yahia (Ed.), *Postharvest physiology and biochemistry of fruits and vegetables* (405–425). Elsevier.

Mahajan, S., & Tuteja, N. (2005). Cold, salinity and drought stresses: An overview. *Archives of Biochemistry and Biophysics, 444*, 139–158.

Mantri, N., Patada, V., Penna, S., Ford, R., & Pang, E. (2012). Abiotic stress responses in plants: Present and future. In P. Ahmad & M. N. V. Prasad (Eds.), *Abiotic stress responses in plants: Metabolism, productivity and sustainability* (pp. 1–19). Springer.

Mittler, R. (2017). ROS are good. *Trends in Plant Science, 22*, 11–19.

Niazi, Z., Razavi, F., Khademi, O., & Aghdam, M. S. (2021). Exogenous application of hydrogen sulfide and γ-aminobutyric acid alleviates chilling injury and preserves quality of persimmon fruit (*Diospyros kaki*, cv. Karaj) during cold storage. *Scientia Horticulturae, 285*, 110198.

Palma, F., Carvajal, F., Jiménez-Muñoz, R., Pulido, A., Jamilena, M., & Garrido, D. (2019). Exogenous γ-aminobutyric acid treatment improves the cold tolerance of zucchini fruit during postharvest storage. *Plant Physiology and Biochemistry, 136*, 188–195.

Park, M.-H., Sangwanangkul, P., & Choi, J.-W. (2018). Reduced chilling injury and delayed fruit ripening in tomatoes with modified atmosphere and humidity packaging. *Scientia Horticulturae, 231*, 66–72.

Paull, R., & Chen, N. J. (2000). Heat treatment and fruit ripening. *Postharvest Biology and Technology, 21*, 21–37.

Phornvillay, S., Pongprasert, N., Wongs-Aree, C., Uthairatanakij, A., & Srilaong, V. (2019). Exogenous putrescine treatment delays chilling injury in okra pod (*Abelmoschus esculentus*) stored at low storage temperature. *Scientia Horticulturae, 256*, 108550.

Pongprasert, N., Sekozawa, Y., Sugaya, S., & Gemma, H. (2011). A novel postharvest UV-C treatment to reduce chilling injury (membrane damage, browning and chlorophyll degradation) in banana peel. *Scientia Horticulturae, 130*, 73–77.

Rai, A., Kumari, K., & Vashistha, P. (2022). Umbrella review on chilling injuries: Postharvest issue, cause, and treatment in tomato. *Scientia Horticulturae, 293*, 110710.

Ranty, B., Aldon, D., Cotelle, V., Galand, J. P., Thuleau, P., & Mazars, C. (2016). Calcium sensors as key hubs in plant responses to biotic and abiotic stresses. *Frontiers in Plant Science, 7*, 327.

Rasouli, M., Saba, M. K., & Ramezanian, A. (2019). Inhibitory effect of salicylic acid and *Aloe vera* gel edible coating on microbial load and chilling injury of orange fruit. *Scientia Horticulturae, 247*, 27–34.

Rehman, R. N. U., Malik, A. U., Khan, A. S., Hasan, . M. U., Anwar, R., Ali, S. & Haider, M. W. (2021). Combined application of hot water treatment and methyl salicylate mitigates chilling injury in sweet pepper (*Capsicum annnuum* L.) fruits. *Scientia Horticulturae, 283*, 110113.

Rehman, M., Singh, Z., & Khurshid, T. (2019). Nitric oxide fumigation alleviates chilling injury and regulates fruit quality in sweet orange stored at different cold temperature. *Australian Journal of Crop Science, 13*, 1975–1982.

Reiter, R. J., Tan, D. X., Zhou, Z., Cruz, M. H. C., Fuentes-Broto, L., & Galano, A. (2015). Phytomelatonin: Assisting plants to survive and thrive. *Molecules, 20*, 7396–7437.

Rodrigues, C., Gasper, P. D., Simões, M. P., Silva, P. D., & Andrade, L. P. (2020). Review on techniques and treatment toward the mitigation of the chilling injury of peaches. *Journal of Food Processing and Preservation, 46*, e14358.

Ruan, J., Li, M., Jin, H., Sun, L., Zhu, Y., Xu, M., & Dong, J. (2015). UV-B irradiation alleviates the deterioration of cold-stored mangoes by enhancing endogenous nitric oxide levels. *Food Chemistry, 169*, 417–423.

Sayyari, M., Castillo, S., Valero, D., Díaz-Mula, H. D., & Serrano, M. (2011). Acetyl salicylic acid alleviates chilling injury and maintains nutritive and bioactive compounds and antioxidant activity during postharvest storage of pomegranates. *Postharvest Biology and Technology, 60*, 136–142.

Seo, J., Yi, G., Lee, J. G., Choi, J. H., & Lee, E. J. (2020). Seed browning in pepper (*Capsicum annuum* L.) fruit during cold storage is inhibited by methyl jasmonate or induced by methyl salicylate. *Postharvest Biology and Technology, 166*, 111210.

Sevillano, L., Sanchez-Ballesta, M. T., Romojaro, F., & Flores, F. B. (2009). Physiological, hormonal and molecular mechanisms regulating chilling injury in horticultural species. Postharvest technologies applied to reduce its impact. *Journal of the Science of Food and Agriculture, 89*, 555–573.

Shan, Y., Li, F., Lian, Q., Xie, L., Zhu, H., Li, T., Zhang, J., Duan, X., & Jiang, Y. (2022a). Role of apyrase-mediated eATP signal in chilling injury of postharvest banana fruit during storage. *Postharvest Biology and Technology, 187*, 111874.

Shan, Y., Zhang, D., Luo, Z., Li, T., Qu, H., Duan, X., & Jiang, Y. (2022b). Advances in chilling injury of postharvest fruit and vegetable: Extracellular ATP aspects. *Comprehensive Reviews in Food Science and Food Safety, 21*, 4251–4273.

Shao, X., Zhu, Y., Cao, S., Wang, H., & Song, Y. (2013). Soluble sugar content and metabolism as related to the heat-induced chilling tolerance of loquat fruit during cold storage. *Food and Bioprocess Technology, 6*, 3490–3498.

Shi, J., Zuo, J., Zhou, F., Gao, L., Wang, Q., & Jiang, A. (2018). Low-temperature conditioning enhances chilling tolerance and reduces damage in cold-stored eggplant (*Solanum melongena* L.) fruit. *Postharvest Biology and Technology, 141*, 33–38.

Si, J., Fan,Y.-Y., Liu, Z.-L., Wei, W., Xiao, X.-M., Yang, Y.-Y., Shan, W., Kuang, J.-F., Lu, W.-J., Fan, Z.-Q., Li, L.-L., & Chen, J.-Y. (2022). Comparative transcriptomic analysis reveals the potential mechanism of hot water treatment alleviated-chilling injury in banana fruit. *Food Research International, 157*, 111296.

Siboza, X., Bertling, I., & Odindo, A. O. (2017). Enzymatic antioxidants in response to methyl jasmonate and salicylic acid and their effect on chilling tolerance in lemon fruit [*Citrus limon* (L.) Burm. F.]. *Scientia Horticulturae, 225*, 659–667.

Singh, S. P., & Singh, Z. (2013). Dynamics of enzymatic and non-enzymatic antioxidants in Japanese plums during storage and lethal temperatures. *LWT-Food Science and Technology, 50*, 562–568.

Sossi, M. L., Valle, E. M., & Boggio, S. (2022). Reversible changes in galactolipid saturation level and head group composition are associated with tolerance to postharvest chilling in tomato fruit. *Journal of the Science of Food and Agriculture, 102*, 531–539.

Taghipour, L., Rahemi, M., Assar, P., Mirdehghan, S. H., & Ramezanian, A. (2021). Intermittent warming as an efficient postharvest treatment affects the enzymatic and non-enzymatic responses of pomegranate during cold storage. *Journal of Food Measurement and Characterization, 15*, 12–22.

Tao, J., Zuo, J., Watkins, C. B., Bai, C., He, X., Liu, S., Han, L., Zhao, X., Liu, Y., Li, J. & Zheng, Y. (2024). Low storage temperature affects quality and volatile compounds in fresh tomatoes. *Food Chemistry*, 460, 140400.

Valero, D., Martínez-Romero, D., & Serrano, M. (2002). The role of polyamines in the improvement of the shelf life of fruit. *Trends in Food Science and Technology, 13*, 228–234.

Vicente, A. R., Martínez, G. A., Chaves, A. R., & Civello, P. M. (2006). Effect of heat treatment on strawberry fruit damage and oxidative metabolism during storage. *Postharvest Biology and Technology, 40*, 116–122.

Wang, C. Y. (1994). Chilling injury of tropical horticultural commodities. *HortScience, 29*, 986–988.

Wang, C. Y. (2010). Alleviation of chilling injury in tropical and subtropical fruits. *Acta Horticulture, 864*, 267–274.

Wang, L., Chen, S., Shao, J., Zhang, C., Mei, L., Wang, K., Jin, P., & Zheng, Y. (2022). Hydrogen sulfide alleviates chilling injury in peach fruit by maintaining cell structure integrity via regulating endogenous H_2S, antioxidant and cell wall metabolisms. *Food Chemistry, 391*, 133283.

Wang, J., Lv, M., Li, G., Jiang, Y., Fu, W., Zhang, L., & Ji, S. (2018). Effect of intermittent warming on alleviation of peel browning of 'Nanguo' pears by regulation energy and lipid metabolisms after cold storage. *Postharvest Biology and Technology, 142*, 99–106.

Wang, S.-Y., Shi, X.-C., Liu, F.-Q., & Laborda, P. (2021). Effects of exogenous methyl jasmonate on quality preservation of postharvest fruits: A review. *Food Chemistry, 353*, 129482.

Wang, W., Vinocur, B., & Altman, A. (2003). Plant responses to drought, salinity and extreme temperatures: Towards genetic engineering for stress tolerance. *Planta, 218*, 1–14.

Wei, C., Ma, L., Cheng, Y., Guan, Y., & Guan, J. (2019). Exogenous ethylene alleviates chilling injury of 'Huangguan' pear by enhancing the proline content and antioxidant activity. *Scientia Horticulturae, 257*, 108671.

Wills, R. B. H., & Golding, J. B. (2016). *Physiological disorders*. In E. B. H. Wills & J. B. Golding (Eds.), *Postharvest: An introduction to the physiology and handling of fruit and vegetables* (pp. 153–169). CABI.

Yahia, E. M., Carrillo-López, A., & Sañudo, A. (2019). Physiological disorders and their control. In E. M. Yahia (Ed.), *Postharvest technology of perishable horticultural commodities* (pp. 499–527). Elsevier.

Yamawaki, K. (2021). Chilling injury. In N. Yamauch & Y. Imahori (Eds.), *Engeiriyougaku* (pp. 215–222). Buneido.

Zhang, W., Cao, J., Fan, X., & Jiang, W. (2020a). Applications of nitric oxide and melatonin in improving postharvest fruit quality and the separate and crosstalk biochemical mechanisms. *Trends in Food Science and Technology, 99*, 531–541.

Zhang, T., Che, F., Zhang, H., Pan, Y., Xu, M., Ban, Q., Han, Y., & Rao, J. (2017). Effect of nitric oxide treatment on chilling injury, antioxidant enzymes and expression of the CmCBF1 and CmCBF3 genes in cold-stored Hami melon (*Cucumis melo* L.) fruit. *Postharvest Biology and Technology, 127*, 88–98.

Zhang, W., & Jiang, H. (2019). UV treatment improved the quality of postharvest fruits and vegetables by inducing resistance. *Trends in Food Science and Technology, 92*, 71–80.

Zhang, W., Jiang, H., Cao, J., & Jiang, W. (2021a). Advances in biochemical mechanisms and control technologies to treat chilling injury in postharvest fruits and vegetables. *Trends in Food Science and Technology, 113*, 355–365.

Zhang, X., Ma, M., Ye, B., Liu, L., & Ji, S. (2021b). Calcium ion improves cold resistance of green peppers (*Capsicum annuum* L.) by regulating the activity of protective enzymes and membrane lipid composition. *Scientia Horticulturae, 277*, 109789.

Zhang, J., Wu, Z., Ban, Z., Li, L., Chen, C., Kowaleguet, M. G., Chen, F., Fei, L., & Wang, L. (2020b). Exogenous polyamines alleviate chilling injury of jujube fruit (*Zizyphus jujuba* Mill). *Journal of Food Processing and Preservation, 44*, e14746.

Zhang, W., Zhao, H., Jiang, H., Xu, Y., Cao, J., & Jiang, W. (2020c). Multiple 1-MCP treatment more effectively alleviated postharvest nectarine chilling injury than conventional one-time 1-MCP treatment by regulating ROS and energy metabolism. *Food Chemistry, 330*, 127256.

Zhang, H., Zhou, S., Pristijono, P., Golding, J. B., Yang, H., Chen, G., & Li, Y. (2021c). Role of AOX in low-temperature conditioning induced chilling tolerance in sweetpotato roots. *Scientia Horticulturae,288*, 110365.

Zhao, Y., Song, C., Brummell, D. A., Qi, S., Lin, Q., & Duan, Y. (2021a). Jasmonic acid treatment alleviates chilling injury in peach fruit by promoting sugar and ethylene metabolism. *Food Chemistry, 338*, 128005.

Zhao, Y., Tang, J., Song, C., Qi, S., Lin, Q., Cui, Y., Ling, J., & Duan, Y. (2021b). Nitric oxide alleviates chilling injury by regulating the metabolism of lipid and cell wall in cold-storage peach fruit. *Plant Physiology and Biochemistry, 169*, 63–69.

Zuo, X., Cao, S., Zhang, M., Cheng, Z., Gao, T., Jin, P., & Zheng, Y. (2021). High relative humidity (HRH) storage alleviates chilling injury of zucchini fruit by promoting the accumulation of proline and ABA. *Postharvest Biology and Technology, 171*, 111344.

Zhou, D., Li, R., Liu, Y., Zhao, N., & Tu, K. (2021). Hot air treatment inhibits chilling injury in peaches by regulating metabolism of sugar and organic acid. The *Journal of Horticultural Science and Biotechnology, 96*, 392–399.

Zhou, D., Sun, Y., Li, M., Zhu, T., & Tu, K. (2019). Postharvest hot air and UV-C treatments enhance aroma-related volatiles by simulating the lipoxygenase pathway in peaches during cold storage. *Food Chemistry, 292*, 294–303.

Zhu, Y., Huang, Q., Pan, Y., Zhang, Z., Yuan, R., & Nie, Y. (2022). Abnormal behavior of chilling injury in postharvest papaya fruit in associated with sugar metabolism. *Journal of Food Science, 87*, 919–928.

CHAPTER 8

Calcium-Related Physiological Disorders

BABAK MADANI[1*], AMIN MIRSHEKARI[2], AHMAD SHAKERARDEKANI[3], FARZAD AZADSHAHRAKI[4], ABOLFAZL GOLSHAN TAFTI[4], and SHOKROLLAH HAJIVAND[5]

[1]*Horticultural Crops Research Department, Natural Resources Research and Education Center of Chaharmahalobakhtiari, Agricultural Research Education and Extension Organization, Shahrekord, Iran*

[2]*Department of Agronomy and Plant Breeding, Faculty of Agriculture, University of Yasouj, Yasouj, Iran*

[3]*Yazd Agricultural and Natural Resources Research and Education Center, AREEO, Yazd, Iran*

[4]*Agricultural Engineering Research Institute, Agricultural Research Education and Extension Organization, Karaj, Iran*

[5] *Department of Genetics and Breeding, Temperate Fruits Research Center, Horticultural Research Center, Agricultural Research Education and Extension Organization, Karaj, Iran*

*Corresponding author

ABSTRACT

Increasing production of horticultural crops has led to the use of calcium to improve their quality. The objective of this chapter was to review recent information on postharvest disorders that are related to the calcium.

Physiological Disorders of Fruits and Vegetables: Causes and Remedies.
Ghulam Khaliq, Sajid Ali, and Mohammed Wasim Siddiqui (Eds.)

DOI: 10.1201/9781998511334-8

We focused on fruit cracking, chilling injury (CI), browning, bitter pit, cork spot, lenticel breakdown, and superficial scald. Moreover, we focused on vegetables blossom-end rot, tip-burn, cracking, and CI. Future studies should be done about gene expression and mode of calcium penetration into fruit and vegetables at cellular and molecular levels and relation with postharvest disorders.

8.1 INTRODUCTION

The significance of calcium (Ca) in the plants nutrition has been recognized. Insufficient calcium content causes physiological disorders in horticultural crops including fruits and vegetables (Kwanele et el., 2019). Calcium is a key mineral nutrient that can affect postharvest crop quality (Madani et al., 2014a, b). Ca maintains cell wall integrity that causes longer storage life (Domoto, 1998). Moreover, it can delay ripening, reduce respiration, increase vitamin C content and firmness (Shear, 1974; Madani et al., 2015). Also, Ca affects resistance to the diseases and pests (Madani et al., 2016; Autio and Bramlage, 2018). However, Shear (1974) and Ferguson (1989) clarified postharvest disorders of horticultural crops that are related to the calcium deficiency and stated that disorders are due to ineffective distribution of calcium in horticultural crops (Masarirambi et al., 2011). Calcium transport is restricted to the xylem (Mengel and Kirkby, 1982; Madani et al., 2014a, b). Calcium moves slowly through xylem exchange and it is not transported from leaves to fruits, or from leaf to leaf, disruptions in this flow can lead to calcium deficiency (Curry, 1998; Kwanele et al., 2019; Nxumalo et al., 2019). So, this chapter focused on recent advances on reducing postharvest disorders of horticultural crops using calcium.

8.2 FRUIT

8.2.1 CRACKING

Cracking is defined as rupturing, or bursting of the skin. It is commonly linked to the rainfall events that take place during ripening. This disorder is characterized by cracks on the skin of the fruits after raining. Over the last few years, mechanisms of susceptibility to this disorder have been studied.

Cracking during maturation depends on cultivars and cultivation conditions (Ramteke et al., 2017; Shakerardekani et al., 2022).

Ca deficiency caused cracking of fruits like pomegranate (Davarpanah et al., 2018) and litchi (Martínez Bolaños et al., 2017). As a macro-element Ca improves cell membranes integrity. Also, it works as a secondary messenger (Yu et al., 2020; Fischer et al., 2021). Cracking is a preharvest disorder for cherry growers. Hydrocooling of cherry fruit with calcium reduced cracking and increased fruit firmness (Wang and Long, 2015). Davarpanah et al. (2018) studied effects of foliar calcium sprays on the quality of pomegranate. Foliar calcium treatment decreased fruit cracking. Draie and Aboras (2021) indicated the percentage of cracked pomegranate fruits decreased in three times spraying calcium. Also, Parthiban et al. (2022) noted that foliar application of Ca and boric acid decreased grape berry cracking (3.46%). Furthermore, immersion of grape berries in calcium solution prevented fruit cracking by promoting cell wall strengthening (Yu et al., 2020).

8.2.2 CHILLING INJURY

When fruits are exposed to low temperature, they typically develop chilling injury (CI) symptoms like skin browning, and pitting. Lipid peroxidation and highly reactive oxygen species (ROS) are associated with CI in fruits. The increase in the antioxidant system of horticultural crops is related to the chilling tolerance due to the prevention of excessive accumulation of ROS, increase in antioxidants like phenolic compounds and antioxidant enzymes. Superoxide dismutase (SOD) reacts with superoxide anion radicals to produce O_2 and H_2O_2, whereas catalase (CAT), peroxidase (POD), or ascorbate peroxidase (APX) convert H_2O_2 into O_2 and H_2O. CAT, POD, SOD, and APX activities are related to stress tolerance (Jiao et al., 2018; Mirshekari et al., 2020).

López-López et al. (2018) showed that hot water and Ca decreased CI of mango by enzymatic and non-enzymatic antioxidant systems, lowering malondialdehyde (MDA), increasing flavonoids, phenols, and carotenoids. Li et al. (2020a, b, c) studied the effects of Ca on CI of loquat fruit. Calcium treatment decreased CI and preserved the membrane integrity and reduced MDA content. Also, Ca increased adenosine triphosphate (ATP) content, polyamine oxidase (PAO), and proline (Li et al., 2020a, b, c).

Also, calcium was effective in alleviating CI of the banana during ripening (Jiao et al., 2018). Elbagoury et al. (2021) stated that the application of methyl jasmonate or calcium increased total phenolic compound contents (TPC) and total antioxidant activity of bananas and decreased CI. Ca decreased PPO and CI of avocado (Barrientos-Priego et al. 2016). Also, the effects of Ca on CI of peach fruit were studied. Calcium treatment maintained fruit quality with decreasing CI (Gang et al., 2015).

8.2.3 BROWNING

Browning is attributed to the phenolic compounds oxidation with polyphenol oxidase (PPO) activity. Decrease in PPO activity decreased browning (Lee et al., 1990). Furthermore, fruit browning processes are caused by ROS accumulation and reduction in enzymatic antioxidant enzymes (SOD, CAT, POD, and APX) (Dong et al., 2015; Zhi et al., 2017). Internal browning (IB) is a postharvest disorder of pineapple. Youryon et al. (2018) studied the effects of calcium for the reduction of IB in "Queen" pineapple. They indicated efficiency of CaGlu and $CaCl_2$ for IB reduction by decreasing browning enzymes and increasing phenols. Aril browning is an important physiological disorder in pomegranates, under low-temperature storage. Tadayon (2021) evaluated effects of foliar nutrition including calcium on aril browning in "Rabab" pomegranate and found that calcium decreased browning index. Furthermore, peel browning appears during ambient temperature (Zhang et al., 2019). Calcium restrained the appearance of peel browning on the pears and decreased electrolyte leakage and MDA. Also, PPO and PLD activities were lower using Ca (Zhang et al., 2019). Moreover, Ca treatment decreased pericarp browning of "Nanguo" pears with increasing GABA biosynthesis in the fruit (Li et al., 2020). Rapid fruit browning and softening has been limited marketability of fresh-cut banana slices. The investigation showed that the usage of calcium retained higher total antioxidant activity and TPC with lower browning, PPO and POD activities (Mirshekari et al., 2017).

8.2.4 BITTER PIT

Bitter pit (BP) is an apple postharvest disorder during storage which causes dark and corky tissue below the skin, accompanied by water-soaking around the calyx with dark spots around skin. BP has been related to three factors:

1—genetic, 2—orchard management, 3—climatic conditions. For example, fruits are more sensitive to the BP disorder in hot dry summers than cooler summers (Nxumalo et al., 2019). Calcium deficiency has been involved in BP of apples. Liu et al. (2021) studied mineral contents of "Honeycrisp" apple fruit and expression genes that were related to the Ca transport. The gene expression was associated with water-soluble Ca content negatively which increased BP (Liu et al., 2021). Spraying Ca during growing season is a common method for decreasing BP. However, its efficacy is dependent on several factors including applications number, Ca sources, and time of application (Wójcik and Borowik, 2013). Amarante et al. (2020) indicated that post bloom spraying with ProCa decreased apple BP (Amarante et al., 2020). Torres et al. (2017) noted that postharvest Ca dips controlled BP better than preharvest treatments, and soil applications. However, Ca sprays with postharvest Ca dip were the best method for BP reduction (Torres et al., 2017). Also, dipping fruit in Ca increased Ca content of fruit with reducing BP (Torres et al., 2017). Moreover, Gago et al. (2016) indicated that postharvest dips of apple fruits in calcium decreased the BP and maintained higher lightness with firmer fruit (Gago et al., 2016).

8.2.5 CORK SPOT, LENTICEL BREAKDOWN, AND SUPERFICIAL SCALD

Cork spot appears in pears and apples with grayish corky lesions beneath the fruit skin, and a rough fruit surface. This disorder begins in the early pears fruit developmental stage and continues to the harvest time, with large economic losses in industry. However, its mechanisms are not well understood. Cui et al. (2021) evaluated effects of nitrendipine (NI) (Ca uptake inhibitor) in the occurrence of cork spot disorder with Ca. They indicated that NI increased cork spot disorder and reduced fruit calcium content (Cui et al., 2021). Micro-cracks on the apple fruit surface are effects of lenticels on apple (Curry, 2003; Singh et al., 2021). Lenticels can be appeared as separate tissue deterioration and caused lenticel breakdown (LB) (Turketti et al., 2012). However, its mechanism is not completely understood (Tessmer et al., 2016; Singh et al., 2021). LB mostly happens in susceptible "Red Delicious" and "Gala" apples. Ca increased LB.

Superficial scald is a physiological disorder that happens in pear fruit and apple after low-temperature storage (Lurie and Watkins, 2012). Symptoms include browning or black color of the exocarp with necrosis of the epidermis (Busatto et al., 2018). Scald is resulted from the production of α-farnesene (Zhao et al., 2016). Calcium reduced scald development in "Chili" pear fruit (Li et al. 2020) (Table 8.1).

TABLE 8.1 Postharvest Disorders of Fruits Treated with Calcium.

Fruit			
Disorder Type	**Fruit Type**	**Results**	**References**
Cracking	Cherry	$CaCl_2$ (0.2–0.5%) reduced cracking and increased fruit firmness.	Wang and Long (2015)
	Pomegranate	Nano-Ca fertilizer at 1% $CaCl_2$ decreased fruit cracking, resulting in increases in the fruit yield.	Davarpanah et al. (2018)
		Increasing the number of times spraying with calcium (0, 1000, 2000 mg L^{-1}) up to four times reduced the percentage of pomegranate fruit cracking. The percentage of cracked fruits reached 73.96, 26.49, 19.81, and 15.74% in treatments of 0, 1, 2, and 3 spraying times, respectively.	Draie and Aboras (2021)
	Grape	Foliar application of chelated EDTA calcium 0.2% + boric acid 0.1% recorded the minimum berry cracking (3.46%).	Parthiban et al. (2022)
		Immersion of grape berries in $CaCl_2$ solution prevented Xiangfei grape cracking.	Yu et al. (2020)
Chilling injury	Mango	SOD and APX showed higher activity with Ca treatment.	López-López et al. (2018)
	Loquat fruit	Calcium chloride reduced chilling injury.	Li et al. (2020)
	Bananas	2% calcium chloride was effective in alleviating CI.	Jiao et al. (2018)
		4% calcium chloride decreased chilling injury.	Elbagoury et al. (2021)
	Avocado	Calcium nitrate decreased PPO and chilling injury.	Barrientos-Priego et al. (2016)
	Peach	$CaCl_2$ treatment decreased chilling injury and increased quality.	Gang et al. (2015)

TABLE 8.1 *(Continued)*

		Fruit	
Disorder Type	**Fruit Type**	**Results**	**References**
Browning	Pineapple	Ca infiltration decreased internal browning.	Youryon et al. (2018)
	Pomegranate	Ca decreased the browning index.	Tadayon (2021)
	Pear	2% $CaCl_2$ treatment restrained the appearance of peel browning.	Zhang et al. (2019)
		Calcium treatment decreased pericarp browning.	Li et al. (2020)
	Banana	Ca decreased browning.	Mirshekari et al. (2017)
Bitter pit	Apple	Post bloom spraying with ProCa (319 mg L^{-1}) with reapplication 20 days later reduced K, N, and Mg while it increased Ca of fruit, and hence decreased "Catarina" and "Fuji" apple BP.	Amarante et al. (2020)
		Preharvest Ca sprays and postharvest calcium dip reduced bitter pit.	Torres et al. (2017)
		Postharvest Ca application for "Golden Delicious" apples decreased bitter pit.	Gago et al. (2016)
Cork spot	Apple	Nitrendipine increased cork spot disorder.	Cui et al. (2021)
		Effects were not positive	Singh et al. (2021)
Superficial scald	Pear	$CaCl_2$ decreased scald development in polyethylene bag.	Li et al. (2020)

8.3 VEGETABLES

8.3.1 BLOSSOM-END ROT (BER) AND TIP-BURN

Blossom-end rot (BER) is a Ca physiological disorder because of its relationship with Ca deficiency (De Freitas et al., 2014). Recent investigations have indicated that cellular localized Ca deficiency can trigger BER. Its indications begin with plasmolysis of cells, and breakdown of membrane (De Freitas and Mitcham, 2012). Factors that cause the development of BER include genetic or environmental conditions and insufficient Ca translocation to the fruit. Fruit expansion dilutes Ca concentration and increases sensitivity to BER during ripening (Ikeda et al., 2017). Tomato plants spraying with Ca as $CaCl_2$ or $CaCO_3$ increased the marketable yield and shelf life and decreased BER (El-Tantawy and Mahmoud, 2016). Cultivar and environmental factors influence BER in tomato, for example, medium-sized "Cindy Sweet," and large-sized cultivar, "Momotaro Fight." Also, moderate Ca application to the tomato plants decreased BER (Reitz et al., 2021).

Tip-burn is described by necrotic tips on younger leaves, making crops unsuitable for sale. Tip-burn is a common issue in lettuce in nutrient film technique (NFT) in commercial greenhouses (Samarakoon et al., 2019). An insufficient Ca provided to the younger leaf tissues, can cause inner leaf tip-burn (Barta and Tibbitts, 2000; White and Broadley, 2003). Moreover, high humidity can limit Ca uptake. Vertical airflow fans can increase transpiration and remove tip-burn (Frantz et al., 2004; Gilliham et al., 2011). Foliar spray of Ca minimized tip burn (Samarakoon et al. 2020).

8.3.2 CRACKING

Cracking is a disorder in the fruits or vegetables which is related to the climate (temperature, rainfall, solar radiation, and air humidity), soil conditions (biological, physical, and chemical properties and fertility), fertilizers (components, chemicals, and bacteria), and pruning (Matas et al., 2004; Li et al., 2017). Root cracking is a physiological disorder that notably reduces commercial value of radish. Yu et al. (2019) identified two candidate genes, RsCDPK and RsANNAT, which encoded proteins involved in Ca signal transduction in the roots of the cracking-tolerant "835" radish cultivar.

8.3.3 BROWNING AND CHILLING INJURY

Browning of the cut stem is one of the most important changes in lettuce during postharvest. Wounding iceberg lettuce stem tissue induces phenylalanine ammonia-lyase (PAL) activity and soluble phenolic components accumulation. The browning process reduces marketability of vegetables (Sapers and Hicks, 1989; Fujita et al., 1991). Once tissue is damaged by cutting, slicing, or pulping, the formation of brown pigments happens. The rate of browning in vegetables is governed by the phenolic content and active PPO and pH, and O_2 availability within the tissue (Rageh, 2018; Mirshekari et al., 2019). The main step in enzyme at browning is the phenolic compounds oxidation by PPO in the presence of oxygen (Karrayan and Aydemir, 2001). Therefore, it is very important to prevent these reactions. Rageh (2018) tested calcium as antibrowning agents and indicated that stem cut lettuce dipped in calcium solution affected PPO, reduced its activity, maintained total phenolic content, and inhibited browning on the cut surface (Rageh, 2018). Furthermore, Feng et al. (2022) indicated that calcium chloride treatment was suitable for decreasing browning of fresh-cut luffa. Moreover, calcium treatment caused these effects: low hydrogen peroxide (H_2O_2) and MDA accumulation, higher SOD, CAT, and PAL activities, lower PPO activity, upregulating SOD, CAT, and PAL genes (Feng et al., 2022).

The symptoms of CI in vegetables differ and are related to the cultivar type, and environmental conditions (Al Shoffe, 2018). The symptoms CI include: chloroplasts and mitochondria expansion and decrease in size and number of starch grains (Zhang et al., 2021) and loss of flavor and aroma and failure to ripen, peel, or pulp browning (Wang, 1994; Lurie and Crisosto, 2005; Azadshahraki et al., 2018). Quick deterioration of vegetables is a usual symptom following CI (Wang, 1994; Zhang et al., 2021). The effects of Ca on green bell pepper were investigated. The highest MDA and leakage, and the lowest calcium content were measured in the control. The results reveal that calcium decreased the CI during storage and can be commercially used to make better shelf life (Bagnazari et al., 2018) (Table 8.2).

TABLE 8.2 Postharvest Disorders of Vegetables Treated with Calcium.

Vegetables			
Postharvest Disorder	**Vegetable Type**	**Results**	**References**
Blossom end rot	Tomato	Spraying tomato plants spray with Ca sources lowered BER% after 10 days of storage period.	El-Tantawy and Mahmoud (2016)
		Moderate application of Ca decreased BER severity.	Reitz et al. (2021)
Tip-burn	Lettuce	Foliar spray of Ca, controlled tip burn.	Samarakoon et al. (2020)
Cracking	Radish	The QTL "RCr1" was in charge of root cracking.	Yu et al. (2019)
Browning	Lettuce	Stem cut lettuce dipped in hot calcium chloride 5% solution reduced browning for 10 days.	Rageh (2018)
	Luffa	1% $CaCl_2$ treatment decreased browning.	Feng et al. (2022)
Chilling injury	Bell pepper	Calcium chloride decreased the chilling injury.	Bagnazari et al. (2018)

8.4 CUT FLOWER

8.4.1 STEM BENDING

The incidence of stem bending is an important postharvest problem for cut flowers, which decreases vase life. In general, stem bending of cut flowers is related to the poor lignification during stem elongation, which reduces the rigidity of the vascular tissues in the xylem. Thus, decreases minerals and water transport to the flowers and leads to the stem bending (Soe et al., 2022). Gerbera flowers (*Gerbera jamesonii* cv. Tamara) vase life is short because of stem bending (Perik et al., 2014). Bending is hypothesized to be due to the following reasons:1- xylem blockage by bacteria, 2- lack of cell wall stiffness, 3- lack of osmotic pressure. Application of Ca is an effective method for increasing vase life of gerbera (Perik et al., 2014; García-González et al., 2022).

8.4.2 CHILLING INJURY

Cut flowers show CI symptoms, such as color changes, at low temperatures. These are influenced by cultivar, species, and environmental factors (Paull, 1987). With chilling stress development, these dysfunctions conduct more CI symptoms (Saltveit and Morris, 1990; Ketrodsakul et al., 2015) which are related to the cultivars and species, and environmental factors. The increase in ROS is an early response to the CI. Cut flowers of anthurium have a limited vase life when kept at low temperature and show CI. Ketrodsakul et al. (2015) studied effects of calcium on CI of Anthurium. The results showed increasing LOX activity and lower vase life of flowers for control. However, Ca treatment increased vase life (Ketrodsakul et al., 2015) (Table 8.3).

TABLE 8.3 Postharvest Disorders of Cut Flowers Treated With Calcium.

Cut Flower			
Postharvest Disorder	**Flower Type**	**Results**	**References**
Stem bending	Gerbera	Ca delayed stem bending.	Perik et al. (2014)
		Ca reduced stem bending.	García-González et al. (2022)
Chilling injury	Anthurium	Cut flowers hold Ca had lower chilling injury.	Ketrodsakul et al. (2015)

8.5 CONCLUSION

Calcium plays a vital role in maintaining the structural and physiological integrity of fruits and vegetables, and its deficiency is directly linked to major postharvest disorders such as cracking, chilling injury, browning, bitter pit, cork spot, lenticel breakdown, superficial scald, blossom-end rot, tip-burn, and stem bending. Through its functions in cell wall stabilization, membrane integrity, and stress resistance, calcium applications via preharvest sprays, postharvest dips, or foliar treatments have been shown to reduce these disorders and improve crop quality, storage life, and marketability. However, the effectiveness of calcium treatments varies with crop type, application method, and environmental conditions, indicating the need for further molecular and physiological research to optimize calcium use in postharvest management.

KEYWORDS

- **calcium**
- **chilling injury**
- **cork spot**
- **cracking**
- **disorders**
- **epidermis**
- **stem bending**

REFERENCES

Al Shoffe, Y. (2018). In *Susceptibility and expression of chilling injury*. Longman.

Amarante, C. V. T., Steffens, C. A., de Freitas, S. T., Silveira, J. P. G., Denardi, V., & Katsurayama, J. M. (2020). Post bloom spraying apple trees with prohexadione-calcium and gibberellic acid affects vegetative growth, fruit mineral content and bitter pit incidence. *Acta Horticulturae, 275*, 193–200.

Autio, W. R., & Bramlage, W. J. (2018). Foliar calcium sprays for apples; U. Mass Fact sheet F-19R.

Azadshahraki, F., Jamshidi, B., & Mohebbi, S. (2018). Postharvest melatonin treatment reduces chilling injury and enhances antioxidant capacity of tomato fruit during cold storage. *Advances in Horticultural Science, 32*(3), 299–309.

Bagnazari, M., Saidi, M., Mohammadi, M., Khademi, O., & Nagaraja, G. (2018). Pre-harvest CaCl2 and GA3 treatments improve postharvest quality of green bell peppers (*Capsicum annum* L.) during storage period. *Scientia Horticulturae, 240*, 258–267.

Barrientos-Priego, A. F., Martínez-Damián, M. T., Vargas-Madríz, H., & Lázaro-Dzul, M. O. (2016). Effect of preharvest calcium spraying on ripening and chilling injury in 'Hass'(*Persea americana* Mill.) avocado. *Revista Chapingo Serie Horticultura, 22*(3), 145–159.

Barta, D. J., & Tibbitts T. W. (2000). Calcium localization and tipburn development in lettuce leaves during early enlargement. *Journal of the American Society for Horticultural Science, 125*, 294–298.

Busatto, N., Farneti, B., Commisso, M., Bianconi, M., Iadarola, B., Zago, E., Ruperti, B., Spinelli, F., Zanella, A., Velasco, R., Ferrarini, A., Chitarrini, G., Vrhovsek, U., Delledonne, M., Guzzo, F., Costa, G., & Costa, F. (2018). Apple fruit superficial scald resistance mediated by ethylene inhibition is associated with diverse metabolic processes. *Plant Journal, 93*, 270–285.

Cui, Z., Wang, N., Li, D., Wang, R., & Ma, C. (2021). Nitrendipine-treatment increases cork spot disorder incidence in pear 'Akituki'(Pyrus pyrifolia Nakai.) by altering calcium distribution inside the fruit. *Plants, 10*(5), 994.

Curry, E. A. (1998). Physiology of braeburn maturity and disorders: A discussion. Postharvest Information Network, WSU Tree Fruit Research & Extension Centre.

Curry, E. (2003). Factors associated with lenticels breakdown. Washington State University-Tree Fruit Research and Extension Center Postharvest Information Network.

Davarpanah, S., Tehranifar, A., Abadía, J., Val, J., Davarynejad, G., Aran, M., & Khorassani, R. (2018). Foliar calcium fertilization reduces fruit cracking in pomegranate (*Punica granatum* cv. Ardestani). *Scientia Horticulturae, 230*, 86–91.

De Freitas, S. T., McElrone, A. J., Shackel, K. A., & Mitcham, E. J. (2014). Calcium partitioning and allocation and blossom-end rot development in tomato plants in response to whole-plant and fruit-specific abscisic acid treatments. *Journal of Experimental Botany, 65*, 235–247.

De Freitas, S. T., & Mitcham. E. J. (2012). Factors involved in fruit calcium deficiency disorders. *Horticultural Reviews*, *40*, 107–146.

Domoto, P. (1998). Calcium helps prevent "bad apples. *The Iowa Horticulture, 14*(4), 16–18.

Dong, Y., Liu, L., Zhao, Z., Zhi, H., & Guan, J. (2015). Effects of 1-MCP on reactive oxygen species, polyphenol oxidase activity, and cellular ultra-structure of core tissue in 'Yali' pear (*Pyrus bretschneideri* Rehd.) during storage. *Horticulture Environment and Biotechnology, 56*(2), 207–215.

Draie, R., & Aboras, A. (2021). Efficacy of foliar spraying with calcium on the fruit cracking of the French pomegranate variety. *International Research Journal of Innovations in Engineering and Technology, 5*(5), 43.

Elbagoury, M. M., Turoop, L., Runo, S., & Sila, D. N. (2021). Regulatory influences of methyl jasmonate and calcium chloride on chilling injury of banana fruit during cold storage and ripening. *Food Science and Nutrition, 9*(2), 929–942.

El-Tantawy, E. M., & Mahmoud, M. I. (2016). Tomato fruit marketability, blossom-end rot incidence and quality as affected by different calcium sources and time of application under calcareous soil. *Journal of Plant Product*, *7*(5), 475–488.

Feng, Y., Feng, C., Wang, Y., Gao, S., Sun, P., Yan, Z., & Zhu, Q. (2022). Effect of $CaCl_2$ treatment on enzymatic browning of fresh-cut luffa (Luffa cylindrica). *Horticulture, 8*(6), 473.

Ferguson, I. B., & Watkins, C. B. (1989). Bitter pit in apple fruit. *Horticultural Reviews, 11*, 289–355.

Fischer, G., Balaguera-López, H. E., & Álvarez-Herrera, J. Causes of fruit cracking in the era of climate change: A Review. *Agronomia Colombiana, 39*(2), 196–207.

Frantz, J. M., Ritchie, G., Cometti, N. N., Robinson, J., & Bugbee. B. (2004). Exploring the limits of crop productivity: Beyond the limits of tip-burn in lettuce. *Journal of the American Society for Horticultural Science, 129*, 331–338.

Fujita, S., Tono, T., & Kawahara, H. (1991). Purification and properties of polyphenol oxidase in head lettuce (*Lactuca sativa*). *Journal of the Science of Food and Agriculture, 55*, 643–651.

Gago, C. M., Guerreiro, A. C., Miguel, G., Panagopoulos, T., da Silva, M. M., & Antunes, M. D. (2016). Effect of Calcium chloride and 1-MCP (Smartfresh™) postharvest treatment on 'Golden Delicious' apple cold storage physiological disorders. *Scientia Horticulturae, 211*, 440–448.

Gang, C., Li, J., Chen, Y., Wang, Y., Li, H., Pan, B., & Odeh, I. (2015). Synergistic effect of chemical treatments on storage quality and chilling injury of honey peaches. *Journal of Food Processing and Preservation, 39*(6), 1108–1117.

García-González, A., Soriano-Melgar, L. D. A. A., Cid-López, M. L., Cortez-Mazatán, G. Y., Mendoza-Mendoza, E., Valdez-Aguilar, L. A., & Peralta-Rodríguez, R. D. (2022). Effects of calcium oxide nanoparticles on vase life of gerbera cut flowers. *Scientia Horticulturae, 291*, 110532.

Gilliham, M., Dayod, M., Hocking, B. J., Xu, B., Conn, S. J., Kaiser, B. N., Leigh, R. A., & Tyerman, S. D. (2011). Calcium delivery and storage in plant leaves: Exploring the link with waterflow. *Journal of Experimental Botany, 62*(7), 2233–2250.

Ikeda, H., Shibuya, T., Nishiyama, M., Nakata, Y., & Kanayama Y. (2017). Physiological mechanisms accounting for the lower incidence of blossom-end rot in tomato introgression line IL8-3 fruit. *The Horticulture Journal, 80*, 327–333.

Jiao, W., Xi, Y., Cao, J., Fan, X., & Jiang, W. (2018). Regulatory effects of CaCl2, sodium isoascorbate, and 1-methylcyclopropene on chilling injury of banana fruit at two ripening stages and the mechanisms involved. *Journal of Food Process Preservating, 42*(2), 134–142.

Karrayan, D., & Aydemir, T. (2001). Partial purification and characterization of polyphenol oxidase from peppermint (Menthapiperita). *Food Chemistry, 74*(2), 147–154.

Ketrodsakul, A., Choosung, P., Penchaiya, P., & Buanong, M. (2015). Effect of calcium on chilling injury of Anthurium'Angel'cut flower. In *III International Conference on Quality Management in Supply Chains of Ornamentals.*

Kwanele, A., Nxumal, O., Christinah, M., & Michael, T. M. (2019). Calcium-related postharvest physiological disorders of fruits and vegetables in eswatini: A review. *Current Journal of Applied Science and Technology,* 1–10.

Lee, C. Y., Kagan, V., Jaworski, A. W., & Brown, S. K. (1990). Enzymic browning in relation to phenolic compounds and polyphenoloxidase activity among various peach cultivars. *Journal of Agricultural and Food Chemistry, 38*(1), 99–101.

Li, J., & Chen, J. (2017). Citrus fruit-cracking: Causes and occurrence. *Horticultural Plant Journal, 3*, 255–260.

Li, Q., Cheng, C. X., Zhang, X. F., Wang, C. H., & Yang, S. L. (2020a). Preharvest bagging and postharvest calcium treatment affects superficial scald incidence and calcium nutrition during storage of 'Chili'pear (*Pyrus bretschneideri*) fruit. *Postharvest Biology and Technology, 163*, 111149.

Li, Z., Wang, L., Xie, B., Hu, S., Zheng, Y., & Jin, P. (2020b). Effects of exogenous calcium and calcium chelant on cold tolerance of postharvest loquat fruit. *Scientia Horticulturae, 269*, 109391.

Li, J., Zhou, Q., Zhou, X., Wei, B., Zhao, Y., & Ji, S. (2020c). Calcium treatment alleviates pericarp browning of 'nanguo'pears by regulating the GABA shunt after cold storage. *Frontiers in Plant Science, 11*, 580986.

Liu, J., Jiang, Z., Qi, Y., Liu, Y., Ding, Y., Tian, X., & Ren, X. (2021). MdCAX affects the development of the 'Honeycrisp'bitter pit by influencing abnormal Ca distribution. *Postharvest Biology and Technology, 171*, 111341.

López-López, M. E., López-Valenzuela, J. Á., Delgado-Vargas, F., López-Angulo, G., Carrillo-López, A., Ayón-Reyna, L. E., & Vega-García, M. O. (2018). A treatment combining hot water with calcium lactate improves the chilling injury tolerance of mango fruit. *HortScience, 53*(2), 217–223.

Lurie, S., & Crisosto, C. H. (2005). Chilling injury in peach and nectarine. *Postharvest Biology and Technology, 37*(3), 195–208.

Lurie, S., & Watkins, C. B. (2012). Superficial scald, its etiology and control. *Postharvest Biology and Technology, 65*, 44–60.

Madani, B., Mirshekari, A., & Yahia, E. (2016). Effect of calcium chloride treatments on calcium content, anthracnose severity and antioxidant activity in papaya fruit during ambient storage. *Journal of the Science of Food and Agriculture, 96*(9), 2963–2968.

Madani, B., Mohamed, M. T. M., Biggs, A. R., Kadir, J., Awang, Y., Tayebimeigooni, A., & Shojaei, T. R. (2014a). Effect of pre-harvest calcium chloride applications on fruit calcium level and post-harvest anthracnose disease of papaya. *Crop Protection, 55*, 55–60.

Madani, B., Mohamed, M. T. M., Watkins, C. B., Kadir, J., Awang, Y., & Shojaei, T. R. (2014b). Preharvest calcium chloride sprays affect ripening of Eksotika II'papaya fruits during cold storage. *Scientia Horticulturae, 171*, 6–13.

Madani, B., Wall, M., Mirshekari, A., Bah, A., & Mohamed, M. T. M. (2015). Influence of calcium foliar fertilization on plant growth, nutrient concentrations, and fruit quality of papaya. *HortTechnology, 25*(4), 496–504.

Martínez Bolaños, M., Martínez Bolaños, L., Guzmán Deheza, A., Gómez Jaimes, R., & Reyes Reyes, A. L. Calcio y ácido giberélico en el bretado de frutos de litchi (*Litchi chinensis* Soon.) cultivar Mauritius. *Revista Mexicana de Ciencias Agrícolas, 8*(4), 837–848.

Masarirambi, M. T., Mhazo, N., Oseni, T. O., & Shongwe, V. D. (2011). Physiological disorders of Brassicas/Cole crops found in Swaziland. *African Journal of Plant Science, 5*(1), 8–14.

Matas, A. J., Cobb, E. D., Paolillo, D. J., & Niklas, K. J. (2004). Crack resistance in cherry tomato fruit correlates with cuticular membrane thickness. *HortScience, 39*, 1354–1358.

Mengel, E. K., & Kirkby, E. A. (1982). Calcium. In: *Principles of plant nutrition.* International Potash Institute.

Mirshekari, A., Madani, B., Golding, J. B. (2017). Suitability of combination of calcium propionate and chitosan for preserving minimally processed banana quality. *Journal of the Science of Food and Agriculture, 97*(11), 3706–3711.

Mirshekari, A., Madani, B., & Golding, J. B. (2019). Aloe vera gel treatment delays postharvest browning of white button mushroom (*Agaricus bisporus*). *Journal of Food Measurement and Characterization, 13*(2), 1250–1256.

Mirshekari, A., Madani, B., Yahia, E. M., Golding, J. B., & Vand, S. H. (2020). Postharvest melatonin treatment reduces chilling injury in sapota fruit. *Journal of the Science of Food and Agriculture, 100*(5), 1897–1903.

Nxumalo, K. A., Matsuane, C., & Masarirambi, M. T. (2019). Calcium-related post-harvest physiological disorders of fruits and vegetables in eswatini: A review. *Current Journal of Applied Science and Technology, 33*(6), 1–10.

Parthiban, S., Indirani, R., Subbiah, A., Saraswathy, S., & Nireshkumar, N. (2022). Effect of calcium, boron and micronutrient formulations on berry cracking in grapes var. Muscat Hamburg. *Madras Agricultural Journal, 108*, 7–9.

Paull, R. E. (1987). Effect of storage duration and temperature on cut anthurium flowers. *HortScience, 22*(3), 459–460.

Perik, R. R., Razé, D., Ferrante, A., & van Doorn, W. G. (2014). Stem bending in cut Gerbera jamesonii flowers: Effects of a pulse treatment with sucrose and calcium ions. *Postharvest Biology and Technology, 98*, 7–13.

Rageh, M. A. (2018). Effect of hot calcium chloride and acetic acid solutions or hot water treatments on inhibition of total microbial counts and browning in lettuce stem cut during cold storage. *Annals of Agricultural Science, 56*(3), 709–720.

Ramteke, S. D., Urkude, V., Parhe, S. D., & Bhagwat, S. R. (2017). Berry cracking; Its causes and remedies in grapes-A review. *Trends in Bioscience, 10*(2), 549–556.

Reitz, N. F.; Shackel, K. A., & Mitcham, E. J. (2021). Differential effects of excess calcium applied to whole plants vs. excised fruit tissue on blossom-end rot in tomato. *Scientia Horticulturae, 290*, 110514.

Saltveit, M. E., & Morris, L. L. (1990). Overview on chilling injury of horticultural crops. *Chilling Injury of Horticultural Crops, 1*, 3–15.

Samarakoon, U. C., Fyffe, C., Bale, J., Ling, P., Basnagala, S., Donley, N., & Altland. J. (2019). Effect of electrical conductivity on the productivity and nutrient uptake of *Lactuca sativa* L. grown using nutrient film technique (NFT). *Acta Horticulturae, 1266*, 137–144.

Samarakoon, U., Palmer, J., Ling, P., & Altland, J. (2020). Effects of electrical conductivity, pH, and foliar application of calcium chloride on yield and tipburn of *Lactuca sativa* grown using the nutrient–film technique. *HortScience, 55*(8), 1265–1271.

Sapers, G. M., & Hicks, K. B. (1989). Inhibition of enzymatic browning in fruits and vegetables. In *J. J.* Jen (Ed.), *Quality factors of fruits and vegetables*. ACS.

Shakerardekani, A., Esmaeilpour, A., Hosseini, S. M., & Saberi, N. (2022). Evaluation of quality of pistachio fruit produced under shade/ net houses and without shade/ net houses and its sensory evaluation. Pistachio Research Center. Report No. 61933.

Shear, C. B. (1974). Interaction of calcium and nitrogen and time of calcium availability in relation to the development of apple disorders. In *Proceedings 7th International Colloquium Plant Analysis Fertility Problems, Hanover* (pp. 427–436). German Society Plant Nutrition.

Singh, V., Gamrasni, D., Parimi, P., Kochanek, B., Naschitz, S., Zemach, H., & Friedman, H. (2021). Postharvest calcium treatment of apple fruit increased lenticel breakdown and altered cuticle structure. *Postharvest Biology and Technology, 171*, 111331.

Soe, M. T., Naing, A. H., Kim, S. R., & Kim, C. K. (2022). Characterizing the effects of different chemicals on stem bending of cut snapdragon flower. *Plant Methods, 18*(1), 1–10.

Tadayon, M. S. (2021). Effect of foliar nutrition with calcium, boron, and potassium on amelioration of aril browning in pomegranate (Punica granatum cv.'Rabab'). *The Journal of Horticultural Science and Biotechnology, 96*(3), 372–382.

Tessmer, M. A., Appezzato-da-Gl´oria, B., & Antoniolli, L. R. (2016). Influence of growing sites and physicochemical features on the incidence of lenticel breakdown in 'Gala' and 'Galaxy' apples. *Scientia Horticulturae, 205*, 119–126.

Torres, E., Recasens, I., Lordan, J., & Alegre, S. (2017). Combination of strategies to supply calcium and reduce bitter pit in 'Golden Delicious' apples. *Scientia Horticulturae, 217*, 179–188.

Turketti, S., Curry, E., & L¨otze, E. (2012). Role of lenticel morphology, frequency and density on incidence of lenticel breakdown in 'Gala'apples. *Scientia Horticulturae, 138*, 90–95.

Wang, C. Y. (1994). Chilling injury of tropical horticultural commodities. *HortScience, 29*(9), 986–988.

Wang, Y., & Long, L. E. (2015). Physiological and biochemical changes relating to postharvest splitting of sweet cherries affected by calcium application in hydrocooling water. *Food Chemistry, 181*, 241–247.

White, P. J., & Broadley M. R. (2003). Calcium in plants. *Annals of Botony, 92*(4), 487–511.

Wójcik, P., & Borowik, M. (2013). Influence of preharvest sprays of a mixture of calcium formate, calcium acetate, calcium chloride and calcium nitrate on quality and 'jonagold'apple storability. *Journal of Plant Nutrition, 36*(13), 2023–2034.

Youryon, P., Supapvanich, S., Kongtrakool, P., & Wongs-Aree, C. (2018). Calcium chloride and calcium gluconate peduncle infiltrations alleviate the internal browning of Queen pineapple in refrigerated storage. *Horticulture Environment and Biotechnology, 59*(2), 205–213.

Yu, X., Choi, S. R., Chhapekar, S. S., Lu, L., Ma, Y., Lee, J. Y., & Lim, Y. P. (2019). Genetic and physiological analyses of root cracking in radish (*Raphanus sativus* L.). *Theoretical and Applied Genetics, 132*(12), 3425–3437.

Yu, J., Zhu, M., Bai, M., Xu, Y., Fan, S., & Yang, G. (2020). Effect of calcium on relieving berry cracking in grape (*Vitis vinifera* L.) 'Xiangfei'. *Peer Journal, 8*, e9896.

Zhang, W., Jiang, H., Cao, J., & Jiang, W. (2021). Advances in biochemical mechanisms and control technologies to treat chilling injury in postharvest fruits and vegetables. *Trends in Food Science and Technology, 113*, 355–365.

Zhang, L., Wang, J. W., Zhou, B., Liu, Y. F., Xia, X. L., Xiao, Z. G., & Ji, S. J. (2019). Calcium inhibited peel browning by regulating enzymes in membrane metabolism of 'Nanguo'pears during post-ripeness after refrigerated storage. *Scientia Horticulturae, 244*, 15–21.

Zhao, J., Xie, X., Shen, X., & Wang, Y. (2016). Effect of sunlight-exposure on antioxidants and antioxidant enzyme activities in' d'Anjou' pear in relation to superficial scald development. *Food Chemistry, 210*, 18–25.

Zhi, H. H., Liu, Q. Q., Dong, Y., Liu, M. P., & Zong, W. (2017). Effect of calcium dissolved in slightly acidic electrolyzed water on antioxidant system, calcium distribution, and cell wall metabolism of peach in relation to fruit browning. *The Journal of Horticultural Science and Biotechnology, 92*(6), 621–629.

CHAPTER 9

Water Loss and Humidity Physiological Disorders

YASAMAN ESMAEILI[1*] and HILMA SABET[2]

[1]*Department of Food Science and Technology, Isfahan (Khorasgan) Branch, Islamic Azad University, Isfahan, Iran*

[2]*Department of Horticulture, Faculty of Agriculture, Isfahan (Khorasgan) Branch, Islamic Azad University, Iran*

*Corresponding author

ABSTRACT

Water loss and humidity-related physiological disorders are among the most significant factors limiting the postharvest quality and shelf life of fruits and vegetables. Even after harvest, horticultural commodities remain biologically active and continue to lose water through transpiration, leading to progressive deterioration in appearance, texture, freshness, and nutritional value. This chapter provides a comprehensive overview of the mechanisms of water loss, the role of relative humidity, and their direct and indirect effects on the development of physiological disorders during postharvest handling, storage, and marketing. The interaction of humidity with temperature, air circulation, commodity characteristics, cultivar differences, and stage of maturity is examined to explain variability in disorder, incidence and severity. Special emphasis is placed on the balance between minimizing water loss and avoiding condensation and microbial proliferation under high-humidity conditions. Furthermore, the chapter reviews practical postharvest strategies to manage water loss and humidity-related disorders, including humidity-controlled cold storage, modified and controlled atmosphere packaging, edible coatings, waxing, and improved handling practices.

Physiological Disorders of Fruits and Vegetables: Causes and Remedies.
Ghulam Khaliq, Sajid Ali, and Mohammed Wasim Siddiqui (Eds.)

DOI: 10.1201/9781998511334-9

9.1 INTRODUCTION

Consuming fruits and vegetables has been linked to many health advantages for humans. Consumption and production of minimally processed foods have become very popular in recent years. Consumers increasingly demand freshly cut fruits and vegetables due to their desire for fresh and natural items and lifestyle changes. However, the fresh-cut business has confronted the issue of preserving the fresh-like characteristics of fresh-cut food items over a more extended storage period (Riva et al., 2020). Fruit and vegetable products are an essential part of the human diet. Vegetables and fruits are actively respiring and continuously use oxygen and produce carbon dioxide throughout respiration. In the process of respiration, substrates, such as triglycerides, peptides, and organic compounds, in addition to carbohydrates, are metabolized. Once substrates and carbohydrates are metabolized, it becomes difficult to replace them since the vegetable or fruit is isolated from the plant (El-Sayed et al., 2020). The quality of color, taste, volume, and nutritional content continue to decline. Water loss is a crucial factor in the deterioration of fruits and vegetables, as it may lead to nutritional loss, softening, flaccidity, shrinkage, and wilting (Lufu et al., 2020). The overall quality of fruit is tied to climatic circumstances and administration throughout the production phase; consequently, it is vital to apply various approaches to forestall microbiological degradation and lessen the biochemical and physiological alterations answerable for postharvest degeneration. Research on sophisticated packaging methods, for instance, the content of atmosphere gas, is necessary to modify the metabolic conditions and to increase the quality and shelf life of fruits (Arroyo et al., 2020; El-Sayed and El-Sayed, 2020).

The effects of water loss on fresh food vary according to the kind of product. Recent knowledge on the crucial limits of water loss, especially fruit, is scarce. However, the water-loss threshold at which some plants become unusable. With a water loss of 5–10% of their original weight, most fruits and vegetables become unmarketable because wilting and shriveling ensue, resulting in a reduction in quality (Bartz and Brecht, 2002; Gunny et al., 2024). In addition to causing loss of marketable weight, considerable water loss causes browning, reduction of fruit structure and taste (Caleb et al., 2013), enhanced aging, cold injury sensitivity, and cell wall breakdown. Although the indirect and direct impacts of water loss on the reliability of harvested produce are pretty apparent, the process of water loss is a complicated procedure influenced by numerous factors that include the biological, biochemical, and physicomechanical characteristics of the yield as well as the production, harvest, and postharvest treatments and handling. Current research has mostly focused on postharvest challenges while disregarding pre-harvest and harvest variables. Recent studies indicate that pre-harvest and harvest factors significantly contribute to postharvest water loss (Lobos et al., 2016; Tyagi et al., 2017). Variations in orchard procedures, environment, and season have a significant influence on the postharvest qualities of a fresh product.

Currently, the effects of pre-harvest and harvest factors on the loss of moisture of fruit and vegetables during postharvest handling are little studied. Second, the existing data on the postharvest factors influencing fruit water loss over preservation must be pooled to aid in the selection of management approaches. Despite extensive research on water-loss management devices and postharvest quality conservation, there is no comprehensive evaluation of the pre-harvest, harvest, and postharvest components that influence water loss under storage settings. Recognizing the multi-stage and multi-level elements that contribute to water loss in the fresh fruit supply chain can open new study opportunities.

9.2 CAUSES OF WATER LOSSES FROM FRUITS AND VEGETABLES

9.2.1 PRE-HARVEST FACTORS

Water loss from fruits and vegetables often begins even before harvest, as several preharvest factors significantly influence the susceptibility of produce to dehydration. One major factor is the environmental conditions under which crops are grown, high temperatures, low relative humidity, and strong winds can increase transpiration rates, leading to greater moisture loss during growth and making tissues more prone to postharvest dehydration. Irrigation practices also play a vital role; inconsistent or insufficient water supply can impair normal physiological development and reduce the water-holding capacity of tissues. Additionally, the canopy structure and leaf area index (LAI) affect the microclimate around the fruit, influencing airflow and exposure to sunlight, both of which impact water retention. The development of the cuticle, a waxy layer that acts as a barrier to water loss, is influenced by both genetics and growing conditions; poorly developed cuticles result in higher transpiration rates.

Nutritional status, particularly calcium and boron levels, affects cell wall strength and membrane integrity, with deficiencies increasing vulnerability to water loss. Furthermore, varietal or genetic differences mean some cultivars naturally have better-developed protective structures or more efficient water retention mechanisms. Lastly, preharvest pest and disease pressure can damage the epidermis or compromise protective barriers, predisposing fruits and vegetables to increased moisture loss after harvest.

9.2.2 HARVEST TECHNIQUE AND PHYSICAL DAMAGE

Due to the increased demand for fruits and vegetables throughout the world, horticulture methods have changed from basic labor-intensive to mechanical systems to assist planting and harvesting (Li and Thomas, 2014). Increased mechanization enhances fresh produce's susceptibility to injury during growth and harvest (Shewfelt and Prussia, 2022). Pressure, vibration, and compression forces from mechanized harvesting can damage fruit texture and must be regulated (Li and Thomas, 2014). Damage to the peel of fresh produce before or after harvest increases peel permeability, moisture flux, and loss during preservation as a result of the disorientation of surface structures (Shewfelt and Prussia, 2022). Abu-Goukh and Elshiekh (2008) examined two grapefruit (cv. Foster) harvest tactics. Fruit collected using the traditional technique (hook on a pole) lost 51% more moisture than fruit collected with the improved method (ringed long cloth sleeve on a pole) (Abu-Goukh and Elshiekh, 2008).

9.2.3 POSTHARVEST FACTORS

Fruits and vegetables are known as nutrient-dense foods. However, these things are incredibly perishable and would not be fit for human consumption if not appropriately managed after harvesting. Moreover, as a result of the globalization of business, fresh horticultural products play an essential role in international trade. Ultimately, lengthy deliveries and distributions may increase the likelihood of significant losses. As a result, precise processes and attention to detail have been stressed for the postharvest handling of fresh items. Postharvest technology aims to preserve the maximum quality of fruits and vegetables, particularly their appearance, taste, nutritional content, and texture. This is done to reduce food waste from harvest time to consumption and improve food safety (Jafarzadeh et al., 2021). Due to their high moisture capacity, nutritional density, and active metabolisms, all fresh fruits and vegetables are prone to dryness, mechanical damage, environmental stress, and pathological breakdown. These classifications significantly reduce the preservation time of vegetables and fruits and result in a discernible decline after harvest (Romanazzi et al., 2017).

9.2.4 EFFECTS OF TEMPERATURE

Temperature is one of the most influential post-climacteric influences on fruit water loss. Vegetables are often kept at low temperatures to prolong their shelf life. Low-moisture storage causes vegetable shrinkage, whereas excessive humidity stimulates microbial growth on the surface. Improper storage temperature raises the metabolic rate of tomatoes, reducing their shelf life (Sharma et al., 2019).

The preceding experiments on pomegranate fruit (cv. Wonderful) revealed that storage temperature shows an impact on the water loss amount of samples during 150 days. Pomegranate samples lost 27.7 and 45.7% of their moisture content in different temperature conditions (5 and 7.5°C at 92% relative humidity (RH), respectively) (Wang, 2016).

Throughout 84 days of storage, it was observed considerably more moisture loss for pomegranate (cv. Ruby and Bhagwa) kept at 10°C, 92% RH compared with fruit held at 5°C, 92% RH and 7°C, 92% RH. They attributed this to changes in respiratory rate (Fawole and Opara, 2013). Similar findings have been discovered for numerous fruits and vegetables, including strawberries (Bovi et al., 2018), pomegranates (Lufu et al., 2019), apples (Cofelice et al., 2019), and table grapes (Pereira et al., 2017). High temperature and low RH together enhance fruit respiration and transpiration, which causes more water loss. After harvest, the fruit must be quickly precooled, and the appropriate temperature and humidity must be preserved throughout the cold storage chain (storage, delivery, and commercialization) (Lufu et al., 2019). To enhance the postharvest quality of sweet cherries, the impact of near-freezing temperature (NFT) storage was examined by analyzing postharvest features, biological components, and total antioxidant activity. While preserving delicious cherries, NFT also prevented weight loss and physical problems (such as pitting and pedicel browsing). The SSC in fruits at NFT was ~ 1.2–1.4 times greater than that at 0°C in both varieties of the sweet cherry after 80 days of cold storage. Moreover, NFT storage improved antioxidant properties and delayed the peak value of fruit antioxidant properties (Zhao et al., 2019).

TABLE 9.1 Impact of Relative Humidity (RH) on Water Loss of Fresh Fruit Samples.

Fruit Type	Storage Temperature (°C)	RH%	Water Loss %	Reference
Figs (Ficus carica L.)	20	80	3.9	Lentzou et al. (2021)
	20	98	2.6	
Pomegranate (cv. Wonderful)	25	77	6.6	Lufu et al. (2020)
	25	82	16.6	
Pomegranate (cv. Wonderful)	20	65	29.1	Mukama et al. (2019)
	20	95	5.8	
Feijoa (cv. Quimba)	18	50	7.1	Castellanos et al. (2016)
	18	70	5.6	
Strawberries	4	76	3	Bovi et al. (2018)
	4	86	5	
	4	96	8	
	12	76	7	
	12	86	7	
	12	96	10	
	20	76	16	
	20	86	20	
	20	96	14	
Pears (Pyrus communis L., Kontoula)	20	95	39	Xanthopoulos et al. (2017)
	0	95	10	

9.2.5 EFFECTS OF HUMIDITY

Limited information exists on how humidity affects water loss in fruit. Earlier research suggests that storage humidity may impact water loss. Minor variations in RH may affect fresh product water loss at a constant temperature (Hussein et al., 2020a). When fresh produce is stored for an extended period, the quality can deteriorate, resulting in a loss of weight. Even though transpiration is the primary process, respiration can still play a considerable role in mass loss. This is doubly apparent under higher humidity conditions than usual, where the rate of transpiration is nearly nonexistent (Table 9.1).

Lentzou et al. (2021) studied transpiration and respiratory water loss in figs (*Ficus carica* L.) at 0, 10, and 20°C and 45.64, 80.22, and 98.65% RH and in stroll cold storage areas. At 20°C and saturating, respiration contributes to 3.9% of the relative water loss owing to moisture pressure difference. In general, respiration contributes to 1.5, 2.1, and 2.6% of the water loss resulting from water vapor pressure at 0, 10, and 20°C (Lentzou et al., 2021).

In another study, the influence of the treatment of nanocomposites on the quality attributes of apples kept at 4°C was evaluated by calculating the weight loss, pH and total titratable acidity, total phenolic amount, and general fruit appearance throughout preservation. Based on the results, coating installation is anticipated to reduce moisture loss. By establishing a high humidity level on the cut apples' surface, the coating succeeded in limiting water loss (Cofelice et al., 2019). Mukama et al. (2019) assessed the pomegranate fruit's (Wonderful) quality degradation throughout precooling and simulating shelf situations. The impacts of RH within the cold environment, polyliner within the packing, and stacked orientation have been evaluated on loss in the quality of the fruit. The shelf life experiment revealed the significance of room humidification for maintaining the quality of the fruit. Keeping pomegranates in a room with a 95% RH minimized weight loss and best-preserved fruit color, firmness, size, and biochemical quality features (Mukama et al., 2019). According to Lufu et al. (2019), respiration and transpiration have a role in the total mass loss of fresh pomegranate fruit. This study examined the effects of temperature range, RH, and storage period. According to the findings, RH had a significant impact on mass loss and transpiration rate than temperature (Lufu et al., 2019).

9.2.6 EFFECTS OF STORAGE DURATION

Vegetables and fruits remain to undertake respiration and transpiration, depleting their sugar and water reserves without being replenished; hence, they continue to lose firmness and mass, culminating in shriveling and shrinking (Ariffin, 2019). Recent research investigated the impacts of bruising impact injuries on the postharvest physiological reaction, physical quality, and oxidative capabilities of the pomegranate fruit. According to the results, during 12-week cold storage, damaged pomegranate fruit lost 40% of its weight (Hussein et al., 2020b). It was demonstrated that after 120 days of preservation, the size of bruising on the pomegranate affected the cheeks with a high impact energy (1390 mJ) increased by 5% (Mohammad Shafie et al., 2015). Consequently, conditions of storage place, such as temperature, RH, and atmospheric air showed different effects on water loss in fruits during storage time.

9.2.7 EFFECTS OF CONTROLLED ATMOSPHERE STORAGE

Modified atmospheres (MA) relate to using environments with variable gas content than ordinary air, which is quietly sustained through product respiration and restricted gas diffusion. Controlled atmospheres (CA) relate to the usage of atmospheres that are distinct from normal air but are accurately and actively controlled. MA and CA can be used to preserve various foods and remarkably fresh horticulture products during packaging, transportation, and storage. Widespread use, particularly for pears and apples for extended storage durations (Yahia et al., 2019). Utilizing controlled atmosphere storage (CAS) in conjunction with temperature regulation has been a standard method for preserving fresh produce' freshness and prolonging its shelf life.

In contrast to regulating air (oxygen and carbon) concentrations, the efficiency of CAS in preserving the quality of the fruit could be accomplished by controlling humidity. Consequently, alterations in other characteristics, such as the physio-mechanical properties of fruit during storage in CAS, may also depend on the conditions mentioned (Hussein et al., 2015; Ju et al., 2025). After two months of storage, the researcher observed lower damage in CAS-stored cranberries compared with the high frequency of bruises in regular atmospheric air. On the other side, increased cranberry losses were caused by high amounts of super-atmospheric O_2 and CO_2. Under typical harvest and storage circumstances, firmer fruits are much more susceptible to damage (Li et al., 2016).

Modified humidity (MH) wrappers preserve tiny fruit freshness. Litchis, sweet cherries, strawberries, blueberries, Chinese bayberries, apricots, loquats, and cherry tomatoes lost 1.2–4.5 times more water in conventional clamshells than in MH clamshells. Quality characteristics of packaged fruits were typically better retained in MH clamshells, particularly those vulnerable to water stress, such as shriveling, desiccation-induced discoloration and/or drying of cherry pedicels, strawberry calyx, litchi carpels, cherry tomato peel shrinking, and blueberry and strawberry softening. Overall, MH clamshells decreased water loss, preserved fruit quality, and prolonged small fruit storage life without altering the package environment and causing deterioration (Bai et al., 2019).

9.3 METHODS TO REDUCE WATER LOSS IN FRUITS AND VEGETABLES

Decreasing the transpiration coefficient or the vapor pressure deficit minimizes water loss. The transpiration coefficient may be decreased by using any sort of packing, waxing fruit, avoiding mechanical damage, or cutting leaves to lower the leaf surface area (Holcroft, 2015).

9.3.1 *EFFECTS OF TEMPERATURE AND TRANSPORTATION*

Transport distance influences vibration, which has negative impacts, and transportation is sometimes the most expensive aspect of the marketing chain. The mode of transport for fresh fruits and vegetables is dictated by the product's distance, perishability, and cost. The shelf life of fresh food mostly relies on temperature fluctuations during shipping and storage. Temperature influences the freshness of stored items. Temperature affects the rate of biological responses. Thus, it greatly affects product quality. Initial quality, moisture, water loss, ambient gas content, mixed loads, physical damage, stress, and transit circumstances all impact product quality (surface road conditions, time, and so forth). Transporting produce may cause physical, chemical, and microbiological harm. Thus, a complete food safety and quality program must address transportation management. Fresh fruit and vegetables have different safe-transporting needs. Transport fundamentals are the same regardless of technique. It is important to load and unload carefully, and overheating must be prohibited. The product should be well covered, transportation duration should be brief, and water losses should be kept to a minimum (Elik et al., 2019).

In a recent research, the effects of transportation and storage on the physical properties of tomatoes, such as weight loss and hardness were analyzed. The greatest decrease in weight and texture was observed in tomatoes that were transported to the longest distance and kept at 22°C. According to the findings, an increase in transit distance and storage temperature causes greater alterations in the physical properties of tomatoes (Al-Dairi et al., 2021).

Another research studied the quality variations of Ponkan fruit throughout postharvest transport and shelf life under different storage periods and temperatures in order to determine optimal transportation conditions. The continual drop in titratable acidity was increased by the high transport temperature. The findings demonstrated that longer storage time and greater transit temperature increased the degradation rate (Cao et al., 2019).

9.3.2 EFFECTS OF PACKAGING TREATMENT

One of the primary problems with packed vegetables and fruits is condensing, which results in more free water. In contrast to the permeability to water vapor of the packaging, the respiratory and transpiration of the fresh product increase the relative moisture inside the box (Bovi and Mahajan, 2017; Esmaeili et al., 2022) (Figure 9.1). Fresh product water loss may be affected by the kind of container and the placement of the fruit within the package. It is well known that chitosan-based coatings may

FIGURE 9.1 Different techniques of applying edible coatings on fresh fruits.

function as a food preservative. It may be utilized as an elevated macromolecule film-forming agent. Chitosan coatings and films offer superior mechanical qualities and are selectively permeable to gases (O_2 and CO_2) (Xing et al., 2020). However, chitosan-based films and coatings have permeability to water vapor, which restricts their application in humid situations where moisture transfer control is needed (Xu et al., 2018).

9.4 EDIBLE COATINGS

Considering the food coatings, it is possible to avoid dehydration and the microbial population of fresh items (Khaliq et al., 2023; Khan et al., 2025; Lal et al., 2025). In reality, edible coatings put on the surface of the food may extend its storage life by minimizing oxidation processes, gas exchange, solubility migration, and water transmission, and avoiding or reducing physiological disorders (Al-Tayyar et al., 2020; Esmaeili et al., 2022; Ul Haq et al., 2024). Consequently, edible films, which often comprise food supplements or antimicrobial compounds, such as spices, minerals, aromas, coloring agents, and anti-browning agents, are gaining importance as crucial components to mitigate the degrading effects of vegetable and food preparation (El-Sayed, 2020; El-Sayed and Youssef, 2019; Esmaeili et al., 2021) (Table 9.2). The efficiency of the outer coating is highly influenced by the features of the substance; consequently, appropriate edible coatings are selected while paying attention to the chemical and physical qualities of the biopolymer's elements and taking into account the mechanical characteristics and film permeability (Al-Tayyar et al., 2020; Esmaeili et al., 2022; Ul Haq et al., 2024). Due to the potential of edible films to minimize water loss, scent loss, water permeability by the food component, and oxygen absorption, they are recommended as feasible options for preserving fresh vegetables and fruit. Strawberry was coated with a bio-nanocomposite composed of sodium alginate and nano-ZnO created. According to their findings, nano-ZnO greatly enhanced the film's moisture resistance, and consequently, decreased strawberry weight loss. At the conclusion of preservation, untreated fruits demonstrated more weight reduction than coated fruits (20 days) (Emamifar and Bavaisi, 2020). Zhang et al. (2020) demonstrated that 2% chitosan coatings enhanced with 4% extract of banana peels produced a homogeneous and compact construction with good tensile strength and outstanding antioxidant properties. According to the attachment of -OH and -CHO groups of phenolics to the chitosan structure, the humidity and water vapor permeability of these films decreased, hence limiting their interaction with water molecules (Zhang et al., 2020).

Recent research by Basiak et al. (2019) showed that treating a starches and starch-whey proteins edible film to the interface of single plums at high transpiration potentials may significantly enhance the overall resistance in the water vapor route by 60–75%. The results revealed that edible films might decrease the fruit's transpiration loss, which is connected to the limitation of alterations in the intercellular gaps' internal gas environment (Basiak et al., 2019; Ali et al., 2025). Coating compositions use essential oil (EO) extracts from tea tree, thyme, clove, and bergamot. During refrigerated temperature, they enhance antibacterial effects and decrease water loss and respiratory. Some oils can also extend the shelf life of fruits and vegetables (Hasheminejad and Khodaiyan, 2020). Few studies have described the application of EO nanoparticles in edible coatings or films

TABLE 9.2 Some of Recent Applications of Edible Coatings on Fruits.

Edible Coating	Fruit Type	Main Properties	Reference
Alginate and gellan coatings	Fresh-cut processed pear	Edible coatings improved fresh-cut pears' water vapor resistance and ethylene generation	Oms-Oliu et al. (2008)
Alginate and gellan-based coatings	Fresh-cut papaya	Firmness, ascorbic acid retention, and water loss were improved	Tapia et al. (2008)
Cassava starch and sodium alginate	Fresh-cut mango	Coatings slowed breathing. Cassava starch coatings prevented weight loss and preserved mechanical and color qualities.	Chiumarelli and Hubinger (2014)
Chitosan	Fresh-cut papaya	Chitosan slowed water loss effectively.	Chien et al. (2013)
Water-in-oil emulsion and chocolate-based coatings	Apple slices	Efficient in keeping apple slices from browning and minimizing water leakage.	Khan et al. (2014)
Alginate and gellan-based edible coating	Fresh-cut pineapples	Significantly reduced weight loss and breathing rate	Azarakhsh et al. (2012)

and its significance for minimally processed food systems in recent years. Acevedo-Fani et al. (2015) have created edible films built on sodium alginate. They used nanoemulsions of EOs, lemongrass (LG), thyme (TH), and sage (SG), to increase the transparency, flexibility, and resistance to water vapor. According to the findings, SG-EO nanoemulsion-formed films displayed increased transparency, moisture resistance, and elasticity (Acevedo-Fani et al., 2015).

9.5 LIPID-BASED COATINGS

Due to the existence of small holes and high solubility and diffusivity, the use of lipids as a covering solvent provides excellent moisture barriers but weak barrier properties. Included among lipid molecules are glycerides, waxes, and resins. As protective coatings, acetylated monoglycerides, beeswax, carnauba wax, natural oil, paraffin, vegetable oil, and surfactants are widely used. Compared with polysaccharides and protein-based films, lipid-based coatings have superior compatibility with other coating agents and barrier qualities. Due to their oily surface and rancidity, lipid-based layers have negative organoleptic qualities (Sharma et al., 2019). At the conclusion of 56 days of preservation at 4°C and 80% RH, surface waxed oranges (Valencia) with various types of waxes considerably reduced water loss to between 4.0 and 5.8% Compared to the 10.4% recorded in uncoated samples (Motamedi et al., 2018).

9.6 EFFECTS OF CHEMICAL TREATMENT

Numerous additional postharvest variables, such as the application of a chemical treatment, affect the water loss rate in fruit. The impact of pre-harvest spraying of $CaCl_2$ and chitosan, singly and in combination, on the freshness characteristics and storage stability of peaches kept at 0 ± 1°C was investigated. Ten days before harvest, plants were sprayed with 0.5% or 1% chitosan individually or in conjunction with 1 and 2% $CaCl_2$. Results indicated that pre-harvest treatment of 2% $CaCl_2$ + 1% chitosan was most efficient in reducing loss of weight (%) and deterioration (%) (Gayed et al., 2017). Li et al. (2016) demonstrate that post-harvesting treatment of "Yali" pears with 10 mg L^{-1} of 1-methylcyclopropene (1-MCP) utilizing compounds that have the capacity to alter fruit texture without harming the fruit's sensory quality might minimize fruit water loss. The reduced damage amount was attributable to the increased firmness of pears treated with 1-MCP (Li et al., 2016). Compared with untreated fruit, dipping apples (cv. Red Delicious) in $CaCL_2$ solution (0.5–2.0%) alone or in conjunction with gamma irradiation (0.4 k Gy) considerably reduced water loss at the conclusion of a 90-day storage regime at 2°C, 90% RH (Hussain et al., 2012). At the end of 60 days at 2°C, 90% RH, water loss was dramatically reduced by the administration of putrescine pressure-infiltration in treated pomegranate fruit (cv. Mollar de Elche) to around 13%, compared with ~ 17% in untreated fruit (Mirdehghan et al., 2007).

9.7 CONCLUSION

Water loss and humidity-related physiological disorders represent a major constraint to maintaining the quality, marketability, and postharvest life of fruits and vegetables. Low relative humidity environments accelerate transpiration, causing rapid dehydration and associated disorders such as shriveling, wilting in leafy vegetables, and increased susceptibility to mechanical damage and decay. Conversely, excessively high humidity, although effective in reducing water loss, can disrupt normal physiological processes, promote condensation, impair gas exchange, and favor the development of disorders such as surface pitting, lenticel breakdown, peel injury, and water soaking. High humidity also creates conditions conducive to microbial growth, further compounding postharvest losses. The severity of water loss and humidity-related disorders is strongly influenced by commodity type, cultivar, maturity stage, cuticle characteristics, surface-to-volume ratio, and storage duration. Adoption of appropriate technologies such as humidity-controlled cold storage, modified and controlled atmosphere packaging, use of moisture-retentive packaging materials, edible coatings, waxing, and careful handling practices plays a critical role in minimizing water loss and preventing humidity-induced disorders. Equally important is commodity-specific optimization, as the ideal relative humidity range varies widely among fruits and vegetables and must be tailored to their physiological requirements. Understanding the mechanisms of water loss and the role of humidity in postharvest physiology is essential for developing effective strategies to reduce physiological disorders and postharvest losses. Improved management of humidity and transpiration not only extends shelf life but also preserves quality, nutritional value, and consumer appeal. Future advancements in storage technologies, sensor-based monitoring, and cultivar selection for improved water retention will further enhance the sustainable management of water loss and humidity-related physiological disorders in fruits and vegetables.

KEYWORDS

- **fresh fruits**
- **nanoemulsions**
- **post-climacteric**
- **postharvest**
- **relative humidity**
- **shelf life**
- **water loss**

REFERENCES

Abu-Goukh, A., & Elshiekh, F. A. (2008). Effect of waxing and fungicide treatment on quality and storability of grapefruits. *Gezira Journal of Agricultural Science, 6*, 31–42.

Acevedo-Fani, A., Salvia-Trujillo, L., Rojas-Grau, M. A., & Martin-Belloso, O. (2015). Edible films from essential-oil-loaded nanoemulsions: Physicochemical characterization and antimicrobial properties. *Food Hydrocolloids, 47*, 168–177.

Al-Dairi, M., Pathare, P. B., & Al-Yahyai R. Effect of postharvest transport and storage on color and firmness quality of tomato. *Horticulture, 7*, 163.

Ali, S., Khaliq, G., Iftikhar, A., Mir, S. A., & Iqbal, S. (2024). Nanoemulsions-based edible coatings for postharvest quality preservation of fruits and vegetables. In *Sustainable Postharvest Technologies for Fruits and Vegetables* (pp. 213-226). CRC Press.

Al-Tayyar, N. A., Youssef, A. M., & Al-Hindi, R. R. (2020). Edible coatings and antimicrobial nanoemulsions for enhancing shelf life and reducing foodborne pathogens of fruits and vegetables: A review. *Sustainable Materials and Technology, 26*, e00215.

Ariffin, S. H. (2019). Effect of processing on quality and shelf life of leafy vegetables. University of Birmingham.

Arroyo, B. J., Bezerra, A. C., Oliveira, L. L., Arroyo, S. J., De Melo, E. A., & Santos, A. M. P. (2020). Antimicrobial active edible coating of alginate and chitosan add ZnO nanoparticles applied in guavas (Psidium guajava L.). *Food Chemistry, 309*, 125566.

Azarakhsh, N., Osman, A., Ghazali, H., Tan, C., & Mohd Adzahan, N. (2012). Optimization of alginate and gellan-based edible coating formulations for fresh-cut pineapples. *International Food Research Journal, 19*(1), 279–285.

Bai, J., Baldwin, E., Tsantili, E., Plotto, A., Sun, X., Wang, L., Kafkaletou, M., Wang, Z., Narciso. J., & Zhao, W. (2019). Modified humidity clamshells to reduce moisture loss and extend storage life of small fruits. *Food Packaging Shelf Life, 22*, 100376.

Bartz, J. A., & Brecht, J. K. (2002). In *Postharvest physiology and pathology of vegetables*. CRC Press.

Basiak, E., Linke, M., Debeaufort, F., Lenart, A., & Geyer, M. (2019). Dynamic behaviour of starch-based coatings on fruit surfaces. *Postharvest Biology and Technology, 147*, 166–173.

Bovi, G. G., & Mahajan, P. V. (2017). Regulation of humidity in fresh produce packaging. In *Reference module in food science.*

Bovi, G. G., Rux, G., Caleb, O. J., Herppich, W. B., Linke, M., Rauh, C., & Mahajan, P. V. (2018). Measurement and modelling of transpiration losses in packaged and unpackaged strawberries. *Biosystems Engineering, 174*, 1–9.

Caleb, O. J., Mahajan, P., Al-said, F. A., & Opara, U. L. (2013). Transpiration rate and quality of pomegranate arils as affected by storage conditions. *CyTA-Journal of Food, 11*, 199–207.

Cao, J., Wang, C., Xu, S., Chen, Y., Wang, Y., Li, X., & Sun, C. (2019). The effects of transportation temperature on the decay rate and quality of postharvest Ponkan (Citrus reticulata Blanco) fruit in different storage periods. *Scientia Horticulturae, 247*, 42–48.

Castellanos, D. A., Herrara, D. R., & Herrara, A. O. (2016). Modelling water vapour transport, transpiration and weight loss in a perforated modified atmosphere packaging for feijoa fruits. *Biosystems Engineering, 151*, 218–230.

Chien, P.-J., Lin, H.-R., & Su, M.-S. (2013). Effects of edible micronized chitosan coating on quality and shelf life of sliced papaya. *Food and Nutrition Sciences, 04*(09), 9–13.

Chiumarelli, M., & Hubinger, M. D. (2014). Evaluation of edible films and coatings formulated with cassava starch, glycerol, carnauba wax and stearic acid. *Food Hydrocollied, 38*, 20–27.

Cofelice, M., Lopez, F., & Cuomo, F. (2019). Quality control of fresh-cut apples after coating application. *Foods, 8*, 189.

El-sayed, S. M. (2020). Use of spinach powder as functional ingredient in the manufacture of UF-Soft cheese. *Heliyon, 6*, e03278.

El-sayed, S. M., & El-sayed, H. S. (2020). Production of UF-soft cheese using probiotic bacteria and Aloe vera pulp as a good source of nutrients. *Annals of Agricultural Science, 65*, 13–20.

El-sayed, S. M., Ibrahim, O. A., & Kholif, A. M. (2020). Characterization of novel Ras cheese supplemented with Jalapeno red pepper. *Journal of Food Processing and Preservation, 44*, e14535.

El-sayed, S. M., & Youssef, A. M. (2019). Potential application of herbs and spices and their effects in functional dairy products. *Heliyon, 5*, e01989.

Elik, A., Yanik, D. K., Istanbullu, Y., Guzelsoy, N. A., Yavuz, A., & Gogus, F. (2019). Strategies to reduce post-harvest losses for fruits and vegetables. *Strategies, 5*, 29–39.

Emamifar, A., & Bavisi, S. (2020). Nanocomposite coating based on sodium alginate and nano-ZnO for extending the storage life of fresh strawberries (Fragaria× ananassa Duch.). *Journal of Food Measurement and Characterization, 14*, 1012–1024.

Esmaeili, Y., Zamindar, N., & Mohammadi, R. (2022). The effect of polypropylene film containing nano-hydroxyapatite on Physicochemical and microbiological properties of button mushrooms (Agaricus bisporus) under Modified atmosphere packaging. *Journal of Food Measurement and Characterization, 17*(1), 1–14.

Esmaeili, Y., Zamindar, N., Paidari, S., Ibrahim, S. A., & Mohammadi Nafchi, A. (2021). The synergistic effects of aloe vera gel and modified atmosphere packaging on the quality of strawberry fruit. *Journal of Food Processing and Preservation, 45*, e16003.

Fawole, O. A., & Opara, U. L. (2013). Effects of storage temperature and duration on physiological responses of pomegranate fruit. *Industrial Crops and ProdUction, 47*, 300–309.

Gayed, A. A. N. A., Shaarawi, S. A. M. A., Elkhishen, M. A., & Elsherbini, N. R. M. (2017), Pre-harvest application of calcium chloride and chitosan on fruit quality and storability of 'Early Swelling'peach during cold storage. *Ciência e Agrotecnologia, 41*, 220–231.

Gunny, A. A. N., Gopinath, S. C., Ali, A., Wongs-Aree, C., & Salleh, N. H. M. (2024). Challenges of postharvest water loss in fruits: Mechanisms, influencing factors, and effective control strategies–A comprehensive review. Journal of Agriculture and Food Research, 101249.

Hasheminejad, N., & Khodaiyan, F. (2020). The effect of clove essential oil loaded chitosan nanoparticles on the shelf life and quality of pomegranate arils. *Food Chemistry, 309*, 125520.

Holcroft, D. (2015). Water relations in harvested fresh produce: PEF White Paper No. 15-01. *The Postharvest Education Foundation (PEF).*

Hussein, Z., Caleb, O. J., & Opara, U. L. (2015). Perforation-mediated modified atmosphere packaging of fresh and minimally processed produce—A review. *Food Packaging Shelf Life, 6*, 7–20.

Hussein, Z., Fawole, O. A., & Opara, U. L. (2020a). Harvest and postharvest factors affecting bruise damage of fresh fruits. *Horticultural Plant Journal, 6*, 1–13.

Hussein, Z., Fawole, O. A., & Opara, U. O. (2020b). Effects of bruising and storage duration on physiological response and quality attributes of pomegranate fruit. *Scientia Horticulturae, 267*, 109306.

Hussain, P., Meena, R., Dar, M., & Wani, A. (2012). Effect of post-harvest calcium chloride dip treatment and gamma irradiation on storage quality and shelf-life extension of Red delicious apple. *Journal of Food and Science Technology, 49*, 415–426.

Jafarzadeh, S., Nafchi, A. M., Salehabadi, A., Oladzad-Abbasabadi, N., & Jafari, S. M. (2021). Application of bio-nanocomposite films and edible coatings for extending the shelf life of fresh fruits and vegetables. *Advances in Colloid and Interface Science, 291*, 102405.

Ju, W., Lee, J. H., Do, E., Ko, D. Y., Jee, E., Chang, M. S., et al. (2025). Controlled atmosphere storage enhances Korean melon shelf life and quality: A comparative metabolic analysis with reefer container export. *Postharvest Biology and Technology, 219*, 113238.

Khaliq, G. H. U. L. A. M. (2015). Effects of edible coatings enriched with calcium chloride on physiological, biochemical and quality responses of mango (*Mangifera indica* L. cv. Choke Anan) fruit during cold storage. Universiti Putra Malaysia, Serdang, Selangor, Malaysia.

Khaliq, G., Ali, S., Gapper, N., & Nicola, S. (2023). Recent advances and approaches in the application of elicitors to enhance resistance mechanisms in fresh produce. *Frontiers in Sustainable Food Systems, 6,* 1061079.

Khan, L., Khaliq, G., Ullah, E., Abdul-Rahaman, A., El-Mogy, M. M., Kesba, H. H., Ismail, A. M., & El-Beltagi, H. S. (2025). Guggul gum and clove essential oil synergistically reduced anthracnose disease and modulated physicochemical changes in tomato during storage. *Journal of Stored Products Research, 114,* 102713.

Khan, M. K. I., Cakmak, H., Tavman, Ş., Schutyser, M., & Schroen, K. (2014). Anti-browning and barrier properties of edible coatings prepared with electrospraying. *Innovative Food Science and Emerging Technology, 25,* 9–13.

Lal, S., Khaliq, G., Mastoi, A. H., Irshad, A., Jaffar, M. A., Khan, L., & Muhammad, N. (2025). Effect of *Fagonia indica* and garlic extract on *Alternaria alternata* and post-harvest quality of tomato fruit during storage. *Journal of Crop Health, 77*(6), 1–13.

Lentzou, D., Xanthopoulis, G., Templalexis, C., & Kaltsa, A. (2021). The transpiration and respiration as mechanisms of water loss in cold storage of figs. *Food Research, 5*, 109–118.

Li, Z., & Thomas, C. (2014). Quantitative evaluation of mechanical damage to fresh fruits. *Trends in Food Science and Technology, 35*, 138–150.

Li, J., Yan, J., Ritenour, M. A., Wang, J., Cao, J., & Jiang, W. (2016). Effects of 1-methylcyclopropene on the physiological response of Yali pears to bruise damage. *Scientia Horticulturae, 200*, 137–142.

Lobos, T., Retamales, J., Ortega-Farias, S., Hanson, E., López-Olivari, R., & Mora, M. (2016). Pre-harvest regulated deficit irrigation management effects on post-harvest quality and condition of V. corymbosum fruits cv. Brigitta. *Scientia Horticulturae, 207*, 152–159.

Lufu, R., Ambaw, A., & Opara, U. L. (2019). The contribution of transpiration and respiration processes in the mass loss of pomegranate fruit (cv. Wonderful). *Postharvest Biology and Technology, 157*, 110982.

Lufu, R., Ambaw, A., & Opara, U. L. (2020). Water loss of fresh fruit: Influencing pre-harvest, harvest and postharvest factors. *Scientia Horticulturae, 272*, 109519.

Mirdehghan, S., Rahemi, M., Castillo, S., Martinez-Romero, D., Serrano, M., & Valero, D. (2007). Pre-storage application of polyamines by pressure or immersion improves shelf-life of pomegranate stored at chilling temperature by increasing endogenous polyamine levels. *Postharvest Biology and Technology, 44*, 26–33.

Mohammad Shafie, M., Rajabipour, A., Castro-Garcia, S., Jimenez-Jimenez, F., & Mobli, H. (2015). Effect of fruit properties on pomegranate bruising. *International Journal of Food Properties, 18*, 1837–1846.

Motamedi, E., Nasiri, J., Malidarreh, T. R., Kalantari, S., Naghavi, M. R., & Safari, M. (2018). Performance of carnauba wax-nanoclay emulsion coatings on postharvest quality of 'Valencia'orange fruit. *Scientia Horticulturae, 240*, 170–178.

Mukama, M., Ambaw, A., Berry, T. M., & Opara, U. L. (2019). Analysing the dynamics of quality loss during precooling and ambient storage of pomegranate fruit. *Journal of Food Engineering, 245*, 166–173.

Oms-Oliu, G., Soliva-Fortuny, R., & Martin-Belloso, O. (2008). Edible coatings with antibrowning agents to maintain sensory quality and antioxidant properties of fresh-cut pears. *Postharvest Biology and Technology, 50*, 87–94.

Pereira, E., Silva, R. G. B., Spagnol, W. A., & Silveira Junior, V. (2017). Water loss in table grapes: Model development and validation under dynamic storage conditions. *Food and Science Technology, 38*, 473–479.

Riva, S. C., Opara, U. O., & Fawole, O. A. (2020). Recent developments on postharvest application of edible coatings on stone fruit: A review. *Scientia Horticulturae, 262*, 109074.

Romanazzi, G., Feliziani, E., Baños, S. B., & Sivakumar S. D. (2017). Shelf life extension of fresh fruit and vegetables by chitosan treatment. *Critical Reviews in Food Science and Nutrition, 57*, 579–601.

Rossi Marquez, G., Di Pierro, P., Esposito, M., Mariniello, L., & Porta, R. (2014). Application of transglutaminase-crosslinked whey protein/pectin films as water barrier coatings in fried and baked foods. *Food and Bioprocess Technology, 7*, 447–455.

Sharma, P., Shehin, V., Kaur, N., & Vyas, P. (2019). Application of edible coatings on fresh and minimally processed vegetables: A review. *International Journal of Vegetable Science, 25*, 295–314.

Shewfelt, R. L., & Prussia, S. E. (2022). Challenges in handling fresh fruits and vegetables. In *Postharvest handling.* Elsevier.

Tapia, M., Rojas-Grau, M., Carmona, A., Rodriguez, F., Soliva-Fortuny, R., & Martin-Belloso, O. (2008). Use of alginate-and gellan-based coatings for improving barrier, texture and nutritional properties of fresh-cut papaya. *Food Hydrocollied, 22*, 1493–1503.

Tyagi, S., Sahay, S., Imran, M., Rashmi, K., & Mahesh, S. S. (2017). Pre-harvest factors influencing the postharvest quality of fruits: A review. *Current Journal of Applied Science and Technology, 23*, 12.

Ul Haq, Z., Khaliq, G., Irshad, A., El-Beltagi, H. S., Mahmoud Ismail, A., & El-Mogy, M. M. (2024). Effect of beeswax and lemon grass oil on Alternaria alternata and post-harvest quality of bitter gourd during storage. New Zealand Journal of Crop and Horticultural Science, 1-21.

Wang, D. (2016). Novel primers for increased specificity and sensitivity for the detection of Staphylococcus aureus by real-time LAMP. *CyTA-Journal of Food, 14*, 88–91.

Xanthopoulos, G. T., Templalexis, C. G., Aleiferis, N. P., & Lentzou, D. I. (2017). The contribution of transpiration and respiration in water loss of perishable agricultural products: The case of pears. *Biosystems Engineering, 158*, 76–85.

Xing, Y., Yang, H., Guo, X., Bi, X., Liu, X., Xu, Q., Wang, Q., Li, W., Li; X., & Shui, Y. (2020). Effect of chitosan/Nano-TiO2 composite coatings on the postharvest quality and physicochemical characteristics of mango fruits. *Scientia Horticulturae, 263*, 109135.

Xu, D., Qin, H., & Ren, D. (2018). Prolonged preservation of tangerine fruits using chitosan/montmorillonite composite coating. *Postharvest Biology and Technology, 143*, 50–57.

Yahia, E. M., Fadanelli, L., Matte, P., & Brecht, J. K. (2019). Controlled atmosphere storage. *Postharvest technology of perishable horticultural commodities.* Elsevier.

Zhang, W., Li, X., & Jiang, W. D (2020). Development of antioxidant chitosan film with banana peels extract and its application as coating in maintaining the storage quality of apple. *International Journal of Biological Macromoleculars, 154*, 1205–1214.

Zhao, H., Wang, B., Cui, K., Cao, J., & Jiang, W. (2019). Improving postharvest quality and antioxidant capacity of sweet cherry fruit by storage at near-freezing temperature. *Scientia Horticulturae, 246*, 68–78.

CHAPTER 10

Storage Atmosphere-Related Physiological Disorders

SAJID ALI[1*], AAMIR NAWAZ[1], SAFINA NAZ[1], SHAGHEF EJAZ[1], GHULAM KHALIQ[2], and MUHAMMAD AZAM[3]

[1]*Department of Horticulture, Faculty of Agricultural Sciences and Technology, Bahauddin Zakariya University, Multan, Punjab, Pakistan*

[2]*Department of Horticulture, Faculty of Agriculture, Lasbela University of Agriculture, Water and Marine Sciences, Uthal, Pakistan*

[3]*Institute of Horticultural Sciences, University of Agriculture, Faisalabad, Punjab, Pakistan*

*Corresponding author

ABSTRACT

Fresh produce such as fruits and vegetables contains significantly higher water content. The higher water content makes the fresh produce susceptible to desiccation-induced mass loss after harvest. For this reason, fresh produce is generally stored under low-temperature conditions to inhibit desiccation-induced postharvest losses. Low-temperature storage (LTS) is not always appropriate, as it often leads to the development of certain physiological disorders, such as chilling injury. There are different types of storage atmospheres, including cold storage, modified atmosphere, controlled atmosphere, dynamic controlled atmosphere, hyperbaric, and hypobaric storage. The aforementioned storage types have been found highly effective in the extension of fresh produce storage life along with

Physiological Disorders of Fruits and Vegetables: Causes and Remedies.
Ghulam Khaliq, Sajid Ali, and Mohammed Wasim Siddiqui (Eds.)

DOI: 10.1201/9781998511334-10

maintained quality. At the same time, the said storage atmospheres are not always productive and lead to the development of certain physiological disorders. In this chapter, various storage atmospheres associated with physiological disorders have been summarized.

10.1 INTRODUCTION

Fruits and vegetables contain a high moisture content and tend to desiccate quickly if not stored under appropriate low-temperature conditions after harvest. Different factors affect the storage potential of harvested fruits and vegetables. The storage of fruits and vegetables under certain conditions may lead to the start and development of various physiological disorders (Zhang et al., 2021; Ali et al., 2022). It has been reported that changes in certain physical, physiological, metabolic, biochemical, and molecular aspects start in the harvested fruits and vegetables during postharvest storage (Aghdam and Bodbodak, 2013). The postharvest storage environment has been categorized into different types, and each storage atmosphere has certain merits and demerits depending upon the storage time and storage conditions.

The physiological, biochemical, metabolic, and molecular mechanisms and/or alterations are generally modulated by the temperature, storage environment, and storage atmosphere conditions that subsequently determine the extent of the onset and severity of physiological disorders in the harvested fruits and vegetables (Khaliq et al., 2023; Khan et al., 2025; Lal et al., 2025). The storage of fresh fruits and vegetables under low-temperature conditions has a long conservation history and is still being used in the different stages of the supply chains in the world (Zhang et al., 2021). Cold and/or low-temperature storage (LTS) regulates certain metabolic, biochemical, and physiological cascades, which ultimately result in prolonged storage life. At the same time, it often results in a physiological disorder, chilling injury in various tropical and subtropical fruits and vegetables (Imahori et al., 2008).

The other storage types include modified atmosphere, controlled atmosphere (CA), hyperbaric storage, hypobaric storage, and dynamic controlled atmosphere (DCA). The said storage atmosphere generally results in the extended storage life of stored horticultural produce along with conserved eating quality and overall acceptability. Nevertheless, all of the aforementioned storage atmospheres sometimes result in certain disorders of a physiological nature. The developed physiological disorders may result from higher and/or lower concentrations of oxygen and carbon dioxide in the modified, controlled, and DCA storage conditions (Koukounaras et al., 2019; Lee et al., 2022; Mattheis et al., 2022; Coser et al., 2023). At the same time, the stowage of fruits and vegetables under hyperbaric and hypobaric storage conditions has also been found to regulate the prevalence and occurrence of certain types of disorders which detrimentally influence the overall quality of the stored produce. It is evident that different types of storage atmospheres affect the onset and development of various types of physiological disorders in harvested horticultural produce. So, in this chapter, various physiological disorders development under different types of storage atmospheres have been discussed along with future research perspectives.

10.2 STORAGE ATMOSPHERES

There are different types of storage atmospheres that have been used and are being utilized in the modern-day postharvest science and technology of fruits and vegetables for the purpose of prolonging their storage life potential along with conservation of eating quality. The brief details about the various types of postharvest storage atmospheres have been described below.

10.2.1 SIMPLE COLD STORAGE

The simplest type of storage is cold and/or LTS. In the LTS, the temperature and relative humidity levels are controlled according to the required level. It has been reported that the level of relative humidity and temperature range should be within the optimal and/or tolerant level specifically required for the stored vegetables and fruits. The temperature below the range of tolerance generally leads to the onset and development of certain freezing and chilling-related disorders (Sevillano et al., 2009). The management of relative humidity is a major aspect to control microbial (microorganism) infestations, desiccation, shriveling symptoms, and the manifestation of certain types of physiological disorders in harvested horticultural produce during storage (Neuwald et al., 2019; Zuo et al., 2021). The availability of higher relative humidity and low range of temperature generally results in extended storage potential and conservation of the overall quality in comparison with ambient temperature storage of fresh horticultural produce (Han et al., 2021; ElManawy et al., 2025). The conservation of the quality and extension of the storage life in cold storage generally depends upon the storage duration and temperature range (Lima et al., 2023; Ali et al., 2022a).

10.3 COLD STORAGE-RELATED DISORDERS OF FRESH PRODUCE

The LTS is considered one of the major strategies to slow down senescence and ripening of fresh produce after harvesting. It is important to mention that LTS does not effectively reduce the release of ethylene from the stored produce. Therefore, the increased concentration of the released ethylene may stimulate the senescence of different fresh produce. It has been reported that LTS stimulates cap expansion and stem elongation in mushroom, russet spots in lettuce (Schotsman et al., 2009), bitter or altered taste in carrot (Wills and Golding, 2016), and enhanced ripening along with softening of kiwifruits and persimmons (Thompson, 2016). It is very important to mention that the increased ethylene concentration in the storage rooms under LTS generally results in early and prompt senescence of the stored produce (Neuwald et al., 2019). The increased respiration is also considered imperative during storage of produce as higher CO_2 levels enhance metabolic activity, thereby eventually resulting in enhanced concentration of the simple sugars upon the breakdown of starch contents of a produce (Bachmann and Earles, 2000; Ali et al., 2022b). It is also reported that increased starch hydrolysis along with higher desiccation may lead to mealiness in fruits and leafy vegetables as well as fruits wilting (Neuwald et al., 2019). The major senescence symptoms appear on asparagus spears in the form of lignification or shriveling owing to water loss upon storage under storage conditions of a suboptimal environment (Techavuthiporn and Boonyaritthongchai, 2016). It has also been reported that increased concentrations of CO_2 under LTS lead to the browning of lettuce midrib and apple skin burning (Wills and Golding, 2016). On the other side, LTS also results in the occurrence of chilling injury in various tropical and subtropical horticultural crops. For example, LTS exhibited symptoms of chilling injury on papaya (Tao et al., 2022), aonla (Ali et al., 2022a), banana

(Ali et al., 2022b), and pomegranate (Islam et al., 2022) fruits (Table 10.1). It has also been reported that LTS results in loss of quality and alteration in the production of aroma volatiles in harvested peach fruits owing to chilling injury-induced altered metabolism (Farcuh and Hopfer, 2023). In addition, the browning of banana peel (Ali et al., 2022b), pomegranate husk (Islam et al., 2022), and guava skin discoloration (Ali et al., 2022c) has also been reported as an indicator of LTS disorders (Table 10.1).

TABLE 10.1 Chilling Injury Symptoms and Temperatures of Different Fresh Horticultural Produce.

Produce	Chilling Temperature	Generalized Chilling Injury Symptoms	References
Aonla	6°C	Surface browning and pitting	Ali et al. (2022a)
Apricot	7°C	Flesh browning	Wang et al. (2016)
Avocado	13°C	Flesh browning	Woolf et al. (2020)
Bamboo shoots	5°C	Flesh browning and water-soaked areas	Luo et al. (2012)
Banana	13°C	Peel browning, pitting, and failure to normally ripen	Khademi et al. (2019)
Citrus	10°C	Rind pitting and discoloration	Pérez-Gago and Palou (2020)
Cucumbers	10°C	Surface pitting and dark water-based patches	Yang et al. (2011)
Eggplant	10°C	Pitting, peeling and pulp browning, and seed browning	Huang et al. (2019)
Guava	12°C	Skin discoloration and failure to ripen	Murmu and Mishra (2018)
Hawthorn	13°C	Surface pitting/discoloration	Aghdam et al. (2018)
Kiwifruits	5°C	Brown and blackish peel, grainy tissues, and subcutaneous tissue lignification	Suo et al. (2018)
Lemons	5°C	Skin discoloration and depressions on the fruit surface	Martínez-Hernández et al. (2017)
Loquat	12°C	Flesh woodiness and internal flesh browning	Cai et al. (2006)
Mango	5°C	Scald-like discoloration, pitting and browning on peel, and discoloration	Cantre et al. (2017)
Melons	10°C	Peel browning, discoloration, and pitting on fruit surface	Zhang et al. (2017)

TABLE 10.1 *(Continued)*

Produce	Chilling Temperature	Generalized Chilling Injury Symptoms	References
Mume	10°C	Surface pitting and flesh softening	Endo et al. (2019)
Okra	10°C	Pod skin browning and pitting or translucency	Huang et al. (2012)
Papaya	9°C	Skin pitting, scald, water-soaked flesh, and abnormal or failed ripening	Pan et al. (2017)
Peaches/ Nectarines	8°C	Internal flesh browning, woolly textures, and lack of juiciness	Zhang et al. (2020)
Persimmons	4°C	Skin and flesh browning, gelling of flesh, and juiciness loss	Li et al. (2018)
Pitaya	10°C	Bract browning, withering, and shriveling occurred	Sheng et al. (2021)
Plum	8°C	Flesh browning, gel breakdown, bleeding or reddening, and flavor loss	Wang et al. (2016)
Pomegranate	15°C	Husk browning, husk scald, surface pitting, and increased decay	Pareek et al. (2015)
Rambutan	5°C	Thorns and pericarp browning	Kondo et al. (2001)
Sweet peppers	8°C	Black watery stains, superficial pitting, and browning	Seo et al. (2020)
Tomatoes	12°C	Surface pitting, blotchy coloration, and prompt softening	Aghdam et al. (2019)
Zucchini	10°C	Surface pitting, sunken lesions, and increased softening	Bokhary et al. (2020)

Source: Modified and updated from Zhang et al. (2021).

10.4 MODIFIED ATMOSPHERE PACKAGING

Modified atmosphere packaging (MAP) is a passive or active process of gaseous modification within the package of stored produce. The passive MAP technology only relies on the initial prevailing composition of the gases along with the interaction between fresh produce metabolic activities and permeation of the gas via the material of the packaging, particularly film packaging (Oliveira et al., 2015; Jalali et al., 2017). On the other side, the active approach of MAP technology depending upon the desired gas composition is normally established by incorporating it into the packages for the purpose of achieving quick atmosphere equilibrium (within the packaging material) and reducing metabolic activities (Oliveira et al., 2015; Jalali et al., 2017; Ali et al., 2019a). The technology of MAP is considered commercially applicable as this is a comparatively cheaper approach and does not change the taste of a stored commodity without probable detrimental effects (on the other side, chemical treatments may sometimes alter the taste of treated produce) and is easy to use (Oliveira et al., 2015). The MAP has been used for different fresh commodities, and the specific gas mixture in packages generally depends upon the produce type, storage temperature, and the types of packaging material (Oliveira et al., 2015). The effect of the different gas compositions has also been observed. It has been reported that MAP-induced higher concentrations of CO_2 generally delay and/or inhibit enzymatic reactions based on browning (Ali et al., 2019a) along with prevented anaerobic-based fermentation and reduced anaerobic and aerobic microbes (Vander der Steen et al., 2002; Oliveira et al., 2015).

10.5 MAP-RELATED DISORDERS OF FRESH PRODUCE

The different rates of respiration and ethylene biosynthesis, along with the conditions of the storage, are some of the key issues of MAP technology (Kader, 2002). The drastic changes in the rates of respiration alter the composition of the gases within MAP bags which thereby detrimentally influencing the quality of fruits and vegetables (Thompson, 2016). The higher accumulation of moisture (owing to lower water vapor transmission rates) generally increases the incidence of decay (Mahajan et al., 2014). The brown strains of lettuce, blackheart in potatoes, and internal pear and apple browning, as well as sprouting in root vegetables and tubers such as potatoes, are other major disorders of fresh produce that develop under MAP conditions (Kader, 2002). The abscission based on ethylene action is an imperative constraint in different ornamental plants. The increased ethylene biosynthesis and production often lead to quick senescence and the complete loss of ornamental plant leaves and flowers (Sandhya, 2010). The increased ethylene concentration leads to blasting in tulip bulbs. The low O_2 concentration in MAP along with higher CO_2 levels generally results in reduced metabolic activity in terms of ethylene generation and respiration rate. The earlier reports show that a 1% O_2 concentration in MAP was successful in improving the shelf life of Asiatic hybrid lily bulbs (Legnani et al., 2004). At the same time, increased CO_2 levels within MAP bags may result in increased senescence and leaf yellowing due to enhanced concentration of CO_2 and higher buildup of ethylene (Legnani et al., 2006). Water loss and /or desiccation are the major factors limiting the cut flowers' market life. The desiccation in combination with increased ethylene concentration further aggravates the senescence phenomenon. The water loss generally results in shriveling and wilting desiccation-induced mass loss, prompt abscission, and accelerated senescence. Although these negative aspects can be successfully controlled with the application of MAP technology, the increased ethylene buildup in the MAP bags may also lead to the senescence of stored produce (Sandhya, 2010).

10.6 HYPOBARIC AND HYPERBARIC ATMOSPHERE STORAGE

Hypobaric storage is a type of storage in which the fresh produce is stored under negative pressure or a partial vacuum. In contrast with CA storage with the normal pressure of the atmosphere, the air pressure reduction, for instance, by 10 MPa is equivalent to the decrease of 2 kPa level of O_2 (Wills and Golding, 2016). The storage facility with a partial vacuum requires air circulation to add water for the purpose of conserving relative humidity and partial O_2 (Thompson, 2016; Neuwald et al., 2019). The advantage of hypobaric conditions of the storage is the removal of ethylene produced in the endogenous tissues of the produce stored (Burg, 1975). The senescence and prompt ripening process of fresh produce are also decreased during storage under a hypobaric environment because the atmospheric pressures decline during storage facility is equivalent to the decrease of partial O_2 (Wills and Golding, 2016; Neuwald et al., 2019). It has been reported that hypobaric storage is very expensive, and the higher cost of construction is the major constraint in its commercial availability. So, hypobaric storage is sometimes used for the transportation of significantly higher-value-associated fresh commodities (Wills and Golding, 2016; Neuwald et al., 2019). Hypobaric storage has been investigated on jujube (Li et al., 2017), pear (Wang et al., 2007), peach (Wang et al., 2015), and kiwifruit (Huan et al., 2021).

Overpressure or hyperbaric is a type of storage, that is, opposite to hypobaric conditions. The quality of apple fruits was maintained under hyperbaric conditions with 15 MPa at 3–4°C even in the presence of certain disease infestations (Neuwald et al., 2019). On the other side, upon the storage of fruits under normal air conditions and environment, the incidence of diseases was significantly increased (Jannasch et al., 1971). The higher pressure storage has been found to reduce freezing temperatures an anticipated stored produce thereby resulting in ice formation (Rostogi, 2013). Very limited literature is available about the hyperbaric storage work on fresh produce probably due to its resemblance with the hypobaric type of storage and needs very expensive facility and appears to be restricted to certain perishable and higher value fresh produce (Wills and Golding, 2016; Neuwald et al., 2019).

10.7 PHYSIOLOGICAL DISORDERS UNDER HYPOBARIC AND HYPERBARIC ATMOSPHERE STORAGE

The hypobaric storage conditions generally result in the removal of ethylene which can help in the reduction of certain physiological disorders of different fresh produce types (Neuwald et al., 2019). On the other side, higher ethylene concentration build-up during storage may result in the aggravation of certain physiological disorders (Neuwald et al., 2019). The higher ethylene production and build-up lead to russet spotting in lettuce (Burg, 1975). The increased incidence of soft scald and internal breakdown has been found associated with low O_2 levels as noted in pear and apple (Burg, 2004). The incidence of bitter pit, internal browning, flesh browning, core flesh, and Jonathan spot can be reduced which are induced in response to LTS (Burg, 2004; Neuwald et al., 2019).

Hyperbaric storage is generally not appropriate for fresh produce such as fruits and vegetables. This is because of high pressure-induced damages which are considered irreversible in nature and greater than 100 Mpa (Thompson, 2016). Hyperbaric storage can lead to excessive softening of the flesh as noted in strawberry and apple (Master et al., 2004). In addition, the negative influences of hyperbaric storage have also been reported in terms of softening, ethylene production, and sprouting in certain crops (Thompson, 2016; Neuwald et al., 2019).

10.8 CONTROLLED ATMOSPHERE STORAGE

Controlled atmosphere storage is a type of technology in which the composition of the gases can be manipulated based on the nature and types of the commodities with the purpose of storage life extension along with maintenance of the overall quality of the stored commodity (Ali et al., 2016, 2021). The increased concentrations of CO_2 and reduced levels of O_2 reduce the metabolic activities of the stored commodity (Table 10.2). In addition, the higher level of relative humidity results in delayed loss of turgidity and delayed degradation of chlorophyll pigments along with extended shelf life of the produce as noted in the case of fresh broccoli (Fernández-León et al., 2013). In the same way, CA storage has been found effective in suppressing the decline of bioactive compounds of a produce (Vallejo et al., 2003; Koyuncu et al., 2019). It has also been reported that CA storage inhibits (due to higher CO_2 and lower O_2 availability) the activity of pro-oxidative enzymes and stimulates the activities of antioxidative enzymes, thereby resulting in the reduction of enzymatic browning in harvested litchi fruits (Ali et al., 2016). In another work, CA storage in combination with oxalic and ascorbic acids significantly suppressed the browning degree and soluble quinones content along with higher antioxidative and lower pro-oxidative enzyme activity which thereby resulted in substantially reduced incidence of enzymatic browning in "Gola" litchi fruits during LTS conditions (Ali et al., 2021).

TABLE 10.2 Different Methods of CA Storage and Gas Compositions During Different Atmospheres.

Methods of CA	N_2 (kPa)	O_2 (kPa)	CO_2 (kPa)
SAS	79	21	0.03
Ventilation CA	79	6–18	3–15
CCA	90–92	2–5	2–5
LO	95–97.5	1.5–2	1–3
ULO	96.8–98.7	0.8–1.2	0.5–2
DCS/DCA	>98	<0.8	<1.5

Source: Adapted from Hoehn et al. (2009) and Bodbodak and Moshfeghifar (2016).

SAS, simple (normal) air storage; CA, controlled atmosphere; CCA, conventional CA storage; LO, low oxygen; ULO, ultra-low oxygen; DCS, dynamic controlled atmosphere storage; DCA, dynamic control system.

10.9 CA STORAGE-RELATED DISORDERS OF FRESH PRODUCE

The CA storage utilizes the effects of low O_2 and high CO_2 to regulate the postharvest physiology of fresh produce. There are various positive influences of low O_2 and high CO_2 (Ali et al., 2016, 2021). At the same time, the aforementioned gaseous composition also negatively affects the physiology of fresh produce during postharvest storage (Table 10.3). It has been reported that low O_2 may cause internal flesh discoloration, skin pitting and surface browning in nectarines and peaches (Wang et al., 1990), red heart or meristem browning in cabbage (Berard, 1994), off-flavors and discoloration (Lipton, 1990), hardcore in kiwifruits (Arpaia, 1990), and meristem discoloration in brussel sprout (Lipton and Mackey, 1987). The elevated concentration of CO_2 also leads to aggravated chilling injury incidence in grapefruits (Purvis and Cubbedge, 1990) along with brown stains, brown heart, and leaves injury of lettuce (Schotsman et al., 2009). Similarly, CA storage increases flesh browning as well as core browning (Neuwald et al., 2015) and corking in apples, external and internal browning in mushrooms, necrotic and black midrib as well spotting of cabbage and peel pitting, discoloration and rind collapse in citrus fruits (Schotsman et al., 2009). In contrast to simple CA, dynamic CA storage has been found to suppress the incidence of certain physiological disorders such as apple fruits' superficial scald better (Neuwald et al., 2019).

TABLE 10.3 Selected Physiological Disorders Reported for Different Fresh Horticultural Produce During CA and DCA Storage Systems.

Produce	Physiological Disorder	Storage Type	References
Lettuce	Russet spotting	Hypobaric	Burg (1975)
Pear	Soft scald	Hypobaric	Burg (1975)
Brussels sprout	Meristem discoloration	CA	Lipton and Mackey (1987)
Kiwifruit	Hardcore	CA	Arpaia (1990)
Peach	Internal browning	CA	Wang et al. (1990)
Nectarine	Internal browning	CA	Wang et al. (1990)
Honeydew	Skin discoloration	CA	Lidster et al. (1990)
Cabbage	Meristem browning	CA	Berard (1994)
Loquat	Internal browning	CA	Ding et al. (1999)
Lettuce	Brown strains	MAP	Kader (2002)
Potato	Blackheart	MAP	Kader (2002)
Apple	Internal browning	MAP	Kader (2002)
Pear	Flesh browning	MAP	Kader (2002)
Apple	External CO_2 injury	DCA	Zanella et al. (2004)
Apple	Epidermal purpling	DCA	DeLong et al. (2006)
Pear	Internal browning	DCA	Lafer (2007)
Apple	Corking	CA	Schotsman et al. (2009)

TABLE 10.3 *(Continued)*

Produce	Physiological Disorder	Storage Type	References
Lily	Ethylene-induced senescence	MAP	Legnani et al. (2004)
Apple	Excessive softening	Hyperbaric	Thompson (2016)
Strawberry	Excessive softening	Hyperbaric	Thompson (2016)
European pear	Senescent scald and internal breakdown	CA	Lum et al. (2017)
Apple	Flesh breakdown	CA	Weber et al. (2020)
Apple	Mealiness and flesh breakdown	DCA	Thewes et al. (2021)
Apple	Mealiness and flesh breakdown	DCA	Wendt et al. (2022)

CA, controlled atmosphere; MAP, modified atmosphere packaging; DCA, dynamic controlled atmosphere.

The storage scald is a physiological disorder that occurs in response to a lower concentration of O_2. The lower concentration of O_2 (0.7 kPa) significantly increased the "Fuji" apple storage scald (Brackmann and Bortoluzzi, 1998). Bauchot et al. (1999) reported that storage scald in response to N_2-induced anaerobiosis within 2–4 days in the apple fruit cultivar "Granny Smith." The storage scald incidence-related symptoms appeared within the time period of only a few minutes after shifting the fruits to normal air conditions. The authors believed that low O_2-induced stress of post-anoxic conditions was accountable for browning. There was no evidence of α-farnesene catabolism in the storage scald development. The late storage corking was aggravated in response to low or high O_2 in apple fruits (Sharples and Johnson, 1987). It has been reported that loquat fruits are susceptible to internal browning during storage. The higher CO_2 (12 kPa) conditions resulted in a high incidence of internal browning in loquat irrespective of the available O_2 concentration (Ding et al., 1999).

Under certain conditions, elevated CO_2 or low O_2 also leads to the induction of certain disorders of physiological nature in fresh plant produce (Prange and Delong, 2006). The chlorophyll degradation sometimes stops in response to elevated CO_2 and low O_2 in different fruits including tomato, banana, and cabbage (Eaks, 1956; Prange and Lidster, 1991; Menniti et al., 1997). The elevated CO_2 or low O_2 generally influences the concentration of red pigments, that is, anthocyanins, resulting in color change from red to purple or blue (Prange and Delong, 2006). The low O_2 level affected the color of strawberries and apples, and high CO_2 concentration changed the red coloration in sweet cherry fruit (Lidster et al., 1990). The epidermal darkening was found to correlate with the degradation of red color in the internal tissues of strawberry fruits in response to elevated CO_2 or low O_2 concentrations (Lidster et al., 1990; Holcroft and Kader, 1999). The petal bluing was noted in anthocyanins containing rose petals under the conditions of abnormal CA, that is, low O_2 and elevated CO_2 environment (Prange and Delong, 2006). It was observed that increased concentration of CO_2 aggravated the chilling injury in cucumbers (Eaks, 1956). Similarly, increased CO_2 from 5 to 70 kPa at 5 or 0°C significantly aggravated chilling injury-like symptoms (Kader and Morris, 1975). The scald-like spots were noted under high CO_2 conditions on grapefruit and orange (Lidster et al., 1990) and honeydew melon skin discoloration (Lidster et al., 1990). The aggravated chilling injury symptoms were noted on "Chondrolia" olive fruits in response to extended exposure to higher CO_2 at 5°C storage (Nanos et al., 2002).

Pears and apples are the two major fruits that are stored in the CA environment. The abnormal condition of CA storage sometimes leads to the occurrence of certain physiological disorders (Prange and Delong, 2006). The elevated CO_2 and low O_2 exhibit peculiar internal and external disorders. The development of the aforementioned disorders may vary with the length of CA storage and cultivars. Different apple fruit cultivars exhibit internal flesh breakdown (flesh browning) as noted in "Fuji," "Braeburn," "Pacific Rose," and "Bramley's Seedling." The CA-stored apple fruits show severe internal browning incidence upon transfer (CA stored) to normal air conditions. In some cases, internal browning disorder is elevated CO_2 induced but sometimes low level of O_2 also increases the incidence of internal browning (Lau, 1998; Chung et al., 2005; Volz et al., 1998; Elgar et al., 1998; Johnson et al., 1998; Colgan et al., 1999; Maguire and Mackay, 2003). The brown core or elevated CO_2-induced chilling-related injury, brown heart and pithy brown core (Neuwald et al., 2019), and flesh browning (internal flesh breakdown) are major physiological disorders induced or aggravated by elevated CO_2 concentration (Prange and Delong, 2006; Neuwald et al., 2019).

The scald-like disorder was noted in CA-stored apple fruits on the cultivar "Empire." The said disorder was observed in the fruits of "Empire" apples kept under the storage conditions of normal air (Burmeister and Dilley, 1995). The authors named it a chilling-related disorder since it only developed in CA storage having elevated concentrations of CO_2, and it was more severe at 1°C in contrast with the fruits stored at 3°C. This disorder can be reduced by temperature conditioning under the storage temperature of 3°C at 1.5 to 3 kPa level of O_2 for 3–4 weeks in the absence of CO_2 before CA environment at 3 kPa CO_2 along with 1.5 kPa O_2 (Wang et al., 1997, 2000). Pear fruits cultivars "Blanquilla," "Conference," "Doyenne du Comice," and "Rocha" exhibit core browning, and internal browning of the flesh and cavity under low O_2 concentrations particularly along with the elevated level of CO_2 (Larrigaudière et al., 2001; Pintó et al., 2001; Galvis-Sánchez et al., 2006). In contrast, Veltman et al. (2003) reported that no disorder was observed in "Conference" pear fruits when stored with or without CO_2 in combination with 0 kPa O_2. In another work, internal breakdown disorder was noted in "Doyenne du Comice" when the fruits of pear were kept in CA conditions, and it was correlated with reduced O_2 concentration in contrast with an elevated level of CO_2 (Ma and Chen, 2003). The authors were of the opinion that this disorder was probably misidentified in the form of flesh breakdown which is considered a senescence-related physiological disorder. The significantly reduced incidence of internal breakdown was observed in "Bosc" pear fruits when these were stored in CA combination of 3 kPa CO_2 and 1.5 kPa O_2 in contrast with 1 kPa level of CO_2 (Drake, 1999). The author proposed this type of contrary response to replace a static condition of CA storage with a purge-through (flow-through) system of CA conditions.

General leaf browning is another major senescence-associated physiological disorder in leafy vegetables. This increases due to the catabolism of chlorophyll pigments. It has been reported that elevated concentration of CO_2 significantly increases the incidence of leaf browning of a general nature. The discoloration of stem (vascular) tissues was observed in response to higher CO_2 (10 kPa) levels in Brussels sprouts (Lidster et al., 1990). It has been reported that a lower concentration of O_2 (below 1 kPa) is not appropriate for lettuce storage. A physiological disorder in the form of the water-soaked leaf was noted under CA storage with the level of 1 kPa O_2 (Lidster et al., 1990). In the same way, higher CO_2 is also not good for lettuce. The higher concentration of CO_2, that is, 2 kPa along with reduced O_2 availability generally leads to brown spot disorder in lettuce (Lidster et al., 1990; Prange and Harrison 1993; Varoquaux et al., 1996; Kim et al., 2005). This type of disorder can be reduced with proper management. It has been described that a 12 h delay before exposure to CO_2 substantially reduces brown stains/spots in Romaine lettuce (Kim et al., 2005). Different fruits show varying types of physiological disorders. For example, "Hayward" kiwifruits show two major types of physiological disorders under the conditions of CA storage (Lidster et al., 1990). The hard core has been found to be associated with elevated concentrations of CO_2 in the CA environment. On the other side, the white core has been found to be associated with CA containing a higher concentration of ethylene. On the other side, the white core does not occur under ethylene-free CA or during the storage of normal air composition (Lidster et al., 1990).

Among the different factors, maturity is considered an imperative aspect during the development of brown skin or skin discoloration during low O_2 concentration CA storage. Fruits of bananas when stored under the atmosphere of pure N_2 (0 kPa O_2) for the period of three days right after the harvesting exhibited an extended period of green life (delayed ripening and discoloration) with the lowest incidence of brown peel disorder under low O_2 storage of CA conditions. The receptacle tissue and bract blackening of artichokes only occur under CA storage. This is not clear whether the said disorder is owing to elevated CO_2 along with lower O_2 or due to chilling injury (Lidster et al., 1990). Flesh browning is a imperative disorder of Asian pear fruits. It has been reported that lower concentration O_2 (<3.0 kPa) normally causes the browning of flesh in "Asian" pear fruits. On the other side, the lower O_2 (<1.0 kPa) level results in discoloration of epidermal in "Asian" pears (Lidster et al., 1990). The florets of cauliflower are prone to low O_2-induced discoloration, along with altered flavor and odor of cooked curd. The higher CO_2 injury has been found associated with mushy and grey cooked curd of cauliflower (Lidster et al., 1990). The low concentration of O_2 and elevated level of CO_2 have been found to cause cracked epidermal tissues and water-soaked areas on the surface of honeydew melons. These physiological disorders occur if the concentration of CO_2 is higher than 10 kPa and O_2 is lower than 3 kPa during storage (Lidster et al., 1990). Skin pitting or browning has been found to be associated with lower O_2 (0 kPa) in nectarine and peach fruits (Lidster et al., 1990). The extent of the said physiological disorder varies with the cultivars of peaches and nectarines. The information about elevated CO_2 level-induced injury is either not reported yet or is very limited and needs future research on this aspect for peach and nectarine fruits (Lidster et al., 1990). It has been reported that low O_2-based injury can be reduced by using CO_2 (5 kPa) in the nectarine fruits cultivar "Late le Grand" (Lidster et al., 1990). The early harvested "Thompson Seedless" grapes exhibit berry browning incidence upon exposure of the berries to elevated CO_2 (10 kPa) levels (Crisosto et al., 2002). The elevated CO_2-induced browning of "Kyoho" grapes has also been reported (Deng et al., 2006).

The apple fruits cultivar "Empire" is prone to chilling injury during CA (CO_2 2 kPa and O_2 3 kPa) storage. The chilling injury is manifested as symptoms of flesh browning. This probably occurs owing to increased accumulation of internal ethylene and ethylene production. It has been reported that "Empire" apple fruits showed higher flesh browning when kept under sole CA storage, but it significantly reduced in response to 1-MCP treatment (Jung and Watkins, 2011). The fruits of "Hass" avocadoes are susceptible to fermentative behavior under CA conditions. The CA storage resulted in fermentative reactions in the harvested "Hass" avocadoes (Valle-Guadarrama, 2011). The incidence of senescent scald and internal breakdown was significantly higher in normal air-stored "Swiss Bartlett" and "Cold Snap" European pear fruits (Table 10.3). The incidence was substantially lower in CA (2.5 or 18 kPa O_2 combined with 2 kPa CO_2)-stored fruits, but it was relatively higher in contrast with CA combined with 1-MCP treatment (Lum et al., 2017). The incidence of black speck, superficial scald, and pithy brown core in "d'Anjou" pears was higher in simple CA storage with a 1 kPa level of O_2 but was lower in contrast with the control (Table 10.3) (Guo et al., 2020). The "Braeburn" apple fruits show flesh breakdown during postharvest storage. The incidence of flesh breakdown was higher in simple CA storage with 0.7 CO_2 kPa and 1.2 O_2 kPa gaseous compositions in comparison with the dynamic CA (Weber et al., 2020).

10.10 DYNAMIC CONTROLLED ATMOSPHERE STORAGE

The initial low O_2 stress and CA storage are some of the main postharvest strategies for the storage life extension of pome and other fruit types. These storage types have been known for reduced respiration rate and ethylene generation, which ultimately help to extend the storage life of climacteric and other fruits (Mditshwa et al., 2018; Delele et al., 2019). These technologies are used to benefit from low O_2 concentration (Table 10.2) for the storage life extension. However, the low O_2-associated technologies also show negative influences on the stored produce under storage, such as the development of some physiological disorders and off-flavor, as reported in apple fruits (Imahori et al., 2005; Wright et al., 2015). Therefore, to enhance the advantages or efficacy of low O_2 technology, dynamic controlled atmosphere storage (DCA) has been established for the management of various off-flavor incidence and physiological disorders reduction (Weber et al., 2015; Wright et al., 2015). The beneficial effects of DCA were reviewed in earlier work (Prange et al., 2012; Reference). The use of DCA technology has been reported on different fresh horticultural produce such as apples (Mditshwa et al., 2018) and pears (Delele et al., 2019).

10.11 PHYSIOLOGICAL DISORDERS UNDER DCA STORAGE

In addition to certain positive effects, DCA also shows some negative effects on the stored produce (Table 10.3). The use of DCA is highly suitable for the prevention of various physiological disorders in various fruits and vegetables. The negative effects of DCA in the form of low O_2 injuries have been reported in apple fruits. The low O_2 in DCA showed epidermal purpling in "Delicious" fruits of apple (DeLong et al., 2006). The reports are contradictory in this regard as no low O_2 injury was noted on "Cortland" apple fruits under DCA with the same storage conditions. This difference could be attributed to the different genetic natures and physiological responses of apples (DeLong et al., 2006). In contrast to low O_2 injury, CO_2 injuries appear in the form of physiological disorder as internal browning manifested by the formation of cavities in the DCA-kept apples (Zanella et al., 2004; Lafer, 2007). In some severe cases, partially sunken areas/spots or brown lesions of rough nature can be found as an external injury on the produce (Table 10.3). The CO_2 injury was noted in Granny Smith" apple fruits stored under DCA (Zanella et al., 2004). In the same way, Lafer (2007) observed external CO_2 injury in 8 months DCA-stored "Braeburn" apple fruits. The increased internal CO_2 accumulation in stored produce sometimes leads to the development of flesh cavities under long-term DCA storage, as noted in "Fuji Suprema" apples (Weber et al., 2017). The flesh breakdown of apples was noted in the "Braeburn" cultivar under DCA-RQ storage (Weber et al., 2020). In the same way, flesh breakdown and mealiness were noted in "Galaxy" apples during storage under DCA environment (Wendt et al., 2022).

10.12 POSTHARVEST TREATMENTS FOR PHYSIOLOGICAL DISORDERS CONTROL

Different types of treatments have been reported for the management of physiological disorders in various fruits and vegetables. It has been reported that the efficacy of the applied treatments generally depends upon the types of produce and storage conditions. The management strategies include various physical and chemical-based treatment applications under cold, MAP, CA, and DCA storage atmosphere conditions. The details are given below.

Various treatments were tested and investigated for the control of chilling injury in fresh horticultural produce during LTS conditions (Table 10.4). One of the traditional and oldest methods is the use of hot water treatment (HWT). Kantakhoo et al. (2022) used HWT for 10 min at 45°C for the mitigation of chilling injury in eggplant fruits. The authors reported a significant reduction in the chilling injury incidence of eggplant stored at 1 and 10°C for 15 days (Kantakhoo et al., 2022). The 8 mmol L^{-1} phenylalanine was applied to reduce chilling injury in mangoes. It was observed that the said treatment significantly suppressed the symptoms of chilling injury at 7 and 10°C for 21 days. The phenylalanine application regulated the phenylpropanoid pathway and inhibited lipid peroxidation which subsequently suppressed the symptoms of chilling injury in mangoes (Patel et al., 2023). The application of nitric oxide donor sodium nitroprusside (10 μmol L^{-1} for 10 min) regulated cell wall and lipid metabolism in LTS-stored peaches at 0°C for 35 days. The treated peaches exhibited significantly lower lipid degradation and showed conserved cell wall polysaccharides which eventually showed inhibited chilling injury symptoms (Zhao et al., 2021). Brassinolide (6 μmol L^{-1})-based treatment chilling injury symptoms (indicated as rib-edge darkening) in carambola fruits at 4°C for 20 days (Duan et al., 2022). It has been reported that banana fruits have a chilling sensitivity below 13°C. The salicylic acid-based treatment consisting of 1 mmol L^{-1} significantly suppressed chilling injury occurrence at 5°C for 10 days (Khademi et al., 2019). The γ-aminobutyric acid (GABA) is one of the important osmolytes with great chilling injury tolerance induction. The 10 mmol L^{-1} GABA application markedly suppressed chilling injury symptoms in the treated aonla fruits stored for 24 days at 5°C (Ali et al., 2022a). The application of glycine betaine (10 mmol L^{-1}) was found to be appropriate in suppressing the symptoms of chilling injury on hawthorns stored at 1°C for 20 days (Razavi et al., 2018). The application of *Aloe vera* gel coating suppressed the enzymatic browning in harvested litchi fruits (Ali et al., 2019b). Pomegranate fruits are also prone to chilling-induced damage under LTS conditions. The application of 15 mmol L^{-1} 24-epibrassinolide markedly suppressed husk browning with maintained eating quality for 84 days at 5°C conditions (Islam et al., 2022).

Different methods have been reported for the management of physiological disorders of fresh horticultural produce under CA conditions (Table 10.4). The CA storage regulated sugars and energy metabolism in "Yulu" flat peach fruits thereby significantly reducing the internal browning index during storage at 1°C for 30 days (Zhou et al., 2022). The nitric oxide application at the rate of 10 μL L^{-1} reduced internal browning in "Laetitia" plums at 1.5°C for 55 days under CA storage (Steffens et al., 2022). The 5 μL L^{-1} NO lowered mealiness and pulp browning in CA-stored "Baigent" apples stored for 7 months at 1.5°C (Coser et al., 2023). The 1 μL L^{-1} 1-MCP (SmartFresh®) application markedly reduced calyx-end flesh and stem-end flesh tissue browning in "Empire" apples under CA storage of 40 weeks at 3.3°C (Lee et al., 2022). The application of 30 μL L^{-1} 1-MCP as a preharvest and postharvest combination significantly suppressed the leather blotch, bitter pit, greasiness, core browning, soft scald, vascular browning, senescent breakdown, and CO_2 injury in "Honeycrisp" apples during storage for 20 weeks at 0.5°C (Al Shoffe et al., 2021).

TABLE 10.4 Management of Physiological Disorders in Different Fresh Horticultural Produce Under Varying Storage Types and Conditions.

Produce	Disorder	Chemical/ Treatment	Dose and Storage	References
Eggplant	Chilling injury	HWT	45°C for 10 min and stored 15 days at 1 and 10°C under LTS	Kantakhoo et al. (2022)
Mango	Chilling injury	Phenylalanine	8 mmol L^{-1} kept at 7 and 10°C for 21 days under LTS	Patel et al. (2023)
Peach	Chilling injury	NO	10 μmol L^{-1} and stored at 0°C for 35 days under LTS	Zhao et al. (2021)
Carambola	Chilling injury	Brassinolide	6 μmol L^{-1} and stored at 4°C for 20 days under LTS	Duan et al. (2022)
Banana	Chilling injury	Salicylic acid	1 mmol L^{-1} and stored at 5°C for 10 days under LTS	Khademi et al. (2019)
Aonla	Chilling injury	γ-aminobutyric acid	5 mmol L^{-1} and stored at 5°C for 2 days under LTS	Ali et al. (2022a)
Hawthorns	Chilling injury	Glycine betaine	10 mmol L^{-1} and stored at 1°C for 20 days under LTS	Razavi et al. (2018)
Pomegranate	Chilling injury	24-Epibrassinolide	15 μmol L^{-1} and stored at 5°C for 84 days under LTS	Islam et al. (2022)
Plum	Internal browning	NO	10 μL L^{-1} at 1.5°C for 55 days under CA	Steffens et al. (2022)

TABLE 10.4 *(Continued)*

Produce	Disorder	Chemical/ Treatment	Dose and Storage	References
Apple	Mealiness	NO	5 μL L^{-1} at 1.5°C for 210 days under CA	Coser et al. (2023)
Apple	Calyx and stem ends browning	1-MCP	1 μL L^{-1} at 3.3°C for 280 days under CA	Lee et al. (2022)
Apple	Leather blotch, bitter pit, greasiness, core browning, soft scald, vascular browning, senescent breakdown, and CO_2 injury	1-MCP	30 μL L^{-1} at 0.5°C for 140 days CA under CA	Al Shoffe et al. (2021)

HWT, hot water treatment; NO, nitric oxide; 1-MCP, 1-methylcyclopropene; CA, controlled atmosphere; LTS, low (cold)-temperature storage.

10.13 CONCLUSION

In conclusion, physiological disorders have been found to be diverse in the harvested fresh horticultural produce during postharvest storage. The physiological disorders generally result in significantly higher quantitative and qualitative losses depending upon their nature and produce type. Various types of atmospheres such as simple cold storage, modified atmosphere, DCA, CA, hypobaric, and hyperbaric storage sometimes lead to severe physiological disorders. Therefore, some appropriate and eco-friendly treatments and/or approaches should be considered in combination with the above-listed storage atmospheres for the purpose of reducing the occurrence of physiological disorders and maintaining the overall eating quality of the fresh produce. Some pre- and postharvest treatments should be carefully investigated on crop by crop as well as genotype-based nature. The influence of some preharvest treatments in combination with postharvest pre-storage treatments should also be explored. It is also proposed that some breeding-oriented strategies may also be investigated which help to genetically improve the tolerance of specific fresh can produce to postharvest physiological disorders.

KEYWORDS

- **cold storage**
- **hypobaric atmosphere**
- **modified atmosphere**
- **physiological disorders**
- **storage atmosphere**

REFERENCES

Aghdam, M. S., & Bodbodak, S. (2013). Physiological and biochemical mechanisms regulating chilling tolerance in fruits and vegetables under postharvest salicylates and jasmonates treatments. *Scientia Horticulturae, 156*, 73–85. https://doi.org/10.1016/j.scienta.2013.03.028

Aghdam, M. S., Mahmoudi, R., Razavi, F., Rabiei, V., & Soleimani, A. (2018). Hydrogen sulfide treatment confers chilling tolerance in hawthorn fruit during cold storage by triggering endogenous H_2S accumulation, enhancing antioxidant enzymes activity and promoting phenols accumulation. *Scientia Horticulturae, 238*, 264–271.

Aghdam, M. S., Moradi, M., Razavi, F., & Rabiei, V. (2019). Exogenous phenylalanine application promotes chilling tolerance in tomato fruits during cold storage by ensuring supply of NADPH for activation of ROS scavenging systems. *Scientia Horticulturae, 246*, 818–825.

Al Shoffe, Y., Nock, J. F., Zhang, Y., & Watkins, C. B. (2021). Physiological disorder development of "Honeycrisp" apples after pre-and post-harvest 1-methycyclopropene (1-MCP) treatments. *Postharvest Biology and Technology, 182*, 111703.

Ali, S., Khan, A. S., Malik, A. U., & Shahid, M. (2016). Effect of controlled atmosphere storage on pericarp browning, bioactive compounds and antioxidant enzymes of litchi fruits. *Food Chemistry, 206*, 18–29. https://doi.org/10.1016/j.foodchem.2016.03.021

Ali, S., Khan, A. S., Malik, A. U., Anjum, M. A., Nawaz, A., & Shah, H. M. S. (2019a). Modified atmosphere packaging delays enzymatic browning and maintains quality of harvested litchi fruit during low temperature storage. *Scientia Horticulturae, 254*, 14–20. https://doi.org/10.1016/j.scienta.2019.04.065

Ali, S., Khan, A. S., Nawaz, A., Anjum, M. A., Naz, S., Ejaz, S., & Hussain, S. (2019b). Aloe vera gel coating delays postharvest browning and maintains quality of harvested litchi fruit. *Postharvest Biology and Technology, 157*, 110960.

Ali, S., Khan, A. S., Malik, A. U., Anwar, R., Anjum, M. A., Nawaz, A., & Naz, S. (2021). Combined application of ascorbic and oxalic acids delays postharvest browning of litchi fruits under controlled atmosphere conditions. *Food Chemistry, 350*, 129277. https://doi.org/10.1016/j.foodchem.2021.129277

Ali, S., Khan, A. S., Anjum, M. A., Ejaz, S., Khaliq, G., Hussain, S., Ahmad, R., & Saleem, M. S. (2022). Pre-and postharvest physiological disorders of citrus fruit. In Citrus Production (pp. 371-390). CRC Press.

Ali, S., Anjum, M. A., Nawaz, A., Ejaz, S., Anwar, R., Khaliq, G. et al. (2022a) Postharvest γ-aminobutyric acid application mitigates chilling injury of aonla (*Emblica officinalis* Gaertn.) fruit during low temperature storage. *Postharvest Biology and Technology, 185*, 111803. https://doi.org/10.1016/j.postharvbio.2021.111803

Ali, S., Nawaz, A., Naz, S., Ejaz, S., Maqbool, M., Siddiqui, M. H. et al. (2022b). Hydrogen sulfide mitigates chilling injury of postharvest banana fruits by regulating γ-aminobutyric acid shunt pathway and ascorbate–glutathione cycle. *Frontiers in Plant Science, 13*.

Ali, M. H., Khan, A. S., Jaskani, M. J., Anwar, R., Ali, S., Malik, A. U., & yyub, S. (2022c). Pre-storage application of L-arginine mitigates chilling injury and maintains quality of Sandhuri guava fruit. *Journal of Food Processing and Preservation, 46*(3), e16405.

Arpaia, M. L. (1990). Kiwifruit. In P. D. Lister, G. D. Blanpied, and R. K. Prange (Eds.), *Controlled atmosphere disorders of commercial fruits and vegetables* (pp. 41–44). Agriculture and Agri-Food Canada.

Bachmann, J., & Earles, R. (2000). Postharvest handling of fruits and vegetables, national center for appropriate technology. https://www.wnc.edu./files/department/ce/sci/postharvesthandling.pdf.

Bauchot, A. D., Reid, S. J., Ross, G. S., & Burmeister, D. M. (1999). Induction of apple scald by anaerobiosis has similar characteristics to naturally occurring superficial scald in "Granny Smith" apple fruit. *Postharvest Biology and Technology, 16*, 9–14.

Berard, L. S. (1994). Storage disorders of cabbage. In R. J. Howard, J. A. Garland, and W. L. Seaman (Eds), *Diseases and pests of vegetable crops in Canada* (pp. 110–112). Canadian Phytopathology Society and Entomology Society of Canada.

Bodbodak, S., & Moshfeghifar, M. (2016). Advances in controlled atmosphere storage of fruits and vegetables. In *Eco-friendly technology for postharvest produce quality* (pp. 39–76). Academic Press.

Bokhary, S. U. F., Wang, L., Zheng, Y., & Jin, P. (2020). Pre-storage hot water treatment enhances chilling tolerance of zucchini (*Cucurbita pepo* L.) squash by regulating arginine metabolism. *Postharvest Biology and Technology, 166*, 111229.

Brackmann, A., & Bortoluzzi, G. (1998). Influence of temperature, CA conditions and relative humidity on physiological disorders of "Fuji" apples. *Acta Horticulturae, 464*, 369–374.

Burg, S. P. (1975). Hypobaric storage and transportation of fresh fruits and vegetables. In N. F. Haard and D. K. Salunkhe (Eds.), *Postharvest biology and handling of fruits and vegetables* (pp. 172–188). The AVI Publishing CO., Inc..

Burg, S. P. (2004). *Postharvest physiology and hypobaric storage of fresh produce*. CAB International.

Burmeister, D. M., & Dilley, D. R. (1995). A scald-like controlled atmosphere storage disorder of Empire apples—A chilling injury induced by CO_2. *Postharvest Biology and Technology, 6*, 1–7.

Cai, C., Xu, C., Shan, L., Li, X., Zhou, C., Zhang, W., & Chen, K. (2006). Low temperature conditioning reduces postharvest chilling injury in loquat fruit. *Postharvest Biology and Technology, 41*(3), 252–259.

Cantre, D., Herremans, E., Verboven, P., Ampofo-Asiama, J., Hertog, M. L., & Nicolaï, B. M. (2017). Tissue breakdown of mango (*Mangifera indica* L. cv. Carabao) due to chilling injury. *Postharvest Biology and Technology, 125*, 99–111.
Chung, H. S., Moon, K. D., Chung, S. K., & Choi, J. U. (2005). Control of internal browning and quality improvement of "Fuji" apples by stepwise increase of CO_2 level during controlled atmosphere storage. *Journal of the Science of Food and Agriculture, 85*, 883–888.
Colgan, R. J., Dover, C. J., Johnson, D. S., Pearson, K., & Delayed, C. A. (1999). and oxygen at 1 kPa or less control superficial scald without CO_2 injury on Bramley's Seedling apples. *Postharvest Biology and Technology, 16*, 223–231.
Coser, W., Steffens, C. A., Lugaresi, A., Buss, M., Gularte, P. S., Fenili, C. L. et al. (2023). Treatment of "Baigent" apple fruit with nitric oxide in controlled atmosphere storage reduces decay and physiological disorders. *Postharvest Biology and Technology, 196*, 112159. https://doi.org/10.1016/j.postharvbio.2022.112159
Crisosto, C. H., Garner, D., & Crisosto, G. (2022). High carbon dioxide atmospheres affect stored "Thompson seedless" table grapes. *HortScience, 37*(7), 1074–1078.
Delele, M. A., Bessemans, N., Gruyters, W., Rogge, S., Janssen, S., Verlinden, B. E., & Nicolai, B. M. (2019). Spatial distribution of gas concentrations and RQ in a controlled atmosphere storage container with pear fruit in very low oxygen conditions. *Postharvest Biology and Technology, 156*, 110903. https://doi.org/10.1016/j.postharvbio.2019.05.004
DeLong, J. M., Harrison, P. A., & Prange, R. K. (2006, Feb). Chlorophyll fluorescence-based low-O_2 CA storage of organic "Cortland" and "Delicious" apples. *I International Symposium on Organic Apple and Pear, 737*, 31–37.
Deng, Y., Wu, Y., & Li, Y. (2006). Physiological responses and quality attributes of "Kyoho" grapes to controlled atmosphere storage. *LWT—Food Science Technology, 39*(6), 584–590.
Ding, C. K., Chachin, K., Ueda, Y., Imahori, Y., & Kurooka, H. (1999). Effects of high CO_2 concentration on browning injury and phenolic metabolism in loquat fruits. *Journal of the Japanese Society for Horticultural Science, 68*, 275–282.
Duan, W., Mekontso, F. N., Li, W., Tian, J., Li, J., Wang, Q., & Xu, X. (2022). Alleviation of postharvest rib-edge darkening and chilling injury of carambola fruit by brassinolide under low temperature storage. *Scientia Horticulturae, 299*, 111015.
Eaks, I. L. (1956). Effect of modified atmosphere on cucumbers at chilling and non-chilling temperatures. *Proceedings of the American Society for Horticultural Science, 67*, 473–478.
Elgar, H. J., Burmeister, D. M., & Watkins, C. B. (1998). Storage and handling effects on a CO_2 -related internal browning disorder of "Braeburn" apple. *HortScience, 33*, 719–722.
ElManawy, A. I., Maratab, A., Ghazal, A. F., Li, F., Li, X., & Zhang, X. (2025). Shelf-life assessment techniques for fruit and vegetables: Recent trends and future prospects. Postharvest Biology and Technology, 226, 113521.
Endo, H., Ose, K., Bai, J., & Imahori, Y. (2019). Effect of hot water treatment on chilling injury incidence and antioxidative responses of mature green mume (*Prunus mume*) fruit during low temperature storage. *Scientia Horticulturae, 246*, 550–556.
Farcuh, M., & Hopfer, H. (2023). Aroma volatiles as predictors of chilling injury development during peach (*Prunus persica* (L) Batsch) cold storage and subsequent

shelf-life. *Postharvest Biology and Technology, 195*, 112137. https://doi.org/10.1016/j.postharvbio.2022.112137

Fernández-León, M. F., Fernández-León, A. M., Lozano, M., Ayuso, M. C., & González-Gómez, D. (2013). Different postharvest strategies to preserve broccoli quality during storage and shelf life: Controlled atmosphere and 1-MCP. *Food Chemistry, 138*(1), 564–573. https://doi.org/10.1016/j.foodchem.2012.09.143

Galvis-Sánchez, A. C., Fonseca, S. C., Gil-Izquierdo, Á., Gil, M. I., & Malcata, F. X. (2006). Effect of different levels of CO_2 on the antioxidant content and the polyphenol oxidase activity of "Rocha" pears during cold storage. *Journal of the Science of Food and Agriculture, 86*, 509–517.

Guo, J., Wei, X., Lü, E., Wang, Y., & Deng, Z. (2020). Ripening behavior and quality of 1-MCP treated d'Anjou pears during controlled atmosphere storage. *Food Container, 117*, 107364. https://doi.org/10.1016/j.foodcont.2020.107364

Han, J. W., Zuo, M., Zhu, W. Y., Zuo, J. H., Lü, E. L., & Yang, X. T. (2021). A comprehensive review of cold chain logistics for fresh agricultural products: Current status, challenges, and future trends. *Trends in Food Science and Technology, 109*, 536–551. https://doi.org/10.1016/j.tifs.2021.01.066

Hoehn, E., Prange, R. K., & Vigneault, C. (2009). Storage technology and applications. In M. Y. Elhadi (Ed.), *Modified and controlled atmospheres for the storage, transportation, and packaging of horticultural commodities* (pp. 16–22). CRC Press.

Holcroft, D. M., & Kader, A. A. (1999). Carbon-dioxide-induced changes in color and anthocyanin synthesis of stored strawberry fruit. *HortScience, 34*, 1244–1248.

Huan, C., Li, H., Jiang, Z., Shen, S., & Zheng, X. (2021). Effect of hypobaric treatment on off-flavour development and energy metabolism in "Bruno" kiwifruit. *LWT—Food Science Technology, 136*, 110349.

Huang, Q., Qian, X., Jiang, T., & Zheng, X. (2019). Effect of eugenol fumigation treatment on chilling injury and CBF gene expression in eggplant fruit during cold storage. *Food Chemistry, 292*, 143–150.

Huang, S., Li, T., Jiang, G., Xie, W., Chang, S., Jiang, Y., & Duan, X. (2012). 1-Methylcyclopropene reduces chilling injury of harvested okra (*Hibiscus esculentus* L.) pods. *Scientia Horticulturae, 141*, 42–46.

Imahori, Y., Takemura, M., & Bai, J. (2008). Chilling-induced oxidative stress and antioxidant responses in mume (*Prunus mume*) fruit during low temperature storage. *Postharvest Biology and Technology, 49*(1), 54–60. https://doi.org/10.1016/j.postharvbio.2007.10.017

Imahori,Y., Uemura, K., Kishioka, I., Fujiwara, H., Tulio, A., Ueda, Y., & Chachin, K. (2005). Relationship between low-oxygen injury and ethanol metabolism in various fruits and vegetables. *Acta Horticulturae, 682*, 1103–1108.

Islam, M., Ali, S., Nawaz, A., Naz, S., Ejaz, S., Shah, A. A., & Razzaq, K. (2022). Postharvest 24-epibrassinolide treatment alleviates pomegranate fruit chilling injury by regulating proline metabolism and antioxidant activities. *Postharvest Biology and Technology*, *188*, 111906. https://doi.org/10.1016/j.postharvbio.2022.111906

Jalali, A., Seiiedlou, S., Linke, M., & Mahajan, P. (2017). A comprehensive simulation program for modified atmosphere and humidity packaging (MAHP) of fresh fruits and vegetables. *Journal of Food Engineering, 206*, 88–97. https://doi.org/10.1016/j.jfoodeng.2017.03.007

Jannasch, H. W., Eimhjellen, K., Wirsen, C. O., & Farmanfarmaian, A. (1971). Micorbial degradation of organic matter in deep sea. *Science, 171*, 672–675.

Johnson, D. S., Dover, C. J., & Colgan, R. J. (1998). Effect of rate of establishment of CA conditions on the development of CO_2-induced injury on Bramley's Seedling apples. *Acta Horticulturae, 464*, 351–356.

Jung, S. K., & Watkins, C. B. (2011). Involvement of ethylene in browning development of controlled atmosphere-stored "Empire" apple fruit. *Postharvest Biology and Technology, 59*(3), 219–226. https://doi.org/10.1016/j.postharvbio.2010.08.019

Kader, A. A. (2002). Modified atmospheres during transport and storage. In A. A. Kader (Ed.), *Postharvest technology horticultural crops* (pp. 135–144). University of California.

Kader, A. A., & Morris, L. L. (1975). Amelioration of chilling injury symptoms on tomato fruits. *HortScience, 10*, 324.

Kantakhoo, J., Ose, K., & Imahori, Y. (2022). Effects of hot water treatment to alleviate chilling injury and enhance phenolic metabolism in eggplant fruit during low temperature storage. *Scientia Horticulturae, 304*, 111325.

Khademi, O., Ashtari, M., & Razavi, F. (2019). Effects of salicylic acid and ultrasound treatments on chilling injury control and quality preservation in banana fruit during cold storage. *Scientia Horticulturae, 249,* 334–339.

Khaliq, G., Ali, S., Gapper, N., & Nicola, S. (2023). Recent advances and approaches in the application of elicitors to enhance resistance mechanisms in fresh produce. *Frontiers in Sustainable Food Systems, 6,* 1061079.

Khan, L., Khaliq, G., Ullah, E., Abdul-Rahaman, A., El-Mogy, M. M., Kesba, H. H., Ismail, A. M., & El-Beltagi, H. S. (2025). Guggul gum and clove essential oil synergistically reduced anthracnose disease and modulated physicochemical changes in tomato during storage. *Journal of Stored Products Research, 114,* 102713.

Kim, J.G,. Luo, Y., Saftner, R. A., & Gross, K. C. (2005). Delayed modified atmosphere packaging of fresh-cut romaine lettuce: Effects on quality maintenance and shelf-life. *Journal of the American Society for Horticultural Science, 130*, 116–123.

Kondo, S., Posuya, P., & Kanlayanarat, S. (2001). Changes in physical characteristics and polyamines during maturation and storage of rambutans. *Scientia Horticulturae, 91*, 101–109.

Koukounaras, A., Siomos, A. S., Gerasopoulos, D., & Papachristodoulou, M. (2019). Active modified atmosphere package induced a new physiological disorder of minimally processed romaine lettuce leaves. *Food Packaging and Shelf Life, 22*, 100411. https://doi.org/10.1016/j.fpsl.2019.100411

Koyuncu, M. A., Erbas, D., Onursal, C. E., Secmen, T., Guneyli, A., & Sevinc Uzumcu, S. (2019). Postharvest treatments of salicylic acid, oxalic acid and putrescine influences bioactive compounds and quality of pomegranate during controlled atmosphere storage. *Journal of Food Science and Technology, 56*(1), 350–359.

Lafer, G. (2007, Nov). Storability and fruit quality of "Braeburn" apples as affected by harvest date, 1-MCP treatment and different storage conditions. *International Conference on Ripening Regulation and Postharvest Fruit Quality, 796,* 179–184.

Lal, S., Khaliq, G., Mastoi, A. H., Irshad, A., Jaffar, M. A., Khan, L., & Muhammad, N. (2025). Effect of *Fagonia indica* and garlic extract on *Alternaria alternata* and post-harvest quality of tomato fruit during storage. *Journal of Crop Health, 77*(6), 1–13.

Larrigaudière, C., Pintó, E., Lentheric, I., & Vendrell, M. (2001). Involvement of oxidative processes in the development of core browning in controlled atmosphere pears. *The Journal of Horticultural Science and Biotechnology, 76*, 157–162.

Lau, O. L. (1998). Effect of growing season, harvest maturity, waxing, low O_2 and elevated CO_2 on flesh browning disorders in "Braeburn" apples. *Postharvest Biology and Technology, 14*, 131–141.

Lee, J., Leisso, R., Rudell, D. R., & Watkins, C. B. (2022). 1-Methylcyclopropene differentially regulates metabolic responses in the stem-end and calyx-end flesh tissues of "Empire" apple during long-term controlled atmosphere storage. *Postharvest Biology and Technology, 192*, 112018. https://doi.org/10.1016/j.postharvbio.2022.112018

Legnani, G., Watkins, C. B., & Miller, W. B. (2004). Low oxygen affects the quality of Asiatic hybrid lily bulbs during simulated dry-sale storage and subsequent forcing. *Postharvest Biology and Technology, 32*(2), 223–233. https://doi.org/10.1016/j.postharvbio.2003.10.008

Legnani, G., Watkins, C. B., & Miller, W. B. (2006). Tolerance of dry-sale lily bulbs to elevated carbon dioxide in both ambient and low oxygen atmospheres. *Postharvest Biology and Technology*, *41*(2), 198–207. https://doi.org/10.1016/j.postharvbio.2006.03.014

Li, J., Bao, X., Xu, Y., Zhang, M., Cai, Q., Li, L., & Wang, Y. (2017). Hypobaric storage reduced core browning of Yali pear fruits. *Scientia Horticulturae, 225*, 547–552. https://doi.org/10.1016/j.scienta.2017.07.031

Li, X., Yang, H., & Lu, G. (2018). Low-temperature conditioning combined with cold storage inducing rapid sweetening of sweetpotato tuberous roots (*Ipomoea batatas* (L.) Lam) while inhibiting chilling injury. *Postharvest Biology and Technology, 142*, 1–9.

Lidster, P. D., Blanpied, G. D., & Prange, R. K. (Eds.). (1990). *Controlled-atmosphere disorders of commercial fruits and vegetables*. Agriculture and Agri-Food Canada.

Lima, F. T. D., Martins, L. P., Toscano, B. L. M. D., de Sousa Costa, E., de Souza, W. F. C., de Lucena, F. A. et al. (2023). Cold damage affects the quality of noni fruits (*Morinda citrifolia* L.). *Biocatalysis and Agricultural Biotechnology,* 102566. https://doi.org/10.1016/j.bcab.2022.102566

Lipton, W. J. (1990). In P. D. Lister, G. D. Blanpied, and R. K. Prange (Eds.), *Controlled atmosphere disorders of commercial fruits and vegetables* (pp. 31–32). Agriculture and Agri-Food Canada.

Lipton, W. J., & Mackey, B. E. (1987). Physiological and quality responses of Brussel sprouts to storage in controlled atmospheres. *Journal of the American Society for Horticultural Science, 112*, 491–496.

Lum, G. B., DeEll, J. R., Hoover, G. J., Subedi, S., Shelp, B. J., & Bozzo, G. G. (2017). 1-Methylcylopropene and controlled atmosphere modulate oxidative stress metabolism and reduce senescence-related disorders in stored pear fruit. *Postharvest Biology and Technology, 129*, 52–63. https://doi.org/10.1016/j.postharvbio.2017.03.008

Luo, Z., Wu, X., Xie, Y., & Chen, C. (2012). Alleviation of chilling injury and browning of postharvest bamboo shoot by salicylic acid treatment. *Food Chemistry, 131*(2), 456–461.

Ma, S. S., & Chen, P. (2003). Storage disorder and ripening behavior of "Doyenne du Comice" pears in relation to storage conditions. *Postharvest Biology and Technology, 28*, 281–294.

Maguire, K. M., & Mackay B. A. (2003). Controlled atmosphere induced internal browning disorder of "Pacific Rose" apples. *Acta Horticulturae, 600*, 281–284.

Mahajan, P. V., Caleb, O. J., Singh, Z., Watkins, C. B., & Geyer, M. (2014). Postharvest treatments of fresh produce. *Philosophical Transactions of the Royal Society A, 372*, 1–19.

Martínez-Hernández, G. B., Artés-Hernández, F., Gómez, P. A., Bretó, J., Orihuel-Iranzo, B., & Artés, F. (2017). Postharvest treatments to control physiological and pathological disorders in lemon fruit. *Food Packaging and Shelf Life, 14*, 34–39.

Master, A. M., Krebbers, B., Van Den Berg, R. W., & Barteles, P. V. (2004). Advancements of high pressure sterilization on quality of food products. *Trends in Food Science and Technology, 15*, 79–85.

Mattheis, J. P., Felicetti, D. A., & Rudell, D. R. (2022). "d'Anjou" pear metabolism during ultra-low O_2, low CO_2 controlled atmosphere storage reflects disorder outcome. *Postharvest Biology and Technology, 185*, 111781. https://doi.org/10.1016/j.postharvbio.2021.111781

Mditshwa, A., Fawole, O. A., & Opara, U. L. (2018). Recent developments on dynamic controlled atmosphere storage of apples—A review. *Food Packaging and Shelf Life, 16*, 59–68. https://doi.org/10.1016/j.fpsl.2018.01.011

Menniti, A. M., Maccaferri, M., & Folchi A. (1997). Physio-pathological responses of cabbage stored under controlled atmospheres. *Postharvest Biology and Technology, 10*, 207–212.

Murmu, S. B., & Mishra, H. N. (2018). Selection of the best active modified atmosphere packaging with ethylene and moisture scavengers to maintain quality of guava during low-temperature storage. *Food Chemistry, 253*, 55–62.

Nanos, G. D., Kiritsakis, A. K., & Sfakiotakis, E. M. (2002). Preprocessing storage conditions for green 'Conservolea'and 'Chondrolia'table olives. Postharvest Biology and Technology, 25(1), 109-115.

Neuwald, D. A., Saquet, A. A., & Klein, N. (2019). Disorders during storage of fruits and vegetables. *Postharvest Physiological Disorders in Fruits and Vegetables*, 89–110.

Neuwald, D. A., Streif, J., & Kittemann, D. (2015). CO_2 skin burning of "Cameo" apples during CA storage: Reasons and possibilities to reduce the disorder. *Acta Horticulturae*, 711–716.

Oliveira, M., Abadias, M., Usall, J., Torres, R., Teixidó, N., & Viñas, I. (2015). Application of modified atmosphere packaging as a safety approach to fresh-cut fruits and vegetables—A review. *Trends in Food Science and Technology, 46*(1), 13–26. https://doi.org/10.1016/j.tifs.2015.07.017

Pan, Y. G., Yuan, M. Q., Zhang, W. M., & Zhang, Z. K. (2017). Effect of low temperatures on chilling injury in relation to energy status in papaya fruit during storage. *Postharvest Biology and Technology, 125*, 181–187.

Pareek, S., Valero, D., & Serrano, M. (2015). Postharvest biology and technology of pomegranate. *Journal of the Science of Food and Agriculture, 95*(12), 2360–2379.

Patel, M. K., Fanyuk, M., Feyngenberg, O., Maurer, D., Sela, N., Ovadia, R. et al. (2023). Phenylalanine induces mango fruit resistance against chilling injuries during storage at suboptimal temperature. *Food Chemistry, 405*, 134909.

Pérez-Gago, M. B., & Palou, L. (2020). Subtropical fruits: Citrus. In *Controlled and modified atmospheres for fresh and fresh-cut produce* (pp. 411–419). Academic Press.

Pintó, E., Lentheric, I., Vendrell, M., & Larrigaudière, C. (2001). Role of fermentative and antioxidant metabolisms in the induction of core browning in controlled-atmosphere pears. *Journal of the Science of Food and Agriculture, 81*, 364–370.

Prange, R. K., & Lidster, P. D. (1991). Controlled atmosphere and lighting effects on storage of winter cabbage. *Canadian Journal of Plant Science, 71*, 263–268.

Prange, R. K., & DeLong, J. M. (2006). Controlled-atmosphere related disorders of fruits and vegetables. *Stewart Postharvest Review, 5*(7), 1–10.

Prange, R. K., & Harrison, P. A. (1993). Effect of controlled atmosphere and humidity on postharvest physiology of buttercup winter squash, Cucurbita maxima Duch. hybrid "Sweet Mama". In *Proceedings of the 6th International Controlled Atmosphere Research Conference* (pp. 759–766). Ithaca.

Prange, R. K., Wright, A. H., DeLong, J. M., & Zanella, A. (2012, June). History, current situation and future prospects for dynamic controlled atmosphere (DCA) storage of fruits and vegetables, using chlorophyll fluorescence. *VII International Postharvest Symposium, 1012*, 905–915.

Purvis, A. C., & Cubbedge, R. H. (1990). In P. D. Lister, G. D. Blanpied, and R. K. Prange (Eds.), *Controlled atmosphere disorders of commercial fruits and vegetables* (pp. 36–38). Agriculture and Agri-Food Canada.

Razavi, F., Mahmoudi, R., Rabiei, V., Aghdam, M. S., & Soleimani, A. (2018). Glycine betaine treatment attenuates chilling injury and maintains nutritional quality of hawthorn fruit during storage at low temperature. *Scientia Horticulturae, 233*, 188–194.

Rostogi, N. K. (2013). *Recent developments in high pressure processing of foods*. Springer.

Sandhya. (2010). Modified atmosphere packaging of fresh produce: Current status and future needs. *LWT—Food Science Technology, 43*(3), 381–392. https://doi.org/10.1016/j.postharvbio.2003.10.008

Schotsmans, W. C., Prange, R. K., & Binder, B. M. (2009). 1-Methylcyclopropene: Mode of action and relevance in postharvest horticulture research. *Horticultural Reviews, 35*, 263–313.

Seo, J., Yi, G., Lee, J. G., Choi, J. H., & Lee, E. J. (2020). Seed browning in pepper (*Capsicum annuum* L.) fruit during cold storage is inhibited by methyl jasmonate or induced by methyl salicylate. *Postharvest Biology and Technology, 166*, 111210.

Sevillano, L., Sanchez-Ballesta, M. T., Romojaro, F., & Flores, F. B. (2009). Physiological, hormonal and molecular mechanisms regulating chilling injury in horticultural species. Postharvest technologies applied to reduce its impact. *Journal of the Science of Food and Agriculture, 89*(4), 555–573. https://doi.org/10.1002/jsfa.3468

Sharples, R. S., & Johnson, D. S. (1987). Influence of agronomic and climatic factors on the response of apple fruit to controlled atmosphere storage. *HortScience, 22*, 763–766.

Sheng, K., Wei, S., Mei, J., & Xie, J. (2021). Chilling injury, physicochemical properties, and antioxidant enzyme activities of red pitahaya (*Hylocereus polyrhizus*) fruits under cold storage stress. *Phyton, 90*(1), 291.

Steffens, C. A., Santana, G. R. O., do Amarante, C. V. T., Antonovviski, J. L., Miqueloto, T., Anami, J. M., & Fenili, C. L. (2022). Treatment with nitric oxide in controlled atmosphere storage to preserve the quality of "Laetitia" plums. *LWT—Food Science Technology, 158*, 113033.

Suo, J., Li, H., Ban, Q., Han, Y. E., Meng, K., Jin, M., & Rao, J. (2018). Characteristics of chilling injury-induced lignification in kiwifruit with different sensitivities to low temperatures. Postharvest Biology and Technology, 135, 8-18.

Tao, S., Zhu, Y., Pan, Y., Zhang, Z., & Huang, L. (2022). Enhancement of respiratory metabolism of the pentose phosphate pathway (PPP) strengthens the chilling tolerance of postharvest papaya fruit stored at 1°C. *Postharvest Biology and Technology, 191*, 111988. https://doi.org/10.1016/j.postharvbio.2022.111988

Techavuthiporn, C., & Boonyaritthongchai, P. (2016). Effect of prestorage short-term anoxia treatment and modified atmosphere packaging on the physical and chemical changes of green asparagus. *Postharvest Biology and Technology, 117*, 64–70. https://doi.org/10.1016/j.postharvbio.2016.01.016

Thewes, F. R., Thewes, F. R., Both, V., Schultz, E. E., Berghetti, M. R. P., Ludwig, V., & Brackmann, A. (2021). Static× dynamic controlled atmosphere: Impacts of aerobic and anaerobic metabolism on physiological disorders and overall quality of "Royal Gala" apples. *LWT—Food Science Technology, 141*, 110922.

Thompson, A. K. (2016). *Fruit and vegetable storage, hypobaric, hyperbaric and controlled atmosphere*. Springer Briefs in Food, Health and Nutrition.

Valle-Guadarrama, S., Morales-Cabrera, M., Peña-Valdivia, C. B., Mora-Rodríguez, B., Alia-Tejacal, I., Corrales-García, J., & Gómez-Cruz, A. (2013). Oxidative/fermentative behavior in the flesh of "Hass" avocado fruits under natural and controlled atmosphere conditions. *Food and Bioprocess Technology, 6*(1), 272–282. https://doi.org/10.1007/s11947-011-0747-8

Vallejo, F., Tomás-Barberán, F., & García-Viguera, C. (2003). Health-promoting compounds in broccoli as influenced by refrigerated transport and retail sale period. *Journal of Agricultural and Food Chemistry, 51*(10), 3029–3034. https://doi.org/10.1021/jf021065j

Van der Steen, C., Jacxsens, L., Devlieghere, F., & Debevere, J. (2002). Combining high oxygen atmospheres with low oxygen modified atmosphere packaging to improve the keeping quality of strawberries and raspberries. *Postharvest Biology and Technology, 26*(1), 49–58. https://doi.org/10.1016/S0925-5214(02)00005-4

Varoquaux, P., Mazollier, J., & Albagnac, G. (1996). The influence of raw material characteristics on the storage life of fresh-cut butterhead lettuce. *Postharvest Biology and Technology, 9*, 127–139.

Veltman, R. H., Lenthéric, I., Van der Plas, L. H. W., & Peppelenbos, H. W. (2003). Internal browning in pear fruit (*Pyrus communis* L. cv conference) may be a result of a limited availability of energy and antioxidants. *Postharvest Biology and Technology, 28*, 295–302.

Volz, R. K., Biasi, W. V., Grant, J. A., & Mitcham, E. J. (1998). Prediction of controlled atmosphere-induced flesh browning in "Fuji" apple. *Postharvest Biology and Technology, 13*, 97–107.

Wang, C. S., Liang, L. S., & Wang, G. X. (2007). Effects of low pressure on quality of DongZao jujube fruit in hypobaric storage. *Food Science, 28*(2), 335–339.

Wang, C. Y., Watada, A. E., & Anderson, R. E. (1990). Peaches and nectarines. In P. D. Lister, G. D. Blanpied, and R. K. Prange (Eds.), *Controlled atmosphere disorders of commercial fruits and vegetables* (pp. 47–49). Agriculture and Agri-Food Canada.

Wang, J., Pan, H., Wang, R., Hong, K., & Cao, J. (2016). Patterns of flesh reddening, translucency, ethylene production and storability of "Friar" plum fruit harvested at three maturity stages as affected by the storage temperature. *Postharvest Biology and Technology, 121*, 9–18.

Wang, J., You, Y., Chen, W., Xu, Q., Wang, J., Liu, Y. et al. (2015). Optimal hypobaric treatment delays ripening of honey peach fruit via increasing endogenous energy status and enhancing antioxidant defence systems during storage. *Postharvest Biology and Technology, 101*, 1–9. https://doi.org/10.1016/j.postharvbio.2014.11.004

Wang, Z., Kossitrakun, M., & Dilley, D. R. (2000). Temperature and atmosphere regimens to control a CO_2-linked disorder of "Empire" apples. *Postharvest Biology and Technology, 18*, 183–189.

Wang, Z., Cao, J., & Jiang, W. (2016). Changes in sugar metabolism caused by exogenous oxalic acid related to chilling tolerance of apricot fruit. *Postharvest Biology and Technology, 114*, 10–16.

Wang, Z., Kossitrakun, M., & Dilley, D. R. (1997). The effect of acclimatization of fruits on the control of CO_2 -linked disorder of "Empire" apples. *Proceedings of the 7th International Controlled Atmosphere Research Conference, 2*, 193–197.

Weber, A., Brackmann, A., Both, V., Pavanello, E. P., Anese, R. D. O., & Thewes, F. R. (2015). Respiratory quotient: Innovative method for monitoring "Royal Gala" apple storage in a dynamic controlled atmosphere. *Scientia Agricola, 72*, 28–33.

Weber, A., Neuwald, D. A., Kittemann, D., Thewes, F. R., Both, V., & Brackmann, A. (2020). Influence of respiratory quotient dynamic controlled atmosphere (DCA–RQ) and ethanol application on softening of Braeburn apples. *Food Chemistry, 303*, 125346. https://doi.org/10.1016/j.foodchem.2019.125346

Weber, A., Thewes, F. R., de Oliveira Anese, R., Both, V., Pavanello, E. P., & Brackmann, A. (2017). Dynamic controlled atmosphere (DCA): Interaction between DCA methods and 1-methylcyclopropene on "Fuji Suprema" apple quality. *Food Chemistry*, *235*, 136–144.

Wendt, L. M., Soldateli, F. J., Both, V., Thewes, F. R., Ludwig, V., Berghetti, M. R. P. et al. (2022). Effects of ethylene control and dynamic controlled atmosphere storage on "galaxy" apple quality. *Journal of Plant Growth Regulation,* 1–18.

Wills, R., & Golding, J. (2016). *Postharvest: An introduction to the physiology and hnadling of fruit and vegetables*, 6th edn. CAB International.

Woolf, A. B., Arpaia, M. L., Defilippi, B. G., & Bower, J. P. (2020). Subtropical fruits: Avocados. In *Controlled and modified atmospheres for fresh and fresh-cut produce* (pp. 389–397). Academic Press.

Wright, A. H., Delong, J. M., Arul, J., & Prange, R. K. (2015). The trend toward lower oxygen levels during apple (*Malus* × *domestica* Borkh) storage. *The Journal of Horticultural Science and Biotechnology, 90*(1), 1–13.

Yang, H., Wu, F., & Cheng, J. (2011). Reduced chilling injury in cucumber by nitric oxide and the antioxidant response. *Food Chemistry, 127*(3), 1237–1242.

Zanella, A., Cazzanelli, P., Panarese, A., Coser, M., Chistè, C., & Zeni, F. (2004, June). Fruit fluorescence response to low oxygen stress: Modern storage technologies compared to 1-MCP treatment of apple. In *V International Postharvest Symposium 682* (pp. 1535–1542).

Zhang, T., Che, F., Zhang, H., Pan, Y., Xu, M., Ban, Q. et al. (2017). Effect of nitric oxide treatment on chilling injury, antioxidant enzymes and expression of the CmCBF1 and

CmCBF3 genes in cold-stored Hami melon (*Cucumis melo* L.) fruit. *Postharvest Biology and Technology, 127*, 88–98.

Zhang, W., Jiang, H., Cao, J., & Jiang, W. (2021). Advances in biochemical mechanisms and control technologies to treat chilling injury in postharvest fruits and vegetables. *Trends in Food Science and Technology, 113*, 355–365. https://doi.org/10.1016/j.tifs.2021.05.009

Zhang, W., Zhao, H., Jiang, H., Xu, Y., Cao, J., & Jiang, W. (2020). Multiple 1-MCP treatment more effectively alleviated postharvest nectarine chilling injury than conventional one-time 1-MCP treatment by regulating ROS and energy metabolism. *Food Chemistry, 330*, 127256.

Zhao, Y., Tang, J., Song, C., Qi, S., Lin, Q., Cui, Y. et al. (2021). Nitric oxide alleviates chilling injury by regulating the metabolism of lipid and cell wall in cold-storage peach fruit. *Plant Physiology and Biochemistry, 169*, 63–69.

Zhou, H., Zhang, X., Su, M., Du, J., Li, X., Zhang, M. et al. (2022). Controlled atmosphere storage alleviates internal browning in flat peach fruit by regulating energy and sugar metabolisms. *Plant Physiology and Biochemistry, 186*, 107–120.

Zuo, X., Cao, S., Jia, W., Zhao, Z., Jin, P., & Zheng, Y. (2021). Near-saturated relative humidity alleviates chilling injury in zucchini fruit through its regulation of antioxidant response and energy metabolism. *Food Chemistry, 351*, 129336. https://doi.org/10.1016/j.foodchem.2021.129336

CHAPTER 11

Physiological and Biochemical Aspects of Physiological Disorders

HARJINDER KAUR[1], AB WAHEED WANI[1*], ATUL B. KHALANGRE[1], KHAN JABROOT J.[1], PALLVI VERMA[1], SAFURA NABI[2], MERAJ AHMAD[3], ANIS A. MERZA[1], GLOMERAZ ABDI[4], JATINDER SINGH[1], and RAHUL R. RODGE[1]

[1]*Department of Horticulture, Lovely Professional University, Jalandhar, Punjab, India*

[2]*Department of Fruit Science, SKUAST-Kashmir, Jammu and Kashmir, India*

[3]*Department of Soil Water and Engineering Lovely Professional University, Jalandhar, Punjab, India*

[4]*Department of Biotechnology, Persian Gulf Research Institute, Persian Gulf University, Bushehr, Iran*

*Corresponding author

ABSTRACT

A physiological disorder is an abnormal growth pattern or abnormal external or internal conditions of fruits as a result of unfavorable environmental conditions, which are caused by variations in the normal state of temperature, light, moisture, nutrients, harmful gases, and an insufficient supply of growth regulators. These disorders include disorders such as chilling injury, freezing injury, heat injury, fruit cracking, and mineral imbalance. The physiological and biochemical aspects of physiological disorders are one of the main aspects where the quality of horticulture fruits has been

Physiological Disorders of Fruits and Vegetables: Causes and Remedies.
Ghulam Khaliq, Sajid Ali, and Mohammed Wasim Siddiqui (Eds.)

DOI: 10.1201/9781998511334-11

determined. The pre-and postharvest disorders are the main parameters to judge the quality of fruits and vegetables. These disorders not only reduce the marketable value but also decrease the yield of horticulture products. Splitting or fruit cracking, which is seen on the flesh of fruit and its skin at the pre-harvest and postharvest phases, is one of the major disorders that significantly restrict fruit quality and quantity. The analysis of all disorders leads the good management strategies and it helps in the reduction of losses in the horticulture sector. Some of the environmental factors also create disorders in the fruits such as low-temperature causes chilling and freezing injury. This results in symptoms such as tissue death and collapse, necrosis, soaking of water, pitting, flavor/aroma loss, and increased decay susceptibility. Bleaching, uneven ripening, excessive softening, surface burning or scalding, and desiccation are a few signs of heat injury. Various biochemical constituents of plants, that is, minerals and their concentration, affect the plant growth which results in different disorders as per the particular mineral (i.e., chlorosis, necrotic spots, leaf margin burning, stunting, fruit drop, etc.). To mitigate these disorders and improve the fruits and vegetables quality, a study on physiological and biochemical components of these disorders is needed. So, in this chapter, we are going to discuss the physiological and biochemical aspects of physiological disorders that happen in the fruit and vegetable plants during pre and postharvest conditions.

11.1 INTRODUCTION

Physiological disorders of plants are generally defined as issues caused by unfavorable environmental factors and incorrect cultural practices that interfere with normal plant growth and development. Rather than being brought on by fungi, bacteria, viruses, or insects, physiological disorders are instead brought on by physiological and/or biochemical factors. Physiological disorders can develop prior to or following harvest. Numerous physiological disorders start before harvest and frequently do not manifest until after harvest, particularly during storage. According to Milind and Ladaniya (2008), the most significant nonpathological issues that face modern fruit growers are physiological disorders, which are brought on by the physiological processes of the fruit tissues malfunctioning as a result of abiotic stresses such as high or low temperature, relative humidity, moisture/water stress, chemicals, and nutrient excesses or deficiencies. The physiological and nutritional abnormalities have a higher impact on the yield and quality of fruits. Numerous physiological disorders, such as fruit drop, uneven bearing, fruit cracking, and so on, have an impact on fruit production and its potential quality yield. According to Campostrini et al. (2008), disruptions in plant metabolic activities are virtually always caused by external conditions, which growers may find challenging to control in some circumstances. According to Du Chatenet et al. (2000), several fruit species exhibit physiological abnormalities that are brought on by various circumstances. On the basis of the underlying causes, the physiological abnormalities in diverse fruit crops can be divided into distinct groups. Due to improper use of organic manures, the adoption of ultra-high density planting techniques, the use of inappropriate root stocks, such as citrus granulation in mandarin when grafted on Jatti Khatti rootstock, and uneven fertilizer application, micronutrient-related disorders in commercial orcharding develop. Micronutrient deficits must be found before visual signs manifest in order to maintain fruit quality and excellent yield. Technical words such as bronzing (the development of a bronze or copper hue), chlorosis (loss of chlorophyll), die-back (collapse of the developing tip affecting the younger leaves), and so on are used to describe physiological and nutritional disorders in crops. The causes of physiological disorders as well as management strategies to maintain high quality for better market value by preventing these disorders as well as the biochemical and physiological elements of these disorders have received specific attention in this chapter.

11.2 LOW-TEMPERATURE DISORDERS

11.2.1 CHILLING INJURY

Low temperatures can affect a lot of tropical or subtropical fruits, vegetables, and ornamentals (Paull, 1990). These crops suffer damage when exposed to chilling temperatures below 10–15°C (50–59°F), but above their freezing points (Lyons, 1973; Wang, 1990). Some crops grown in temperate climates are also vulnerable to chilling damage (Bramlage and Meir, 1990). The threshold temperatures for those temperate crops are typically lower, at 5°C (41°F). The tissues become weaker as a result of being unable to carry out normal metabolic functions at these freezing temperatures. For instance, banana fruits that are exposed to CI display pitting on the peel surface, the loss of aroma, and abnormal ripening when exposed to temperatures below 13°C. When it comes to papaya, low temperatures used improperly can result in water-soaked flesh and scorching skin (Zou et al., 2014). Cucumber is extremely vulnerable to chilling injury (CI), primarily manifesting as dehydration and significant surface sunken areas. Therefore, chilling damage reduces the commercial profitability of horticultural products by having a negative impact on their quality. Products that are chilled frequently continue to look good even after being kept at low temperatures. Once they are exposed to warmer temperatures, however, the signs of a chilling injury start to show up quickly. Both exposure duration and temperature have an impact on the onset of chilling injury. In some commodities, the effects of chilling build up over time. The overall chilling effects that occur in cold storage are a result of low temperatures experienced during transportation or even in the field just prior to harvest (Park et al., 2018).

11.2.1.1 CHILLING INJURY SYMPTOMS

Depending on how sensitive the commodity is to CI, chilling injury symptoms frequently manifest after keeping the product at room temperature for a sufficient amount of time after exposure to chilling temperatures. These might include the next:

1. The fruit does not ripen. In fridge, a mature green tomato kept for a few days is a good example of it.
2. Fruit softening, which affects all sensitive crops.
3. Guavas kept at 8°C or less for a few days will lose flavor and aroma the easiest.
4. Increased decay sensitivity in all susceptible crops.
5. Skin browning or internal browning (IB), as in bananas.
6. The production of off flavors in various fruits and vegetables.
7. Locations that are wet.
8. Pitting.

11.2.1.2 FACTORS AFFECTING THE SYMPTOM DEVELOPMENT OF CHILLING INJURY

1. Time: In unfavorable temperatures: The severity of the symptoms increases with the length of exposure. However, crops can bounce back after brief exposure to dangerous temperatures.
2. Temperature: When temperatures are unsafe, the severity of the symptoms increases the lower the temperature.

3. Until the removal of produce to ambient temp erature symptoms may not appear.
4. Chilling injury is cumulative. The amount of time spent in dangerously low temperatures prior to harvest adds up to postharvest exposure. Most of the time, low temperatures inhibit the reactions that lead to symptom development, but ambient temperatures cause them to advance and manifest.

11.2.1.3 FACTORS THAT AFFECT CHILLING INJURY

The following two types of factors affect chilling injury:

11.2.1.3.1 Internal Factors

1. Genetics: The degree of chilling sensitivity can vary between species and, in some cases, even between cultivars of the same species.
2. Maturity level: The ability to feel cold gets less sensitive as fruit ages and ripens.

11.2.1.3.2 External Factors

1. Adaptation: The temperatures exposure that is low but safe for brief periods of time.
2. For instance, two days in the air at 38°C or 10 min in water at 53°C may make some fruits more resistant to CI.
3. Alternate heating: This could make people more resistant to CI, and it is thought to encourage the detoxification of the toxic substances that cause the injury.
4. Excessive relative humidity while being stored. This lessens the rate of water evaporation and pitting development.
5. Controlled and modified (MA) atmospheres (CA): It has been demonstrated that MA/CA, particularly high CO_2 concentrations, can reduce CI in a variety of fruits, including avocados, mangoes, and grapefruit.
6. A few fungicides: When applied to grapefruit, thiabendazole has been shown to reduce CI.

More specific information about chilling injury is provided for each product. Many of the goods that are prone to CI are listed in Table 11.1, along with threshold temperatures and a few symptoms.

TABLE 11.1 Chilling Injury of Fruits and Vegetables.

Fruits and Vegetables	Safe Temperature (Approx.)		Chilling Symptoms
	°C	°F	
Apples—certain cultivars	2–3	36–38	Brown core, soggy breakdown, internal browning (IB), and soft scald
Asparagus	0–2	32–33	Dull, grey-green, and limp tips
Avocados	4.5–13	40–55	Greyish-brown discoloration of flesh
Bananas, green or ripe	11.5–13	53–56	Dull color when ripened
Beans (lima)	1–4.5	34–40	Rusty brown specks, spots, or areas
Beans (snap)	7	45	Pitting and russeting
Cranberries	2	36	Rubbery texture, red flesh
Cucumbers	7	45	Pitting, water-soaked spots, decay
Eggplants	7	45	Surface scald, alternaria rot, blackening of seeds
Guavas	4.5	40	Pulp injury, decay
Grapefruit	10	50	Scald, pitting, watery breakdown
Jicama	13–18	55–65	Surface decay, discoloration
Lemons	11–13	52–55	Pitting, membranous staining, red blotch
Limes	7–9	45–48	Pitting, turning tan with time
Mangos	10–13	50–55	Greyish scaldlike discoloration of skin, uneven ripening
Melons			
Cantaloups	2–5	36–41	Pitting, surface decay
Honeydew	7–10	45–50	Reddish-tan discoloration, pitting,
Cassava	7–10	45–50	surface decay, failure to ripen
Crenshaw and Persian	7–10	45–50	Same as above but no discoloration
Watermelons	4.5	40	Same as above but no discoloration
			Pitting, objectionable flavor
Okra	7	45	Discoloration, water-soaked areas, pitting, decay
Olives, fresh	7	45	IB

TABLE 11.1 *(Continued)*

Fruits and Vegetables	Safe Temperature (Approx.)		Chilling Symptoms
	°C	°F	
Oranges, California and Arizona	3	38	Pitting, brown stain
Papayas	7	45	Pitting, failure to ripen, off-flavor, decay
Peppers, sweet	7	45	Sheet pitting, alternaria rot on pods and calyxes, darkening of seed
Pineapples	7–10	45–50	Dull green when ripened
Pomegranates	4.5	40	Pitting, external and IB
Potatoes	3	38	Mahogany browning (Chippewa and Sebago), sweetening
Pumpkins and hard shell squashes	10	50	Decay, especially alternaria rot
Sweet potatoes	13	55	Decay, pitting, internal discolorations; hardcore when cooked
Tamarillos	3–4	37–40	Surface pitting, discoloration

11.2.2 FREEZING INJURY

When goods are kept below their freezing points either before or after harvest, freezing can happen, which can cause the tissue to collapse. At temperatures between 2.2 and 1.7°C, some fruits and vegetables, such as potatoes, freeze. Potatoes also freeze at 1.8 and 1.7°C, cucumbers at 0.9 and 0.8°C, and lettuce at 0.6 and 0.3°C. The disturbance caused by freezing frequently resulted into the tissues collapsing, which may go in front to the complete product loss. Goods freeze as a result of the cells' subsequent dehydration brought on by the condensation of water into crystals of ice both inside and between the cells. The cell walls can also be penetrated by expanding ice crystals. In order to reduce the size of the ice crystals and maintain the quality of the frozen product, in food processing facilities rapid blast freezers are used that create frozen fruits and vegetables. If they are defrosted gradually and for a short time period, some vegetables like cabbage, can recover from freezing. Keep fresh horticultural products above freezing at all times for the best protection against freezing damage. Storage conditions should be as low as possible but just above freezing for goods that are resistant to chilling damage.

Freezing injury happens when ice crystals develop within the tissues. The most typical sign of a frozen injury is a wet appearance. Upon thawing, tissues that have been frozen-damaged typically lose their rigidity and become mushy. Different fresh fruits and vegetables have varying degrees of resistance to freezing damage. Others are irreparably harmed by even a slight freezing, while some products can be repeatedly frozen and thawed with little to no damage. All fruits and vegetables can be categorized into three groups based on how sensitive they are to freezing: the most susceptible, which are liable to suffer serious harm from even one light freezing; the moderately susceptible, which can withstand one to two light freezings, and the least susceptible, which can withstand a number of light freezings without significantly suffering. For instance, apples would sustain severe damage after just a few hours at −7 to −10°C (19.4–14°F), despite only being exposed for a few days to temperatures near or below freezing point.

Depending on the fruit or vegetable type, different types are more susceptible to freezing damage. For instance, leafy lettuce is extremely vulnerable to freezing damage, whereas cabbage and kale can withstand several light freezings without suffering significant damage. For instance, potatoes, which are extremely prone to freezing damage, can be carefully warmed up to −3.9°C (25°F), which is about 3°C (5.4°F) below their freezing point, without showing any signs of freezing (Hruschka et al., 1961). Plant tissues are extremely susceptible to yellowing when frozen, so it is best to avoid handling goods until they have warmed up. A trade-off must be made in choosing an appropriate thawing temperature. Fast thawing harms tissues, whereas extremely slow thawing, such as at 0–1°C (32–33.8°F), allows ice to remain in the tissues for an excessively long time and causes injury. According to studies on the rate of thawing, most commodities sustain the least damage when thawing at 4°C (39.2°F) (Lutz, 1936). Despite the fact that some fruits and vegetables can tolerate freezing to some extent, frozen goods frequently have a shorter shelf life and are more prone to microbial invasion. For instance, frozen apples are softer than fresh apples, and frozen carrots are more susceptible to rotting than fresh ones. As a result, it is preferable to keep fresh produce away from freezing temperatures. Many of the goods that are susceptible to freezing injury are listed in Table 11.2.

TABLE 11.2 Fresh Fruits and Vegetables' Resistance to Freezing Degradation.

S. No.	The Most Prone Crops	Comparatively Prone Crops	Least Prone Crops
1	*Prunus armeniaca*	*Brassica oleracea var. italic*	*Brassica oleracea var. capitata*
2	*Asparagus officinalis*	*Brassica oleracea var. botrytis*	*Beta vulgaris*
3	*Persea Americana*	*Malus domestica*	*Phoenix dactylifera*
4	*Musa spp.*	*Daucus carota subsp. sativus*	*Brassica oleracea var. gemmifera*
5	*Phaseolus vulgaris*	*Citrus × paradise*	*Brassica rapa*
6	*Prunus persica*	*Vaccinium subg. Oxycoccus*	*Brassica oleracea L. var. acephala*
7	*Solanum tuberosum*	*Vitis vinifera*	*Pastinaca sativa L.*
8	*Cucurbita pepo*	*Citrus × sinensis*	*Brassica oleracea*
9	*Solanum lycopersicum*	*Pyrus communis* L.	*Tragopogon porrifolius*
10	*Ipomoea batatas*	*Pisum sativum*	*Brassica napus Napobrassica*
11	*Cucumis sativus*	*Spinacia oleracea*	
12	*Solanum melongena*	*Cucurbita pepo*	

Source: Adapted and modified from Wang and Wallace (2004).

11.2.3 MECHANISM EMPLOYED BY SALICYLATES AND JASMONATES FOR ALLEVIATING OF CHILLING INJURY

In postharvest, the main technique used to extend the fruits and vegetables shelf life while maintaining their quality is low temperature storage (LTS). LTS lowers the rate of respiration and slows the spread of fungi. However, fruits and vegetables belongs to tropical and subtropical areas are vulnerable to chilling injury and low temperature treatment. When fruits or vegetables are frozen-damaged, they develop unfavorable quality changes like abnormal ripening, pitting, or browning. When the effects are particularly bad, the produce suffers a lot, which significantly lowers its final market value and results in significant financial losses. When salicylates and jasmonates are used to treat fruits and vegetables, they can reduce the activity of PLD, PLC, and LOX enzymes, improve unSFA/SFA ratio (possibly by increasing FAD gene expression), and maintain ATP and AEC levels while also (1) alleviating CI; (2) increasing the expression and accumulation of HSPs; (3) increasing the activity of the immune system; (4) increasing the arginine pathway, which resulted in the buildup of signaling molecules such as polyamines, nitric oxide (NO), proline, and GABA, all of which are essential for chilling tolerance; (5) causing changes in the activities of the PAL and PPO enzymes, and (6) activating the CBF pathway.

11.2.3.1 ENHANCING MEMBRANE INTEGRITY

Through its effects on the arrangement of integral membrane proteins, membrane permeability, and transmembrane transport activity, membrane fluidity has the capacity to regulate membrane function (Los and Murata, 2004). Increased membrane fluidity improves CI tolerance by reducing the membrane's propensity to transition from flexible liquid crystalline to solid gels (Los and Murata, 2004). According to Cao et al. (2009), a MeJA treatment of 10 M for 24 h at 20°C significantly reduced chilling injury in loquat fruit. Internal browning was reduced and fruit firmness was maintained as chilling injury was lessened. As a result, CI development was linked to a decline in the unSFA/SFA ratio. They discovered that in peach fruit, the MeJA treatment increased ATP and ADP levels while decreasing AMP levels, which resulted in a higher level of AEC. The meJA treatment increased the activity of several enzymes necessary for cells to produce ATP, including succinic dehydrogenase (SDH), H+-ATPase, Ca^{2+}-ATPase, and cytochrome C oxidase (CCO).

11.2.3.2 ENHANCING ANTIOXIDANT ACTIVITY

According to Yang et al. (2012), inducing endogenous SA biosynthesis and successfully reducing the effects of CI were achieved in peach fruits by postharvest treatment with 1.0 mM SA for 10 min, either alone or in combination treatment with ultrasound (40 kHz, 10 min). They hypothesized that when ultrasound and SA were used together, both treatments worked more effectively to treat CI. By reducing the activity of the enzyme that causes oxidation of AA and, consequently, the loss of this essential antioxidant molecule, AA oxidase (AAO), in sweet pepper postharvest treatment with SA and $CaCl_2$ helps in maintaining the high levels of AA (Rao et al., 2011). AA content is essential for the antioxidant capacity due to the preservation it is also crucial for its anti-browning property, the postharvest treatment's reduction of AAO enzyme activity was helpful for maintaining nutritional and organoleptic quality (Rao et al., 2011). According to Cao et al. (2010),

SA application (1 mM for 5 min) and hot air treatment (38°C for 12 h) both helped to reduce CI in peach fruit. Superoxide radicals are produced by LOX, and once the SOD enzyme gets involved, they can be converted to H_2O_2. The activities of the enzymes CAT, APX, and GR can scavenge H_2O_2 (Mittler, 2002). Hot air combined with SA applied to peaches increases SOD enzyme activity while decreasing LOX activity, increasing the SOD/LOX ratio, which lowers superoxide levels. Increasing CAT, APX, and GR enzyme activity lowers H_2O_2 levels (Cao et al., 2010).

11.2.3.3 ENHANCING HSPS ACCUMULATION

HSPs are a family of proteins that respond to stress and its molecular weights lie between 15 and 115 kDa. Chaperonins (HSP60s), HSP90s, HSP70s, HSP100s, and HSPs with less molecular weight, which is also known as small HSPs. There are the five HSP families that have been discovered (sHSPs). Chaperone activity (that is independent of ATP) is exhibit small HSPs (sHSPs), which have low molecular weights between 15 and 42 kDa, in contrast to HSPs with higher molecular weights (Gusev et al., 2002).

11.2.3.4 ENHANCING ARGININE PATHWAYS

As a metabolically versatile amino acid, arginine is essential not only as a protein building block but also as a precursor for the biosynthesis of signaling molecules such as proline, NO, and polyamines (putrescine, spermidine, and spermine), which may increase tolerance to CI (Gao et al., 2009; Morris, 2009; Jubault et al., 2008). Arginase, arginine decarboxylase (ADC), and nitric oxide synthase (NOS) are three essential enzymes that catabolize arginine (Morris, 2009). Ornithine, a precursor for the biosynthesis of proline and polyamines, is produced by the enzyme arginase when arginine is converted to it.

11.2.3.5 ACTIVATION OF CBF PATHWAY

Fruits and vegetables' ability to withstand CI is greatly influenced by CBF, a transcription factor, because it ecognizes and binds to the C-repeat/DRE DNA regulatory elements (Zhao et al., 2009). MeJA treatment increased the expression of the MaMYC2a and MaMYC2b genes, which both encode isoforms of the basic helix-loop-helix (bHLH) transcription factor MYC2, during the later cold storage (7°C) in fruit of banana. According to Cheng et al. (2011), Fernández-Calvo et al., (2011), and Niu et al., (2011), MYC2 is a crucial regulator in the activation of the JA response. However, the researchers found that MeJA treatment had little effect on MaICE1 gene expression, whose by product induces the CBF regulon in plants' response to cold stress (Chinnusamy et al., 2003, Zhao et al., 2013).

11.2.3.6 ALTERATION IN PAL AND PPO ENZYMES ACTIVITIES

It is common knowledge that the following things occur when fruits and vegetables are kept in a chilling environment: (1) there is a loss of selective membrane permeability; (2) a CI effect results in an increase in the total phenols that accumulate in vacuoles; (3) PPO activity increases in the cytoplasm, which causes internal or flesh browning; and (4) accumulated phenolic compounds in vacuoles leak into the cytoplasm because a vacuole ruptures (Sevillano et al., 2009). The chilling injury effect in plum fruits was significantly diminished by SA treatment (1.5 mM for 10 min), according to research by Luo et al. (2011). Lu et al. (2011) hypothesized that it is possible that a decline in PPO and PAL activities caused the observed IB reduction in pineapple fruit treated with PSS and PSI. Additionally, pineapple treated with PSI and PSS contains high concentrations of AA, an antioxidant molecule well known for its ability to prevent browning, which may help lessen IB. As a result, another advantageous side effect of PSS and PSI treatments is maintaining AA levels (Lu et al., 2011). The authors claim that SA treatment decreased chilling injury in bamboo shoots by postponing the enzymatic reactions that lead to browning (PPO and PAL). To lessen the effects of CI, loquat fruit was treated with ASA at a 1 mM concentration for 5 min at pH 3.5 after harvest. This outcome was escort by a decrease in lignification's, leakage of electrolyte, and internal and external browning (Cai et al., 2006).

11.3 HIGH TEMPERATURE

11.3.1 HEAT INJURY

In the field, heat damage can occur during harvest. When horticultural commodities are exposed to high temperatures, such as when they are in direct sunlight or are being stored or transported without refrigeration, this can significantly raise their temperature and result in significant losses of qualitative and quantitative traits. Breaks in the cold chain may result in extremely high temperatures and heating, which may cause severe stress and harm. Temperatures above the optimum speed up metabolic process, shorten fruit life, and cause other significant qualitative and quantitative losses. Due to the fact that these temperatures prevent ethylene synthesis, several fruits will not ripen after being exposed to 35–40°C. The tissue will be killed by extremely high temperatures. Reactive oxygen species (ROS) can build up as a result of the steady-state flux of metabolites being disrupted. In response to heat stress, plants modify their proteome, metabolome, liposome, and transcriptome. These adjustments result in a brand new study state equilibrium of metabolic processes, which can aid in the organism's adjustment to the new temperature as well as help it cope with other types of stress, such as low temperatures (Lauxmann et al., 2014). For a variety of commodities, high temperature treatments have been developed and integrated into the postharvest handling chain, in order to reduce pathogen inoculum, lessen the symptoms of chilling injury, and control insect pests. Some commercially available insect control methods require the fruit's core temperature to reach a specific temperature, which could require a few minutes in water or several hours in the air. Treatments that involve hot water dips for 3–20 min at 48–56°C help to prevent decay. There have also been studies done on hot water rinsing and brushing (HWB) treatments, which only last 10–25 s but can reach 63°C. Because a high temperature stress can alter plant tissues and affect various physiological and biochemical processes, these treatments may have effects on the fruits or vegetables that are different from those intended for their intended use (Cai et al., 2006).

11.4 PHYSIOLOGICAL DISORDERS RELATED TO MINERAL IMBALANCE

Deficiency disorders in minerals: Two implications of mineral imbalances in horticultural products can be used to understand changes in postharvest life and physiology of fruit: (1) the result of a loss of homeostasis in the structural and metabolic processes of fruit, which has an impact on quality of fruit; and (2) sensitivity of diseases, which can speed up infection and decay biologically. On the other hand, when the imbalances exceed the vital level of adaptability, neither physiological nor morphological responses are adequate to stop metabolic disorders. Horticultural products may exhibit a number of disorders caused by nutrient deficiencies before or after harvest.

11.4.1 DISORDERS DUE TO CALCIUM AND BORON DEFICIENCIES

Blackheart, bitter pit, blossom-end rot fruit cracking, and cork spot are a few of the common disorders brought on by a Ca deficiency. Other disorders linked to calcium deficiency include blackheart of celery, brown heart of escarole, and blossom-end rot in watermelons. In order for the components of the cell membrane, cell wall, cytosol, and vacuole to function properly, Ca is a crucial nutrient for plants. It is needed for a number of metabolic and structural processes. The transpiration stream carries calcium through the xylem after being absorbed by the roots, along with other nutrients. Since calcium is comparatively immobile in plants, certain physiological and environmental conditions can impair its movement (Barta et al., 2000). A short-circuit in its distribution could quickly impact developing cells. As boron serves as a crucial micronutrient for carrying out tasks such as strengthening cell walls and sugar transport, there is still some correlation between this mineral and calcium. Calcium deficiency is partially caused by a lack of calcium in the soil or by the condition of the plants, which may have varying demands for calcium concentration. According to plant concentration supplies, calcium is a macronutrient that may occasionally be temporarily unavailable to developing tissues or phloem distribution. Additionally, some data indicates that environmental factors that cause stress, such as high humidity, watering, and high temperatures, can exacerbate the symptoms of calcium deficiency. Blossom-end rot, cracking, cork, spongy tissue, and bitter pit are among the physiological disorders linked to Ca imbalance in fleshy fruits. Leafy vegetables are more prone to tip burn, brown heart, and blackheart disorders (Bakeer, 2016). Some physiological disorders of different fruits are listed in Table 11.3.

TABLE 11.3 Physiological Disorders of Some Fruits.

Disorder	Principal Affected Commodity
Fruit cracking	Tomatoes, husk tomatoes, apricots, pomegranates, lychee, chillis, prunes, peach and plum
Cork (spot like)	Apple, pear
Spongy tissue	Mango
Tip-burn and IB	Lettuce, cabbage, Brussels sprouts, escarole, celery, and cauliflower
Hollow stem	Tomatoes, broccoli
Blossom end rot	Peppers, squash, melon, cucumbers, and tomatoes
Bitter pit	Apples
Blackheart	Cabbage, celery
Pre Harvest Factors Associated with High Bitter Pit Incidence	References
Low fruit Ca at harvest	Ferguson and Watkins (1989), De Freitas et al. (2010)
High fruit Mg and K at harvest	Ferguson and Watkins (1989); De Freitas et al. (2015)
Low crop load/low fruit Ca/high fruit K	Ferguson and Watkins (1992), Volz et al. (1993), Khadivi-Khub et al. (2010)
Large fruit volume and weight/low fruit Ca	Ferguson and Watkins (1989), Huang et al. (2008)
Poor pollination/low seed number/low fruit Ca	Bramlage et al. (1990), Tomala and Dilley (1990), Brookfield et al. (1996), Volz et al., (1996), Islam et al. (2017)
Low spur/bourse leaf area/low fruit Ca	Jones and Samuelson (1983), Proctor and Palmer (1991), Volz et al. (1994, 1996)
Fruiting wood age and terminal or lateral positions	Schumacher et al. (1979); Volz et al. (1994), Khadivi-Khub et al. (2010)
High position on tree/low fruit Ca	Ferguson and Triggs (1990); De Freitas et al. (2010)

11.4.2 FRUIT CRACKING

Surface or tissue cracking can occur at a variety of levels depending on environmental factors, varietal susceptibilities, and nutritional deficiencies, according to this disorder. Fruit cracking is one of the main quality disorders, and research on the status of the nutrient calcium has shown that it plays a role in the physiological process of cellular adhesion loss. Weather patterns and germplasm tolerant to cracking must be combined in order to control this mineral deficiency. Increases in structural minerals and crack prevention have not been achieved by simply incorporating mineral amendments. Sweet cherries, plums, apricots, apples, litchi, pomegranates, citrus fruits, bananas, avocados, grapes, persimmons, peaches, tomatoes, pistachios, and other fruits are frequently affected by fruit cracking disorder (Opara, 1996; Simon, 2006; Khadivi-Khub, 2009).

11.4.2.1 APPLE

Some cultivars of commercial apples exhibit more fruit cracking. Commercial cultivars "Golden Russet," Golden Delicious, Fuji," and "Gala" are vulnerable to fruit cracking (Sadamori, 1963; Opara, 1996; Andrews et al., 1999; O'Rourke et al., 2003). Fruit cracking in apple is influenced by the following factors:

11.4.2.1.1 Fruit Development

According to the signs, there are two types of fruit cracking in apples. Internal ring-cracking (IRC) is one of them and develops at the stem joint of fruit. Stem-End Splitting develops in the cavity of stem (Opara, 1996). So, water absorption-induced cell growth may have boosted IRC or SES. A fruit's uneven cell enlargement can lead to microscopic cell splits and induce IRC. IRC occurs just below the epicarp and appears to be caused by unevenness between the enlargement of cells for the epicarp and mesocarp (Opara, 1996). Additionally, apples are more likely to crack if they are grown away from a tree and exposed to the sun (Verner, 1957).

11.4.2.1.2 Gene Expression

Different researchers have found that some genes involved in the cell wall contribute to apple fruit cracking. Cell wall proteins called expansins have the potential to promote wall thinning and loosening of cell wall during cell expansion under expandable stress (McQueen-Mason and Cosgrove, 1994; Cosgrove, 2000). In peach, expansins from hypocotyls were isolated (Hayama et al., 2001), tomato (Brummell et al., 1999), and cucumber (Cosgrove, 1999) showed a correlation with fruit development.

Six expansin genes have also been found in apples, and during fruit development their expression patterns were identified (Wakasa et al., 2003). Expansins may contribute to fruit cracking because they increase the extensibility of the cell wall and encourage cell growth. MdEXPA3 transcript levels in the mesocarp at 30 days after full bloom increased, peaked at 95 DAFB, then began to decline, paralleling the rate of fruit growth, according to Kasai et al. (2008). As a result, early fruit cracking symptoms are present when mesocarp *MdEXPA3* expression is higher than that of the epicarp. Due to insufficient levels of growth stimulating expansins, epicarp cells may not be able to follow the mesocarp cells expansion in such circumstances. If this is the case, an imbalance in the levels of expansin-dependent tissue growth is what causes IRC to manifest (Kasai et al., 2008).

11.4.2.2 CITRUS

Citrus fruit cracking can begin in the middle of the summer, but it usually starts in the fall. The fruits of "Navel" oranges, "Valencia," "Hamlin," "Murcott," "Tangelos," and a number of varieties of mandarins with a tighter skin are the most prone to cracking. Splitting is less frequent in grapefruit, "Dancy" tangerines, and "Temple" fruits. Citrus fruit's fruit blossom end, which is the skin's most fragile area, is where cracks typically first appear. The split may be shallow and short or wide and deep. Additionally, only a small percentage of cracks reach the inner center, making partial cracking more common. The cracked area of the fruit can become infected by organisms which can then spread disease that results in early fruit falling from trees and partial rotting. Citrus fruit cracking is influenced by the following factors.

11.4.2.2.1 Environment

Fruit splitting is closely related to extreme fluctuations in temperature, soil moisture, humidity, and fertilizer levels. It is believed that a combination of the aforementioned factors, as opposed to a single agent, causes fruit to crack. Therefore, it is important to always provide trees with enough water and nutrients (García-Luis et al., 1994).

11.4.2.2.2 Growth Regulators

Fruit cracking is influenced by the time at which gibberellic acid (GA3) is applied. Citrus fruit cracking is increased by GA3 application during flowering, but it is reduced soon after the end of the June drop, (Agusti et al., 2002). The best results were obtained when GA3 and 2,4-D were applied together (Monselise and Costo, 1985; Aliviela et al., 1994), which greatly diminished fruit splitting. In addition, Monselise and Costo (1985) discovered that spraying calcium nitrate just as cells begin to enlarge significantly decreased the proportion of citrus fruit that cracked.

11.4.2.3 SWEET CHERRY

There are three known varieties of fruit cracking in sweet cherries: Three different types of splits are known as lateral splitting: (1) cracks around the stem end on the cavity that are semicircular or circular; (2) small splits present at the apical end of the fruit; and (3) numerous, frequently on the fruit sides deep splits are present (Simon, 2006; Khadivi-Khub, 2009). Many days before fruit maturity and harvest, in cherry, at the base and apical end of fruit, there may be tiny, fine splits visible. The market value of fruits is reduced by these tiny cracks present at an early stage of development of fruit. The fungi (brown rot and grey mould) can enter through the large, deep side cracks, which are frequently followed by severe attacks. Fruit quality will be impacted by fungus attacks, making them unsellable. The following factors can cause cherry to crack:

11.4.2.3.1 Fruit Size

It is a common misconception that larger sweet cherry fruits will crack more easily than smaller ones. According to studies, cultivars with large fruits are more likely to crack than those with small fruits (Sekse, 2005; Khadivi-Khub et al., 2007). Way (1967) demonstrated that, for the same cultivar of sweet cherry, fruits from trees loaded with heavy crops crack less than those from trees with light crops.

11.4.2.3.2 Fruit Firmness

Firm-fleshed cherry cultivars are thought to be more susceptible to fruit cracking than soft-fleshed cultivars. In cherry, it seems paradoxical that cultivars with firm-fleshed fruit are more prone to cracking of fruits than those with soft-fleshed flesh. soft, such as "Arak," "Mashhad," "Mesh-kinshahr," "MehaliKaraj," and "Hamedan," do not exhibit any cracking, but cultivars with firm flesh fruit, such as "Soorati-Lavasn," "Rafat," "Ghazvin," "Maremot," and "Siah-Daneshkade," exhibit significant cracking (Khadivi-Khub et al., 2007).

11.4.2.3.3 Fruit Shape

In sweet cherries, shape of fruit is a crucial determinant of cracking in fruits (Beyer et al., 2002; Sekse et al., 2001; Sekse, 2005; Khadivi-Khub, 2007). Researcher believed that when fruits are at a right angle to the shape's curvature, they can break more easily. Cherry fruit cracking can be significantly influenced by the fruit's shape (Khadivi-Khub et al., 2007). It means that if a cultivar has a kidney, blocky, or heart shape and is genetically predisposed to moderate susceptibility to splitting (based on cuticle thickness and sugar content), it may also be susceptible to stem end type splitting (Simon, 2006; Khadivi-Khub, 2007).

11.4.2.3.4 Fruit Development

Most sweet cherry cultivars are believed to become more susceptible to cracking as they mature more fully, but there are some instances where this relationship does not hold true. The reason for this deviation is that climatic conditions have an impact on the susceptibility of fruits to its cracking which is prior to the start of fruit sampling or picking (Christensen, 1996; Yamaguchi et al., 2004).

11.4.2.3.5 Cuticular Properties

According to Belmans et al. (1990), a significant relationship between cuticle thickness and resistance was seen in 13 cultivars. By putting the fruit's surface in contact with liquid water or a lot of water. Knoche and Peschel (2006) showed that the cuticular membrane developed micro cracks.

11.4.2.3.6 Water-Retaining Capacity of the Fruit Pulp

It is hypothesized that cracking in dwarf rootstocks happens as a result of stress on their smaller roots (Knoche and Peschel, 2009; Hovland and Sekse, 2004; Khadivi-Khub, 2009).

11.4.2.3.7 Genetic

According to reports, some SSR and RAPD markers have a strong correlation with fruit cracking, specifically the *BPPCT004197* SSR locus and *TIBMBA-09521* RAPD marker (Khadivi-Khub, 2007, 2011). Additionally, research was done to find quantitative trait loci (QTL) for sweet cherry cracking resistance (Quero-Garcia et al., 2012). They stated that distinct linkage groups were found for the QTL of cracking is resistance in each fruit zone. This finding demonstrates that different genetic mechanisms contribute to fruit cracking in various fruit regions (Quero-Garcia et al., 2012).

11.4.2.3.8 Environment

The rate at which fruit cracks depends significantly on temperature. Overall, the rate of fruit cracking increases linearly as temperature rises from 10 to 40°C. Additionally, many other factors, including the permeability of cell walls and biochemical processes within cells, are influenced by temperature (Simon, 2006). Additionally, how susceptible sweet cherries are to fruit cracking depends on the amount and type of rainfall during the maturing and ripening stages of fruit. (Christensen, 1996; Simon, 2006).

11.4.2.4 LITCHI

In many commercial orchards, cracking in litchi fruit is a serious problem. In litchi fruit, if there is higher levels of GA and ABA in the skin, aril, and seed, as well as endogenous growth regulator concentrations, are all directly related to splitting (Menzel, 1984). Among the causes of fruit cracking in litchi are:

11.4.2.4.1 Gene Expression

Yong et al. (2006) used the cracking-resistant cultivars "Huaizhi" and "Nuomici" to measure the expression profiles of two expansin genes from litchi fruit at various fruit development stages in order to determine the relationship between fruit splitting and gene expression (the cracking-susceptible cultivar). By using RT-PCR and RACEPCR techniques, two full-length cDNAs called *LcExp1* and *LcExp2* that encode expansin and have 1087 and 1010 base pairs respectively were isolated from expanding fruit. They noted that in the pericarp of "Huaizhi," the *LcExp1* mRNA increased, whereas in the pericarp of "Nuomici," the mRNA could be detected at the stage of rapid fruit growth, then increased slightly, and finally kept almost constant. Peaked at the end of the growth phase (80 days after anthesis), *LcExp2* mRNA accumulated similarly in the fruit arils of "Nuomici" and "Huaizhi," but only the pericarp of "Huaizhi," not the pericarp of "Nuomici," accumulated *LcExp2*. Their research demonstrates a direct relationship between the expansion of two genes (expansin) in the pericarp of litchi and fruit growth and cracking (Huang et al., 2008).

11.4.2.4.2 Fruit Development

Cracking and fruit development are closely related. The rapid growth phase of the aril occurs 60 days after blooming and lasts until harvest, which is the critical period for fruit cracking. Before the cracking of fruits, there is raise in diameter of fruit typically exceeds 1.00 mm/day, usually cracks after that in 1–2 days. The peel feels pressure from the "spurt of fruit growth," but only ruptures under excessive pressure. This varies depending on the cultivar and type of fruit on the tree.

11.4.2.4.3 Peel Morphology

The part of the litchi pericarp that is most likely to crack is the mesocarp because of its size. Litchi cultivars differ significantly from one another in terms of structure. It has been established that the features of peel crackles serve as significant indices for classifying cultivars according to their varying susceptibilities to crack. When compared to the resilient "Huaizhi," crackles for the simple "Nuomici" are wider and shallower. The development of the peel has been hampered, as evidenced by the smaller and sharper tubercles of the cracked pericarp than those of the unbroken peel. According to Rab and Haq (2012), the pulp and skin characteristics of litchi fruit can indicate how susceptible the fruit is to cracking.

11.4.2.4.4 Environment

There is a correlation between the vapor pressure deficit and daily fruit cracking rate (−0.90), relative humidity (0.90), temperature (−0.80), and sunshine hours (−0.70). Fruit cracking can occur in the summer when there is a hot wind. Currently, some commercial orchards accept applications of organic fertilizer and lime during the winter. In addition to pruning and bagging, irrigation is also crucial. Fruit cracking is decreased during the stage of fruit pit hardening by spraying with boron 0.40%.

11.4.2.5 TOMATO

One of the most widely consumed commercial vegetables, tomatoes are prone to fruit cracking. There are a number of theories put forth for tomato fruit cracking, including:

11.4.2.5.1 Genetic Differences

Despite the fact that there are clear cultivar differences in fruit cracking susceptibility (Abbott et al., 1986), it is not yet clear what the physiological causes of these variations are. Although breeders have made great strides in improving modern tomato cultivars such as crack resistance, the genetically crack resistance are poorly understood. According to Walter (1967), there is no study has found that inheritance of cracking resistance is straightforward. Different gene families are probably in charge of regulating the concentric, burst, and radial types of cracking. Even when only one type of cracking is considered, direct selection of resistance is not easy due to the fact that numerous genes appear to be involved, low heritability, and environmental influence is significant (Cuartero et al., 1981; Islam et al., 2017).

11.4.2.5.2 High Differences Between Day and Night Temperatures

Bulky fruits such as tomatoes should be expected to shrink during night-time fruit cooling (Corey and Tan, 1990). This unenthusiastic pressure may draw moisture from the plant itself as well as condense or other moisture that is on the fruit surface. Positive pressures develop inside the fruit as it warms up in the sun and stretching the gas impermeable as the fruit gets bigger, the tomato skin extends outward (Corey and Tan, 1990). The more stress the skin is likely to endure, the greater the day/night difference.

11.4.2.5.3 High Humidity

Increased cracking is linked to high humidity or shifts between night and daytime humidity. High temperatures and high humidity both contribute to tomato fruit cracking in greenhouse environments (Byari, 1984). Increases in water and gas pressure may be related to how humid conditions affect tomato fruit cracking. Ninety-seven percent of tomato skin is impermeable to gas (Corey and Tan, 1990).

11.4.2.6 CORK (SPOT-LIKE) DISORDER

More severe cases occur in susceptible cultivars of late-maturing apples and pears. It is regarded as a harmful flesh effect frequently linked to calcium and boron deficiency. Apples with cork disorder have quality issues (like dry and withered tissue) as well as spots that develop a bitter taste and may affect the size of the fruit. However, it is also thought that cork spots considerably multiply on larger fruit. Following fruit cutting, the condition may manifest as interior lesions that are brown in color, corky hard, and made of dead tissue that faces the calyx, as well as cracking in more severe conditions. The blemishes on the peel might merely be a figment of external harm. The disorder can also appear in some pear cultivars (Hayama et al., 2017; Matsumoto et al., 2018).

11.4.2.7 SPONGY TISSUE

Fruit cutting reveals the disorder even though there are no visible symptoms on the outside of the fruit. The pulp nearer the seed develops ripening changes (texture and color), turning the tissue hard and less-colored pulp. Depending on whether air has been added to the matrix or not, the color can change from a faint yellow to a brownish black. Due to its significance in maintaining the integrity of cell walls, calcium deficiency has been proposed as one of the candidates causing spongy defects. Therefore, in addition to Ca deficiency, there may be other causes for this internal breakdown disorder in mangoes. The biochemical and physiological processes involved in the breakdown may be significantly regulated by environmental factors. Triazole fungicides and growth retardants have helped to control the problem to some extent by harvesting fruits that are nearly fully ripe (Huang et al., 2008).

11.4.2.8 TIP BURN AND INTERNAL BROWNING

Tip burn symptoms can be seen in the developing tissues and appear at the leaf margins (i.e., lettuce's heart leaves). A longitudinal incision is required to assess the severity of symptoms in the disorder known as IB, which is often characterized by a necrosis of tissues between the growth point and the exterior. Failures to distinguish between minerals based on symptomatology have been discovered, and this internal disorder is also associated with the deficiency of Boron (B). The tissues affected may change color from brown to black and may also be influenced hormonally by ethylene. In addition, while lettuce experiences large visible affected tip burns from desiccation during storage, Brussels sprouts and cabbage do not exhibit IB until after harvest (Bakeer, 2016).

11.4.2.9 BITTER PIT

Apples with bitter pit have a disorder that affects their appearance. The condition is characterized by dark skinned, circular depressions that affect the fruits postharvest quality. A localized calcium deficiency has been linked to the disorder during at the time of cell division and elongation of cell that take place 4–6 weeks after anthesis. Decrease in Ca levels noticed due to mineral imbalance at this stage and an increase in potassium and magnesium concentration, which may affect the permeability of cell membranes and cause progressive cell death (Ferguson and Triggs, 1990). At the end of the fruit calyx, symptoms of the bitter pit are more common and are correlated with dry summers. The primary factor affecting the disorder is varietal sensibility; even before harvest cultivars like Honey crisp, Golden Delicious, Granny Smith, and Gala may show symptoms. Internal breakdown, storage scalds, and Jonathan spot are additional conditions that can affect apples while they are being stored (De Freitas et al., 2010, 2015).

11.4.2.10 BLOSSOM-END ROT

Blossom-end rot is one of the disorders that cause significant monetary crops losses like watermelons and tomatoes, especially in susceptible varieties. It is a serious condition that has an antagonistic effect on the fruit quality and is linked to uneven calcium distribution and absorption during times of high mineral demand. The specific rotting indication found on the underside of developing or ripe fruit is known as end rot. The lesion starts out as a small, depressed lesion, then it gets bigger and becomes an oxidized dark spot. Blossom-end rot can be prevented using a special control strategy called prevention during pre-harvest management of nutrition. Similar factors that affected the severity of the bitter pit disorder have also been discussed previously (Taylor and Salvadore, 2004; Ho and White, 2005).

11.4.2.11 BLACKHEART

Celery has a condition called blackheart, which exhibits symptoms similar to tipburn. It can be identified by the development of necrotic tissue at the tip of young leaves and the progression of necrosis from brown to black throughout the leaf. Blackheart symptoms are more common when translocation of Ca occurs to young, actively growing tissues; treating older leaves with foliar calcium does not alleviate the condition. Ca is sprayed in the form of chloride or nitrate salts may be effective in some control situations, but new Ca-chelated presentations may enhance mineral absorption and translocation (Pardossi et al., 1999). Mineral Overflow The central role that nitrogen plays in protein conformation explains imbalances brought on by improper nitrogen management. The negative effects of excess nitrogen on tomatoes include changes in herbivore defence mechanisms, decreased fruit quality, decreased acidity, decreased volatile production, and decreased sugar content, all of which have an impact on insect pest infestation. Other minerals that, when present in high concentrations, affect the chemical composition of fruit include phosphorus and potassium. Flavor has been linked to high phosphorus and potassium levels. Phosphorus likely supports the accumulation of sugar and the supremacy of sweetness over organic acid concentration by activating synthesis and energy flow. Access to potassium causes vitamin C and other organic acids to accumulate (Raid, 2004).

11.5 PHYSIOLOGICAL DISORDERS RELATED TO GASES

Some horticultural products require atmospheric modifications and control, often involving low oxygen and high levels of carbon dioxide (CO_2), for storage, transportation, and packaging. Fresh horticultural products can withstand a range of oxygen and carbon dioxide concentrations. Fresh horticultural commodities suffer damage and losses when oxygen and carbon dioxide levels are too low or too high for long enough periods of time.

11.5.1 LOW-OXYGEN DISORDERS

Brown color areas on the skin which is fluctuating from distinct tiny patches to areas larger in size of the surface of fruit, are the symptoms shown externally of an injury with low oxygen levels. These signs resemble the ones listed for scald. Inside the fruit, there are sections that are brownish corky and occasionally have cavities. Sometimes the same symptoms appear on the inside and outside. Purpling of the flushed areas of the skin, bleaching or scorching of the skin, and fermentative odors and tastes are further symptoms. Late fruit harvest, the emergence of a low-oxygen atmosphere a week following fruit harvest, a gradual drop in atmospheric oxygen levels, and sluggish cooling of the product can all enhance the risk of injury. The incidence of low-oxygen injury may increase if carbon dioxide is added to the low-oxygen environment. Different species and cultivars of the same species are more sensitive to and more tolerant of low oxygen environments than other species (Zanella et al., 2004). Fruits like citrus are particularly susceptible to harm from low oxygen levels. Oranges kept at zero degrees for a period of 12 weeks in an atmosphere with no oxygen display scald-like signs on the rind of fruits, and juice stored in environments with oxygen levels less than 5% may have off flavors due to the build-up of ethanol and acetaldehyde. Observations indicate that Northern Spy and Empire apples may be more resistant to low- oxygen damage than McIntosh and Delicious apples, which may be more susceptible. Aeration may aid tissue recovery and assist get rid of some of the unpleasant flavors and odors if low-oxygen injury has already started but is not as severe (Corey and Tan, 1990).

11.5.2 HIGH-OXYGEN DISORDERS

Some commodities can be damaged by extremely high levels of super-atmospheric oxygen (O_2), which is greater than the 20–21 kPa typically detected in unrestricted air. Apples from the variety Bramley Seedling show browning of the skin; flesh as well as mealy flesh after being kept at 4°C in 100 kPa O_2. When stored at 100 kPa O_2 for more than three months, one of apple variety Granny Smith develop high concentrations of ethanol and take on a bronzed coloration. When exposed for more than 5 days to oxygen level of 80 or 100 kPa, green tomatoes at maturity develop spots which are dark brown in color on their skin, the severity depends on the length of exposure. The intensification of ethylene-related disorders on some vegetables is another consequence of the super-atmospheric oxygen levels. Environments with high oxygen levels also increase ethylene production and the lettuce disorder known as russet spotting caused by ethylene (Wang et al., 2005; Prange and DeLong, 2006).

11.5.3 HIGH CARBON DIOXIDE DISORDERS

This damage frequently befalls as a result of atmospheric exposure with the levels of CO_2 above what the end product can tolerate. Packed produce in containers which are gas-tight as well as stored produce, conveyed, or packaged in specially designed modified/controlled atmosphere systems may contain the injury. The sensitivity to and tolerance to CO_2 varies between species and cultivars as well. For instance, the majority of berries are very tolerant of CO_2, whereas lettuce is very sensitive. Cortland, Golden Delicious, Granny Smith, Cox's Orange Pippin, McIntosh, Jonathan, Rome Beauty, Northern Spy are among the apple cultivars that are vulnerable to carbon dioxide damage. Newton Wonder is a very delicate cultivar that can suffer internal harm from 3 kPa CO_2. On the other hand, Worcester Pear main is a hardy cultivar that can withstand pressures of CO_2 of up to 13 kPa. Most fresh horticultural commodities generally suffer harm from higher CO_2 levels more than 20 kPa (De Corato et al., 2017). These CO_2 concentrations may result in tissue breakdown, acetaldehyde and ethanol buildup, as well as unfavorable odors. Reducing carbon dioxide injury during modified and controlled environment storage, transport, and packaging requires a thorough understanding of postharvest produce practices (such as harvesting, handling and storage of fruits, their packaging and transportation) (Assis et al., 2001).

11.5.4 ETHYLENE DISORDERS

Fruits and vegetables may experience physiological disorders due to ethylene. Certain disorders are frequently observed when handling fresh horticultural products, such as isocoumarin accumulation-induced different disorders like bitter taste of carrots, russet spotting in lettuce, uneven degreening in citrus, yellowing induced in fruits and vegetables. Disorders caused by ethylene are frequently the result of exposure to ethylene at extremely low levels. It has been demonstrated that exogenously use of ethylene concentrations more than 0.3 ppm are sufficient to cause the formation of isocoumarin in carrot root discs. Due to its ability to inhibit the action of ethylene, carbon dioxide can reduce the amount of isocoumarin. Russet spotting, which appears as tiny spots which are reddish-brown in color on the midribs of leaves of lettuce, is brought on by ethylene exposure. In lettuce kept at 5°C, levels of ethylene as low as 1 ppm cause rusty spots to appear. It has been also demonstrated that using 1-Methylcyclopropene (1-MCP), an activity of ethylene inhibitor, can stop some disorders induced by ethylene like broccoli yellowing (Martínez-Romero et al., 2007; Zhang et al., 2012).

11.6 CONCLUSION

Numerous different physiological disorders can cause significant qualitative and quantitative losses in fresh horticultural commodities, particularly in the case of CI. Effective pre- and postharvest handling is more important for preventing physiological disorders than applying control measures after harvest. The development of physiological disorders may be prevented by using knowledge of the atmosphere gas composition, plant mineral nutrition that the commodity can tolerate while in use, and the protected temperature for storage and transport. A proper treatment or treatments could extend the shelf life of sensitive commodities, allowing them to remain on the market for longer. These chemical or physical substitues are believed to fortify the cellular anti-oxidative system and the integrity of the cell membrane, which may help control certain disorders like CI.

KEYWORDS

- **fruit cracking**
- **heat injury**
- **low temperature**
- **mineral imbalance**
- **physiological disorders**

REFERENCES

Abbott, J. D., Peet, M. M., Willits, D. H., Sanders, D. C., & Gough, R. E. (1986). Effects of irrigation frequency and scheduling on fruit production and radial fruit cracking in greenhouse tomatoes in soil beds and in a soil-less medium in bags. *Scientia Horticulturae, 28*(3), 209–217.

Agusti, M., Martinez-Fuentes, A., & Mesejo, C. (2002). Citrus fruit quality. Physiological basis and techniques of improvement. *Agrociencia, 6*(2), 1–16.

Ali, S., Khan, A. S., Anjum, M. A., Ejaz, S., Khaliq, G., Hussain, S., Ahmad, R., & Saleem, M. S. (2022). Pre-and postharvest physiological disorders of citrus fruit. In Citrus Production (pp. 371-390). CRC Press.

Aliviela, V., Zaragoza, S., Primo-Millo, E., & Agusti M. (1994). Hormonal control of splitting in "Nova" mandarin fruit. *Journal of Horticultural Science, 69*(6), 969–973.

Andrews, P., Collier, M., Fahy, D., & Evans, R. (1999). Gala stem-end splitting and internal ring cracking. *Good Fruit Grower, 50*, 20–23.

Assis, J. S., Maldonado, R., Muñoz, T., María, I., Escribano, C. M. (2001). Effect of high carbon dioxide concentration on PAL activity and phenolic contents in ripening cherimoya fruit. *Postharvest Biology and Technology, 23*(1), 33–39.

Bakeer, S. M. (2016). Effect of ammonium nitrate fertilizer and calcium chloride foliar spray on fruit cracking and sunburn of Manfalouty pomegranate trees. *Scientia Horticulturae, 209*, 300–308.

Belmans, J., Keulemans, Y., Debarsy, T., & Bronchart. R. (1990). Influence of sweet cherry epidermal characters on the susceptibility to fruit cracking. *Proceedings of the International Horticultural Congress, 637*.

Beyer, M., Hahn, R., Peschel, S., Harz, M., & Knoche, M. (2002). Analysing fruit shape in sweet cherry (*Prunus avium* L.). *Scientia Horticulturae, 96*(1–4), 139–150.

Bramlage, W. J., & Meir, S. (1990). Chilling injury of crops of temperate origin. *Chilling Injury of Horticultural Crops, 37*–49.

Bramlage, W. J., Weis, S. A., & Greene, D. W. (1990). Observations on the relationships among seed number, fruit calcium, and senescent breakdown in apples. *Horticultural Science, 25*(3), 351–353.

Brookfield, P. L., Ferguson, I. B., Watkins, C. B., & Bowen, J. H. (1996). Seed number and calcium concentrations of "Braeburn" apple fruit. *Journal of Horticultural Science, 71*(2), 265–271.

Brummell, David A., Mark H., Harpster, Pamela Dunsmuir. (1999). Differential expression of expansin gene family members during growth and ripening of tomato fruit. *Plant Molecular Biology, 39*(1), 161–169.

Byari, S. H. (1984). *Physiological, genetic, and breeding studies of tolerance to high temperature and high relative humidity in tomatoes (Lycopersicon Esculentum Mill.)*. The University of Nebraska-Lincoln.

Cai, C., Li, X., & Chen, K. (2006). Acetylsalicylic acid alleviates chilling injury of postharvest loquat (*Eriobotrya japonica* Lindl.) fruit. *European Food Research and Technology, 223*(4), 533–539.

Campostrini, E., Valdevino Pommer, C., & Yamanishi, O. K. (2008). Environmental factors causing physiological disorders in papaya plants. *II International Symposium on Papaya, 851*, 453–458.

Cao, S., Hu, Z., Zheng, Y., & Lu, B. (2010). Synergistic effect of heat treatment and salicylic acid on alleviating internal browning in cold-stored peach fruit. *Postharvest Biology and Technology, 58*(2), 93–97.

Cao, S., Zheng, Y., Wang, K., Jin, P., & Rui, H. (2009). Methyl jasmonate reduces chilling injury and enhances antioxidant enzyme activity in postharvest loquat fruit. *Food Chemistry, 115*(4), 1458–1463.

Du Chatenet, C., Latché, A., Olmos, E., Ranty, B., Charpenteau, M., Ranjeva, R., Pech, J.-C., & Graziana, A. (2000). Spatial-resolved analysis of histological and biochemical alterations induced by water-soaking in melon fruit. *Physiologia Plantarum, 110*(2), 248–255.

Cheng, Z., Sun, L., Qi, T., Zhang, B., Peng, W., Liu, Y., & Xie, D. (2011). The bHLH transcription factor MYC3 interacts with the jasmonate ZIM-domain proteins to mediate jasmonate response in Arabidopsis. *Molecular Plant Pathology, 4*(2), 279–288.

Chinnusamy, V., Zhu, J.-K., & Sunkar, R. (2003). Gene regulation during cold stress acclimation in plants. In *Plant stress tolerance* (pp. 39–55). Humana Press.

Christensen, J. V. (1996). Rain-induced cracking of sweet cherries: Its causes and prevention. *Cherries,* 297–327.

Corey, K. A., & Tan, Z.-Y. (1990). Induction of changes in internal gas pressure of bulky plant organs by temperature gradients. *Journal of the American Society for Horticultural Science, 115*(2), 308–312.

Cosgrove, D. J. (1999). Enzymes and other agents that enhance cell wall extensibility. *Annual Review of Plant Physiology and Plant Molecular Biology, 50*(1), 391–417.

Cosgrove, D. J. (2000). Loosening of plant cell walls by expansions. *Nature, 407*(6802), 321–326.

Cuartero, J., Palomares, G., Balasch, S., & Nuez. F. (1981, May 18–21). Tomato fruit cracking under plastic-house and in the open air. II. General and specific combining abilities. In *Genetics and breeding of tomato: Proceedings of the meeting of the Eucarpia Tomato Working Group, Avignon-France, Versailles*. Institut national de la recherche agronomique.

De Corato, U., Salimbeni, R., De Pretis, A., Avella, N., & Patruno, G. (2017). Antifungal activity of crude extracts from brown and red seaweeds by a supercritical carbon dioxide technique against fruit postharvest fungal diseases. *Postharvest Biology and Technology, 131*, 16–30.

de Freitas, S. T., do Amarante, C. V. T., Labavitch, J. M., & Mitcham, E. J. (2010). Cellular approach to understand bitter pit development in apple fruit. *Postharvest Biology and Technology, 57*(1), 6–13.

de Freitas, S. T., do Amarante, C. V. T., & Mitcham, E. J. (2015). Mechanisms regulating apple cultivar susceptibility to bitter pit. *Scientia Horticulturae, 186*, 54–60.

Domínguez, E., Fernández, M. D., López Hernández, J. C., Pérez Parra, J., España, L., Heredia, A., & Cuartero, J. (2012). Tomato fruit continues growing while ripening, affecting cuticle properties and cracking. *Physiologia Plantarum, 146*(4), 473–486.

Ferguson, I. B., & Triggs. C. M. (1990). Sampling factors affecting the use of mineral analysis of apple fruit for the prediction of bitter pit. *New Zealand Journal of Crop and Horticultural Science, 18*(2–3), 147–152.

Ferguson, I. B., & Watkins. C. B. (1989). *Bitter pit in apple fruit*. Horticultural Reviews.

Ferguson, I. B., & Watkins, C. B. (1992). Crop load affects mineral concentrations and incidence of bitter pit in "Cox's Orange Pippin" apple fruit. *Journal of the American Society for Horticultural Science, 117*(3), 373–376.

Fernández-Calvo, P., Chini, A., Fernández-Barbero, G., Chico, J.-M., Gimenez-Ibanez, S., Geerinck, J., Eeckhout, D. et al. (2011). The *Arabidopsis* bHLH transcription factors MYC3 and MYC4 are targets of JAZ repressors and act additively with MYC2 in the activation of jasmonate responses. *The Plant Cell, 23*(2), 701–715.

Gao, H.-J., Yang, H.-Q., & Wang, J.-X. (2009). Arginine metabolism in roots and leaves of apple (Malus domestica Borkh.): The tissue-specific formation of both nitric oxide and polyamines. *Scientia Horticulturae, 119*(2), 147–152.

García-Luis, A., Duarte, A. M. M., Porras, I., Garcia-Lidon, A., & Guardiola, J. L. (1994). Fruit splitting in "Nova" hybrid mandarin in relation to the anatomy of the fruit and fruit set treatments. *Scientia Horticulturae, 57*(3), 215–231.

Gusev, N. B., Bogatcheva, N. V., & Marston, S. B. (2002). Structure and properties of small heat shock proteins (sHsp) and their interaction with cytoskeleton proteins. *Biochemistry (Moscow), 67*(5), 511–519.

Hayama, H., Shimada, T., Ito, A., Yoshioka, H., & Kashimura, Y. (2001). Changes in the levels of mRNAs for putative cell wall-related genes during peach fruit development. *Scientia Horticulturae, 91*(3–4), 239–250.

Hayama, H., Mitani, N., Yamane, T., Inoue, H., & Kusaba S. (2017). Characteristics of cork spot like disorder in Japanese pear "Akizuki" and "Oushuu". *Horticulture Research, 16*(1), 79–87.

Ho, L. C., & White, P. J. (2005). A cellular hypothesis for the induction of blossom-end rot in tomato fruit. *Annals of Botany, 95*(4), 571–581.

Louise Hovland, K., & Sekse, L. (2004). Water uptake through sweet cherry (*Prunus avium* L.) fruit pedicels: Influence of fruit surface water status and intact fruit skin. *Acta Agriculturae Scandinavica: Section B. Soil and Plant Science, 54*(2), 91–96.

Hruschka, H. W. (1961). Seed potato productivity after cooling, supercooling, or freezing. No. 507. US Department of Agriculture, Agricultural Marketing Service, Market Quality Research Division.

Huang, X.-M., Wang, H.-C., Zhong, W.-L., Yuan, W.-Q., Lu, J.-M., & Li, J.-G. (2008). Spraying calcium is not an effective way to increase structural calcium in litchi pericarp. *Scientia Horticulturae, 117*(1), 39–44.

Islam, M. N., Hasanuzzaman, A. T. M., Zhang, Z.-F., Zhang, Y., & Liu, T.-X. (2017). High level of nitrogen makes tomato plants releasing less volatiles and attracting more Bemisia tabaci (Hemiptera: Aleyrodidae). *Frontiers in Plant Science, 8*, 466.

Jones, H. G., & Samuelson, T. J. (1983). Calcium uptake by developing apple fruits. II. The role of spur leaves. *Journal of Horticultural Science, 58*(2), 183–190.

Jubault, M., Hamon, C., Gravot, A., Lariagon, C., Delourme, R., Bouchereau, A., & Manzanares-Dauleux, M. J. (2008). Differential regulation of root arginine catabolism and polyamine metabolism in clubroot-susceptible and partially resistant Arabidopsis genotypes. *Plant Physiology, 146*(4), 2008–2019.

Kasai, S., Hayama, H., Kashimura, Y., Kudo, S., & Osanai, Y. (2008). Relationship between fruit cracking and expression of the expansin gene MdEXPA3 in "Fuji" apples (Malus domestica Borkh.). *Scientia Horticulturae, 116*(2), 194–198.

Khadivi-Khub, A. (2009). *Pomology*. Agriculture Education Press. (In Farsi).

Kiran, P. R., Jadhav, P., Avinash, G., Aradwad, P., TV, A., Bhardwaj, R., & Parray, R. A. (2024). Detection and classification of spongy tissue disorder in mango fruit during ripening by using visible-near infrared spectroscopy and multivariate analysis. Journal of Near Infrared Spectroscopy, 32(4), 140-151.

Khadivi-Khub, A. (2011). *Evaluation of genetic diversity in Cerasus subgenus using morphological and SSR markers*. PhD diss., Ph. D. Thesis, University of Tehran.

Khadivi-Khub, A., Khanali, M., Sahraroo, A., & Yavari, A. (2010). Physical properties of tangerine. *American-Eurasian Journal of Agricultural & Environmental Sciences, 3*, 216–220.

Khadivi-Khub, A., Zamani, Z., & Bouzari, N. (2007). Study of fruit cracking and doubled fruit in some sweet cherry cultivars. In *Fifth horticulture congress*. Shiraz, Iran (in Farsi).

Knoche, M., & Peschel, S. (2006). Water on the surface aggravates microscopic cracking of the sweet cherry fruit cuticle. *Journal of the American Society for Horticultural Science, 131*(2), 192–200.

Knoche M., & Peschel, S. (2009). Studies on water transport through the sweet cherry fruit surface: VI. Effect of hydrostatic pressure on water uptake. *Journal of Horticultural Science and Biotechnology, 77*, 609–614.

Milind, S., & Ladaniya. (2008). *Citrus fruit: Biology, technology and evaluation* (pp. 1–4). Elsevier.

La Spada, P., Dominguez, E., Continella, A., Heredia, A., & Gentile, A. (2024). Factors influencing fruit cracking: An environmental and agronomic perspective. Frontiers in Plant Science, 15, 1343452.

Lauxmann, M. A., Borsani, J., Osorio, S., Lombardo, V. A., Budde, C. O., Bustamante, C. A., Monti, L. L. et al. (2014). Deciphering the metabolic pathways influencing heat and cold responses during post-harvest physiology of peach fruit. *Plant Cell Environment, 37*(3), 601–616.

Los, D. A., & Murata, N. (2004). Membrane fluidity and its roles in the perception of environmental signals. *Biochimica et Biophysica Acta (BBA)—Biomembranes. 1666*(1–2), 142–157.

Lu, P.-L., & Lin, C.-H. (2011). Physiology of fruit cracking in wax apple (*Syzygium samarangense*). *Botanica Orientalis: Journal of Plant Science, 8*, 70–76.

Luo, Z., Chen, C., & Xie, J. (2011). Effect of salicylic acid treatment on alleviating postharvest chilling injury of "Qingnai" plum fruit. *Postharvest Biology and Technology, 62*(2), 115–120.

Luo, Z., Wu, X., Xie, Y., & Chen, C. (2012). Alleviation of chilling injury and browning of postharvest bamboo shoot by salicylic acid treatment. *Food Chemistry, 131*(2), 456–461.

Lutz, J. M. (1936). The influence of rate of thawing on freezing injury of apples, potatoes and onions. *Proceedings of the American Society for Horticultural Science, 33*, 227–233.

Lyons, J. M. (1973). Chilling injury in plants. *Annual Review of Plant Biology, 24*(1), 445–466.

Martínez-Romero, D., Bailén, G., Serrano, M., Guillén, F., Valverde, J. M., Zapata, P., Castillo, S., & Valero, D. (2007). Tools to maintain postharvest fruit and vegetable quality through the inhibition of ethylene action: A review. *Critical Reviews in Food Science and Nutrition, 47*(6), 543–560.

Matsumoto, K., Kobayashi, T., Kougo, T., Fujita, T., Sato, S. & Moriguchi, T. (2018). Prevention of new cork spot-like physiological disorder in "Kurenainoyume" apples by pre-harvest fruit bagging. *Journal of Horticulture, 87*(2), 174–183.

McQueen-Mason, S., & Cosgrove, D. J. (1994). Disruption of hydrogen bonding between plant cell wall polymers by proteins that induce wall extension. *Proceedings of the National Academy of Sciences, 91*(14) 6574–6578.

Meir, S., Philosoph-Hadas, S., Lurie, S., Droby, S., Akerman, M., Zauberman, G., Shapiro, B., Cohen, E., & Fuchs, Y. (1996). Reduction of chilling injury in stored avocado, grapefruit, and bell pepper by methyl jasmonate. *Canadian Journal of Botany, 74*(6), 870–874.

Menzel, C. M. (1984). The pattern and control of reproductive development in lychee: A review. *Scientia Horticulturae, 22*(4), 333–345.

Mihrete, T. B., Workie, M. A., & Bogale, F. A. (2024). Applications of biotechnology for enhancing the shelf life of horticultural crops. Journal of Sustainable Agriculture and Environment, 3(3), e12121.

Mittler, R. (2002). Oxidative stress, antioxidants and stress tolerance. *Trends in Plant Science, 7*(9), 405–410.

Monselise, S. P., & Costo, J. (1985). Decreasing splitting incidence in "Murcott" by 2, 4-D and calcium nitrate. *Alon Hanotea, 39*, 731–733.

Morris, Jr, S. M. (2009). Recent advances in arginine metabolism: Roles and regulation of the arginases. *British Journal of Pharmacology, 157*(6), 922–930.

Nazari, J., Nabigol, A., Rasouli, M., & Aghdam, M. S. (2024). Exogenous dopamine ameliorates chilling injury of banana fruits during cold storage. Scientific Reports, 14(1), 25802.

Niu, Y., Figueroa, P., & Browse, J. (2011). Characterization of JAZ-interacting bHLH transcription factors that regulate jasmonate responses in Arabidopsis. *Journal of Experimental Botany, 62*(6), 2143–2154.

O'Rourke, D., Janick, J., & Sansavini, S. (2003). World apple cultivar dynamics. *Chronica Horticulturae, 43*, 10–13.

Opara, L. U. (1996). Some characteristics of internal ring-cracking in apples. *Fruit Var Joural, 50, 260–262.*

Pardossi, A., Bagnoli, G., Malorgio, F., Campiotti, C. A., & Tognoni, F. (1999). NaCl effects on celery (*Apium graveolen*s L.) grown in NFT. *Scientia Horticulturae, 81*(3), 229–242.

Park, M.-H., Sangwanangkul, P., & Choi, J.-W. (2018). Reduced chilling injury and delayed fruit ripening in tomatoes with modified atmosphere and humidity packaging. *Scientia Horticulturae, 231*, 66–72.

Paull, R. E. (1990). Chilling injury of crops of tropical and subtropical origin. *Chilling Injury of Horticultural Crops, 17*, 36.

Prange, R. K., & DeLong, J. M. (2006). Controlled-atmosphere related disorders of fruits and vegetables. *Stewart Postharvest Review, 5*(7), 1–10.

Proctor, J. T. A., & Palmer, J. W. (1991). The role of spur and bourse leaves of three apple cultivars on fruit set and growth and calcium content. *Horticultural Science, 26*(6), 789F.

Quero-Garcia, J., Campoy, J. A., Joly, J., Tauzin, Y., Rosyara, U., Iezzoni, A., & Dirlewanger, E. (2012). QTL detection for fruit weight, fruit firmness and fruit cracking tolerance in sweet cherry. *Plant and Animal Genome Conference, 20*, 2012.

Rab, A., & Haq, I.-U. (2012). Irrigation intervals affect physicochemical quality attributes and skin cracking in litchi fruit. *Turkish Journal of Agriculture and Forestry, 36*(5), 553–562.

Raid, R. N. (2004). Celery diseases and their management. In *Diseases of fruits and vegetables* (Vol. I, pp. 441–453). Springer.

Rao, T. V. R., Gol, N. B., & Shah, K. K. (2011). Effect of postharvest treatments and storage temperatures on the quality and shelf life of sweet pepper (*Capsicum annum* L.). *Scientia Horticulturae, 132*, 18–26.

Sadamori, S. (1963). New apple variety "Fuji". *Bulletin of Horticultural Research Station, Japanese Series (C), 1*, 1–6.

Schumacher, R., Fankhauser, F., & Stadler. W. (1979). Influence of shoot growth, average fruit weight and daminozide on bitter pit. *Symposium on Mineral Nutrition and Fruit Quality of Temperate Zone Fruit Trees, 92*, 83–92.

Sekse, L. (2005). Fruit cracking in sweet cherries-some recent advances. *V International Cherry Symposium, 795*, 615–624.

Sekse, L., Bjerke, K. L., & Vangdal, E. (2001). Fruit cracking in sweet cherries-an integrated approach. *IV International Cherry Symposium. 667*, 471–474.

Sevillano, L., Sanchez-Ballesta, M. T., Romojaro, F., & Flores, F. B. (2009). Physiological, hormonal and molecular mechanisms regulating chilling injury in horticultural species. Postharvest technologies applied to reduce its impact. *Journal of the Science of Food and Agriculture, 89*(4), 555–573.

Simon, G. (2006). Review on rain induced fruit cracking of sweet cherries (*Prunus avium* L.), its causes and the possibilities of prevention. *International Journal of Horticultural Science, 12*(3), 27–35.

Taylor, M., Salvadore, D., & Locascio, J. (2004). Blossom-end rot: A calcium deficiency. *Journal of Plant Nutrition, 27*(1), 123–139.

Verner, L. (1957). Procedure for determining resistance of sweet cherry varieties to fruit cracking. *Fruit Var Horticultural Digging, 12*, 3–4.

Volz, R. K., Ferguson, I. B., Bowen, J. H., & Watkins, C. B. (1993). Crop load effects on fruit mineral nutrition, maturity, fruiting and tree growth of "Cox's Orange Pippin" apple. *Journal of Horticultural Science, 68*(1), 127–137.

Volz, R. K., Ferguson, I. B., Hewett, E. W., & Woolley, D. J. (1994). Wood age and leaf area influence fruit size and mineral composition of apple fruit. *Journal of Horticultural Science, 69*(2), 385–395.

Volz, R. K., Stuart Tustin, D., & Ferguson, I. B. (1996). Pollination effects on fruit mineral composition, seeds and cropping characteristics of "Braeburn" apple trees. *Scientia Horticulturae, 66*(3–4), 169–180.

Wakasa, Y., Hatsuyama, Y., Takahashi, A., Sato, T., Niizeki, M., & Harada, T. (2003). Divergent expression of six expansin genes during apple fruit ontogeny. *Eur. Journal of Horticultural Science, 68*(6), 253–259.

Walter, J. M. (1967). Hereditary resistance to disease in tomato. *Annual Review of Phytopathology, 5*(1), 131–160.

Wang, C. Y. (1990). *Chilling injury of horticultural crops*. CRC Press, p. 313.

Wang, C. Y., & Wallace, H. A. (2004). Chilling and freezing injury. The commercial storage of fruits, vegetables, and florist and nursery stocks (p. 66). Agriculture handbook.

Wang, Y.-S., Tan, S.-P., & Xu, Y. (2005). Effects of high oxygen concentration on pro-and anti-oxidant enzymes in peach fruits during postharvest periods. *Food Chemistry, 91*(1), 99–104.

Way, R. D. (1967). *Cherry varieties for New York State*. Cornell Extension Bulletin 1197, New York State College of Agriculture.

Yamaguchi, M., Sato, I., Takase, K., Watanabe, A., & Ishiguro, M. (2004). Differences and yearly variation in number and size of mesocarp cells in sweet cherry (*Prunus avium* L.) cultivars and related species. *Journal of the Japanese Society for Horticultural Science, 73*(1), 12–18.

Yang, Z., Cao, S., Zheng, Y., & Jiang, Y. (2012). Combined salicyclic acid and ultrasound treatments for reducing the chilling injury on peach fruit." *Journal of Agricultural and Food Chemistry, 60*(5), 1209–1212.

Yong, W., Wangjin, L., Jianguo, L., & Yueming, J. (2006). Differential expression of two expansin genes in developing fruit of cracking-susceptible and-resistant litchi cultivars. *Journal of the American Society for Horticultural Science, 131*(1), 118–121.

Zanella, A., Cazzanelli, P., Panarese, A., Coser, M., Chistè, C., & Zeni, F. (2004). Fruit fluorescence response to low oxygen stress: Modern storage technologies compared to 1-MCP treatment of apple. *V International Postharvest Symposium, 682*, 1535–1542.

Zhang, Z., Tian, S., Zhu, Z., Xu, Y., & Qin, G. (2012). Effects of 1-methylcyclopropene (1-MCP) on ripening and resistance of jujube (Zizyphus jujuba cv. Huping) fruit against postharvest disease. *LWT—Food Science Technology, 45*(1), 13–19.

Zhao, D. Y., Shen, L., Bei Fan, K. L., Liu, M., Yu, M., Zheng, Y., Ding, Y., & Sheng, J. P. (2009). Physiological and genetic properties of tomato fruits from 2 cultivars differing in chilling tolerance at cold storage. *Journal of Food Science, 74*(5), C348–C352.

Zhao, M.-L., Wang, J.-N., Shan, W., Fan, J.-G., Kuang, J.-F., Wu, K.-Q., Li, X.-P. et al. (2013). Induction of jasmonate signalling regulators MaMYC2s and their physical interactions with MaICE1 in methyl jasmonate-induced chilling tolerance in banana fruit. *Plant, Cell & Environments, 36*(1), 30–51.

Zou, Y., Zhang, L., Rao, S., Zhu, X., Ye, L., Chen, W., & Li, X. (2014). The relationship between the expression of ethylene-related genes and papaya fruit ripening disorder caused by chilling injury. PLoS One, 9(12), e116002.

CHAPTER 12

Causes of Postharvest Physiological Disorders

SHAGHEF EJAZ[1*], MUHAMMAD AZAM[2], and BEENISH CHAUDHARY[2]

[1]Department of Horticulture, Faculty of Agricultural Sciences and Technology, Bahauddin Zakariya University, Multan, Pakistan

[2]Institute of Horticultural Sciences, Faculty of Agriculture, University of Agriculture, Faisalabad, Pakistan

*Corresponding author

ABSTRACT

Postharvest physiological disorders of fruits and vegetables represent a major constraint to postharvest quality and marketability, arising from complex interactions among genotype, preharvest environment, and postharvest handling conditions. This chapter comprehensively examines the underlying causes of physiological disorders, emphasizing the roles of genetic susceptibility, environmental stresses, mineral nutrition, cultural practices, and postharvest factors such as temperature, relative humidity, gaseous composition, and light exposure. The discussion integrates preharvest and postharvest perspectives to explain how disruptions in cellular homeostasis, energy balance, membrane integrity, and oxidative metabolism trigger visible disorder symptoms in fresh produce. By synthesizing current knowledge on environmental, physiological, and biochemical mechanisms, this chapter provides a foundation for understanding disorder development and highlights critical control points for minimizing postharvest losses in horticultural commodities.

Physiological Disorders of Fruits and Vegetables: Causes and Remedies.
Ghulam Khaliq, Sajid Ali, and Mohammed Wasim Siddiqui (Eds.)

DOI: 10.1201/9781998511334-12

12.1 INTRODUCTION

Physiological disorders refer to the loss of the functionality of a plant tissue that is neither as a result of pathogen (fungi, bacteria, virus, etc.) invasion nor by mechanical injury. Physiological disorders can have an important role, depending on genotype and environmental conditions. Physiological disorders are abnormal nonpathological changes in plant tissues expressed in response to the interaction between genotype and environment. These disorders may be reversible or irreversible, depending on time and extent/intensity (Kumar and Kumar, 2016). However, most of them are nonreversible once they have occurred.

Crop plants, however, have been reported to develop a wide range of physiological disorders, suggesting that mechanisms leading to physiological disorders have also been selected during plant breeding for human purposes. Changes in genetic regulation can determine plant susceptibility to physiological disorders. Indeed, different genotypes, tissues in the same genotype, and cells in the same tissue can have different susceptibility to physiological disorders. On the other hand, environment stresses can have a great impact on the development of physiological disorders in plants.

Indeed, elevated stress levels may explain the greater susceptibility of crop genotypes to physiological disorders. In this context, it is possible that most physiological disorders develop at the point where cell stress levels overcome the capacity of the biological systems to control stress. In other words, physiological disorders may develop when plant cells are no longer able to maintain normal homeostasis, leading to cell death and the development of disorder symptoms. Generally, there are various preharvest and postharvest aspects of the environment such as temperature or insufficient mineral nutrition that directly or indirectly cause physiological disorders.

12.2 PREHARVEST FACTORS INFLUENCING POSTHARVEST DISORDERS

Many factors contribute to the definition of quality. Quality includes intrinsic and extrinsic factors which can change according to the consumer's memories and expectations. Intrinsic factors include external traits such as shape, color, size, freedom from defects, sweetness, aroma, texture, flavor, and nutritional values. While extrinsic factors include production, packaging types, sustainability, and consumer acceptability. After harvest, quality can only be maintained; it cannot be improved. Preharvest factors are essential for ensuring the production of high-quality horticultural commodities. Many preharvest factors, such as weather and agricultural practices, have a significant impact on the postharvest quality attributes and physiology and shelf life of fresh horticultural commodities. External and internal quality characteristics are determined by genetic material, while agronomic and cropping conditions, such as environmental conditions, mineral nutrition, chemical treatments, and pest management, have direct and indirect effects on these characteristics. The use of different growing techniques, manipulation of environmental conditions (e.g., light, temperature, and carbon dioxide) and nutrients, irrigation, pests, and other factors can have significant direct and indirect effects on the quality, physiology, and shelf life of horticultural commodities after harvest. Soil management, fertilization, irrigation, and agricultural practices affect mineral composition, storage capacity, and susceptibility to a variety of diseases. Commonly, the components of several nutrients, including N, P, K, Ca, and Mg, are used as a basis for evaluating the storage potential of certain commodities, such as apples. Chemical treatments, such as growth regulators, can alter fruit characteristics and susceptibility to particular storage disorders, and may contribute to the variable storage capacity of certain fruits. Consequently, it is crucial to comprehend which preharvest factors influence the quality characteristics of horticultural products and how they affect their postharvest shelf life, storability, and rate of deterioration.

12.3 ENVIRONMENTAL AND GENOTYPE-ASSOCIATED FACTORS

Postharvest physiological disorders in horticultural crops are influenced by a combination of environmental and genotype-related factors. Environmental conditions such as temperature extremes, relative humidity, light intensity, and soil fertility significantly affect the incidence and severity of postharvest physiological disorders. High temperatures can accelerate senescence and increase susceptibility to disorders like sunscald, while low temperatures may lead to chilling injury. Nutrient imbalances, especially calcium deficiency, contribute to disorders such as bitter pit in apples and blossom-end rot in tomatoes. Water stress, either drought or over-irrigation can disturb physiological balance and promote disorders like fruit cracking or internal browning. Harvesting at improper maturity stages and mechanical injuries during handling further exacerbate physiological disorder development. On the other hand, genotype-related factors, including genetic susceptibility, fruit morphology, metabolic rate, nutrient uptake efficiency, and cuticle characteristics, play a critical role in determining a crop's resilience or vulnerability to postharvest disorders. Varieties with poor calcium translocation, thinner cuticles, or lower antioxidant capacities are generally more prone to postharvest physiological disorders.

12.3.1 GENOTYPES/CULTIVARS

The quality of the crop after harvest can be significantly impacted by selecting the best crop variety, genotype, or cultivar. This is because some of them can handle being chilled, are more firm, resistant to fungal diseases or insects, and can maintain their quality for a longer period of time, making them preferable for long-term preservation (Hewett, 2006; Ali et al., 2025). Due to genetic variations, different cultivars, varieties, or genotypes of the same fruit have unique characteristics. Because of this, the genes are expressed differently, which results in variations in the type and activity of enzymes and metabolites between genotypes. How these factors impact the quality of fruits and vegetables has been the subject of numerous research. For instance, Tribute strawberries are among the tiniest varieties, and the weight of strawberries has varied between cultivars (in certain cases, by more than five times). Butternut and Winter Luxury squash are smaller than Violina squash (Tyagi et al., 2017).

With a few notable exceptions, several studies have revealed that there are not many differences in the weight of olives between varieties. However, fruit differs significantly in terms of symmetry and surface roughness (Barman et al., 2015). Different quince cultivars range slightly in fruit size as well as in other quality characteristics such as pectin content, measurable acidity, and soluble solids content. Out of all citrus fruits, oranges have the largest total soluble solids. These solids typically fall between 9% and 14%. The total soluble solids content of different squash cultivars does not differ significantly. Tetsu kabuto and Red Kury squash varieties, in contrast, have up to 62 times as much starch as other squash cultivars. Packham Triumph and Beurre d's Anjou pears are not as hard when plucked, although they both have the same levels of total soluble solids and starch (Shewfelt and Henderson, 2003).

Additionally, there are significant variations across strawberry cultivars in terms of total soluble solids (4.8–10.9%) and titratable acidity (0.5–0.8%). The amounts of fructose, glucose, and sugar in different strawberry cultivars vary greatly; 1700–4240, 1225–3260, and 780–1600 mg/100 g, respectively, are the ranges for these sugars. These various sugar content levels result in significant sweetness variations. The titratable acidity and total soluble solids of kiwifruit typically range between 7.1 and 15.2 g/L and 15 and 8%, respectively, depending on the cultivar. Fruits vary in their starch content and in the production and utilization of amylases, which influences the quantity of total soluble solids and titratable acidity they contain. The amylase isoforms in less sweet fruits are less active (Brasil and Siddiqui, 2018).

The cultivar/genotype has a significant impact on the firmness of the fruit. This is probably due to variations in perspiration capacity and in the activity/biosynthesis of the enzymes that break down structural polysaccharides (such as pectin methylesterase, polygalacturonase, xylanase, galactosidase, and cellullase). The breakdown of structural polysaccharides has a greater impact on this quality attribute than starch degradation, although starch breakdown also influences fruit firmness. Different fruit cultivars, variations, or genotypes have wildly varied colors, which are brought on by various pigment concentrations (Prasad et al., 2018; Ro et al., 2024).

Other crucial components of fruits and vegetables that vary greatly depending on cultivar, variety, or genotype are minerals, fragrance compounds, and vitamins. Mandarin juice, for instance, contains between 21 and 600 mg/L of vitamin C, depending on the variety, cultivar, or genotype. There is not much variation in the amount of vitamin C across various varieties of potatoes and pome fruits, unlike citrus fruits. Different types of fruits and vegetables are thought to have varying quantities of vitamin C due to sugar metabolism. The cultivar has a significant impact on how much tocopherol an avocado contains. It has been shown that variations in how chlorophyll is broken down among cultivars are responsible for the varying levels of tocopherols in fruits and vegetables. A component in the production of tocopherols is the phytol chain found in chlorophyll. After being selected, fruits and vegetables are susceptible to a variety of physical issues, primarily as a result of exposure to extremes in temperature as well as oxygen or carbon dioxide (Gudkovskiy et al., 2015).

12.3.2 ROOTSTOCKS

Grafting is the process of attaching pieces of two or more plants so that they can grow together. The merged plant's scion is at the top, while its rootstock is at the bottom. The rootstock typically withstands situations that are detrimental for plant growth, such as saline soil, low or high water levels, and low or high water levels. The scion, on the other hand, produces a lot of high-quality fruit. Therefore, grafted plants are far better than plants that grow from their own roots in many ways. Twenty of the top 25 fruit and nut crops are occasionally grafted, and more than 70 woody perennial crop species that produce edible fruits are cultivated on rootstocks (Fallik and Ilic, 2014).

The rootstock has a significant impact on the productivity and quality of a number of horticultural products. A plant's ability to endure adverse conditions (such as poor soil, drought, or pests) and its ability to absorb nutrients are both changed by rootstock, which has a significant impact on crop yield. Additionally, it alters the physiological, hormonal, and biochemical equilibrium between the base and the top, which has an impact on final size, fruit set, growth, and flowering. The rootstock alters the leaf area, which impacts how efficiently stomata function and how well plants absorb CO_2. Thus, photosynthesis and agricultural yield are increased. In the shoots of stress-sensitive grafted plants, some signals from the roots, including abscisic acid and cytokinin, safeguard leaf photosynthesis. In some instances, this results in an increase in the amount of cytokinin, which aids in the formation of chloroplasts and the production of chlorophyll, and a decrease in the amount of abscisic acid in the xylem sap, which improves Rubisco's performance (Edelstein, 2004).

The quality of a plant's fruits alters as a result of grafting. When the fruit ripens, how hard the flower buds are and when the flowers blossom are all influenced by the rootstocks. When the fruit ripens has a significant impact on the fruit's quality. Compared to plants that develop from their own roots, grafted plants often absorb more water and minerals. This strengthens the plants and provides them with more nutrients so they can produce fruit. The quality of the fruit can vary depending on how robust the shoots are. The amount of dry matter that matures into fruit is decreased by rootstock vigor. The appearance of the root system, such as its lateral and vertical growth, which improves its ability to absorb water and minerals, has typically been related to the effect of the rootstock on the mineral content of the plant's leaves, stems, and fruits. But the amount of particular nutrients in the fruit is also influenced by the scion. Since fruit and vegetative organ competition for photosynthetic products (sugars) is typically lower, dwarfing rootstocks can transport more of these substances to fruits. Less plant growth facilitates light penetration into tree canopies, enhancing photosynthesis and fruit quality (Benkeblia et al., 2011).

On the other hand, the quality of the fruit is diminished by excessive canopy shade. Fruit size, color, sugar accumulation, firmness, taste, smell, phytochemical concentration, and antioxidant capacity all tend to decline. Although rootstock has an impact on chemical composition and antioxidant capacity, it is frequently impossible to predict how a typical scion will react to rootstock vigor. Rootstocks have an impact on more than just vigor, as evidenced by fruits from rootstocks with comparable growth vigor but varied nutritional characteristics (Roupheal et al., 2010).

The number of carotenoids, particularly lutein and carotene, tends to increase when melons are grafted onto pumpkin-hybrid rootstocks. The output is greatly increased when Birgah eggplant is grafted onto *S. paniculatum*, Msa 2/2 E7, and 460 CAL eggplant hybrids; however, the rootstock has no effect on the fruit's quality. Cherry trees can be made less vigorous by using dwarfing rootstocks, which will boost the fruit's quality. But these rootstocks frequently result in overcropping, which reduces the leaf-to-fruit ratio, and results in smaller fruits, and slows vegetative growth (Davis et al., 2018).

Some rootstocks do not significantly affect the size, number of locules, pericarp thickness, or firmness of the fruit. The majority of rootstocks do, however, alter the fruit's chemical composition. In general, the quality of cherries, peaches, and nuts such as almonds and walnuts is not greatly impacted by the rootstock. Rootstock typically has little-to-no impact on the uniformity, color, or shape of almond fruit. Grafting occasionally results in fruit that has a poor flavor or is less resilient to issues after being picked. For instance, Seckel pears grown on the same rootstock produce large, high-quality fruit but Anjou pears grafted onto *Pyrus betulifolia* rootstock frequently have internal necrotic areas (also known as "cork spot") in the flesh. Some plum and prune rootstocks may increase the likelihood that the fruit may develop brown spots, shrinking stem ends, and gum marks (Yuri et al., 2019).

12.3.3 CLIMATE VARIABLE INFLUENCING POSTHARVEST DISORDERS

Based on where they originated, where they thrive, and where they can generate the most, horticultural products are separated into tropical, subtropical, and temperate zones. This classification is based on the environment and weather, as well as proof that these factors have an impact on the fruit's quality.

12.3.4 LIGHT

The length of the day and night, which is determined by genes, affects how the plants behave. Therefore, plants' physiology is altered by light, which has an impact on the quality of the fruits they produce. The photoreceptors of the plants pick up the various electromagnetic spectrum wavelengths, which is how they get light. Photoreceptors transmit signals that cause the plant to alter in both form and function, allowing it to better adapt to its surroundings. These alterations include flowering, the formation of certain fragrance compounds, the germination of seeds, and the coloration of fruits and vegetables. The quantity and caliber of light affect the signal's intensity. It is also well recognized that the fruit's quality is significantly influenced by its elevation above sea level. This is due to the fact that the light is stronger the higher you are (Usall et al., 2016).

By adjusting the tree's canopy or by utilizing hail nets, fruit trees can receive more light. The black hail nets significantly alter the quantity and quality of light, which has an impact on how physiologically plants and fruits function. But employing nets of various colors can help to partially solve this hail nets issue. White and red hail nets typically improve the fruit compared to black hail nets. The amount of light that reaches the fruit is influenced by the color of the hail net, mostly through altering how much ultraviolet, visible, or far-red light is absorbed. The total soluble solids, color, firmness, and rate of ripening of some fruits, including apples, peaches, plums, and blueberries, can all be influenced by the color of the hail net (Madakadze and Kwaramba, 2004).

The amount of light that a fruit tree receives might vary depending on its canopy. Fruit with a smaller canopy receives more sunlight and soon takes on the color of mature fruit. It typically smells better as well. When exposed to additional sunlight, many fruits, including olives, peaches, grapes, mangoes, grapes, and apples, exhibited these impacts. The concentrations of chlorophylls, carotenoids, anthocyanins, fragrance compounds, total polyphenols, and total soluble solids varied with the amount of sunlight. Apples can receive more sunlight by altering the tree's canopy, which alters how esters, which give apples their scent, are created. Methoxypyrazines are produced less when Cabernet Sauvignon grapes are exposed to increased sunlight. Red wines with this herbaceous aroma (similar to green bell pepper) are of poorer grade (Mditshwa et al., 2017).

12.3.5 SOIL TYPES

How plants grow and develop is influenced by a variety of factors, including structure, water, and the quantity of nutrients in the soil. Because roots cannot expand in compacted soil, plants have a tougher time drawing water and nutrients from the ground. As a result, plants wilt more frequently, grow slowly, and produce inferior fruits and vegetables. In comparison to clay soils and soils with a lot of organic matter, sandy soils and soils with little organic matter are less able to hold water and nutrients. Plants struggle to grow and develop as a result since they do not receive enough water and nutrients. When there is a drought or not enough nutrients in the soil, plants frequently become unwell. However, consuming too much of some necessary nutrients can also lead to issues. This implies that plants require the proper ratio and balance of nutrients to maintain healthy cell metabolism and homeostasis (Kumar and Kumar, 2016).

Plant growth and development are significantly influenced by the soil structure, water content, and nutritional content. Compacted soils prevent root growth, which restricts plants' ability to absorb water and nutrients from the soil, leading to stunted growth, poor fruit and vegetable quality and yield, and increased susceptibility to disease. Sandy and low organic matter soils hold onto water and nutrients less effectively than clay and high organic matter soils, which restricts plant growth and development because of the plants' insufficient access to water and nutrients. A higher occurrence of physiological diseases in plants has frequently been linked to dry spells and nutrient deficiency in the soil. On the other side, too much of some necessary nutrients can also cause problems, indicating that a plant needs the right amounts of all its nutrients to sustain healthy cell metabolism and homeostasis (Prusky, 2011).

12.3.6 TEMPERATURES

Due to its impact on all of the phenological stages of plants, temperature is one of the most significant environmental elements influencing the development, production, and quality of horticultural products (e.g., dormancy, sprouting, blooming, and growth). Preharvest temperature has a significant impact on fruit yield, color, scent, and the production of carbohydrates because it affects photosynthesis, maturity, and ripening.

Prior to and following harvest, temperatures have an impact on the likelihood that horticultural goods will contract specific disorders. It is established that high preharvest temperatures have an impact on a product's ability to withstand low or high temperatures after harvest. Preharvest temperature influences the ripening rate, with effects varying among CAM, C3, and C4 plants. This classification of plants is based on how the temperature affects how well they fix CO_2. In C3 and C4 plants, the mesophyll and bundle-sheath cells fix CO_2 during the day. Phosphoenolpyruvate carboxylase is a nighttime CO_2 fixation enzyme found in mesophyll cells of CAM plants. The type of plant determines the ideal temperatures for fruit growth and ripening. For instance, C3 plants require temperatures of 20–30°C, whereas C4 and CAM plants require temperatures of 40°C or more (Yahia, 2019).

The C3 class of plants includes those that produce fruits like grapes, apples, peaches, oranges, pears, bananas, mangoes, cherries, tomatoes, guavas, strawberries, limes, and prunes. These plants are extremely susceptible to high temperatures because they cause a rise in photorespiration, energy loss, and the release of CO_2 rather than its fixation to produce cell wall components and store sugars and sucrose in fruits. These plants produce fruits that are slightly stressed before harvest, which improves them afterward, slows down their ripening afterward, and increases their tolerance to cold conditions afterward. On the other side, these plants' fruits frequently burn or scald before being harvested. On the other side, low temperatures lead to the fruit freezing and the cells bursting, which detracts from the fruit's appearance after it has been picked because the wounds sometimes turn black. When the temperature is really low, some grape varieties have significantly more phenolic chemicals. With the help of preharvest methods including canopy control, watering, and growing plants in greenhouses, the ideal temperature balance between cold and warm can be achieved during the ripening process. These plants require the ideal temperature conditions to produce high-quality fruits. The amount of total soluble solids, fragrance compounds, and color in various kinds of plants' fruits is influenced by the preharvest temperatures. For instance, exposing strawberries to temperatures as high as 30°C for a week before to harvest aids in the development of their color but reduces the quantity of pelagic acid, ascorbic acid, and total soluble solids present in the fruit. Tomatoes do not change color in the 35°C range before harvest, but they do ripen more slowly, which reduces the quality of the fruit once it is picked. Before harvest, bell peppers should be exposed to temperatures of 32°C to improve the pollen quality. This is due to the pollen's greater starch and sugar content, which increases fruit yield and quality. When the temperature is lower before harvest than when it is higher, more delphinidin is produced by merlot grapes. Low temperatures aid in giving grapes their color since delphinidin is what creates the primary anthocyanin pigment in grapes. The lowest overnight temperatures, which can activate the gene expression of chalcone synthetase, are similarly related to the color of the fruit. The formation of anthocyanins in fruits including grapes, grapefruits, oranges, strawberries, and red apples depends heavily on this enzyme (Aked, 2002).

12.3.7 ATMOSPHERIC HUMIDITY

Typically, the cells and tissues of fruits and vegetables contain at least 90% water. The plants require water to develop, sprout, bloom, and bear fruit (i.e., cellular division and expansion). Because environment relative humidity has an impact on the fruit tangency at harvest, it has a significant impact on how fruit looks. How well fruit can survive being subjected to poor relative humidities after harvest depends on its original tangency. When it rains during the ripening process, the environment's high relative humidity increases, promoting the growth of larger and better-quality fruits and vegetables. However, the relationship between temperature and watering and relative humidity is intricate (Kader, 2014).

In some instances, such as with tomatoes and citrus fruits, high relative humidity just prior to harvest weakens the peels, resulting in the fruit cracking and a reduction in quality. The impact of ambient relative humidity on the quality of fruits and vegetables after harvest is well documented. Information on how this element alone influences the quality of fruits and vegetables after harvest is few, though. The impact of relative humidity is influenced by other factors, including the cultivar, the soil, the ambient temperature, the quantity of light, the amount of solar radiation, the amount of rain, and how the canopy is controlled. For instance, feijoa and grapes exposed to relative humidity levels greater than 70% during blooming or ripening may be more susceptible to fungal diseases after harvest, primarily from Botrytis, and may therefore produce less fruit. In some circumstances, a high relative humidity can make fruits have a high internal osmotic pressure. Cellular rupture, the release of carbohydrates, a rise in the likelihood of fungal infections following harvest, and a decline in quality are all effects of this. The amount of irrigation water that crops receive is influenced by the relative humidity of the surrounding air. A relative humidity of about 80% makes the pollen more viable, which results in more fruit, in the case of passion fruit. The Penman-Monteith equation, the most crucial equation for determining evapotranspiration, contains variables such as solar radiation, temperatures, and ambient humidity, among others. Relative humidity is a crucial component of this equation (Hussein et al., 2014).

12.3.8 WIND/STORMS/CYCLONES DURING PRODUCTION PHASE

Research has shown that the wind can harm fresh fruit in both obvious and subtle ways. Strong wind defoliation results in tiny fruits, poor color, and more surface faults such as friction marks on kiwifruits, wind scab on European prunes, and wind rub on persimmons. Cutting back the leaves on Crimson seedless grape vines led to an increase in berry juice, fruit hardness, total soluble solids (TSS), and a decrease in acidity (Abd El-Razek et al., 2010). The "Dominico Harton" plantain fruit contained greater total sugar, less pulp ash, and more patches of dead tissue on the peel when strong wind speeds occurred 30–45 days after flowers blossomed (Parra et al., 2001). A combination of defoliation rates of 1 fruit per 4, 10, and 20 leaves with fruit thinning of 0, 60, and 80% lowered fruit weight and TSS in a study on how wind damage impacts the quality of persimmons in Korea. However, if 60 and 80% thinning was performed without defoliation, fruit weight and TSS increased. These experiments demonstrate how the wind has an impact on both the outside and interior of fruits (Choi et al., 2005).

Fruits that have external blemishes are graded after harvest. However, mechanical disturbances that do not directly harm the tissue can still have an impact on plants or some parts of them. Although vibration and shaking were mentioned as examples of mechanical perturbations, stroking, bending, flexing, and rubbing were all discovered to have an impact on plant growth and development. Although the mechanism of mechanical disruption is unclear, we do know that it has an immediate impact, and that recovery takes time (resuming normal growth takes days). However, disturbance significantly alters the production of gibberellic acid and raises the production of ethylene. After the first sense is detected, both ethylene and gibberellic acid undergo a sequence of processes. It is assumed that a fast shift in membrane permeability and bioelectrical potential are among the first signs of a response to a disruption. Wind is a common source of mechanical stress that results in both physical damage and mechanical disturbances (Kays and Paull, 2004).

12.4 POSTHARVEST FACTORS AFFECTING PHYSIOLOGICAL DISORDERS

To maintain quality and extend the marketing window of fresh fruits and vegetables, senescence related metabolic activities are often inhibited during storage. This is generally achieved by lowering the storage temperature, which can be combined with low O_2 and high CO_2 levels—named as controlled atmosphere (CA) storage—to further slowdown the rate of respiration and ethylene production. Ultimately, these conditions delay the onset of senescence-related physiological disorders. On the other hand, during long term storage, too much fluctuations or improperly low temperature and oxygen and carbon dioxide partial pressures lead to physiological disorders in certain temperate fruits (Thompson, 2016).

Generally, storage specifications vary from one species to other species or even among cultivars. Sometimes, geographic location also determines the storage recommendation for a commodity. During storage, various postharvest treatments are used to extend the postharvest life of various fruits and vegetables. However, if improper treatment any disrupt metabolism and cause physiological disorder of the treated commodity. Generally, overexposure to some treatment leads to the accumulation of reactive oxygen species. Resultantly, oxidative stresses damage cellular membranous structure and interrupt ATP production leading to energy imbalance (Pedreschi et al., 2015; Mditshwa et al., 2023). These injuries initiate secondary reactions and visible disorder, such as, russet spotting in lettuce, blackheart in potato, and browning of tissues. Accumulation of off-flavors and odors, water-soaking, and blackening in various fruits and vegetables (Pedreschi et al., 2015 and Kader 2002).

12.5 GASEOUS COMPOSITION

Low O_2 and high CO_2 levels are used to extend the storage life of horticultural commodities. However, increased O_2 level within fruit or on surface causes over-accumulation of free radicals that, ultimately, damage tissues of fruits and vegetables (Fridovich, 1986). Horticultural produces from different species and their developmental stages respond differently to O_2 toxic levels. Improper storage conditions cause abiotic stress to stored commodities. For example, within pome fruits, abiotic stresses caused by suboptimal partial pressures promote the development including flesh browning and dry internal cavities. During oxidative stress caused by hypoxia or elevated CO_2 levels, there is either accumulation of certain metabolites such as gamma-aminobutyrate or reduction in some metabolites such as ascorbate and glutathione. Such stressful environment may also cause altered ATP production and energy imbalance (Pedreschi et al., 2015; Mditshwa et al., 2023). High oxygen-atmospheres/hyperbaric atmospheres, where O_2 level is $\geq$21 kPa, interact with postharvest physiology and directly or indirectly influence quality of stored fresh horticultural produces either by altering the rate of production of CO_2 and C_2H_4.

Ethylene, a colorless and odorless gas, is a natural hormone. Climacteric fruits, such as apple and banana, and vegetables synthesize ethylene during ripening. Ethylene have been used in agriculture owing to its ability to impart positive and negative effects on ripening of climacteric fruits. On the other hand, ethylene is not required for nonclimacteric fruits. However, it has been observed that in nonclimacteric fruits and vegetables, ethylene exposure reduces their postharvest life by conferring negative effects by developing physiological disorders and enhancing senescence-related activities (Martínez-Romero et al., 2007). Ethylene causes physiological disorders such as chilling injury in persimmon and avocado, russet spotting in lettuce, superficial scald in pear and apple and internal browning in pear and peach (Martínez-Romero et al., 2007). Certain storage conditions such as low temperature, higher CO_2 and low O_2 inhibit the biosynthesis of ethylene. Ethylene can also have undesirable changes in fruits such as skin damage and premature ripening (Kahramanoğlu, 2017).

12.6 TEMPERATURE

Low temperature storage, the most commonly used technique for long term storage, is also a leading cause of chilling injury in several tropical and subtropical horticultural stored produces. Generally, the symptoms of chilling injury appear after the stored produce is transferred from cold storage to ambient or shelf conditions. Physical signs of chilling injury include discoloration, internal breakdown, surface pitting, loss of textural quality and higher disease incidence. There are several other chilling injury symptoms that appear depending on the type of commodity, storage and thawing temperatures. These symptoms range from slow, uneven or failed ripening process; accelerated rate of respiration and ethylene production; increased membrane disintegration and lipid peroxidation, and thus, early senescence; reduced postharvest life and loss of texture and flavor (Khaliq et al., 2023).

Under chilling conditions, cell membranes lose semipermeability due to a transition from the flexible liquid-crystalline state to a solid-gel structure. During membrane degradation, increase in the peroxidation and saturation degree of fatty acid occurs along with disintegration of phospholipids and galactolipids. On the other hand, ratio between the sterol and phospholipid rises ratio, this all decreasing the fluidity and efficiency of cellular membranes. Prolonged exposure to chilling temperatures disintegrates cell membranes leading to leakage of ions and metabolites and intracellular fluids (Sharom et al., 1994; Tomita et al., 2023).

Chilling injury where, on one hand, directly affect cellular functions, it also leads to oxidative stress caused by the over-production of reactive oxygen species. The accumulation of reactive oxygen species boosts peroxidation of membrane lipids, hence, exponentially increasing the damage to cellular structures (Sevillano et al., 2009). The genotype plays important role in tolerating chilling temperatures. Cao et al. (2011) observed that chilling tolerance in the fruits of loquat cultivar "Qingzhong" was attributed to higher levels of linolenic and linoleic acids and lower concentration of palmitic and stearic acid as compared to the fruits of chilling sensitive cultivar "Fuyang." Further, chilling-tolerant cultivar had higher superoxide dismutase and catalase activities and lower lipoxygenase activity that scavenge H_2O_2 and O_2.

12.7 HUMIDITY

Loss of moisture from the surface of a commodity is not only a loss of marketable weight (quantitative loss) but also a loss of marketable quality in terms of appearance, due to shrinking and wilting, and loss of edible quality, due to flesh softening and texture deterioration (Kader, 1992).

During storage, if the temperature and air flow is kept constant, then water loss is directly proportional to relative humidity of the ambient environment. Low relative humidity during storage increases certain physiological disorders, such as, shriveling and skin browning.

Leafy vegetables have large surface area and, therefore, have high rate of water loss that makes them very perishable commodities. Therefore, relative humidity had great effect on storage life of leafy vegetables. For example, if the relative humidity of cold stores falls below optimum levels, that is, 95–98%, transpiration rate exponentially increases and, consequently, significant water loss and resultant physiological disorders deteriorate commercial quality of the stored commodities to a greater extent. Agüero et al. (2011) reported the deterioration of marketable quality lettuce heads stored at 70–72% (low relative humidity) and suggested to prevent dehydration of freshly harvested vegetables by regulating relative humidity. As a rule of thumb, if freshly harvested fruits and vegetables are stored at very high relative humidity (>90%), these commodities maintain their physical, edible, cosmetic and nutritional quality by preserving appearance, weight, texture, and flavor. In contrast, low relative humidity causes loss of crispness, wilting, softening, and shriveling (Arah et al., 2015).

12.8 LIGHT

Plant growth and development critically require light for several key physiological functions. However, even after harvest, light may play important role in maintaining marketable quality of fresh fruits and vegetables. For example, potatoes are stored in the dark because exposure to light triggers photosynthesis and accumulation of chlorophyll turn them green. The green pigment in potato skin contains the toxin, alkaloid solanine, seriously deteriorating potato quality. Similarly, for better leaf color retention and low browning index, leafy vegetables are preferred to be stored under darkness. In minimally processed leafy vegetables, tissue browning is quite common and major physiological disorder that is detected immediately with plain sight and thus results in rejection of the product by the consumers. Light has been found to significantly influence the browning index and color (hue angle) of leafy vegetables during storage (Zappia et al., 2018). Exposing the spinach leaves to continuous light during storage resulted in leaf wilting, although the light-exposed leaves had more bioactive compounds (Lester et al., 2010). During exposure to natural or fluorescent light, stomata remain open which is strongly and positively linked with the loss of moisture. Noichinda et al. (2007) found the similar relation between stomatal opening and loss of fresh weight, when Chinese kale (*Brassica oleracea* var. *alboglabra*) was exposed to relatively low light levels (21.8 μmol m^{-2} s^{-1}) during storage at 1°C and ≥95% relative humidity under.

12.9 CONCLUSION

Postharvest physiological disorders in fruits and vegetables remain a persistent challenge due to their multifactorial nature and the intricate physiological and biochemical processes involved in their development. This chapter highlights that these disorders arise when environmental or management-induced stresses exceed the plant tissue capacity to maintain cellular homeostasis, leading to metabolic imbalance, membrane degradation, and oxidative damage. Both preharvest factors such as genotype, climate, soil conditions, and cultural practices and postharvest conditions particularly temperature, humidity, atmospheric composition, and light play decisive roles in determining disorder incidence and severity. A thorough understanding of these interacting factors is essential for developing effective mitigation strategies. Future advancements integrating genomics, proteomics, and metabolomics will be instrumental in elucidating disorder-specific mechanisms and in designing cultivar- and commodity-specific management approaches to reduce postharvest losses and improve fresh produce quality.

KEYWORDS

- **climate**
- **disorder**
- **fresh produce**
- **genotype**
- **preharvest**
- **postharvest**

REFERENCES

Abd El-Razek, E., Treutter, D., Saleh, M. M. S., El Shammaa, M., Amira Fonad, A., Abdel-Hamid, N., & Abou-Rawash, M. (2010). Effect of defoliation and fruit thinning on quality of 'Crimson Seedless' grapes. *Journal of Agricultural and Biological Science*, *6*(3), 289–295.

Agüero, M. V., Ponce, A. G., Moreira, M. R., & Roura S. I. (2011). Lettuce quality loss under conditions that favor the wilting phenomenon. *Postharvest Biology and Technology*, *59*(2), 124–131.

Ahn, S. Y., Kim, S. A., Choi, S. J., & Yun, H. K. (2015). Comparison of accumulation of stilbene compounds and stilbene related gene expression in two grape berries irradiated with different light sources. *Horticulture, Environment, and Biotechnology*, *56*, 36–43.

Aked, J. (2002). Maintaining the postharvest quality of fruits and vegetables. In W. Jongen (Ed.), *Fruit and vegetable processing: Improving quality* (pp. 119–149). Woodhead Publishing, Cambridge.

Ali, S., Khaliq, G., Iftikhar, A., Mir, S. A., & Iqbal, S. (2025). Nanoemulsions-based edible coatings for postharvest quality preservation of fruits and vegetables. In Sustainable Postharvest Technologies for Fruits and Vegetables (pp. 213-226). CRC Press.

Arah, I. K., Amaglo, H., Kumah, E. K., & Ofori, H. (2015). Preharvest and postharvest factors affecting the quality and shelf life of harvested tomatoes: A mini review. *International Journal of Agronomy*, *2015*, 1–6.

Barman, K., Ahmad, M. S., & Siddiqui, M. W. (2015). Factors affecting the quality of fruits and vegetables: Recent understandings. In M. W. Siddiqui (Ed.), *Postharvest biology and technology of horticultural crops: Principles and practices for quality maintenance* (pp. 1–50). Waretown, NJ: Apple Academic Press

Benkeblia, N., Tennant, D. P. F., Jawandha, S. K., & Gill, P. S. (2011). Preharvest and harvest factors influencing the postharvest quality of tropical and subtropical fruits. In E. M. Yahia (Ed.), *Postharvest biology and technology of tropical and subtropical fruits* (pp. 112–142e). Woodhead Publishing.

Brasil, I. M., & Siddiqui, M. W. (2018). Postharvest quality of fruits and vegetables: An overview. In M. W. Siddiqui (Ed.), *Preharvest modulation of postharvest fruit and vegetable quality* (pp. 1–40). Amsterdam: Elsevier.

Cao, S., Yang, Z., Cai, Y., & Zheng Y. (2011). Fatty acid composition and antioxidant system in relation to susceptibility of loquat fruit to chilling injury. *Food Chemistry*, *127*, 1777–1783.

Choi, T., Park, D., & Kang, S. (2005). Effect of thinning degrees on fruit characteristics and carbohydrate accumulations of persimmon trees defoliated in early autumn. *Acta Horticulturae*, *685*, 105–112.

Davis, A. R., Perkins-Veazie, P., Hassell, R., Levi, A., King, S. R., & Zhang, X. (2008). Grafting effects on vegetable quality. *Horticultural Science*, *43*(6), 1670–1672.

Edelstein, M. (2004, March). Grafting vegetable-crop plants: Pros and cons. In *VII International Symposium on Protected Cultivation in Mild Winter Climates: Production, Pest Management and Global Competition 659* (pp. 235–238).

Fallik, E., & Ilic, Z. (2014). Grafted vegetables-the influence of rootstock and scion on postharvest quality. *Folia Horticulturae*, *26*(2), 1–12.

Fridovich, I. (1986). Biological effects of the superoxide radical. *Archives of Biochemistry and Biophysics*, *247*, 1–11.

Gudkovskiy, V., Kozhina, L., Balakirev, A., & Nazarov, Y. (2015). The effect of cultivar genotype, harvest date and fruit treatments with ethylene inhibitor on intensity of scald development. *Russian Journal of Horticulture*, *2*(1), 35–48.

Hewett, E. W. (2006). An overview of preharvest factors influencing postharvest quality of horticultural products. *International Journal of Postharvest Technology and Innovation*, *1*(1), 4–15.

Hussein, Z., Fawole, O. A., & Opara, U. L. (2018). Preharvest factors influencing bruise damage of fresh fruits–a review. *Scientia Horticulturae*, *229*, 45–58.

Kader, A. A. (1992). *Postharvest technology of horticultural crops publ. 3311*. Oakland, CA: University of California, Division of Agricultural Resources.

Kader, A. A. (2002). *Postharvest technology of horticultural crops* (pp. 39–48). 3rd Edition. University of California, Agriculture and Natural Resources. Oakland, Publication 3311.

Kader, A. A. (2013). Postharvest technology of horticultural crops-An overview from farm to fork. *Ethiopian Journal of Science and Technology*, *1*, 1–8.

Kahramanoğlu, İ. (2017). Introductory chapter: Postharvest physiology and technology of horticultural crops. *Postharvest Handling*, Intech, Rijeka.

Kays, J. S., & Paull, R. E. (2004). *Postharvest biology*. Athens: Exon Press.

Khaliq, G., Ali, S., Gapper, N., & Nicola, S. (2023). Recent advances and approaches in the application of elicitors to enhance resistance mechanisms in fresh produce. Frontiers in Sustainable Food Systems, 6, 1061079.

Kumar, R., & Kumar, V. (2016). Physiological disorders in perennial woody tropical and subtropical fruit crops-A review. *Indian Journal of Agricultural Sciences*, *86*(6), 703–717.

Lester, G. E., Makus, D. J., & Hodges, D. M. (2010). Relationship between fresh-packaged spinach leaves exposed to continuous light or dark and bioactive contents: Effects of cultivar, leaf size, and storage duration. *Journal of Agricultural and Food Chemistry*, *58*(5), 2980–2987.

Madakadze, R. M., & Kwaramba, J. (2004). Effect of preharvest factors on the quality of vegetables produced in the tropics. In R. Dris, & S. M. Jain (Eds.), *Production practices and quality assessment of food crops* (Vol. 1, pp. 1–36). Springer, Dordrecht.

Martínez-Romero, D., Bailén, G., Serrano, M., Guillén, F., Valverde, J. M., Zapata, P., ... & Valero, D. (2007). Tools to maintain postharvest fruit and vegetable quality through the inhibition of ethylene action: A review. *Critical Reviews in Food Science and Nutrition*, *47*(6), 543–560.

Mditshwa, A., Khaliq, G., Hussein, Z., & Ejaz, S. (2023). Sustainable postharvest management practices for fresh produce. Frontiers in Sustainable Food Systems, 7, 1143759.

Mditshwa, A., Magwaza, L. S., Tesfay, S. Z., & Mbili, N. (2017). Postharvest quality and composition of organically and conventionally produced fruits: A review. *Scientia Horticulturae*, *216*, 148–159.

Noichinda, S., Bodhipadma, K., Mahamontri, C., Narongruk, T., & Ketsa, S. (2007). Light during storage prevents loss of ascorbic acid, and increases glucose and fructose levels in Chinese kale (*Brassica oleracea* var. *alboglabra*). *Postharvest Biology and Technology*, *44*(3), 312–315.

Parra, C. R., Sarmiento, L. M. L., Salinas, D. G. C., & Giraldo, G. G. (2001). Effects of wind and hail on the development and quality of the fruits of Dominico Harton and FHIA-21 Plantain. *Information Musa*, *10*(2), 13–17.

Pedreschi, R., & Lurie, S. (2015). Advances and current challenges in understanding postharvest abiotic stresses in perishables. *Postharvest Biology and Technology, 107*, 77–89.

Prasad, K., Jacob, S., & Siddiqui, M. W. (2018). Fruit maturity, harvesting, and quality standards. In *Preharvest modulation of postharvest fruit and vegetable quality* (pp. 41–69). Academic Press.

Prusky, D. (2011). Reduction of the incidence of postharvest quality losses, and future prospects. *Food Security*, *3*(4), 463–474.

Ro, N., Oh, H., Ko, H. C., Yi, J., Na, Y. W., & Haile, M. (2024). Genome-wide analysis of fruit color and carotenoid content in capsicum core collection. Plants, 13(18), 2562.

Rouphael, Y., Schwarz, D., Krumbein, A., & Colla, G. (2010). Impact of grafting on product quality of fruit vegetables. *Scientia Horticulturae*, *127*(2), 172–179.

Sevillano, L., Sanchez-Ballesta, M. T., Romojaro, F., & Flores, F. B. (2009). Physiological hormonal and molecular mechanisms regulating chilling injury in horticultural species. Postharvest technologies applied to reduce its impact. *Journal of the Science of Food and Agriculture*, *89*, 555–573.

Sharom, M., Willemot, C., & Thompson, J. E. (1994). Chilling injury induces lipid phase changes in membranes of tomato fruit. *Plant Physiology*, *105*, 305–308

Shewfelt, R. L., & Henderson, J. D. (2003). The future of quality. *Acta Horticulturae*, *604*, 49–59.

Thompson, A. K. (2016). Controlled atmosphere storage. In R. W. Hartel (Ed.), *Fruit and vegetable storage. hypobaric, hyperbaric and controlled atmosphere* (pp. 21–36). New York: Springer International Publishing.

Tomita, S., Ando, Y., Watanabe, T., & Nagata, M. (2023). Physical, chemical, and physiological characterization of the reactions preceding chilling injury–induced cell membrane damage in cucumber. Postharvest Biology and Technology, 201, 112349.

Tyagi, S., Sahay, S., Imran, M., Rashmi, K., & Mahesh, S. S. (2017). Pre-harvest factors influencing the postharvest quality of fruits: A review. *Current Journal of Applied Science and Technology*, *23*(1), 1–12.

Usall, J., Ippolito, A., Sisquella, M., & Neri, F. (2016). Physical treatments to control postharvest diseases of fresh fruits and vegetables. *Postharvest Biology and Technology*, *122*, 30–40.

Yahia, E. M., Gardea-Béjar, A., de Jesús Ornelas-Paz, J., Maya-Meraz, I. O., Rodríguez-Roque, M. J., Rios-Velasco, C., Ornelas-Paz, J., & Salas-Marina, M. A. (2019). Preharvest factors affecting postharvest quality. In *Postharvest technology of perishable horticultural commodities* (pp. 99–128). Woodhead Publishing.

Yuri, J. A., Moggia, C., Sepulveda, A., Poblete-Echeverría, C., Valdés-Gómez, H., & Torres, C. A. (2019). Effect of cultivar, rootstock, and growing conditions on fruit maturity and postharvest quality as part of a six-year apple trial in Chile. *Scientia Horticulturae, 253*, 70–79.

Zappia, A., De Bruno, A., Torino, R., Piscopo, A., & Poiana, M. (2018). Influence of light exposure during cold storage of minimally processed vegetables (*Valeriana* sp.). *Journal of Food Quality*, *2018,* 4694793.

CHAPTER 13

Mechanism and Management of Postharvest Physiological Disorders

RACHID EZZOUGGARI[1,2], NAJWA SEDDIQI KALLALI[1,3], ILHAM DEHBI[1,4], KHADIJA BENAMAR[1,5], and RACHID LAHLALI[1*]

[1]*Phytopathology Unit, Department of Plant Protection, National School of Agriculture, Meknes, Morocco*

[2]*Laboratory of Biotechnology, Conservation and Valorization of Natural Resources (LBCVNR), Faculty of Sciences Dhar El Mehraz, Sidi Mohamed Ben Abdallah University, Fez, Morocco*

[3]*Laboratory of Biotechnology and Valorization of Phyto-Resources, Faculty of Sciences, Moulay Ismail University of Meknes, Meknes, Morocco*

[4]*Laboratory of Plant Biotechnology and Molecular Biology, Faculty of Sciences, Moulay Ismail University, Meknes, Morocco*

[5]*Department of Biology, Laboratory of Microbial Biotechnology and Bioactive Molecules, Sciences and Technologies Faculty, Sidi Mohamed Ben Abdallah University, Fez, Morocco*

*Corresponding author

ABSTRACT

Fruits and vegetables are abundant in organic acids, sugars, vitamins, minerals, and other components that contribute to human health, rendering them an indispensable component of the daily diet. The fruits and vegetables sector's impact on the global economy has markedly increased in recent decades, propelled by improvements in living

Physiological Disorders of Fruits and Vegetables: Causes and Remedies.
Ghulam Khaliq, Sajid Ali, and Mohammed Wasim Siddiqui (Eds.)

DOI: 10.1201/9781998511334-13

standards and significant shifts in consumer preferences. Despite their nutritional fruits and vegetables are susceptible to both external and internal damage, primarily stemming from various physiological factors, that lead to deterioration and reduced commercial quality. These diverse lesions arise from metabolic imbalances triggered by conditions such as exposure to extreme temperatures (either low or high), mineral imbalances, or alterations in atmospheric composition, among other factors. Termed physiological disorders, these disruptions are unique in that they do not arise from the action of fungal, bacterial, viral, or insect agents. Some physiological disorders may manifest before harvesting or emerge postharvest. While substantial research efforts have been directed toward developing strategies to mitigate. Although extensive research has focused on mitigating the effects of cold temperatures on fruits and vegetables, this chapter focuses on the primary physiological disorders occurring during the postharvest handling of fruits and vegetables. It explores and addresses the most common postharvest physiological disorders resulting from extreme temperatures, such as chilling injury, freezing, heat injury, and those associated with mineral imbalances, including fruit cracking, cork-like spots, and so forth. This chapter also delves into methods aimed at reducing the incidence of these disorders, emphasizing their frequency, significance, symptoms, physiology, and associated control strategies.

13.1 INTRODUCTION

Due to their high nutritional content, appealing colors, and delightful flavors, fruits and vegetables are among the most coveted horticultural commodities sought by consumers (Khalil et al., 2022) and play a pivotal role in global agricultural production. Statistics indicate that worldwide fruit production in 2021 reached 944 million metric tons (MT), reflecting a 0.61% increase compared to 2019, according to the Food and Agricultural Organization, Statistics Division (FAOSTAT, 2023b). Bananas are the world's leading fruit crop in terms of production volume, with 124 MT, followed by watermelon, apples, oranges, and grapes with 101.63, 93.74, 75.57, and 72.52 million MT, respectively (Statista, 2021).

Bananas (*Musa acuminata*) play a significant role in coastal agroecosystems in tropical and subtropical zones. India, China, and Indonesia are the top three producers of bananas, with an annual production of over 32.8, 11.6, and 8.46 million tons, respectively (FAOSTAT, 2023b). Bananas serve as a staple food for millions of people in tropical regions and are widely consumed globally. Available year-round, they constitute a good source of vitamin C, dietary fiber, and minerals such as potassium (Falcomer et al., 2019).

Watermelon is primarily produced in China, which accounts for nearly 60% of global production approximately 60.86 million MT annually. Watermelon is an excellent source of hydration. With its high water content, it is rich in bioactive compounds, such as cucurbitacin, triterpenes, and alkaloids (Tlili et al., 2011); watermelon also contains carotenoids including lycopene which is the molecule responsible for its red color, this red pigment has the highest antioxidant activity of all dietary antioxidants (Di Mascio et al., 1989). It is also rich in vitamins and minerals, in particular, vitamins B, C, and E, minerals potassium and magnesium, iron, and amino acids (citrulline, arginine, among others) (Romdhane et al., 2017). Apple (*Malus domestica*) production is about 45.98 million MT annually. The other primary producers of apples are Turkey, the United States of America (USA), and Poland, each contributing more than 4 million MT a year; apples provide fiber such as pectin, as well as polyphenols and vitamin C, antioxidants with multiple benefits. They also contribute to our potassium intake (Colin-Henrion, 2008), Brazil is the world's largest producer of orange fruits (*Citrus sinensis* L), with a 21% share and a volume of 16.21 million tons, followed by India with a share of 10 million tons, China in third place (7.5 million tons), and Mexico (4.6 million tons); citrus fruits, which are widely planted around the world, contain vitamins A, E, and B (thiamine, riboflavin, and niacin), minerals and antioxidants, such as flavonoids, vitamin C, phenolic compounds, and carotenoids (Uthman and Garba, 2023). They are still processed into various types of citrus-based products, such as canned fruit, citrus drinks, and citrus jams, and so forth (FAOSTAT, 2023a). In addition to these mentioned fruits, it is important to highlight other fruits that hold significant nutritional importance. For example, berries, such as strawberries, blueberries, and raspberries are rich in antioxidants, vitamins C and K, as well as fiber. Mangos, primarily cultivated in tropical regions, are an important source of vitamin A, vitamin C, and dietary fiber. Avocados, on the other hand, are well known for their high content of healthy fats, which also provide fiber, vitamins B, C, E, and K, as well as minerals like potassium. By incorporating a variety of these fruits into the diet, one can benefit from an extensive range of essential nutrients.

Given their nutritional richness, fruits are an essential part of a healthy diet and lifestyle. As per recommendations of the WHO, they serve as an excellent means to maintain the body's health (Mritunjay and Kumar, 2015). Insufficient fruit intake is associated with various metabolic disorders, particularly cardiovascular diseases, contributing to 11%. Based on the aforementioned facts, it was also predicted that the daily consumption of fruits may protect more than 2.7 million people annually. The therapeutic significance of fruits might also be due to the characteristics of low calories, high nutritional values, dietary fibers, and robust biologically active compounds (Farooq et al., 2020). Generally, fruits contain 70–80% moisture, 1.5% proteins, 13–15% carbohydrates, up to 6% dietary fibers, 500 mg of minerals, and up to 90 mg vitamins, but this composition varies depending on the type of fruit and its variety (Farooq et al., 2020). Despite this, postharvest losses of fruits and vegetables are estimated to be 5–20% in developed countries and 20–50 % in developing countries (Bantayehu, 2017). Losses after harvesting are influenced by several factors, including physical, physiological, mechanical, and hygienic conditions (Khan et al., 2025). Fruits characterized by high metabolic activity and a short shelf life are especially vulnerable. Due to these factors, a significant proportion of loss occurs between the time of harvest and the consumption of the product, such as when it is harvested, dried, and threshed, as well as losses along the chain during storage and transportation. On-farm losses occur during storage when the crop is being stored for self-consumption or while the farmer awaits a selling opportunity or a price rise (Yahaya, 2019).

Losses are generally higher at all stages of the food supply chain. In low-income countries, losses of fresh fruits and vegetables are primarily attributed to poor infrastructure, especially inadequate storage facilities, including refrigerated warehouses. Despite the availability of good storage facilities in high-income countries, losses still occur. They are often due to technical failures, poor control of temperature, relative humidity (RH), oxygen, carbon dioxide, and ethylene, or overstocking (FAO, 2019). During the storage period, there is a possibility of a manifestation of certain physiological disorders.

13.2 POSTHARVEST PHYSIOLOGICAL DISORDERS

Physiological disorders are those not caused by fungal, bacterial, viral, or insect agents but rather result from the loss of cellular integrity as produce responds to stress during storage. Physiological disorders can occur before or after harvest. Several physiological disorders manifest before harvest and typically only become apparent postharvest, especially during storage. The causes of these disorders are diverse and include factors such as temperature (both low and high), mineral imbalance, modified atmospheres (high carbon dioxide and low oxygen), and chemicals, including ethylene. Environmental elements such as wind, hail, and certain farming practices also contribute to these disorders (Kahramanoğlu, 2017; Watkins, 2017). A broad spectrum of physiological disorders has been identified among fruits; nevertheless, they vary depending on the type of crop. It is crucial to note that the specific nature of these disorders can vary significantly depending on the type of cultivation under consideration. Each fruit crop may exhibit its own characteristics and peculiarities in terms of physiological responses to environmental conditions, agricultural practices, and various stresses during growth and storage.

The most prevalent physiological disorder encountered during postharvest is chilling injury, which occurs at low, non-freezing temperatures. Numerous horticultural commodities are susceptible to chilling, a disorder that restricts their postharvest shelf life and leads to significant losses and wastage (Yahia et al., 2019a, b, c). Additionally, various other physiological disorders contribute to substantial deterioration in horticultural products, causing economic losses. Research into these physiological disorders has resulted in the identification of their causal agents, the understanding of their development mechanisms and the formulation of suitable and sustainable control strategies. These approaches encompass the utilization of various chemicals, such as ozone, electrolyzed oxidizing water, edible coatings, 1-methylcyclopropene, modified atmosphere packaging, as well as physical treatments, such as heat application, ultrasound, and UV-C radiation (Pinheiro et al., 2014). Despite the progress made in understanding postharvest physiological disorders in fruit crops, their physiological complexity remains an active area of investigation. This chapter delves into recent insights regarding the occurrence, significance, symptoms, and physiological aspects of postharvest physiological disorders in fruits. Furthermore, it explores various control measures aimed at preventing or mitigating these disorders.

13.3 CHILLING INJURY

13.3.1 STAGE OF OCCURRENCE

Chilling injury is a physiological dysfunction or abnormality that occurs after fruits and vegetables have been exposed to low, non-freezing temperatures for a period, which is completely different from frost injury (0°C, and below) (Khaliq et al., 2023). The low-temperature storage is the best way to preserve quality and increase the shelf life of many horticultural items, making long-distance shipment easier (Rodrigues et al., 2022). However, because of due to the cold sensitivity of certain commodities and their susceptibility to chilling damage, low temperatures above zero and below a specified threshold cannot be used for certain crops, especially tropical or subtropical ones, whose storage conditions are restricted (Albornoz et al., 2022; Yahia et al., 2019b). Furthermore, in comparison to quicker air freight, the globalization of trade has resulted in longer lead times and routes, raising the danger of losses due to cold weather (Aung and Chang, 2023). Cryogenic storage-related postharvest losses might be more than previously reported or measured (Wang, 1994). Since symptoms of cold damage frequently do not show up while the product is still chilled and only do so after it has been moved to ambient temperature, it can be challenging to estimate losses from this type of damage. Since chilling injury CI may cause widespread cellular damage and increase susceptibility to pathogen invasion, a substantial amount of real losses from CI might be unjustly attributed to maturation obstacles in the market or pathogen infection (Wang, 1994). Since temperate fruits tolerate low temperatures better than those of tropical or subtropical origin, which can occur at storage temperatures below 12°C, some are damaged if exposed to low temperatures for too long (Yahia et al., 2019b). CI can occur at any stage of supply and distribution centers, or in the field during growth and development (Mercier et al., 2018). CI builds up and usually manifests itself after 2 days at room temperature after being stored at a low temperature. After being stored at 4°C for 39 days followed by 5 days at 25 °C, the avocado pulp often darkened around the vascular tissues. When stored at room temperature, avocados take a few days to display lesion, but lemons require 60 days at 4°C to produce visible skin sores (Lo'ay and Dawood, 2019). In contrast, banana peels require 2 days at 4°C to begin browning, while the pulp takes 7 days to discolor (Yahia et al., 2019b).

13.3.2 IMPORTANCE

Since cold-sensitive fruits and vegetables cannot be kept at low temperatures, chilling injury is the most crucial postharvest procedure for the proper treatment of perishable horticultural produce, and it has significant economic implications. Because of this, they are often kept in storage for a shorter amount of time, and when compared with items that can withstand cold, their losses both qualitative and quantitative are higher (Yahia et al., 2019b; Zhang et al., 2021). According to (Lukatkin et al., 2012), the temperature to which the crop is exposed, the period of exposure, and the species' susceptibility to low temperature all affect the severity of cold-induced damage. Generally, fruits and vegetables recover from a brief exposure to refrigeration temperature and resume normal operations as soon as the stress caused by the low temperature is removed. On the other hand, if exposure to low temperatures (below a threshold) lasts for an extended length of time, symptoms may manifest, and irreparable damage or metabolic malfunction may result. The species, cultivar, tissue type, metabolic status (active or dormant) of the tissue at the moment chilling stress is applied, and several environmental variables all affect how quickly and how severely observable CI symptoms develop (Wang, 1989; Yahia et al., 2019b).

13.3.3 SYMPTOMS

The difference between a cold injury and a freezing injury is that the former happens above freezing, between 0 and 15°C. When a product is moved to a warmer, non-refrigerated temperature, it frequently develops symptoms of CI. These symptoms may include incomplete or inhibited ripening, excessive or inhibited softening, and a loss of flavor and fragrance (Gierson and Kader, 1986). According to Jackman et al. (1988), specific physiological signs may include skin pitting or depressions, aberrant skin discoloration, surface or interior browning, mealy or dry flesh, tissue deterioration, and fungal or bacterial rots, and ultimately tissue death (Yahia et al., 2019b). Tissues can be affected by various disorders in different ways. They can sometimes occur only in certain parts of the flesh or heart, or they can damage the skin but not the flesh (Franck et al., 2007). A specific crop's symptoms, such as the aberrant development of surface scald in apples, sunken areas in citrus fruits, and chalkiness in peaches and nectarines, are also typically indicative of CI (Yahia et al., 2019b). Although internal or surface browning is the typical indication of CI in apples, the symptoms vary greatly throughout cultivars. For instance, browning manifests as diffuse browning of the firm flesh in "Cripps Pink" apples (Mogollon et al., 2020). The appearance of browning spots on the "Huangguan" pear is accompanied by a reduction in the wax layer and an increase in lignified cells (Wang et al., 2021).

13.3.4 PHYSIOLOGY

CI may manifest itself in stone fruit in several ways, including as a glassy appearance and interior browning of the flesh (Tang et al., 2022). According to Fruk et al. (2014), CI in peaches and nectarines appears as brittle, leathery fruit, dry chalking or wooliness of the flesh, or both, with little to no free juice produced upon biting and chewing. As the CI progresses, interior browning of the flesh may also occur (Sidhu et al., 2023). According to Tremblay and Bélec (2000), microscopic investigations of cold-damaged mealy fruit revealed changed pectin-rich intermediate lamellae between cells and noticeably larger air gaps between them. Because there is much less adhesion between the cells, the cells break apart in clusters as opposed to rupturing and releasing their contents as juice (Brummell, 2006; Toivonen and Brummell, 2008).

Although the exact causes of browning are unknown, they most likely involve a loss of membrane integrity that compromises intracellular integrity and permits polyphenol oxidases in the plastids to combine with phenolic compounds in the vacuole (Bangar et al., 2022). This mixture causes oxidation reactions, which in turn produce melanin, the brown pigment (Zhang et al., 2023). The ethylene released in the fruits as a result of chilling may boost the activity of phenylalanine ammonia-lyase, which may explain the rise in total phenolic content following fruit chilling. It is the essential enzyme in the initial step of carbon flow from primary to secondary metabolism, specifically into the phenylpropanoid pathway. Citrus fruit exposed to chilling damage shows an increase in phenylalanine ammonia-lyase activity, which was followed by an increase in ethylene synthesis (Samad et al., 2016).

13.3.5 CONTROL

To manage CI and extend the cold-sensitive fruits' shelf life, scientists have turned to physical processing, which includes coating, UV irradiation, controlled atmospheres, and temperature treatments, such as low-temperature conditioning, heat treatment (HT), and storage at almost freezing temperatures. Plant hormones and natural elicitors, such as glycine betaine, ethylene, salicylic acid (SA), oxalic acid, hydrogen sulfide, calcium ions, 1-methylcyclopropene (1-MCP), melatonin, and methyl jasmonate are also used in chemical therapy (Zhang et al., 2021). Apples are typically kept in a controlled atmosphere (CA) with higher CO_2 and lower O_2 concentrations. This is frequently done after treating the fruit with 1-MCP (Smart Fresh SM), an ethylene antagonist that prolongs storage life by postponing softening but makes browning symptoms worse.

Nonetheless, the degree of susceptibility to the disorder can be determined by preharvest weather conditions, nutrition, crop maturity, crop load, orchard, and season (Sidhu et al., 2023), indicating a significant interaction between pre- and postharvest weather conditions. Nitric oxide (NO) was used at the start and finish of Baigent's apple fruit storage in CA to preserve a greater overall phenolic component content and decrease the likelihood of flesh browning (Coser et al., 2023).

13.4 FREEZING INJURY

13.4.1 STAGE OF OCCURRENCE

Freezing injury can be observed on fruits and vegetables when they are exposed to very low temperatures below their freezing point (Chaves and Zaritzky, 2018). The occurrence of freezing injury in postharvest fruits and vegetables and its severity depends on many factors related to the preharvest conditions, such as the genomic features of fruits, which have a great impact not only on their growth but also on their cellular metabolism (Yahia et al., 2019c). In this aspect, it was observed that fruits with high levels of soluble solids are less susceptible to freezing lesions than other fruits since no major symptoms of this physiological disorder were observed after their storage at very low temperatures (De Freitas and Pareek, 2019). This physiological disorder can occur during fruits and vegetables storage at freezing temperatures which are usually used to preserve the quality of fruits and vegetables. However, due to the richness of these fresh products in water, they could face several damages attributed to the transformation of water to ice crystals during freezing, which destroys cell walls and leads to the appearance of freezing injury. Furthermore, it is important to mention that this phenomenon can also occur before harvest in the case of freezing environmental conditions surrounding fruit trees, which constitute a stressful factor limiting the development and productivity of temperate fruit trees (Yu and Lee, 2020).

13.4.2 IMPORTANCE

Freezing injury has a negative impact on fruits and vegetables quality since ice crystals formed inside fruits and vegetables at temperatures below their freezing point destroy cellular membranes. These damages to the cellular structure are followed by several deteriorations of the fruit tissue, such as dehydration, and biochemical modifications including protein denaturation, leading to tissue softening and the degradation by microorganisms that penetrate the vulnerable fruit very easily (De Freitas and Pareek, 2019). Overall, quality parameters of fruits and vegetables are impacted (flavor, moisture, taste, safety) due to freezing injury. Hence, economic losses and may pose health risks to consumers.

13.4.3 SYMPTOMS

Symptoms of freezing injuries may differ between fruits and vegetables. Thus, for instance, in bananas, the appearance of different sections in the endocarp separated from each other reflects the symptom of this physiological disorder (De Freitas and Pareek, 2019). In apples, freezing injury is responsible for a softer texture (Saltveit et al., 2016). In watermelon, different amounts of lycopene loss are induced depending on freezing temperature (Fish and Davis, 2003). In mangoes, cell wall destruction is observed, especially after exposure to a slow freezing process (Charoenrein and Owcharoen, 2016). In citrus fruits, freezing injury symptoms are characterized by the modification of fruit color, which becomes lighter than before, due to the loss of fruit juice, resulting therefore in fruit puffiness. However, the symptoms of freezing injuries may sometimes be absent on citrus fruits, since this physiological disorder can occur only on the pericarp without any apparition on peels which freeze generally at lower temperatures than the pericarp. Hence, the freezing injuries may be invisible (Salehi Sardoei et al., 2023).

13.4.4 PHYSIOLOGY

The physiological disorders attributed to freezing injuries lie in the crystallization of fruit tissues, followed by the alteration of their structure, and the loss of their marketability (Mir et al., 2018). Indeed, the crystallization of fruit the liquid in fruit cells at a low temperature leads to an enhancement of the osmotic pressure. Hence, cellular permeability increases, intracellular cell fluids escape, and severe damage to cellular integrity occurs. Alabi et al. (2022) and De Freitas and Pareek (2019) have further detailed the mechanism that occurs during the freezing of fruits and vegetables and have summarized it in the notion of the transport phenomenon. In fact, during freezing, the liquid contained in fruit tissue is converted to ice crystals. This transformation of cells liquid occurs firstly in the extracellular spaces, leading to a higher concentration of solutes (minerals and sugar) inside cells. Hence, under the created osmotic pressure, the intracellular water moves toward the extracellular space, resulting in the growth of extracellular ice crystals. Therefore, great damage to cell walls appears, and the symptoms of freezing injury start to be observed (Alabi et al., 2020).

13.4.5 CONTROL

Several treatments have been demonstrated efficient in preventing freezing injury, namely, osmotic dehydration (Alabi et al., 2022), magnetic field (Kaur and Kumar, 2020), near-freezing temperature (Cui et al., 2019), high-pressure freezing, microwave, and ultrasound-assisted technologies. These techniques not only prevent the formation of ice crystals or reduce their size during freezing but also help to make the freezing process faster. Indeed, during fast freezing, the movement of moisture is minimal and prompt, preserving, therefore, fruits' cellular structure (Alabi et al., 2020). In this context, it was found that the fast freezing of strawberries resulted in less damage to the fruit tissue, since smaller ice crystals were formed (Kobayashi and Suzuki, 2019). Furthermore, it is important to mention that controlling thawing also plays an important role in freezing injury control. Temperature is an essential parameter which must be taken into consideration during the thawing process. It has been reported that fast thawing at high temperatures leads to great tissue damage, whereas thawing at 4°C causes less damage to frozen tissues (De Freitas and Pareek, 2019).

13.5 HEAT INJURY

13.5.1 STAGE OF OCCURRENCE

In the field, heat damage can happen either before or after harvest. Horticultural produce can experience sunlight temperature increases and substantial qualitative and quantitative losses when exposed to high temperatures, such as when exposed to direct sunshine or during unrefrigerated storage or transportation (Joardder and Masud, 2019). For many goods at various stages of the postharvest chain, including the postharvest waiting period, before chilling, before packing, during shipping, during marketing, and so forth, temperature abuse during postharvest handling is a typical problem. Breaks in the cold chain have the potential to result in extremely high temperatures can lead to heat buildup, stress, and injury (Shewfelt and Prussia, 2022).

13.5.2 IMPORTANCE

Physical treatment before storage, including hot water (HW) soaking, short HW rinsing and brushing, hot air (HA) or steam HT, radio frequency irradiation, and so forth, can reduce subsequent decomposition losses, ensure quarantine safety, and maintain product quality during refrigeration and shelf life (Fallik and Ilic', 2019). Despite this, HT may have effects on fruits beyond their intended purpose, as high thermal stress induces changes in plant tissues that affect various physiological and biochemical processes. These processes include inhibition of ethylene production and other processes associated with maturation and senescence, induction of defense against parasite attack, and induction of resistance to other types of stress, including cold stress (Yahia et al., 2019b).

13.5.3 SYMPTOMS

In addition to reducing the fruits and vegetables shelf life and increasing metabolic activity, temperatures over the ideal range also result in significant quantitative and qualitative losses. Heat damage manifests as surface scalding or bleaching, uneven ripening, excessive softening, and drying (Fallik and Ilic', 2019; Yahia et al., 2019a, b, c).

When chlorophyll-containing tissues are subjected to high temperatures and strong solar radiation throughout the growth season, photooxidative stress occurs, resulting in sun damage to fruit (Naschitz et al., 2015). The fruits and vegetables ability to absorb light will be exceeded by too much light energy. Reactive oxygen species (ROS) are created when tissues absorb more light energy than they can use for photosynthetic processes. This process is known as photooxidation, and it results in sunburn symptoms (Morales-Quintana et al., 2020).

13.5.4 PHYSIOLOGY

Heat stress injury may result in increased efflux due to impaired plasma membrane semipermeability or decreased influx due to impairment of active transport systems (Palou and Smilanick, 2019). Because ethylene production is inhibited around 35–40°C (Boller, 2018), many fruits will not mature after being exposed to these temperatures. There will be tissue death at really high temperatures (Saltveit, 2018). Heat stress modifies the efficiency of enzymatic activities in the cell, leading to metabolic imbalance, and impacts the stability of different proteins, membranes, RNA species, and cytoskeletal structures (Firmansyah and Argosubekti, 2020). When the steady flow of metabolites is disturbed, harmful substances like ROS can build up. In response to heat stress, plants rewire their transcriptome, proteome, metabolome, and lipidome (Khan and Mehmood, 2023). The organism can adapt to the new temperature and respond to other forms of stress, such as low temperatures, attributable to these modifications that create a new, stable equilibrium of metabolic activities (Yahia et al., 2019b).

13.5.5 CONTROL

For several reasons, including lowering pathogen inoculum, managing insect pests, and lessening the symptoms of cold damage, high-temperature treatments either in the air or water have been developed and included in the postharvest handling chain for various products (El-Ramady et al., 2015). From many days at 35–40°C in the air to less than a minute in water at 63°C, treatment durations and temperature ranges differ significantly (Fallik and Ilic', 2019; Paull and Chen, 2000). A few commercially available methods of controlling insects include applying steam heat or submerging the fruit in HW until the core temperature reaches a certain point; this process can take several minutes in the case of water or many hours in the case of air (Yahia et al., 2019b). HT at 38–60°C has demonstrated positive results as an eco-friendly method of reducing CI in horticultural produce postharvest (Wang, 1994). In cold storage, HT mostly refers to the application of HA or HW to postharvest horticultural produce for a predetermined amount of time. The growth circumstances, size, maturity, and variety of the horticulture produce determine the treatment's temperature and length. Short HT has less detrimental impact on the quality attributes of fruits and vegetables than lengthy HT, which is why many professionals advise using it. Furthermore, short-term HT can lessen fruit and vegetable postharvest rotting as a useful and affordable technique, increasing the final product's aesthetic appeal (Bokhary et al., 2020). The potential and biochemical processes of HT in the postharvest management of horticultural crops were summarized in a previous report (Aghdam and Bodbodak, 2014). Recent studies have also shown that HT can effectively reduce CI in postharvest fruit, such as mango. The MW treatment of mandarins at a temperature of 50°C, maintained for 2.5 min, results in lower citric acid content, preserves color and total soluble solids (TSS), maintains a high thermal ripeness index, and is more effective in deactivating molds compared with HW treatment (Queb-González et al., 2020).

13.6 FRUIT CRACKING

Fruit cracking, a common physiological disorder, is a major challenge to the horticultural industry, affecting a wide range of fruit crops and resulting in significant economic losses for growers (Ray and Solankey, 2022). This disorder manifests itself in two primary forms: radial cracking, which extends from the stem to the end of the flower, and concentric cracking, which forms rings around the circumference of the fruit (Petit et al., 2017). Environmental conditions are closely linked to the development of fruit cracking. Fluctuations in temperature and humidity, particularly during the development phase of the fruit, are critical determinants of susceptibility to cracking (Li et al., 2021). Water availability is another important factor. Fruit exposed to an inconsistent water supply, especially those that experience a period of drought followed by a sudden flood, tends to expand rapidly (Kumar et al., 2021). As a result of the skin is often unable to withstand the internal pressure, leading to crack formation. Although the primary function of the fruit skin is to protect against pathogens and physical damage, it has its limitations. If the fruit is growing rapidly, the internal turgor pressure can exceed the tensile strength of the skin, resulting in cracks (Yu et al., 2023). At the cellular level, calcium plays a key role in ensuring the strength of cell walls and the stability of cell membranes. A lack of calcium can weaken these structures, making the fruit more susceptible to cracking (Fischer et al., 2021). Aside from environmental and nutritional factors, genetics also plays an important part. Some varieties are inherently more susceptible to cracking due to their genetic makeup. Growers need to make careful selections to minimize losses (Khadivi-Khub, 2015). The role of hormonal imbalances in fruit cracking has also been highlighted in recent studies. Ethylene, a plant hormone, has been observed to affect skin elasticity, and its dysregulation can exacerbate cracking tendencies. The need for holistic management practices is further emphasized by the fact that postharvest handling of fruit, including storage conditions and transport, can influence the severity and incidence of cracking (Seo et al., 2022).

13.6.1 STAGE OF OCCURRENCE

Fruit cracking can occur at various stages of development, from flowering to postharvest. The stage of development and the environmental and physiological factors to which the fruit is exposed often determine the susceptibility and severity of the disease (Fischer et al., 2021). The early stages of fruit development are characterized by rapid cellular changes, making fruit particularly sensitive to both environmental and hormonal cues. The cherry (*Prunus avium*) is a prime example of how phytohormones can greatly influence fruit development and subsequent vulnerability to diseases such as cracking (Wójcik et al., 2013). In sweet cherry, early developmental applications of phytohormones such as abscisic acid (ABA) and methyl jasmonate (MeJA) have been shown to profoundly affect fruit quality and cracking resistance. In particular, significant reductions in the number of mature cracked fruits have been observed when ABA and MeJA are applied preharvest, either independently or in combination, at the fruit set or color change stages (Balbontín et al., 2018). Furthermore, a recent study highlighted the tissue-specific dynamics of phytohormones, including jasmonic acid (JA), SA, and ABA, in the exocarp and mesocarp of sweet cherries during fruit development. The study showed that the exocarp or outer skin of the fruit responds more dynamically to hormones than the mesocarp (Kumar et al., 2016). This difference in behavior between the two tissues suggests that endogenous phytohormones have a key role in the modulation of fruit development and its interaction with microbes. Fruit expansion and growth are a dynamic period characterized by significant physiological changes predisposing fruit to cracking (Li et al., 2022). There are several stages in this phase, each with its own contribution to the development of cracking (Kumar et al., 2016). The onset of fruit expansion is characterized by rapid cell division followed by cell enlargement, leading to an increase in fruit volume. This stage is critical as internal stresses begin to develop within the developing fruit, setting the stage for possible cracks if the skin is insufficiently elastic to accommodate the internal growth (Luo et al., 2021). Rapid cellular turnover is a key determinant. In the case of bananas, the study by Khodabakhshian et al. (2021), which used AFM to determine the structural properties of banana fruit cells, is instructive. The study revealed considerable variation in elastic-mechanical properties at the cellular level. Even among cells from the same region of the fruit, the elastic modulus and stiffness of the cell walls, which indicate the mechanical strength of the fruit, vary considerably (Khodabakhshian et al., 2021). This variability can be attributed to the different growth rates and developmental changes that occur within the fruit. These changes can influence the susceptibility of the fruit to cracking as it matures. As the fruit continues to grow, there is an escalating demand for both water and nutrients. The efficient transport of water through the vascular system is of vital importance at this stage. Disturbances in the water supply such as periods of drought followed by heavy rainfall, can cause sudden changes in turgor pressure, which is a major cause of cracking (Li et al., 2022). At the same time, as internal growth, the development of the skin of the fruit is a critical factor in its resistance to cracking. Skin elasticity, influenced by factors, such as cuticle development, epidermal cell size, and deposition of structural compounds, such as suberin and lignin must keep pace with expanding fruit to prevent cracking. As the fruits approach their ultimate size, the rate at which they expand slows and ripening begins (Alkio et al., 2012).

At this late stage, skin elasticity decreases and additional internal pressure from environmental factors or growth spurts may cause cracking. This risk is particularly pronounced around the stalk and calyx, where the skin is often less flexible and more susceptible to stress (Knoche and Grimm, 2008). When the fruit starts to ripen, its physical properties can change, leading to an increased risk of cracking. The mechanical properties of the skin change considerably during the late stages of fruit development. In the case of sweet cherry (*Prunus avium* L.), the mechanical properties of the skin, such as its elasticity and its ability to withstand stress, are important determinants of the quality of the fruit (Brüggenwirth and Knoche, 2017).

Fruit cracking, as mentioned above, is a detrimental postharvest physiological disorder that compromises the integrity of the fruit skin. This leads to reduced marketability and economic losses. The etiology of fruit cracking is multifactorial and including environmental, nutritional, and genetic factors that interact with each other to predispose the fruit to this condition (Ikram and Nafees, 2020).

13.6.1.1 ENVIRONMENTAL FACTORS

The rapid changes in environmental conditions can stress the fruit's physiological processes, leading to a higher incidence of cracking (Fischer et al., 2021). Temperature extremes can cause differential growth rates in fruit tissues, resulting in cracking, especially when transitioning from a period of drought to a period of heavy rainfall. This phenomenon is well documented for pomegranate, where the sclerenchyma tissue developed during drought do not grow at the same rate as the meristem tissue when the water supply is increased, leading to cracks Sheikh and Manjula (2012) conducted a study on the control of fruit cracking in pomegranate that provides insights into how fluctuations in temperature and water availability can lead to fruit cracking. We also note that a major cause of cracking in pomegranates has been identified as dry heat accompanied by hot winds during the fruit ripening stage. The study indicates that temperatures in excess of 38°C in combination with humidity levels below 60% are conducive to the cracking of the fruit. Large day/night temperature fluctuations, especially when combined with heavy irrigation after a dry spell, also contribute to cracking (Sheikh and Manjula, 2012). In bananas, temperature stress plays a crucial role in the postharvest life and quality of the fruit. Huang et al. (2021) investigated the changes in cuticle morphology, chemical composition, and biosynthetic regulation in response to chilling injury in stored bananas. Significant changes in the chemical composition of the cuticle, indicating a response to thermal stress, were found during low-temperature storage. This response may be related to banana crack occurrence under different environmental conditions. Similarly, a study by Sicilia et al. (2020) on pigmented orange fruit varieties under cold stress showed that temperature variation has a significant impact on physiological processes in oranges. Focusing on anthocyanin biosynthesis in sweet orange fruit, they found that cold temperatures caused an increase in anthocyanin content by inducing related gene expression. This shows how temperature extremes lead to different growth rates in fruit tissues, resulting in stress responses like cracks (Sicilia et al., 2020).

Water stress, both drought and excessive humidity, may result in differential growth rates in the fruit tissue, resulting in cracking. Goodwin et al. (2022) studied the effects of water deficiency on "Cripps Pink" apples and found that insufficient irrigation increased crack formation, highlighting the need for effective water management in orchards (Goodwin et al., 2022). Also, the study of Sheikh and Manjula (2012) in pomegranates shows that when the water supply is increased, the hardened tissues do not continue to grow at the same rate as the meristem tissues. The study also emphasizes the importance of a balanced water supply, noting that maintaining 25% available soil moisture during the summer can reduce fruit cracking (Sheikh and Manjula, 2012). Also, heavy rains after a dry period can cause fruit to crack, especially if the skin of the fruit has been hardened due to previous water stress (El-Bolok and Sawarsan, 2017). The reason for this is that the arils swell rapidly with the water, while the skin is unable to expand fast enough resulting in cracking. Singh et al. (2020b) study of fruit cracks in pomegranates provides a comprehensive review of their causes and management, emphasizing their role as a result of water imbalances and sharp temperature fluctuations during fruit growth and development. In the same context, Sheikh and Manjula (2012) conducted a study on mangosteen trees and investigated the effects of different levels of soil moisture on the yield and growth of the fruit. It was found that unbalanced water supply can negatively affect fruit development and may contribute to cracking, as both severe water stress and excess water supply resulted in increased fruit drop and affected fruit size.

13.6.1.2 CALCIUM, POTASSIUM, AND MAGNESIUM IMBALANCES

Mineral imbalances, especially those involving calcium, potassium, and magnesium, are known to affect the integrity of the fruit (Singh et al., 2020b). The role of calcium in cell wall stability is well documented, and calcium deficiency has been directly linked to increased fruit cracking. The study on "Clementino" mandarins shows that calcium deficiency results in weaker fruit skins that are prone to cracking under stress (Chabbal et al., 2020). There is also evidence that potassium and magnesium are critical for the maintenance of turgor pressure and enzyme activity within the fruit. Research on tomato varieties has shown that these minerals play a role in maintaining the fruit's ability to manage internal pressure without damage, as inadequate levels of these minerals can lead to an increased risk of cracking (Ren et al., 2014). In addition, the review by Singh et al. (2020a, b) on the cracking of pomegranate fruit reinforces the understanding that nutritional deficiencies, particularly in calcium, boron, and potassium, are the major factors that contribute to fruit cracking. The review also indicates that environmental stressors, such as high evapotranspiration, low humidity, and high-temperature variation aggravate these deficits and lead to cracking, especially as the fruit approaches maturity (Singh et al., 2020b). The incidence of fruit cracking can also be influenced by the interaction of different mineral elements. For example, a calcium-to-magnesium imbalance can affect the absorption and utilization of these minerals, causing the fruit's peel to become structurally weak and prone to cracks. The delicate balance required to maintain fruit integrity is highlighted by the effects of different mineral elements on tomato fruit cracking (El-Bolok and Sawarsan, 2017).

13.6.1.3 GENETIC FACTORS

The susceptibility of fruit to cracking is significantly influenced by genetic factors. Inherent characteristics of the skin and cell structure make some varieties more susceptible to cracking (Khadivi-Khub, 2015). The genetic component in physiological disorders suggest that breeding for crack-resistant varieties may be a long-term solution to the problem (Peet, 2009). Similarly, Ikram and Nafees (2020) studied the causes and control of fruit cracking in pomegranates. They found that under the same environmental and cultural conditions, different cultivars exhibit different intensities of cracking. This review suggests the possibility of identifying and manipulating the genes that cause cracking, which has been reported in many fruit crops, to select more resistant plants.

In the same context, genetic factors can be influenced by environmental stresses. A study on the role of microRNAs (miRNAs) in banana heat stress response provides deep insight into the complex molecular mechanisms underlying fruit physiology and response to environmental challenges. This research is especially important for treating and mitigating physiological disorders such as fruit cracking. By identifying the target genes of these temperature-responsive miRNAs and verifying their interactions, the study provides insight into the specific pathways and processes involved in the stress response of bananas, which will greatly enhance our understanding of the mechanisms leading to fruit scarring. Therefore, thermal stress may elicit specific molecular responses, which may result in differential growth rates in fruit tissue, and eventually fruit cracks (Zhu et al., 2019).

13.6.2 IMPORTANCE

Fruit cracking not only reduces the aesthetic appearance of the fruits and vegetables but also severely reduces the commercial value of the fruits and vegetables, resulting in serious economic damage. The visual quality of fruits and vegetables is a primary indicator of marketability, and cracked fruit is frequently considered unfit for sale. In Li's research, the author the author investigated the factors that influence fruit cracking and its detrimental effect on the visual quality of fruit, which significantly reduces its market value (Li, 2010). Cracked fruit skins allow pathogens to enter, which can lead to rapid spoilage and reduce the shelf life of the fruit (Ikram and Nafees, 2020).

Otherwise, Niu and his team provided insight into the molecular mechanisms that regulate fruit skin cracking in *Akebia trifoliata*, a problem that limits its commercial value due to increased susceptibility to pathogens (Niu et al., 2020). Li (2010) identified differentially expressed genes and proteins involved in fruit cracking and that provide targets for improving fruit cracking resistance and reducing pathogen penetration. Taken together, these studies highlight the significant impact of fruit cracking on the commercial value of the produce and the role of cracking in facilitating the entry of pathogens that lead to spoilage. In terms of both the marketability of fresh produce and postharvest losses due to increased spoilage, the economic impact of cracking is significant (Niu et al., 2020). Also, Ozores-Hampton et al. (2013) in their evaluation of tomato cultivars resistant to tomato yellow leaf curl virus, observed a high percentage of unmarketable fruit due to various defects, including radial or concentric cracks. This study underlines the importance of selecting varieties that are not only resistant to diseases but also to physiological disorders such as fruit cracking.

13.6.3 SYMPTOMS

Fruit cracking is characterized by visible splits or cracks in the skin, which may be radial, concentric, or a combination of both. In addition to compromising the aesthetic quality of the fruit, these symptoms decrease market value and increase the risk of pathogen penetration. Olle and Williams (2017) provided a comprehensive review of physiological disorders in tomatoes, including fruit splitting, and describe epidermal splitting around the calyx or stem scarring as a common feature; the authors note that rapid net influx of solutes and water into the fruit, coupled with factors that reduce skin strength and elasticity, can result in cracks. In this regard, the types of fruit cracks, including radial and concentric cracks, are described in detail in a review of physiological disorders in tomatoes by Srinivasulu et al. (2020). Radial cracking develops from the stem end of the fruit, often because of heavy rains or irrigation followed by a long dry period. Concentric cracks are circular cracks that develop around the stem end. These symptoms indicate the underlying physiological stress the fruit is under, which can be aggravated by environmental factors and cultural practices (Li, 2010; Srinivasulu et al., 2020).

13.6.4 PHYSIOLOGY

Fruit cracking is a physiological disorder that can have a significant impact on the quality and yield of a fruit crop. This section focuses on will be on the two main factors that contribute to the phenomenon of rapid water uptake, which leads to fruit swelling and skin tears, and inadequate skin strength due to poor cell wall structure and composition. In order to develop strategies to reduce the risk of fruit cracking and ensure the production of high-quality fruit, an understanding of these factors is critical (Li et al., 2021).

13.6.4.1 RAPID WATER ABSORPTION AND FRUIT SWELLING LEAD TO SKIN CRACKING

The dynamics of water movement within the fruit are closely related to the phenomenon of fruit cracking in sweet cherry cultivars. A significant difference in water potential between the tree and the fruit can cause water to move from the branches and leaves into the fruit during prolonged rainfall before and during harvest, or during periods of high humidity (Borve et al., 2000). The fruit can swell rapidly from this water influx. For example, water uptake initially causes separation of the cuticle from the epidermal wall in "Bing" cherries. This can culminate in cuticle rupture and cracking as the fruit swells further (Glenn and Poovaiah, 1989). Lane et al. (2000) investegated this further, measuring water absorption as a percentage of fruit weight at the point of break. The researchers found that these measurements were related to the fruit's resistance to cracking, with the mesocarp cells of resistant varieties being more squared, which may provide a structural advantage to resist cracking (Balbontín et al., 2013). They also found that the thickness of the skin or the thickness of the epidermis does not explain the resistance. Indeed, this research underscores the importance of understanding fruit water relations, particularly in relation to fruit development and cracking susceptibility (Stojanović et al., 2013). It suggests that the response of fruit to rapid water uptake and the associated risk of skin cracking may be influenced by the structural characteristics of the fruit tissue, including the shape and arrangement of cells (Lane et al., 2000).

13.6.4.2 INADEQUATE SKIN STRENGTH DUE TO POOR CELL WALL STRUCTURE AND COMPOSITION

In order to determine the susceptibility of fruit skins to cracking, their mechanical properties are crucial. In their study, Brüggenwirth and Knoche (2016) examined the mechanical properties of the peels of two sweet cherry cultivars, "Regina" and "Burlat," which differ in their susceptibility to cracking. Using biaxial tensile tests, the researchers found that the less susceptible "Regina" had stiffer skin with a higher modulus and higher fracture pressure than "Burlat." This suggests that Regina has stronger skin capable of withstanding greater internal pressures without breaking. In addition, the contribution of cell turgor and cell wall properties to the mechanical strength of fruit peel was investigated. The researchers determined that the differences in mechanical properties were due to the cell walls, rather than cell turgor, by subjecting the skins to freeze/thaw cycles to disrupt the plasma membranes. Microscopy also confirmed this finding, showing no significant differences in cell size between the two varieties (Brüggenwirth and Knoche, 2016a). However, Regina had a higher mass of cell walls per unit fresh weight and a lower ratio of tangent to radial epidermis cell diameter, suggesting that the physical and possibly chemical properties of the cell walls account for the observed differences in skin mechanical properties and crack sensitivity. The implications of these findings are important for understanding the physiology of fruit cracking. They suggest that cell wall strength and composition may be more critical than cuticle or skin elasticity in the prevention of cracking. This finding can guide breeding efforts to select cultivars with stronger cell walls, potentially reducing the incidence of cracking (Brüggenwirth and Knoche, 2016b).

13.6.5 CONTROL

13.6.5.1 REGULATE WATER SUPPLY AND PREVENT WATER STRESS

To prevent fruit cracking in fruit crops, effective water management is a critical component (Kumar et al., 2017). Fruit cracking is a significant economic concern because it can reduce marketability and increase pathogen susceptibility (Singh et al., 2020a; Li et al., 2021).

Several studies have shown that regulating water supply and preventing water stress are an important strategy for controlling fruit cracking in fruits such as apples. Goodwin et al. (2022) stressed the importance of consistent and sufficient watering, especially for cultivars such as "Cripps Pink," in reducing the risk of cracks caused by water deficiency. The research by Li et al. (2021) further supports this by emphasizing the need for balanced soil moisture, especially during critical growth periods, to prevent rapid cell expansion that leads to cracks. Furthermore, a study on wax apples by Lee (2013), which can be applied to other fruits, demonstrates that tailored soil moisture management based on specific soil types and environmental conditions can significantly reduce fruit cracking.

Blanco et al. (2022) studied the effects of deficit irrigation and crop stress on cherry (*Prunus avium*) physiology. Their study showed that a deficit irrigation strategy that provided 70% of crop evapotranspiration significantly reduced the incidence of rain-induced cherry cracking, highlighting the importance of precise irrigation management (Blanco et al., 2022). Similarly, research conducted by Sandhu and Bal (2013) on cracking in lemon trees (*Citrus limon*), also showed that integrated management, including proper water management, fertilizer application, and preventive spray programs could significantly reduce cracking. In particular, irrigation at a 20% depletion of the available soil moisture in combination with mulching was found to be effective in reducing cracking losses. Recently, the importance of adequate irrigation and the role of micronutrients were highlighted in another study on the causes and management of fruit cracking in pears (*Pyrus* spp.). In addition, a comprehensive understanding of soil moisture, plant water uptake, and the use of plant growth regulators may help develop effective strategies for reducing fruit cracking (Seo et al., 2022).

13.6.5.2 STRENGTHENING CELL WALLS WITH CALCIUM-BASED FERTILIZERS

Calcium is a critical nutrient that is essential for the maintenance of the structural integrity of plant cell walls (Glenn and Poovaiah, 1989). When applied as a calcium-based fertilizer, it has been shown to help prevent fruit cracking, a common physiological disorder in various fruits and vegetables (El-Bolok and Sawarsan, 2017).

The effects of calcium treatments on fruit cracking and cell wall enzyme activity were investigated in navel oranges (*Citrus sinensis*). It was found that calcium content was higher in normal than in cracking fruits, and that calcium hydroxide treatments significantly increased pericarp calcium content, improved bruising resistance, and reduced cracking rate. Treatments also reduced cellulase activities, an enzyme involved in cell wall degradation, suggesting that calcium contributes to fruit integrity by affecting cell wall enzymatic activity (Chen et al., 2014).

Nevertheless, a study by Schumann et al. (2022), focused on the effect of calcium on the swelling of the epidermal cell wall of the anticlinal zone of sweet cherries (*Prunus avium*). They found that calcium chloride treatments decreased cell wall thickness and reduced swelling, which was associated with reduced susceptibility to cracking. This study supports the idea that calcium may reduce crack susceptibility by cross-linking cell wall components and reducing cell wall swelling (Schumann et al., 2022). In the same context, another study by Huai et al. (2022) investigated the effects of calcium on cell wall metabolizing enzymes and the expression of related citrus skin wrinkle genes. They found that foliar application of calcium nitrate significantly reduced the fruit wrinkling rate rate and suppressed the expression of genes encoding cell wall degrading enzymes. This suggests that calcium treatments can enhance skin mechanical properties and thus reduce wrinkling and cracking in fruits (Huai et al., 2022).

13.6.5.3 SELECTIVE BREEDING FOR CRACK-RESISTANT VARIETIES

A proactive and sustainable strategy for mitigating fruit cracking in fruits and vegetables are the selective breeding of resistance to fruit cracking. This approach consists of identifying and selecting genetic traits that impart resistance to cracking, which can then be enhanced through breeding schemes. While direct studies on Rosaceae crops are limited, the principles of selective breeding have been successfully applied in a variety of agricultural contexts and have provided valuable insights for the improvement of fruits and vegetables. In their findings on breeding programs, Doyle et al. (2019) considered the balance between improved production and increased disease risk due to inbreeding. They emphasize the importance of maintaining genetic diversity to prevent increased disorder susceptibility, perhaps analogous to fruit cracking. This suggests that genetic diversity should also be a priority in rosaceous breeding programs to ensure resilience and adaptability to the environment (Doyle et al., 2019). In a similar vein, Súnico et al. (2021) describe the use of synthetic biology-assisted intragenomic strategies to improve fruit quality and pathogen resistance in strawberries (*Fragaria* spp.), a member of the Rosaceae family, by targeting specific genes associated with fruit skin integrity and water regulation, this innovative approach can be adapted for the development of crack-resistant varieties (Súnico et al., 2021). Additionally, a recent 2022 study discusses using wild relatives in soybean breeding to expand the gene pool and introduce traits lost during domestication. This concept can be applied to rosaceous crops by screening wild species of the family for traits that confer resistance to cracking, which can then be introduced into cultivated varieties (Zhuang et al., 2022).

13.7 CORK (SPOT-LIKE)

Corking, a physiological disorder of orchard crops, appears as recessed, corky areas on the skin and flesh of fruit, especially members of the Rosaceae family such as apples and pears (Blevins and Lukaszewski, 1998). This disorder is often associated with a deficiency of essential nutrients, particularly calcium, which is essential for maintaining cell wall structure and function in developing fruit tissues. Cork spot development can be exacerbated by various environmental stresses, including fluctuating soil moisture and temperature, which can disrupt the uptake and transport of calcium and other nutrients within the plant (Brasier, 2008). Managing cork spot requires a multifaceted approach, including soil and foliar calcium applications, careful irrigation practices to maintain consistent soil moisture, and selecting resistant cultivars through breeding programs (Blevins and Lukaszewski, 1998). Advancements in nanotechnology could provide new methods for precision delivery of nutrients such as calcium, possibly reducing nutrient-related disorders and improving fruit integrity and plant resilience (Ghormade et al., 2011). To develop effective strategies to prevent and control cork spots in fruit crops, it is essential to understand the complex interactions between plant nutrition, environmental factors, and genetic predisposition. Research is ongoing into the physiological mechanisms underlying the disorder, providing insights for improving fruit quality and yield (Matsumoto et al., 2018).

13.7.1 STAGE OF OCCURRENCE

Cork spot, particularly in apples, is a physiological disorder that can manifest during multiple developmental stages of the fruit (Matsumoto et al., 2018). It is generally observed at key points in the ripening process of the fruit, although the specific timing may vary. Cork spots can begin to develop during the early stages of fruit growth. A consistent supply of nutrients, especially calcium, is required to ensure proper cell wall development and function during this critical period of rapid fruit expansion and differentiation. Disruptions in the availability of nutrients during this stage can predispose the fruit to cork spot development (Blevins and Lukaszewski, 1998). Indeed, as the fruit matures on the tree, any existing nutrient imbalance or environmental stress may aggravate the development of the cork spot. Ripening involves major physiological changes in fruit, including changes in metabolism and water content, which can further stress fruit tissues if the nutritional supply is inadequate (Turker et al., 2017). Once harvested, the fruit is particularly susceptible to cork spot, especially when stored. Cork spots can develop or worsen due to the lack of a continuous supply of nutrients from the tree and the stress of storage conditions. Careful management is needed to prevent postharvest losses during this critical stage in maintaining fruit quality (Denardi et al., 2023).

13.7.1.1 CALCIUM DEFICIENCY IN FRUIT TISSUE

Fruit tissue calcium deficiency is a recognized contributor to cork spots in several fruits, including apples. Studies have shown a lack of calcium, especially during early development, can significantly increase the risk of these problems. Underscoring the importance of calcium in maintaining fruit integrity, a study by Cui et al. (2021) on "Akituki" pear showed that application of a calcium uptake inhibitor resulted in an increased incidence of cork spot. These results are consistent with what was observed in other fruits where irregular calcium distribution or deficiencies were associated with pip formation. It is also important to note that calcium deficiencies are not solely responsible for this condition. In addition, environmental factors such as light exposure, as Matsumoto et al. (2018) found in their study of "Kurenainoyume" apples, where opaque bags were effective in mitigating corking. This suggests that for effective management and prevention of cork spot-like disorders in orchards, calcium plays a critical role, but a holistic approach that considers all contributing factors is needed.

13.7.1.2 OTHER INFLUENCES ON CALCIUM DEFICIENCY

The role of calcium in fruit development goes beyond that of a nutrient. It is a secondary messenger in cellular signaling and provides structural stability to cell walls and membranes. Moreover environmental and physiological factors, such as water uptake and nutrient transport in the xylem, influence the acquisition of calcium by the roots and its transport to the shoot. These factors highlight the complexity of calcium transport and its critical role in the prevention of disorders such as cork spot (Mayorga-Gómez et al., 2020).

13.7.1.3 UNEVEN NUTRIENT DISTRIBUTION

Nutrient distribution, especially calcium, for preventing corking. Research has shown that uneven calcium distribution is often the cause of cork spot disorder in citrus fruits, particularly orange trees. Methods such as foliar calcium sprays have been studied to address this (Saunt, 2000). In apples, cork spot is often associated with uneven water supply, which affects calcium uptake and distribution (Ferguson, 1989). In watermelons, inconsistent irrigation patterns and calcium transport problems have been associated with cork spots (Robinson and Saúco, 2010). Nutritional disorders are less common in bananas, but can still have an impact on quality, with the emphasis usually being on balanced fertilizer application to maintain an even supply of nutrients (Siddiq, 2012). Furthermore, highlighting the complex interplay between nutrient availability and plant health, marginal leaf scorch in bananas, similar to cork spot, was associated with boron toxicity exacerbated by low calcium availability in Taiwanese fruit crops (Oberly, 1976).

13.7.2 IMPORTANCE

In general, cork does not adversely affect the edibility or the nutritional value of the fruit. However, the development of unsightly spots on the surface of the fruit can significantly affect consumer perception and purchasing decisions (Matsumoto et al., 2018; Tarancón et al., 2021). This has a significant economic impact on the agricultural industry and may significantly reduce the market price of produce (Jijakli and Lepoivre, 2006).

The presence of cork spots on fruits such as apples and pears may cause significant economic losses to the producers. This loss results from the reduced market value of the affected fruits because they do not meet the aesthetic requirements necessary for higher market prices. For high-value crops, where appearance plays an important role in consumer choice, the economic impact is particularly severe (Djekic et al., 2019). Nevertheless, the appearance of cork spots requires a rigorous sorting process in the supply chain, which increases the cost of labor and resources. Blemished fruit is often relegated to lower-priced markets or processed products, resulting in a significant price differential compared with unblemished fruit. This pricing effect ripples throughout the supply chain, impacting growers, distributors, and retailers (Jijakli and Lepoivre, 2006; Njaya, 2014).

Additionally, the appearance of the fruit is a key factor in the consumer's purchase decision. Studies indicate that, irrespective of internal quality or flavor, most consumers are less likely to buy fruit with visible defects such as cork stains. The reason for this behavior is the association of physical defects with a reduction in quality or freshness (Hayati et al., 2023). Consumers associate the presence of cork spots with reduced freshness or lower quality. While this perception is not always accurate, it can discourage consumers from purchasing the fruit in question, which can affect overall sales and consumer satisfaction (Poole et al., 2007). The challenge for the industry is to inform consumers about the nature of these blemishes and their actual effect on the quality of the fruit.

13.7.3 SYMPTOMS

Cork spot disorder is a physiological disorder of fruits and vegetables. It is characterized by a variety of symptoms that primarily affect the epidermal and subepidermal tissues of the fruit necrotic spots on the surface of the fruit, often resulting in significant economic losses due to reduced market value (Blevins and Lukaszewski, 1998).

The earliest sign is when tiny, dark green, sunken spots appear on the fruit skin. These spots gradually increase in size and become brown to black in color. They have a corky texture (Jijakli and Lepoivre, 2006). Spots are generally irregularly shaped and may coalesce to form larger necroses (Kays, 1999).

Deeper below the surface, affected tissues appear brownish-discolored, indicating cellular destruction (Blevins and Lukaszewski, 1998). Histological examination reveals disruption of cell structure and suberization in the affected areas (Klinkowski, 1970). This suberization process, which results in the formation of cork-like tissue, is a defensive response of the plant to stress and injury. It should be noted that in severe cases, the disorder can penetrate deeper into the flesh of the fruit, resulting in an internal browning and a mealy texture (González-Aguilar et al., 2000). This not only affects the aesthetic appearance of the fruit but also affects the taste and texture of the fruit, making it less appetizing (Jijakli and Lepoivre, 2006).

13.7.4 PHYSIOLOGY

Cork spot in fruits and vegetables is a complex physiological disorder state that involves various aspects of plant biology, from nutrient uptake to cell structure and function (Matsumoto et al., 2018). Nevertheless, the physiological study of boron deficiencies in plants (Reed, 1947) highlights the essential role of trace elements like boron in plant health. A deficiency of boron is often associated with the development of cork-like symptoms in fruit crops. Boron insufficiency can cause cell wall instability and altered cell functions that contribute to the formation of corky tissue.

Cork disorder has a major impact on the process by which fruit breathe and produce ethylene. A key plant hormone, plays a central role in fruit ripening and senescence. Altered ethylene levels in fruit affected by cork spots can result in altered ripening patterns and reduced fruit quality (Kader, 1985; Zhang and Cui, 2023). In fact, calcium is essential for the maintenance of cell wall structure and function, and in cork spot disorder, calcium deficiency is a common feature of fruit. In cork spots, calcium deficiency can lead to the weakening of the cell walls, making them susceptible to necrosis and corky tissue formation (Nxumalo et al., 2019). The relationship between calcium and ethylene is complex (Zhang and Cui, 2023). Calcium can modulate the production and action of ethylene in fruit, and any imbalance in calcium nutrition can upset this balance, leading to abnormal ethylene development. Similarly, Al-Ani (1978) well illustrated the relationship between mineral nutrition and cork spot. This research is of crucial importance for the understanding of how mineral imbalances, and especially calcium imbalances, can influence the development of symptoms similar to those of cork disorder.

In addition, the researchers found an altered expression of calcium transport-related genes and an uneven distribution of intracellular calcium in the affected fruits. This suggests that there is a dysfunction of both long-range and transmembrane calcium transport channels, which may be responsible for the uneven distribution of calcium within the fruit and may contribute to the development of cork spot disorder, the same research showed that the size of pores in corked fruits is significantly larger than in healthy fruits, indicating that tissue damage may precede the development of corking. These findings suggest underlying structural defects predisposing the fruit to the disorder (Cui et al., 2021).

13.7.5 CONTROL

13.7.5.1 MANAGING CALCIUM

Managing calcium during fruit development is a multifaceted approach that involves not only applying calcium at critical growth stages but also balancing calcium with other nutrients such as nitrogen. Several studies have shown that the incidence of physiological disorders, such as cork spots in various fruit crops can be significantly reduced by proper calcium nutrition (Nxumalo et al., 2019). In order to prevent calcium-related disorders, calcium can be supplied to the plant through the soil as well as through foliar applications. The most commonly used foliar applications are calcium chloride and calcium nitrate. Because calcium must be available at key stages of fruit development, the timing and frequency of these applications are critical (Rebouças et al., 2015; Bitange et al., 2023).

13.7.5.2 BALANCED FERTILIZATION

Balanced fertilization plays a key role in preventing physiological disorders such as cork spot in orchard preventing physiological disorders such as cork spot on fruit trees (Lather et al., 2021), because it ensures that no single nutrient is dominant within the soil composition, thus avoiding competition between nutritional elements, particularly between calcium and elements such as nitrogen and potassium. Adequate calcium availability is especially important in conjunction with high nitrogen or potassium fertilization, preventing cell wall weakening and maintaining overall plant health (Nava et al., 2022). To minimize the risk of nutrient imbalances and associated disorders, regular soil testing and analysis are recommended to tailor fertilization practices to the specific needs of the crop (Lather et al., 2021).

13.7.5.3 ADEQUATE IRRIGATION

Proper irrigation keeps essential nutrients soluble and available for root uptake. For example, in their study of peppers, Ningoji et al. (2023) demonstrated that sensor-guided irrigation at optimal soil moisture, combined with appropriate fertigation, significantly increased nutrient uptake and fruit yield. These results highlight the importance of maintaining a balanced soil moisture balance, which facilitates nutrient solubility and uptake by plant roots. Furthermore, Nava et al. (2022) highlighted the effectiveness of drip irrigation in the management of saline water and emphasized that appropriate irrigation methods can optimize nutrient availability even in difficult conditions.

Once nutrients are absorbed, an adequate water supply is essential for their transport throughout the plant. Santos et al. (2021) found in their study that optimal soil water tension was crucial for both crop development and nutrient uptake. This indicates that consistent soil moisture not only aids in nutrient uptake but also in their efficient distribution within the plant, which is vital for overall plant health and fruit quality. Furthermore, Vivaldi et al. (2017), in their research on a nectarine orchard irrigated with treated municipal wastewater, observed that the water source significantly affected nutrient uptake and fruit quality. This study illustrates how irrigation water quality can influence the nutrient dynamics within the plant, affecting fruit development and the incidence of disorders like cork spots.

13.8 CONCLUSION

Fruit and vegetables have high nutritional value but are prone to internal and external damage, mostly from several physiological disorders that cause deterioration and lower commercial quality. These various lesions result from metabolic imbalances brought on, among other things, caused by factors such as extreme temperature, mineral imbalances, or changes in the composition of the atmosphere. Physiological disorders cause disturbances that are distinct from those caused by fungal, bacterial, viral, or insect agents. Some physiological disorders appear prior to harvesting, while others develop after harvest. Although much research has been done on ways to lessen the negative effects of cold temperatures on fruits.

KEYWORDS

- **chilling injury**
- **cork**
- **cracking**
- **freezing injury**
- **fruits crops**
- **heat injury**
- **postharvest losses**
- **postharvest physiological disorders**

REFERENCES

Aghdam, M. S., & Bodbodak, S. (2014). Postharvest heat treatment for mitigation of chilling injury in fruits and vegetables. *Food and Bioprocess Technology*, *7*(1), 37–53. https://doi.org/10.1007/s11947-013-1207-4.

Al-Ani, A. M. (1978). Postharvest physiology of "Anjou" pear fruit: Relations between mineral nutrition and cork spot, respiration, and ethylene evolution. *Oregon State University Library*.

Alabi, K. P., Olalusi, A. P., Olaniyan, A. M., Fadeyibi, A., & Gabriel, L. O. (2022). Effects of osmotic dehydration pretreatment on freezing characteristics and quality of frozen fruits and vegetables. *Journal of Food Process Engineering*, *45*(8), e14037. https://doi.org/10.1111/jfpe.14037

Alabi, K. P., Zhu, Z., & Sun, D.-W. (2020). Transport phenomena and their effect on microstructure of frozen fruits and vegetables. *Trends in Food Science & Technology*, *101*, 63–72. https://doi.org/10.1016/j.tifs.2020.04.016

Albornoz, K., Zhou, J., Yu, J., & Beckles, D. M. (2022). Dissecting postharvest chilling injury through biotechnology. *Current Opinion in Biotechnology*, *78*, 102790. https://doi.org/10.1016/j.copbio.2022.102790

Alkio, M., Jonas, U., Sprink, T., van Nocker, S., & Knoche, M. (2012). Identification of putative candidate genes involved in cuticle formation in Prunus avium (sweet cherry) fruit. *Annals of Botany*, *110*(1), 101–112. https://doi.org/10.1093/aob/mcs087

Aung, M. M., & Chang, Y. S. (2023). Cold chain management essentials. In Springer *series in advanced manufacturing* (pp. 45–76). Springer. https://doi.org/10.1007/978-3-031-09567-2_4

Balbontín, C., Ayala, H., Bastías, R. M., Tapia, G., Ellena, M., Torres, C., Yuri, J. A., Quero-García, J., Ríos, J. C., & Silva, H. (2013). Cracking in sweet cherries: A comprehensive review from a physiological, molecular, and genomic perspective. *Chilean Journal of Agricultural Research*, *73*(1), 66–72. https://doi.org/10.4067/S0718-58392013000100010

Balbontín, C., Gutiérrez, C., Wolff, M., & Figueroa, C. R. (2018). Effect of abscisic acid and methyl jasmonate preharvest applications on fruit quality and cracking tolerance of sweet cherry. *Chilean Journal of Agricultural Research*, *78*(3), 438–446. https://doi.org/10.4067/S0718-58392018000300438

Bangar, S. P., Trif, M., Ozogul, F., Kumar, M., Chaudhary, V., Vukic, M., Tomar, M., & Changan, S. (2022). Recent developments in cold plasma-based enzyme activity (browning, cell wall degradation, and antioxidant) in fruits and vegetables. *Comprehensive Reviews in Food Science and Food Safety*, *21*(2), 1958–1978. https://doi.org/10.1111/1541-4337.12895

Bantayehu, M. (2017). Postharvest losses assessment of tropical fruits in the market chain of north western Ethiopia. *Journal of Food Science and Quality Management*, 66, 13–24..

Bitange, N., Chemining'wa, G., Ambuko, J., & Owino, W. (2023). Effect of varied calcium formulations and time of application on postharvest quality and organoleptic acceptability of mango fruits. *African Journal of Food, Agriculture, Nutrition and Development*, *23*(03), 22871–22892. https://doi.org/10.18697/ajfand.118.22485

Blanco, V., Blaya-Ros, P. J., Torres-Sánchez, R., & Domingo, R. (2022). Irrigation and crop load management lessen rain-induced cherry cracking. *Plants*, *11*(23), 3249. https://doi.org/10.3390/plants11233249

Blevins, D. G., & Lukaszewski, K. M. (1998). Boron in plant structure and function. *Annual Review of Plant Physiology and Plant Molecular Biology*, *49*(1), 481–500. https://doi.org/10.1146/annurev.arplant.49.1.481

Bokhary, S. U. F., Wang, L., Zheng, Y., & Jin, P. (2020). Pre-storage hot water treatment enhances chilling tolerance of zucchini (Cucurbita pepo L.) squash by regulating arginine metabolism. *Postharvest Biology and Technology*, *166*, 111229. https://doi.org/10.1016/j.postharvbio.2020.111229

Boller, T. (2018). Ethylene in pathogenesis and disease resistance. In *The plant hormone ethylene* (pp. 293–314). CRC Press. https://doi.org/10.1201/9781351075763-16.

Borve, J., Sekse, L., & Stensvand, A. (2000). Cuticular fractures promote postharvest fruit rot in sweet cherries. *Plant Disease*, *84*(11), 1180–1184. https://doi.org/10.1094/PDIS.2000.84.11.1180

Brasier, C. M. (2008). The biosecurity threat to the UK and global environment from international trade in plants. *Plant Pathology*, *57*(5), 792–808. https://doi.org/10.1111/j.1365-3059.2008.01886.x

Brüggenwirth, M., & Knoche, M. (2016a). Factors affecting mechanical properties of the skin of sweet cherry fruit. *Journal of the American Society for Horticultural Science*, *141*(1), 45–53. https://doi.org/10.21273/JASHS.141.1.45

Brüggenwirth, M., & Knoche, M. (2016b). Mechanical properties of skins of sweet cherry fruit of differing susceptibilities to cracking. *Journal of the American Society for Horticultural Science*, *141*(2), 162–168. https://doi.org/10.21273/JASHS.141.2.162

Brüggenwirth, M., & Knoche, M. (2017). Cell wall swelling, fracture mode, and the mechanical properties of cherry fruit skins are closely related. *Planta*, *245*(4), 765–777. https://doi.org/10.1007/s00425-016-2639-7

Brummell, D. A. (2006). Cell wall disassembly in ripening fruit. *Functional Plant Biology*, *33*(2), 103–119. https://doi.org/10.1071/FP05234

Chabbal, M. D., Yfran-Elvira, M. D. M., & Llarens-Beyer, L. A. (2020). Control of fruit cracking in clementino mandarin plants. *Cultivos Tropicales*, 41(4), e06.

Charoenrein, S., & Owcharoen, K. (2016). Effect of freezing rates and freeze-thaw cycles on the texture, microstructure and pectic substances of mango. *International Food Research Journal*, *23*, 613–620. https://api.semanticscholar.org/CorpusID:99376496

Chaves, A., & Zaritzky, N. (2018). Cooling and freezing of fruits and fruit products. In *Fruit preservation* (pp. 127–180). Springer. https://doi.org/10.1007/978-1-4939-3311-2_6.

Chen, J., Liu, L., Chen, J., & Zhang, H. (2014). Effects of various calcium treatments on fruit cracking and cell wall enzyme activities in navel orange. *Journal of South China Agricultural University*, *35*(6), 29–32.

Colin-Henrion, M. (2008). De la pomme à la pomme transformée: Impact du procédé sur deux composés d'intérêt nutritionnel. Caractérisation physique et sensorielle des produits transformés. *Université d'Angers*.

Coser, W., Steffens, C. A., Lugaresi, A., Buss, M., Gularte, P. S., Fenili, C. L., Miqueloto, T., & Talamini do Amarante, C. V. (2023). Treatment of 'Baigent' apple fruit with nitric oxide in controlled atmosphere storage reduces decay and physiological disorders. *Postharvest Biology and Technology*, *196*, 112159. https://doi.org/10.1016/J.POSTHARVBIO.2022.112159

Cui, Z., Wang, N., Duan, Y., Xu, X., Wang, R., Zhang, S., & Ma, C. (2021). High-resolution microstructure analysis of cork spot disordered pear fruit "Akizuki" (Pyrus pyrifolia Nakai) using X-ray CT. *Frontiers in Plant Science*, *12*. https://doi.org/10.3389/fpls.2021.715124

Cui, K., Zhao, H., Sun, L., Yang, L., Cao, J., & Jiang, W. (2019). Impact of near freezing temperature storage on postharvest quality and antioxidant capacity of two apricot (Prunus armeniaca L.) cultivars. *Journal of Food Biochemistry*, *43*(7), e12857. https://doi.org/10.1111/jfbc.12857

De Freitas, S.T., & Pareek, S. (2019). Postharvest physiological disorders in fruits and vegetables. In *Postharvest physiological disorders in fruits and vegetables* (pp. 3–14). CRC Press.

Denardi, F., Kvitschal, M. V., Argenta, L. C., Couto, M., & Araújo, L. (2023). SCS443 ISADORA: Late ripening apple cultivar with very high fruit storage ability. *Revista Brasileira de Fruticultura*, *45*. https://doi.org/10.1590/0100-29452023161

Di Mascio, P., Kaiser, S., & Sies, H. (1989). Lycopene as the most efficient biological carotenoid singlet oxygen quencher. *Archives of Biochemistry and Biophysics*, *274*(2), 532–538.

Djekic, I., Radivojevic, D., & Milivojevic, J. (2019). Quality perception throughout the apple fruit chain. *Journal of Food Measurement and Characterization*, *13*(4), 3106–3118. https://doi.org/10.1007/s11694-019-00233-1

Doyle, R. W., Lal, K. K., & Virapat, C. (2019). Domestication and genetic improvement: balancing improved production against increased disease risks from inbreeding. *Revue Scientifique et Technique de l'OIE*, *38*(2), 615–628. https://doi.org/10.20506/rst.38.2.3008

El-Bolok, T., & Sawarsan, M. (2017). Effect of calcium and some nutrients on pomegranate fruit cracking under souhag governorate conditions. *Egyptian Journal of Agricultural Sciences*, *68*(3), 305–318. https://doi.org/10.21608/ejarc.2017.212749

El-Ramady, H. R., Domokos-Szabolcsy, É., Abdalla, N. A., Taha, H. S., & Fári, M. (2015). Postharvest management of fruits and vegetables storage. In E. Lichtfouse (Ed.), *Sustainable agriculture reviews* (Vol. 15, pp. 65–152). Springer. https://doi.org/10.1007/978-3-319-09132-7_3

Falcomer, A. L., Riquette, R. F. R., de Lima, B. R., Ginani, V. C., & Zandonadi, R. P. (2019). Health benefits of green banana consumption: A systematic review. *Nutrients*, *11*(6), 1222.

Fallik, E., & Ilic', Z. (2019). Control of postharvest decay of fresh produce by heat treatments; the risks and the benefits. In *Postharvest pathology of fresh horticultural produce* (pp. 521–537). https://doi.org/10.1201/9781315209180-16/CONTROL-POSTHARVEST-DECAY-FRESH-PRODUCE-HEAT-TREATMENTS-RISKS-BENEFITS-ELAZAR-FALLIK-ZORAN-ILIC

FAO. (2019). In *The state of food and agriculture 2019. Moving forward on food loss and waste reduction* (pp. 2–13). FAO.

FAOSTAT. (2023a). FAOSTAT database. Food and Agriculture Organization of the United Nations. https://www.fao.org/faostat/en/#home

FAOSTAT. (2023b). FAOSTAT crops and livestock products. Food and Agriculture Organization of the United Nations. https://www.fao.org/faostat/en/#data/QCL

Farooq, U., Shafi, A., Akram, K., & Hayat, Z. (2020). Fruits and nutritional security. In M. S. Siddiqui, S. A. Raza, & M. Y. Khan (Eds.), *Fruit crops* (pp. 1–12). Elsevier. https://doi.org/10.1016/B978-0-12-818732-6.00001-0

Ferguson, S. (1989). Calcium in plant senescence and fruit ripening. In J. Janick (Ed.), *Horticultural reviews* (Vol. 11, pp. 197–232). Wiley. https://doi.org/10.1002/9781118060841.ch6

Firmansyah, & Argosubekti, N. (2020). A review of heat stress signaling in plants. In *IOP Conference Series: Earth and Environmental Science* (Vol. 484). https://doi.org/10.1088/1755-1315/484/1/012041

Fischer, G., Balaguera-López, H. E., & Álvarez-Herrera, J. (2021). Causes of fruit cracking in the era of climate change. A review. *Agronomía Colombiana*, *39*(2), 196–207. https://doi.org/10.15446/agron.colomb.v39n2.97071

Fish, W. W., & Davis, A. R. (2003). The effects of frozen storage conditions on lycopene stability in watermelon tissue. *Journal of Agricultural and Food Chemistry*, *51*(12), 3582–3585. https://doi.org/10.1021/jf030022f

Franck, C., Lammertyn, J., Ho, Q. T., Verboven, P., Verlinden, B., & Nicolaï, B. M. (2007). Browning disorders in pear fruit. *Postharvest Biology and Technology*, *43*(1), 1–13. https://doi.org/10.1016/J.POSTHARVBIO.2006.08.008

Fruk, G., Cmelik, Z., Jemric, T., Hribar, J., & Vidrih, R. (2014). Pectin role in woolliness development in peaches and nectarines: A review. *Scientia Horticulturae*, *180*, 1–5. https://doi.org/10.1016/j.scienta.2014.09.042

Ghormade, V., Deshpande, M. V., & Paknikar, K. M. (2011). Perspectives for nano-biotechnology enabled protection and nutrition of plants. *Biotechnology Advances*, *29*(6), 792–803. https://doi.org/10.1016/j.biotechadv.2011.06.007

Gierson, D., & Kader, A. A. (1986). Fruit ripening and quality. In *The tomato crop* (pp. 241–280). https://doi.org/10.1007/978-94-009-3137-4_6

Glenn, G. M., & Poovaiah, B. W. (1989). Cuticular properties and postharvest calcium applications influence cracking of sweet cherries. *Journal of the American Society for Horticultural Science*, *114*(5), 781–788. https://doi.org/10.21273/JASHS.114.5.781

González-Aguilar, G. A., Fortiz, J., Cruz, R., Baez, R., & Wang, C. Y. (2000). Methyl jasmonate reduces chilling injury and maintains postharvest quality of mango fruit. *Journal of Agricultural and Food Chemistry*, *48*(2), 515–519. https://doi.org/10.1021/jf9902806

Goodwin, I., McClymont, L., & Green, S. (2022). The effects of water deficits on fruit cracking and sunburn damage in 'Cripps Pink' apple. *Acta Horticulturae*, *1335*, 421–428. https://doi.org/10.17660/ActaHortic.2022.1335.52

Hayati, N. Q., Sayekti, A., Sulistyaningrum, A., Waryat, W., Yusuf, A. T., Wiratno, A. T., Udiarto, B. K., & Puspitasari. (2023). Consumer preferences of vegetables and fruits at farmer's household level on the potential development of horticultural technology innovations in Central Kalimantan's Food Estate (Case Study in Village of Belanti Siam, District of Pandih Batu, Regency of Pulang Pisau, Central Kalimantan).In *IOP Conference Series: Earth and Environmental Science* Vol. 1153(1), 012031. https://doi.org/10.1088/1755-1315/1153/1/012031

Huai, B., Wu, Y., Liang, C., Tu, P., Mei, T., Guan, A., Yao, Q., Li, J., & Chen, J. (2022). Effects of calcium on cell wall metabolism enzymes and expression of related genes associated with peel creasing in Citrus fruits. *PeerJ*, *10*, e14574. https://doi.org/10.7717/peerj.14574

Huang, H., Wang, L., Qiu, D., Zhang, N., & Bi, F. (2021). Changes of morphology, chemical compositions, and the biosynthesis regulations of cuticle in response to chilling injury of banana fruit during storage. *Frontiers in Plant Science*, *12*. https://doi.org/10.3389/fpls.2021.792384

Ikram, S., & Nafees, M. (2020). Causes and control of fruit cracking in pomegranate: A review. *Journal of Global Innovations in Agricultural and Social Sciences*, *8*(4), 183–190. https://doi.org/10.22194/JGIASS/8.920

Jackman, R. L., Yada, R. Y., Marangoni, A., Parkin, K. L., & Stanley, D. W. (1988). Chilling injury: A review of quality aspects. *Journal of Food Quality*, *11*(4), 253–278. https://doi.org/10.1111/j.1745-4557.1988.tb00887.x

Jijakli, M. H., & Lepoivre, P. (2006). State of the art and challenges of post-harvest disease management in apples. In S. A. M. H. Naqvi (Ed.), *Fruit and vegetable diseases* (pp. 59–94). Springer. https://doi.org/10.1007/0-306-48575-3_3

Joardder, M. U. H., & Masud, M. H. (2019). Food preservation in developing countries: Challenges and solutions. In *Food preservation in developing countries: Challenges and solutions* (pp. 1–245). https://doi.org/10.1007/978-3-030-11530-2/COVER

Kader, A. A. (1985). Ethylene-induced senescence and physiological disorders in harvested horticultural crops. *HortScience*, *20*(1), 54–57.

Kahramanoğlu, İ. (2017). Introductory chapter: Postharvest physiology and technology of horticultural crops. In *Postharvest handling* (pp. 1–5). IntechOpen. https://doi.org/10.5772/intechopen.69169

Kaur, M., & Kumar, M. (2020). An innovation in magnetic field assisted freezing of perishable fruits and vegetables: A review. *Food Reviews International*, *36*(8), 761–780. https://doi.org/10.1080/87559129.2019.1683746

Kays, S. J. (1999). Preharvest factors affecting appearance. *Postharvest Biology and Technology*, *15*(3), 233–247. https://doi.org/10.1016/S0925-5214(98)00088-X

Khadivi-Khub, A. (2015). Physiological and genetic factors influencing fruit cracking. *Acta Physiologiae Plantarum*, *37*(1), 1718. https://doi.org/10.1007/s11738-014-1718-2

Khalil, H. A., Abdelkader, M. F. M., Lo'ay, A. A., El-Ansary, D. O., Shaaban, F. K. M., Osman, S. O., Shenawy, I. E., Osman, H.-E. H., Limam, S. A., Abdein, M. A., & Abdelgawad, Z. A. (2022). The combined effect of hot water treatment and chitosan coating on mango (Mangifera indica L. cv. Kent) fruits to control postharvest deterioration and increase fruit quality. Coatings, 12(1), 83. https://doi.org/10.3390/coatings12010083

Khaliq, G., Ali, S., Gapper, N., & Nicola, S. (2023). Recent advances and approaches in the application of elicitors to enhance resistance mechanisms in fresh produce. Frontiers in Sustainable Food Systems, 6, 1061079.

Khan, L., Khaliq, G., Ullah, E., Abdul-Rahaman, A., El-Mogy, M. M., Kesba, H. H., Ismail, A. M., & El-Beltagi, H. S. (2025). Guggul gum and clove essential oil synergistically reduced anthracnose disease and modulated physicochemical changes in tomato during storage. Journal of Stored Products Research, 114, 102713.

Khan, N., & Mehmood, A. (2023). Revisiting climate change impacts on plant growth and its mitigation with plant growth promoting rhizobacteria. *South African Journal of Botany*, *160*, 586–601. https://doi.org/10.1016/j.sajb.2023.07.051

Khodabakhshian, R., Naeemi, A., & Bayati, M. R. (2021). Determination of texture properties of banana fruit cells with an atomic force microscope: A case study on elastic modulus and stiffness. *Journal of Texture Studies*, *52*(3), 389–399. https://doi.org/10.1111/jtxs.12594

Klinkowski, M. (1970). Catastrophic plant diseases. *Annual Review of Phytopathology*, *8*(1), 37–60. https://doi.org/10.1146/annurev.py.08.090170.000345

Knoche, M., & Grimm, E. (2008). Surface moisture induces microcracks in the cuticle of 'Golden Delicious' apple. *HortScience*, *43*(6), 1929–1931. https://doi.org/10.21273/HORTSCI.43.6.1929

Kobayashi, R., & Suzuki, T. (2019). Effect of supercooling accompanying the freezing process on ice crystals and the quality of frozen strawberry tissue. *International Journal of Refrigeration*, *99*, 94–100. https://doi.org/10.1016/j.ijrefrig.2018.11.045

Kumar, S., Lahlali, R., Liu, X., & Karunakaran, C. (2016). Infrared spectroscopy combined with imaging: A new developing analytical tool in health and plant science. *Applied Spectroscopy Reviews*, *51*(6), 466–483. https://doi.org/10.1080/05704928.2016.1157808

Kumar, K., Pinder, R., Jitender, D., Dabas, S. T. K. S., Yadav, B., & Rana, S. (2017). Effect of growth regulators and micronutrients on fruit cracking and fruit yield in pomegranate. *Indian Journal of Agricultural Research*, 51(3), 275–280. https://doi.org/10.18805/ijare.v0iOF.4575

Kumar, M., Singh, V., Jat, R., Ahamad, S., & Kumar, V. (2021). Preharvest fruit bagging for quality improvement in fruit crops: A review. *The Pharma Innovation Journal, 10*, 530–541.

Lane, W. D., Meheriuk, M., & Mckenzie, D.-L. (2000). Fruit cracking of a susceptible, an intermediate, and a resistant sweet cherry cultivar. *Hortscience, 35*(2), 239–242.

Lather, R., Vandana, Tallapragada, S., & Singh, G. (2021). Balanced fertilization for increasing and sustaining fruit quality and productivity-A review. *International Journal of Agricultural Sciences, 17*(2), 704–711. https://doi.org/10.15740/HAS/IJAS/17.2/704-711

Lee, Y. H. (2013). Effect of soil moisture management on the quality of wax apple. *African Journal of Biotechnology, 12*(42), 6107–6112. https://doi.org/10.5897/AJB10.094

Li, Y.-Y. (2010). Advances in research on the factors influencing fruit cracking. *Journal of Huazhong Agricultural* University, 29(4), 123–130.

Li, H., Liu, G., Tian, H., & Fu, D. (2021). Fruit cracking: A review. *Sheng wu gong cheng xue bao = Chinese journal of biotechnology, 37*(8), 2737–2752. https://doi.org/10.13345/j.cjb.200553

Li, Y., Xia, M., He, P., Yang, Q., Wu, Y., He, P., Ahmed, A., Li, X., Wang, Y., & Munir, S. (2022). Developing Penicillium digitatum management strategies on post-harvest citrus fruits with metabolic components and colonization of Bacillus subtilis L1-21. *Journal of Fungi, 8*(1), 80.

Lo'ay, A. A., & Dawood, H. D. (2019). Chilling injury, fruit color maturity stages, and antioxidant enzyme activities of lemon "baladi CV" fruits under cold storage stress. *Scientia Horticulturae, 257*, 108676. https://doi.org/10.1016/j.scienta.2019.108676

Lukatkin, A., Brazaityte, A., Bobinas, C., & Duchovskis, P. (2012). Chilling injury in chilling-sensitive plants: A review. *Agriculture, 99*(2), 111–124.

Luo, H., Ding, Z., Zhang, C., Zhao, L., Bi, R., Chen, Y., Wang, Y., & Hao, Z. (2021). Breeding report of a new big seed, mid-maturing and red peel pomegranate cultivar Yizhouhong. *Journal of Fruit Science, 38*(6). https://doi.org/10.13925/j.cnki.gsxb.20200506

Matsumoto, K., Kobayashi, T., Kougo, T., Fujita, T., Sato, S., & Moriguchi, T. (2018). Prevention of new cork spot-like physiological disorder in 'Kurenainoyume' apples by pre-harvest fruit bagging. *The Horticulture Journal, 87*(2), 174–183. https://doi.org/10.2503/hortj.OKD-117

Mayorga-Gómez, A., Nambeesan, S. U., Coolong, T., & Díaz-Pérez, J. C. (2020). Temporal relationship between calcium and fruit growth and development in bell pepper (Capsicum annuum L.). *HortScience, 55*(6), 906–913. https://doi.org/10.21273/HORTSCI14892-20

Mercier, S., Mondor, M., Villeneuve, S., & Marcos, B. (2018). The Canadian food cold chain: A legislative, scientific, and prospective overview. *International Journal of Refrigeration, 88*, 637–645. https://doi.org/10.1016/j.ijrefrig.2018.01.006

Mir, S. A., Shah, M. A., & Mir, M. M. (2018). Postharvest biology and technology of temperate fruits. Springer. https://doi.org/10.1007/978-3-319-76843-4

Mogollon, M. R., Jara, A. F., Contreras, C., & Zoffoli, J. P. (2020). Quantitative and qualitative VIS-NIR models for early determination of internal browning in 'Cripps Pink' apples during cold storage. *Postharvest Biology and Technology, 161*, 111060. https://doi.org/10.1016/j.postharvbio.2019.111060

Morales-Quintana, L., Waite, J. M., Kalcsits, L., Torres, C. A., & Ramos, P. (2020). Sun injury on apple fruit: Physiological, biochemical and molecular advances, and future challenges. *Scientia Horticulturae*, *260*, 108866. https://doi.org/10.1016/j.scienta.2019.108866

Mritunjay, S. K., & Kumar, V. (2015). Fresh farm produce as a source of pathogens: A review. *Research Journal of Environmental Toxicology*, *9*(2), 59–70. https://doi.org/10.3923/rjet.2015.59.70

Naschitz, S., Naor, A., Sax, Y., Shahak, Y., & Rabinowitch, H. D. (2015). Photo-oxidative sunscald of apple: Effects of temperature and light on fruit peel photoinhibition, bleaching and short-term tolerance acquisition. *Scientia Horticulturae*, *197*, 5–16. https://doi.org/10.1016/j.scienta.2015.11.003

Nava, G., Reisser Júnior, C., Parent, L.-É., Brunetto, G., Moura-Bueno, J. M., Navroski, R., Benati, J. A., & Barreto, C. F. (2022). Esmeralda peach (Prunus persica) fruit yield and quality response to nitrogen fertilization. *Plants*, *11*(3), 352. https://doi.org/10.3390/plants11030352

Ningoji, S. N., Mn, T., Mudalagiriyappa, Vasanthi, K. G., Tulja, S., & Shivaramu, H. S. (2023). Influence of automated sensor-based irrigation and fertigation on fruit yield, nutrient utilization and economics of capsicum (Capsicum annuum L.). *Communications in Soil Science and Plant Analysis*, *54*(15), 2126–2144. https://doi.org/10.1080/00103624.2023.2211608

Niu, J., Shi, Y., Huang, K., Zhong, Y., Chen, J., Sun, Z., Luan, M., & Chen, J. (2020). Integrative transcriptome and proteome analyses provide new insights into different stages of Akebia trifoliata fruit cracking during ripening. *Biotechnology for Biofuels*, *13*(1), 149. https://doi.org/10.1186/s13068-020-01789-7

Njaya, T. (2014). The economics of fruit and vegetable marketing by smallholder farmers in murehwa and mutoko districts in Zimbabwe. *International Journal of Research in Humanities and Social Sciences*, *1*(1), 35–43.

Nxumalo, K. A., Matsuane, C., & Masarirambi, M. T. (2019). Calcium-related post-harvest physiological disorders of fruits and vegetables in Eswatini: A review. *Current Journal of Applied Science and Technology*, *33*(6), 1–10. https://doi.org/10.9734/cjast/2019/v33i630107

Oberly, G. H. (1976). Observations and assessments of nutritional conditions and production practices of some fruit crops. 中國園藝, *22*(3), 97–105.

Olle, M., & Williams, I. H. (2017). Physiological disorders in tomato and some methods to avoid them. *The Journal of Horticultural Science and Biotechnology*, *92*(3), 223–230. https://doi.org/10.1080/14620316.2016.1255569

Ozores-Hampton, M., Stansly, P. A., & McAvoy, E. (2013). Evaluation of round and roma-type tomato varieties and advanced breeding lines resistant to tomato yellow leaf curl virus in Florida. *HortTechnology*, *23*(5), 689–698. https://doi.org/10.21273/HORTTECH.23.5.689

Palou, L., & Smilanick, J. L. (2019). Postharvest pathology of fresh horticultural produce. In *Postharvest pathology of fresh horticultural produce* (pp. 1–824). CRC Press. https://doi.org/10.1201/9781315209180

Paull, R. E., & Chen, N. J. (2000). Heat treatment and fruit ripening. *Postharvest Biology and Technology*, *21*(1), 21–37. https://doi.org/10.1016/S0925-5214(00)00162-9

Peet, M. M. (2009). Physiological disorders in tomato fruit development. *Acta Horticulturae*, *821*, 151–160. https://doi.org/10.17660/ActaHortic.2009.821.16

Petit, J., Bres, C., Mauxion, J.-P., Bakan, B., & Rothan, C. (2017). Breeding for cuticle-associated traits in crop species: Traits, targets, and strategies. *Journal of Experimental Botany*, *68*(19), 5369–5387. https://doi.org/10.1093/jxb/erx341

Pinheiro, J., Goncalves, E. M., & Silva, C. L. M. (2014). Alternative technologies for tomato post-harvest quality preservation. *CABI Reviews*, *2013*, 1–15.

Poole, N. D., Martínez, L. M. C., & Giménez, F. V. (2007). Quality perceptions under evolving information conditions: Implications for diet, health and consumer satisfaction. *Food Policy*, *32*(2), 175–188. https://doi.org/10.1016/j.foodpol.2006.05.004

Queb-González, D. B., Lopez-Malo, A., Sosa-Morales, M. E., & Villa-Rojas, R. (2020). Postharvest heat treatments to inhibit Penicillium digitatum growth and maintain quality of Mandarin (Citrus reticulata blanco). *Heliyon*, *6*(1), 0–7. https://doi.org/10.1016/j.heliyon.2020.e03166

Ray, P. K., & Solankey, S. S. (2022). In *Physiological disorders of horticultural crops: Causes and management*. NIPA. https://doi.org/10.59317/9789394490192

Rebouças, T. N. H., Porto, J. S., Jesus, J. S., & Moraes, M. O. B. (2015). Effects of different nitrogen sources and levels on tomato fruit quality. *Acta Horticulturae*, *1106*, 79–84. https://doi.org/10.17660/ActaHortic.2015.1106.13

Reed, H. S. (1947). A physiological study of boron deficiency in plants. *Hilgardia*, *17*(11), 377–411. https://doi.org/10.3733/hilg.v17n11p377

Ren, X. J., Yang, Z., Qifeng, Y., & GuangHui, P. (2014). Effects of different mineral elements on tomato fruit cracking. *Southwest China Journal of Agricultural Sciences*, *27*(1), 259–262.

Robinson, J., & Saúco, V. (2010). Bananas and plantains (2nd ed.). CABI. https://doi.org/10.1079/9781845936587.0000

Rodrigues, C., Gaspar, P. D., Simões, M. P., Silva, P. D., & Andrade, L. P. (2022). Review on techniques and treatments toward the mitigation of the chilling injury of peaches. *Journal of Food Processing and Preservation*, *46*(8), e14358. https://doi.org/10.1111/jfpp.14358

Romdhane, M. B., Haddar, A., Ghazala, I., Jeddou, K. B., Helbert, C. B., & Ellouz-Chaabouni, S. (2017). Optimization of polysaccharides extraction from watermelon rinds: Structure, functional and biological activities. *Food Chemistry*, *216*, 355–364. https://doi.org/10.1016/j.foodchem.2016.08.056

Salehi Sardoei, A., Alizadeh, A., & Fallah Imani, A. (2023). How that trends in north Iran citrus cultivars production correlate with evastating freeze events. *Agrotechniques in Industrial Crops,* 3(3), 104-110.https://civilica.com/doc/1779598.

Saltveit, M. E. (2018). Respiratory metabolism. *Postharvest Physiology and Biochemistry of Fruits and Vegetables*, *1*, 73–91. https://doi.org/10.1016/B978-0-12-813278-4.00004-X

Saltveit, M., Gross, K. C., & Wang, C. Y. (2016). The commercial storage of fruits, vegetables, and florist and nursery stocks (USDA Agriculture Handbook No. 66). United States Department of Agriculture. https://www.ars.usda.gov/ARSUserFiles/oc/np/CommercialStorage/CommercialStorage.pdf

Samad, M. A., Hashim, S. H., Simarani, K., Yaacob, J. S. (2016). Antibacterial properties and effects of fruit chilling and extract storage on antioxidant activity, total phenolic and anthocyanin content of four date palm (Phoenix dactylifera) Cultivars. *Molecules, 21*(4), 419. https://doi.org/10.3390/molecules21040419

Santos, V. K. S., Silva, E., Rocha, W. W., dos Santos, L. L., & Lima, R. C. (2021). Development and nutrient uptake by physalis under different soil water tensions. *Communications in Soil Science and Plant Analysis*, *52*(6), 576–585. https://doi.org/10.1080/00103624.2020.1862156

Saunt, J. (2000). In *Citrus varieties of the world* (2nd Ed.). Sinclair International Norwich.

Sandhu, S., & Bal, J. S. (2013). Quality improvement in lemon (Citrus limon (L.) Burm.) through integrated management of fruit cracking. *African Journal of Agricultural Research*, *8*(27), 3552–3557. https://doi.org/10.5897/AJAR2013.6876

Schumann, C., Winkler, A., & Knoche, M. (2022). Calcium decreases cell wall swelling in sweet cherry fruit. *Scientific Reports*, *12*(1), 16496. https://doi.org/10.1038/s41598-022-20266-9

Seo, H.-J., Sawant, S. S., & Song, J. (2022). Fruit cracking in pears: Its cause and management—A review. *Agronomy*, *12*(10), 2437. https://doi.org/10.3390/agronomy12102437

Sheikh, M. K., & Manjula, N. (2012). Effect of chemicals on control of fruit cracking in pomegranate (*Punica granatum* L.) var. Ganesh. *Acta Horticulturae*, 940, 131–133. https://doi.org/10.17660/ActaHortic.2012.940.19

Shewfelt, R. L., & Prussia, S. E. (2022). Challenges in handling fresh fruits and vegetables. In *Postharvest handling: A systems approach* (pp. 167–186). https://doi.org/10.1016/B978-0-12-822845-6.00006-3

Sicilia, A., Scialò, E., Puglisi, I., & Lo Piero, A. R. (2020). Anthocyanin biosynthesis and DNA methylation dynamics in sweet orange fruit [Citrus sinensis L. (Osbeck)] under cold stress. *Journal of Agricultural and Food Chemistry*, *68*(26), 7024–7031. https://doi.org/10.1021/acs.jafc.0c02360

Siddiq, M. (Ed.). (2012). In *Tropical and subtropical fruits*. Wiley. https://doi.org/10.1002/9781118324097

Sidhu, R. S., Bound, S. A., & Swarts, N. D. (2023). Internal flesh browning in apple and its predisposing factors—A review. *Physiologia*, *3*(2), 145–172. https://doi.org/10.3390/physiologia3020012

Singh, A., Shukla, A. K., & Meghwal, P. R. (2020a). Fruit cracking in pomegranate: Extent, cause, and management—A review. *International Journal of Fruit Science*, 20(sup3), S1234–S1253. https://doi.org/10.1080/15538362.2020.1784074

Singh, A., Shukla, A. K., & Meghwal, P. R. (2020b). Fruit cracking in pomegranate: Extent, cause, and management – A review. *International Journal of Fruit Science*, *20*(sup3), S1234–S1253. https://doi.org/10.1080/15538362.2020.1784074

Srinivasulu, B., Rao, G. S., & Singh, P. K. (2020). Physiological disorders of tomato and their management. *Journal of Pharmacognosy and Phytochemistry*, *9*(3), 2149–2150.

Statista. (2021). Volume de la production mondiale de fruits dans le monde en 2021. Statista. https://www.statista.com/statistics/673783/volume-of-global-fruit-production

Stojanović, M., Milatović, D., Kulina, M., & Alić-Džanović, Z. (2013). Susceptibility of sweet cherry cultivars to rain-induced fruit cracking in the region of Sarajevo. *Agroznanje* [*Agro-knowledge Journal*], 14(2), 179–186. https://doi.org/10.7251/AGREN1302179S

Súnico, V., Higuera, J. J., Molina-Hidalgo, F. J., Blanco-Portales, R., Moyano, E., Rodríguez-Franco, A., Muñoz-Blanco, J., & Caballero, J. L. (2021). The intragenesis and synthetic biology approach towards accelerating genetic gains on strawberry:

Development of new tools to improve fruit quality and resistance to pathogens. *Plants, 11*(1), 57. https://doi.org/10.3390/plants11010057

Tang, T., Zhang, M., & Mujumdar, A. S. (2022). Intelligent detection for fresh-cut fruit and vegetable processing: Imaging technology. *Comprehensive Reviews in Food Science and Food Safety*, *21*(6), 5171–5198. https://doi.org/10.1111/1541-4337.13039

Tarancón, P., Tárrega, A., González, M., & Besada, C. (2021). External quality of mandarins: Influence of fruit appearance characteristics on consumer choice. *Foods*, *10*(9), 2188. https://doi.org/10.3390/foods10092188

Tlili, I., Hdider, C., Lenucci, M. S., Riadh, I., Jebari, H., & Dalessandro, G. (2011). Bioactive compounds and antioxidant activities of different watermelon (Citrullus lanatus (Thunb.) Mansfeld) cultivars as affected by fruit sampling area. *Journal of Food Composition and Analysis*, *24*(3), 307–314. https://doi.org/10.1016/j.jfca.2010.06.005

Toivonen, P. M. A., & Brummell, D. A. (2008). Biochemical bases of appearance and texture changes in fresh-cut fruit and vegetables. *Postharvest Biology and Technology*, *48*(1), 1–14. https://doi.org/10.1016/j.postharvbio.2007.09.004

Tremblay, N., & Bélec, C. (2000). A synthesis of worldwide approaches for rationalizing nitrogen fertilization of vegetable crops. *HortScience*, 35(3), 422–429. https://DOI: 10.21273/HORTSCI.35.3.514E

Turker, D., Sezer, R., Isik, S., & Ozkan, K. (2017). Quality measurement of fruits and vegetables with near infrared camera. In *25th Signal Processing and Communications Applications Conference (SIU)*, pp. 1–4. https://doi.org/10.1109/SIU.2017.7960341

Uthman, A., & Garba, Y. (2023). Citrus mineral nutrition and health benefits: A review. In *Citrus research - Horticultural and human health aspects*. IntechOpen. https://doi.org/10.5772/intechopen.107495

Vivaldi, G. A., Stellacci, A. M., Vitti, C., Rubino, P., Pedrero, F., & Camposeo, S. (2017). Nutrient uptake and fruit quality in a nectarine orchard irrigated with treated municipal wastewaters. *Desalination and Water Treatment*, *71*, 312–320. https://doi.org/10.5004/dwt.2017.20564

Wang, C. Y. (1989). Chilling injury of fruits and vegetables. *Food Reviews International*, *5*(2), 209–236. https://doi.org/10.1080/87559128909540850

Wang, C. Y. (1994). Chilling injury of tropical horticultural commodities postharvest techniques for reducing CI. *Hortscience*, *29*(9), 986–988.

Wang, Q., Wu, X., Liu, L., Yao, D., Li, J., Fang, J., Chen, X., Zhu, L., Liu, P., Ye, Z., Jia, B., & Heng, W. (2021). Transcriptome and metabolomic analysis to reveal the browning spot formation of 'Huangguan' pear. *BMC Plant Biology*, *21*(1), 1–18. https://doi.org/10.1186/s12870-021-03049-8

Watkins, C. B. (2017). Postharvest physiological disorders of fresh crops. In *Encyclopedia of applied plant sciences* (pp. 315–322). Elsevier. https://doi.org/10.1016/B978-0-12-394807-6.00217-3

Wójcik, P., Akgül, H., Demirtaş, İ., Sarısu, C., Aksu, M., & Gubbuk, H. (2013). Effect of preharvest sprays of calcium chloride and sucrose on cracking and quality of 'Burlat' sweet cherry fruit. *Journal of Plant Nutrition*, *36*(9), 1453–1465. https://doi.org/10.1080/01904167.2013.793714

Yahaya, S. M. (2019). Review of post-harvest losses of fruits and vegetables. *Biomedical Journal of Scientific & Technical Research*, *13*(4). https://doi.org/10.26717/BJSTR.2019.13.002448

Yahia, E. M., Carrillo-López, A., & Bello-Perez, L. A. (2019a). Carbohydrates. In *Postharvest physiology and biochemistry of fruits and vegetables* (pp. 175–205). Elsevier. https://doi.org/10.1016/B978-0-12-813278-4.00009-9

Yahia, E. M., Carrillo-López, A., & Sañudo, A. (2019b). Physiological disorders and their control. In *Postharvest technology of perishable horticultural commodities* (pp. 499–527). https://doi.org/10.1016/B978-0-12-813276-0.00015-8

Yahia, E. M., Gardea-Béjar, A., Ornelas-Paz, J. de J., Maya-Meraz, I. O., Rodríguez-Roque, M. J., Rios-Velasco, C., Ornelas-Paz, J., & Salas-Marina, M. A. (2019c). Preharvest factors affecting postharvest quality. In *Postharvest technology of perishable horticultural commodities* (pp. 99–128). Elsevier. https://doi.org/10.1016/B978-0-12-813276-0.00004-3

Yu, D. J., & Lee, H. J. (2020). Evaluation of freezing injury in temperate fruit trees. *Horticulture, Environment, and Biotechnology*, *61*(5), 787–794. https://doi.org/10.1007/s13580-020-00264-4

Yu, J., Wang, R., Ma, W., Lei, S., Zhu, M., & Yang, G. (2023). Pectate Lyase Gene VvPL1 plays a role in fruit cracking of table grapes. *Journal of Agricultural and Food Chemistry*, *71*(3), 1643–1654. https://doi.org/10.1021/acs.jafc.2c05996

Zhang, X., & Cui, Z. (2023). Review of fruit cork spot disorder of Asian pear (Pyrus spp.). *Frontiers in Plant Science*, *14*, 1211451. https://doi.org/10.3389/fpls.2023.1211451

Zhang, W., Jiang, H., Cao, J., & Jiang, W. (2021). Advances in biochemical mechanisms and control technologies to treat chilling injury in postharvest fruits and vegetables. *Trends in Food Science and Technology*, *113*, 355–365. https://doi.org/10.1016/j.tifs.2021.05.009

Zhang, W., Pan, Y., Jiang, Y., & Zhang, Z. (2023). Advances in control technologies and mechanisms to treat peel browning in postharvest fruit. *Scientia Horticulturae*, *311*, 111798. https://doi.org/10.1016/J.SCIENTA.2022.111798

Zhu, H., Zhao, L., Zhang, X., Foku, J. M., Li, J., Hu, W., & Zhang, H. (2019). Efficacy of Yarrowia lipolytica in the biocontrol of green mold and blue mold in *Citrus reticulata* and the mechanisms involved. *Biological Control*, *139*, 104096. https://doi.org/10.1016/j.biocontrol.2019.104096

Zhuang, Y., Li, X., Hu, J., Xu, R., & Zhang, D. (2022). Expanding the gene pool for soybean improvement with its wild relatives. *aBIOTECH*, *3*(2), 115–125. https://doi.org/10.1007/s42994-022-00072-7

CHAPTER 14

Future Aspects of Postharvest Physiological Disorders

YASIR ABBAS SHAH[1*], MUHAMMAD AFZAAL[1], FARHAN SAEED[1], NOOR AKRAM[2], and ZARGHAM FAISAL[3]

[1]*Department of Food Science, Government College University, Faisalabad, Pakistan*

[2]*Department of Food and Nutrition, Government College University, Faisalabad, Pakistan*

[3]*Institute of Food Science and Nutrition, Bahauddin Zakariya University, Multan, Pakistan*

*Corresponding author

ABSTRACT

It is essential for a nation's political, social, and economic well-being to maintain a level of food production that satisfies the needs of its population at all times while also meeting the demand. The retail and consumption stages of the food supply chain account for the vast majority of food waste in industrialized countries, whereas the manufacturing and processing stages account for enormous amounts of waste in less developed nations. Increasing the amount of research, development, teaching, and training that occurs after harvest may assist in cutting down on losses like this. After the crop has been harvested, the quality of the storage can be significantly impacted, as can the longevity of the storage by improper regulation of temperature and humidity, packaging, and handling. Different innovative strategies can be implemented to reduce postharvest physiological disorders, such as cytokinin applications, ionizing radiation, edible coatings, and the use of colored

Physiological Disorders of Fruits and Vegetables: Causes and Remedies.
Ghulam Khaliq, Sajid Ali, and Mohammed Wasim Siddiqui (Eds.)

DOI: 10.1201/9781998511334-14

shade nets. In this chapter, different factors influencing postharvest quality are discussed, such as temperature, freezing, light, humidity, harvesting practices, handling, packaging, storage, and chemical and biochemical factors. Furthermore, different strategies to improve postharvest quality and reduce physiological disorders are summarized in this chapter.

14.1 INTRODUCTION

Postharvest physiological disorders might happen throughout the supply and handling chain. Before utilization, postharvest losses occur at different stages, including sorting and grading, processing, storage, marketing, and transportation to houses or markets, as well as at home before or after preparation (Strecker et al., 2022). These issues bring about the consumer's unacceptability of the food leading to wasting and discarding the food. Different issues, including high and low temperature, mineral imbalance, synthetic compounds such as ethylene, windy weather including hail, and different agriculture practices, are the causes of these disorders (Yahia et al., 2019). Fruits and vegetables are progressively known as a vital group for eradicating extreme poverty and hunger, improving food security and nutrition, and eventually contributing to the general enhancement of human health and well-being in the twenty-first century (Siddique and Gruber, 2021). Horticultural products are significant because they provide calories, macro- and micronutrients such as carbs, proteins, calcium, phosphorus, iron, magnesium, dietary fiber, and vitamins, as well as hundreds of phytochemicals that are vital for human nutrition and health. However, mishandling and improper treatment of vegetables and fruits during the postharvest phases reduces nutritional quality (Baltazari et al., 2020). Postharvest physiological disorders continue to pose significant challenges to the quality, shelf life, and marketability of horticultural produce. Despite advances in postharvest handling and storage technologies, many disorders still occur due to complex interactions among genetic factors, preharvest conditions, and postharvest environments. The chapter focuses on emerging trends, innovative technologies, and research priorities aimed at improving the understanding, prediction, and management of these disorders. Emphasis is placed on integrative approaches involving cultivar selection, precision postharvest management, non-chemical technologies, and advanced diagnostic tools to minimize losses and ensure sustainable postharvest systems. In this chapter, different physiological disorders and the factors affecting the postharvest quality of fruits and vegetables are discussed with reference to recent literature. Moreover, different strategies to improve postharvest physiological disorders are also summarized.

14.2 POSTHARVEST PHYSIOLOGICAL DISORDERS

Postharvest physiological disorders is one of the major and foremost issues with food security and world hunger in many nations and developing countries (Sisipo et al., 2021; Ali et al., 2022). In the future, fruit and vegetable loss can exceed 50%, making it one of the most serious challenges for feeding the global population (Searchinger et al., 2019). Fruit and vegetable postharvest losses are mostly caused by handling and storage after harvest, processing steps, distribution, and consumption (Dos Santos et al., 2020). To reduce the amount of postharvest loss, it is crucial to utilize the proper postharvest processing, packaging, transportation, and storage techniques. To cut postharvest losses, proper postharvest innovations and technologies must be used (Gardas et al., 2018; Gammage and Marangoni, 2025). Successful management of postharvest stages plays a significant influence in reducing postharvest losses, in addition to postharvest technologies (Rajapaksha et al., 2021; Mditshwa et al., 2023). Postharvest physiological disorders of fruits and vegetables are listed in Table 14.1.

TABLE 14.1 Postharvest Disorders of Fruits and Vegetables.

Disorder	Affected Commodity	Symptoms	References
Bitter pit	Apple	Small sunken pits, scattered browning of flesh	DeBrouwer et al., 2020; Torres et al., 2021
Black pit	Citrus fruits	Light brown spots later become dark brown and black pits	Sharma et al., 2021; Jyothi and Shilpa, 2020
Blackheart	Celery, cabbage	Tip burn, necrosis of leaves	Smith, 2020; Yahia et al., 2019
Blossom-end rot	Tomatoes, peppers, squash, melon, cucumber	Large brown to black dry area at the blossom end, dark almost black discoloration	Reitz and Mitcham, 2021; Al-Saloul et al., 2022
Bulb rot	Onion	Yellow to brown discoloration, softened neck	Khanal et al., 2022; Leach et al., 2020
Cork	Apple, pears	Small green depressions	Singh and Aravind, 2021; Cui et al., 2020

TABLE 14.1 *(Continued)*

Disorder	Affected Commodity	Symptoms	References
Crown rot	Banana	Blackening and softening of tissues, black to brown lesions on skin	Jahan et al., 2019; Murmu and Mishra, 2018
Dry rot	Potato	Sunken and wrinkled brown to black patches	Tiwari et al., 2020; Azil et al., 2021
Fruit cracking	Tomatoes, pomegranates, apricots, peaches, plums	Circular or deep cracks	Singh et al., 2020; Wang et al., 2021
Hollow stem	Broccoli	Small and elliptical cracks forming cavity	Frieman et al., 2021; Sharifi et al., 2020
Sour rot	Citrus fruits, tomato	Slimy and watery mass	François et al., 2022; Yasser et al., 2019
Spongy tissues	Mangoes	White and soft area like sponge	Ram et al., 2020; Singh, 2019
Tip burn and browning	Lettuce, cabbage, celery, cauliflower	Leaf margin turning brown and decaying	Maruo and Johkan, 2020; Wien and Stützel, 2020
Watery white rot	Apple, pear, peach, cherry, tomato, carrot	Small circular spots or blisters, depressed area	Srivastava et al., 2020; Shree and Kumari, 2019

14.2.1 PLANT GROWTH REGULATORS AND PHYSIOLOGICAL DISORDERS

Bioactive compounds such as cytokinins (CKs) belong to a chemically diverse family of plants that act as growth regulators and are used to control postharvest disorders and improve crops (Aremu et al., 2020). Cytokinin interact with other phytohormones and signaling networks to trigger and regulate biological activity at the cellular and organismal levels (Schäfer et al., 2015). As shown by the varied physiological and biochemical roles that occur throughout the life cycles of various species, CKs are thought to have the potential to impact a plant's development in the plant life cycle (embryogenesis to death) (Nguyen et al., 2021). CKs serve direct or indirect roles in plant physiology. Seed germination, flowering, shoot elongation, seed set, proliferation, fruiting, and senescence are regulated (Jameson and Song 2016). CKs have a key role in the production of horticultural crops, where their application affects morphology, nutritional content, harvesting, and yield in many fruit crops (Aremu et al., 2020). Growing data suggests that CKs reduce biotic and abiotic stress. CKs in plants entail sophisticated coordination via a wide range of cross-talk pathways, which regulate various physiological processes, including axillary shoot branching, apical dominance release, lateral root development, and root meristematic cell patterning (Koprna et al., 2016). Postharvest affects fruit production significantly. As fruit harvests are perishable, postharvest losses are common. CKs impact fruit quality and shelf life throughout ripening and senescence in planta and postharvest (Aremu et al., 2020; Yahia et al., 2019).

Although exogenous CKs are excellent plant regulators of ripening and senescence, their use can prevent chlorophyll degradation can successfully prolong senescence and enhance the quality of fruits that contain chlorophyll (Massolo et al., 2014). According to a study, the synthetic cytokinin 6-Benzylaminopurine (BAP), which promotes cell division, can promote plant growth and development. It has been shown that the use of BAP can prevent deterioration and browning while also extending the shelf life of harvested litchi (Zhang et al., 2018).

Phenylurea and gibberellins (GA3) were studied to maintain banana fruit quality and shelf life. Phenylurea and GA3 reduced fruit softening, delayed ethylene and respiration generation rates, and delayed hue value and maximum chlorophyll fluorescence of banana fruit during storage. Ascorbic acid, total phenol, and soluble reducing sugar losses were delayed. The current application led to the postharvest banana fruit quality being maintained, and its shelf life being extended in the most advantageous way (Huang et al., 2014). A cytokinin, benzyl adenine (BA) was used to examine the rind color and fruit quality of harvested calamondin fruit. The calamondin fruit's de-greening was postponed by BA treatment under both light and dark circumstances. Aside from ascorbic acid (AA) concentration, research reveals that treatment with BA after harvest and dark storage preserve the green rind color of calamondin fruit (Kawai et al., 2018).

For the first time, a study showed how 6-benzylaminopurine (BAP), a synthetic cytokinin, inhibits *Monilinia fructicola* invasion of peach fruit. It has been observed by the results that BAP therapy reduced sickness incidence by 63% and lesion breadth by ten times. Moreover, BAP boosted polyphenol oxidase and peroxidase activity, boosting host defenses. Results confirmed the notion that BAP controls *M. fructicola* by inhibiting growth, delaying senescence, and activating protective enzymes (Zhang et al., 2015). Such developments will require a wider range of horticulture fruit crops as novel models to better understand CKs application and their regulatory constraints in developmental pathways. This strategy seems crucial to increasing output and achieving food security and sustainable fruit production along with fewer postharvest disorders.

14.2.2 IONIZING RADIATIONS CONTROL POSTHARVEST DISORDERS

Postharvest disorders cause major losses worldwide. The global trend is toward environmentally friendly, healthier options, as well as treatments to prevent postharvest disorders losses (Xu et al., 2021). The use of pesticides has been the primary strategy for postharvest disease management up until this point. However, the extensive use of chemical fungicides has had unintended consequences, including environmental problems and the emergence of isolates that are fungicide resistant (Raveau et al., 2020). Ionizing radiation has been the subject of investigation as a possible application in a variety of postharvest treatment fields, particularly as a significant potential method for postharvest disease management. This is in response to the increasing demand for non-chemical therapies, such as ionizing radiation (Fan and Song, 2020). In order to control postharvest disorders, a number of technologically advanced procedures, such as irradiation combined with other treatments and created disease resistance, have been developed. These operations aim to improve the quality of fresh commodities and extend their shelf life (Zhang and Jiang, 2019).

Radiation with sufficient energy to break chemical bonds is referred to as ionizing radiation. This type of radiation is also referred to as high-energy particles or electromagnetic radiation (Ashfaq et al., 2020). Radio waves, X-rays, heat waves, ultraviolet light, infrared light, visible light, electron beams, and gamma rays are electromagnetic radiations. However, only electron beams, gamma rays, and X-rays are utilized professionally to preserve fresh products (Trivedi and Dikshit, 2021). To control pathogens in produce after harvest, it is essential to understand how they react physiologically to ionizing radiation. Ionizing radiation can destroy a pathogen's cell membrane and cause it to lose its cytoplasmic nutrition is proof of its antibacterial properties. Ionizing radiation can also harm people directly or indirectly (Ngolong Ngea et al., 2021).

Moreover, pathogens are sensitive to radiation, as this factor determines whether postharvest illnesses can be controlled by irradiation (Table 14.2). An organism's sensitivity is connected with the complication and the processes by which numerous cells can repair radiation damage. In this way, parasites and insects exhibit the greatest sensitivity to irradiation, whilst viruses exhibit the highest resilience (Jeong and Jeong, 2018). However, it is important to take into account the impact of the unavoidable

TABLE 14.2 Irradiation Combined Treatments to Enhance Postharvest Quality.

Crop	Treatment	Outcomes	References
Sweet cherry	Combined effect of UV-B radiation with chitosan coating for 28 days at 4°C	Inhibits weight loss of fruit, maintains stability of firmness, extends shelf life, and prevents undesirable postharvest changes	Abdipour et al., 2020
Lime fruit	Electron beam radiations combined with shellac coating for 30 days stored at 13°C	Maintains chlorophyll content, suppresses activity of chlorophyll degrading enzyme, increases ascorbic acid, and reduces titratable acidity	Pongsri et al., 2021
Pakchoi	Light-emitting diode LED irradiation	Positive respiratory rate, enhanced preservation effect, and extended shelf life	Zhang et al., 2022
Strawberry	UV-B radiations, aloe vera gel, and acetic acid	Lowers decay, maintains content of phenols and vitamins, improves postharvest quality of strawberry	Hosseinifarahi and Radi, 2018
Red raspberries	UV-C radiations along with passive modified atmosphere packaging at 6°C stored for 12 days	Delaying senescence and extending shelf life	Gimeno et al., 2021

mutations brought on by irradiation because they may result in higher, lower, or equivalent levels of pathogenicity (Vijg and Dong, 2020).

For the purpose of determining whether or not radiation is lethal to bacteria, the D10 value is utilized. This number reflects the dose required to render 90% of a population inactive (Ebrahim et al., 2022). To evaluate their radio sensitivity and calculate their D10 values, various species of postharvest pathogens have been exposed to several radiations. Additionally, compared to mature mycelia, young mycelia are more resistant. It is interesting to note that different irradiations cause fungal pathogens to respond to radiation differently, indicating that species-specific radio sensitivity may be connected to how those pathogens adapt to radiation stress (Jeong et al., 2015). According to international norms, radiation treatment has no negative effects on human health. The physical characteristics and nutritional value of fresh produce may be impacted by radiation treatment in addition to plant diseases. Irradiation is often used at dosages below 1 kGy since larger doses significantly alter the product's physical characteristics (Jaleel et al., 2018). Moreover, several international bodies and organizations have all endorsed irradiation as being safe and effective (Jeong and Jeong, 2018). However, this technology will be excellent for preventing postharvest disorders in fruits and vegetables and will eventually be used in export markets.

14.2.3 EDIBLE COATINGS

Stone fruits, which include apricots, nectarines, cherries, almonds, peaches, and plums, belong to the expansive and diverse family Rosaceae, together with raspberries, strawberries, blackberries, pears, apples, and roses (Potter, 2012). Due to their morphology, which includes a thin outer epicarp, an edible mesocarp, a fleshy and hard, lignified stone (endocarp) in the fruit's center that protects the seed, they are known as drupes (Canton et al., 2020). Stone fruits are in high demand due to their nutritional benefits and taste. South Africa, Chile, Spain, and Turkey are major exporters of stone fruit (Boonzaaier, 2015). However, a long handling chain causes export fruit to wilt, rot, lose weight, and over-ripen. The stone fruit industry is dependent on postharvest technologies to achieve maximum earnings from harvesting. Edible coatings reduce moisture loss and gas exchange, slowing down ripening and senescence. They help maintain fruit quality, extend shelf life, and reduce postharvest spoilage (Khaliq, 2015; Choi et al., 2016). Even though these technologies have been proven to be successful in prolonging the shelf life of stone fruit, their use still has certain drawbacks (Thakur et al., 2018). It has been suggested that the use of edible coatings as a postharvest technique can lower quality losses in various stone fruits including plums (Kumar et al., 2017), cherries (Mahfoudhi and Hamdi, 2015), apricots (Zhang et al., 2018), and peaches (Guillén et al., 2013).

Fruit coating has a long history. In China around the twelfth century, oranges and lemons were waxed to prevent moisture loss and enhance visual appeal (Aayush et al., 2022). Commercial use of wax to decrease postharvest fruit and vegetable losses began in 1922. Hydrocolloids have been utilized as packaging films, stabilizers, coating agents, thickeners, texture modifiers, and emulsifiers. Edible coatings improve food presentation and enhance fruit shelf life. Several types of research have shown the possible use of natural gums used in the formulation of edible coatings to improve postharvest qualities (Salehi, 2020). These waxes are detrimental to consumers, creating a demand for natural, sustainable coatings.

However, consumer demand for healthful, quality fruit prompted edible coverings (Ncama et al., 2018; Ul Haq et al., 2024). Many studies claim that applying a coating on stone fruit in conjunction with other postharvest methods can delay ripening and increase shelf life. But by producing a comparable changed environment effect, coatings might be able to replace the requirement for pricey and wasteful packaging materials like HDPE (High-Density Polyethylene) plastic bags (Maringgal et al., 2020; Vázquez-Celestino et al., 2016).

A study was conducted to determine whether or not plum fruit could be kept if a covering of carrageenan, rice starch, and sucrose fatty acid esters (FAEs) was applied to it. Before covering the fruit, many aspects of the film, including its barriers, surface shape, and physical and mechanical properties, were analyzed. Over the course of three weeks, measurements were taken of the fruit's weight loss, respiration rate, color change, ethylene production, firmness, titratable acidity (TA), and soluble solid content (SSC). It is possible to develop the RS-car-FAEs coating for commercial use as an innovative edible covering for the plum fruit market. The coating extends the plum fruit's shelf life and maintains its quality while it is being stored (Thakur et al., 2018).

Investigations were conducted on how chitosan coating affected the Santa Rosa plum's quality and shelf life. The plum fruits received a 2% chitosan-coating treatment. Results show that using chitosan as an edible covering after harvest can preserve plum quality and increase their storage life by up to 35 days when stored at moderate temperatures (Kumar et al., 2017). In the postharvest storage period, edible coatings might be a useful strategy for preventing the ripening of fruits and vegetables. It has been discovered that coating sweet cherries with 10% gum Arabic or almond gum can extend their shelf life and improve their postharvest quality. Furthermore, compared to uncoated control fruit, coatings were able to postpone changes in firmness, weight, soluble solids concentration, TA, and color development (Mahfoudhi and Hamdi, 2015).

Aloe arborescens gel was originally utilized on fruit in a study as an edible covering. *Aloe vera* or *Aloe arborescens* gels were applied to freshly picked peaches and plums, which were then left to mature at 20°C for six days. *Aloe arborescens* gel was originally utilized on fruit as an edible covering. Fruit that had been coated saw a significant delay in quality parameters associated with postharvest ripening of peach and plum. The effects were greater in plums, which had the highest rates of ethylene production (Guillén et al., 2013).

The current state of the technology used to apply edible coatings for postharvest uses in stone fruits should be the primary focus of research to be conducted in the future. Edible coatings play an important role in reducing postharvest physiological disorders by modifying the microenvironment around harvested fruits and vegetables (Ali et al., 2024; Khan et al., 2025). These coatings act as semi-permeable barriers to gases and water vapor, thereby reducing moisture loss, respiration rate, and oxidative stress, which are key factors involved in the development of disorders such as chilling injury, browning, and shriveling (Ali et al., 2022; Khaliq et al., 2023). In addition, edible coatings can help maintain membrane integrity, delay senescence, and preserve firmness and overall quality (Lal et al., 2025). When enriched with bioactive compounds, they may further enhance resistance to physiological stress, making edible coatings an effective and sustainable approach for postharvest disorder management.

14.2.4 USE OF COLORED SHADE NETS

Since fresh fruits and vegetables have exceptionally high nutritional value and numerous positive effects on health, their intake has seen a major uptick in recent years. However, even though climate change affects a variety of biotic and abiotic factors, farmers are unable to produce considerable amounts of high-quality vegetables and fruits (Munaweera et al., 2022). Abiotic disorders in postharvest are probably going to become more common due to the impacts of global warming, which are already beginning to manifest (Andreassen et al., 2021). To maximize plant production and quality, active manipulation of plants is frequently performed (Dueck et al., 2016; Lee et al., 2010).

Using colored shade nets to alter spectral quality can be a physiological technique for altering the microenvironment of a crop and enhancing plant growth along with productivity (Purbey et al., 2019). The productivity, quality, and phytochemical content of cultivated plants have currently improved through the use of photo-selective netting or films in horticulture. It is a technology that can be used in place of traditional methods to shield crops from harmful environmental factors including high sun radiation (Tinyane et al., 2013).

The intense solar radiation and high temperatures of summer affect the postharvest quality as well as the storage life of crops. The utilization of several light-dispersive and reflective chromatic additives within

TABLE 14.3 Photo-Selective Netting System Improves Postharvest Quality.

Color Nets	Application	Findings	References
Shade nets	Green houses	Improves environment for crop growth and enhances crop production	Ahemd et al., 2016
White, green, yellow, and red nets	Pepper cultivars	Enhances fruit protection from sunburn; white and red nets are more effective	Ombodi et al., 2015
Yellow net	Lettuce varieties	Prevents disorders and achieves higher yield	Ntsoane, 2015
Pearl net	Bell pepper crop	Lowers irradiation, air, and soil temperature	Díaz-Pérez et al., 2014
Red, pearl, and yellow nets	Sweet pepper crop	Maintains fruit quality and content of bioactive compounds, improves marketability of crop, and retains storage of aromatic compounds	Selahle et al., 2015
Red net	Sweet pepper crop	Reduces postharvest losses	Santana et al., 2012

the photo-selective nets assists in the transformation of direct light into dispersed light that can travel into the interior of the plant canopy (Sivakuma et al., 2018). The postharvest quality and storage life are both impacted by the excessive sun exposure and high temperatures that summer brings. The use of a variety of light-dispersive and reflecting chromatic additives in photo-selective nets makes it easier for direct light to be converted into scattered light, which can then travel into the inner plant canopy (Table 14.3). It is possible that the use of photo-selective and light-dispersive shade nets could be an alternative method for protecting crops from potentially damaging environmental factors such as excessive sun radiation, heat stress, hail, wind, and flying pests, while simultaneously improving crop quality, yield, and productivity (Ilić et al., 2017, 2018). This approach might reduce postharvest losses by extending produce shelf life. Photo-selective netting is a practical way to adjust the crop microclimate to manage yield, retail and eating quality, and functional aspects of bioactive compounds in vegetables that contribute to human health.

14.3 FUTURE ASPECTS OF POSTHARVEST PHYSIOLOGICAL DISORDERS

Postharvest advances in packaging, storage, transport systems, pests and disease control for market access, senescence control, supply chain optimization, and track and trace systems are required to deliver premium quality products to discriminating markets. Further novel R&D methods are needed to understand the genetic and molecular basis of quality traits like stress resistance, resistance to postharvest diseases, and pests. The growing world population and food quality requirements require a better knowledge of the processes governing physiological abnormalities in diverse crop species to increase agricultural production efficiency by lowering losses. Genotype and environment determine fruit and vegetable disease susceptibility. Studies have sought genetic markers to identify crop genotypes with increased physiological problem susceptibility. The information will aid in improving pre- and postharvest crop management to minimize disorder occurrence. Genomic, transcriptomic, proteomic, and metabolomic processes governing physiological disorders can be exploited to uncover important regulatory genes and genetic markers to drive breeding programs and genome transformation and editing to produce highly resistant plants. From harvest to consumption, both industrialized and developing countries must maintain the quality and security of horticulture perishables. The main problems in developing countries are poor harvesting techniques, postharvest management and transportation, infrastructure, processing and packaging technology, and marketing information. Based on scale, market, and return on investment, there are significant differences in the adoption of these processes within and between countries (Parwez, 2013).

It is, therefore, recommended to encourage the development of simple technologies such as enhanced field packaging methods during harvesting to minimize handling damage, enhanced curing of root and tuber crops, disinfestation, and protection against re-infestation. Well-designed containers that offer protection from environmental stress and include shading features can effectively reduce heat buildup, providing a natural cooling effect that helps slow down produce deterioration and maintain quality. These technologies should be combined with intermediate-level ones, like those that improve cooling procedures (like hydro-cooling), employ low-energy cooling storage techniques, make use of cool transportation routes, and make use of alternative cooling technologies (Sibanda, 2013; Legg et al., 2017). Postharvest losses might also be decreased by using new marketing tactics, managing product better during transit to help retain quality, and paying attention to ripening techniques, such as the slow or quick ripening of fruits. Additionally, it is important to reduce the food supply chain by enabling small farmers to communicate with consumers directly.

In the new era of agricultural innovation and sustainability, addressing postharvest physiological disorders demands a multifaceted and forward-thinking approach. As global food systems face mounting pressures from climate variability, population growth, and supply chain complexity, minimizing postharvest losses through the control of postharvest physiological disorders is more critical than ever. Future research should prioritize the development of predictive models that integrate climatic data, physiological markers, and storage conditions to anticipate disorder onset and progression. Cutting-edge technologies such as artificial intelligence, remote sensing, and non-destructive imaging tools hold great promise in enabling real-time monitoring and early detection of physiological disorders. These digital tools, when combined with Internet of Things (IoT)-based storage systems, can help optimize temperature, humidity, and gas composition in storage environments, reducing the risk of disorder development while maintaining produce quality and safety.

Equally important in the new era is the advancement of genetic and biochemical research to unravel the molecular basis of postharvest physiological disorders and breed crops with enhanced resistance. The integration of genomics, transcriptomics, and metabolomics can identify key genes and pathways involved in stress responses and disorder manifestation, enabling the development of cultivars better adapted to postharvest handling. Furthermore, there is a growing need to design sustainable, eco-friendly postharvest treatments such as the use of natural coatings, biostimulants, and bio-based antioxidants that reduce chemical dependency while extending shelf life. Collaboration between breeders, physiologists, engineers, and supply chain specialists will be essential to develop holistic solutions. Education and capacity building for farmers and stakeholders in proper harvest timing, handling, and storage practices must also be emphasized. As we move forward, the future of postharvest physiological disorders management lies in innovation, integration, and sustainability to ensure a resilient and efficient postharvest system.

The future of managing postharvest physiological disorders in fruits and vegetables lies increasingly in the integration of advanced technologies, particularly artificial intelligence (AI), to address challenges related to food quality, food security, and sustainable development. AI holds immense potential in early detection, diagnosis, and prediction of physiological disorders such as chilling injury, internal browning, fruit cracking, and superficial scald by analyzing complex patterns in visual, spectral, and environmental data that are often undetectable to the human eye. Machine learning algorithms, combined with imaging techniques such as hyperspectral imaging, near-infrared spectroscopy, and computer vision, can enable real-time, non-destructive monitoring of produce during storage, transportation, and retail. Furthermore, AI-driven decision support systems can optimize storage conditions by dynamically controlling temperature, humidity, and atmospheric composition based on predictive models, thus minimizing the risk of disorder development and extending shelf life. These technologies can also be used to sort and grade produce more efficiently, ensuring only high-quality fruits and vegetables reach consumers, thereby reducing postharvest losses. In the broader context, AI applications in postharvest management contribute to global food security by preserving the nutritional value and marketability of horticultural commodities, ultimately reducing waste and enhancing resource-use efficiency. As part of a sustainable development framework, the adoption of AI in postharvest physiology represents a transformative approach that not only mitigates losses but also promotes resilient food systems capable of feeding a growing population under changing climate conditions.

14.4 CONCLUSION

Future advances in the management of postharvest physiological disorders will depend on a deeper integration of fundamental plant physiology with emerging technologies and innovative postharvest practices. This chapter highlights that improved understanding of disorder-specific mechanisms at the molecular, biochemical, and cellular levels will be essential for early diagnosis, accurate prediction, and effective intervention. The application of omics approaches, non-destructive sensing tools, artificial intelligence, and precision storage technologies offers promising opportunities to monitor produce responses in real time and to tailor postharvest conditions according to commodity- and cultivar-specific requirements. In addition, the development of stress-tolerant cultivars, sustainable postharvest treatments, and climate-resilient handling systems will play a critical role in reducing losses under changing environmental conditions. Collectively, these future-oriented strategies underscore the need for interdisciplinary research and closer linkage between science, technology, and industry to ensure improved quality, extended shelf life, and reduced postharvest losses of fruits and vegetables in the years ahead.

KEYWORDS

- **cytokinin**
- **food quality**
- **fruits**
- **physiological disorders**
- **postharvest**
- **radiation**
- **vegetables**

REFERENCES

Aayush, K., McClements, D. J., Sharma, S., Sharma, R., Singh, G. P., Sharma, K., & Oberoi, K. (2022). Innovations in the development and application of edible coatings for fresh and minimally processed Apple. *Food Control*, *141*, 109188.

Abdipour, M., Malekhossini, P. S., Hosseinifarahi, M., Radi, M. (2020). Integration of UV irradiation and chitosan coating: A powerful treatment for maintaining the postharvest quality of sweet cherry fruit. *Scientia Horticulturae, 264*, 109197.

Ahemd, H. A., Al-Faraj, A. A., & Abdel-Ghany, A. M. (2016). Shading greenhouses to improve the microclimate, energy and water saving in hot regions: A review. *Scientia Horticulturae*, *201*, 36–45.

Ali, S., Khaliq, G., Iftikhar, A., Mir, S. A., & Iqbal, S. (2024). Nanoemulsions-based edible coatings for postharvest quality preservation of fruits and vegetables. In *Sustainable postharvest technologies for fruits and vegetables,* (pp. 213–226). CRC Press.

Ali, S., Khan, A. S., Anjum, M. A., Ejaz, S., Khaliq, G., Hussain, S., Ahmad, R., & Saleem, M. S. (2022). Pre-and postharvest physiological disorders of citrus fruit. In *Citrus Production*, (pp. 371–390). CRC Press.

Al-Saloul, N. J., El-Hamarnah, H. A., Lafi, O. I., Radwan, H. I., & Abu-Naser, S. S. (2022). A knowledge based system for cucumber diseases diagnosis. *International Journal of Academic Information Systems Research, 6*(5), 29–45.

Andreassen, H. P., Sundell, J., Ecke, F., Halle, S., Haapakoski, M., Henttonen, H., & Ylönen, H. (2021). Population cycles and outbreaks of small rodents: Ten essential questions we still need to solve. *Oecologia*, *195*(3), 601–622.

Aremu, A. O., Fawole, O. A., Makunga, N. P., Masondo, N. A., Moyo, M., Buthelezi, N., & Doležal, K. (2020). Applications of cytokinins in horticultural fruit crops: Trends and future prospects. *Biomolecules, 10*(9), 1222.

Ashfaq, A., Clochard, M. C., Coqueret, X., Dispenza, C., Driscoll, M. S., Ulański, P., & Al-Sheikhly, M. (2020). Polymerization reactions and modifications of polymers by ionizing radiation. *Polymers*, *12*(12), 2877.

Azil, N., Stefańczyk, E., Sobkowiak, S., Chihat, S., Boureghda, H., & Śliwka, J. (2021). Identification and pathogenicity of *Fusarium* spp. associated with tuber dry rot and wilt of potato in Algeria. *European Foundation for Plant Pathology*, *159*(3), 495–509.

Baltazari, A., Mtui, H., Chove, L., Msogoya, T., Kudra, A., Tryphone, G., & Mwatawala, M. (2020). Evaluation of post-harvest losses and shelf life of fresh mango (*Mangifera indica* L.) in Eastern zone of Tanzania. *International Journal of Fruit Science*, *20*(4), 855–870.

Boonzaaier, J. D. T. L. (2015). *An inquiry into the competitiveness of the South African stone fruit industry* (Doctoral dissertation, Stellenbosch: Stellenbosch University).

Canton, M., Drincovich, M. F., Lara, M. V., Vizzotto, G., Walker, R. P., Famiani, F., & Bonghi, C. (2020). Metabolism of stone fruits: Reciprocal contribution between primary metabolism and cell wall. *Frontiers in Plant Science*, *11*, 1054.

Choi, W. S., Singh, S., & Lee, Y. S. (2016). Characterization of edible film containing essential oils in hydroxypropyl methylcellulose and its effect on quality attributes of 'Formosa'plum (*Prunus salicina* L.). *LWT*, *70*, 213–222.

Cui, Z., Jiao, Q., Wang, R., & Ma, C. (2020). Investigation and analysis of relationship between mineral elements alteration and cork spot physiological disorder of Chinese pear 'Chili'(Pyrus bretschneideri Rehd.). *Scientia Horticulturae*, *260*, 108883.

DeBrouwer, E. J., Sriskantharajah, K., El Kayal, W., Sullivan, J. A., Paliyath, G., & Subramanian, J. (2020). Pre-harvest hexanal spray reduces bitter pit and enhances post-harvest quality in 'honeycrisp'apples (malus domestica borkh.). *Scientia Horticulturae*, *273*, 109610.

Díaz-Pérez, J. C. (2014). Bell pepper (Capsicum annum L.) crop as affected by shade level: fruit yield, quality, and postharvest attributes, and incidence of phytophthora blight (caused by Phytophthora capsici Leon.). *Horticultural Science*, *49*(7), 891–900.

DosSantos,S.F.,Cardoso,R.D.C.V.,Borges,Í.M.P.,Almeida,A.C.E.,Andrade,E.S.,Ferreira, I. O., & Do Carmo Ramos, L. (2020). Post-harvest losses of fruits and vegetables in supply centers in Salvador, Brazil: Analysis of determinants, volumes and reduction strategies. *Journal of Waste Management*, *101*, 161–170.

Dueck, T., Ieperen, W. V., & Taulavuori, K. (2016). Light perception, signalling and plant responses to spectral quality and photoperiod in natural and horticultural environments. *Environmental and Experimental Botany*, *121*, 1–150.

Ebrahim, H., Abou ElNour, S., Hammad, A. A., Abouzeid, M., & Abdou, D. (2022). Comparative effect of gamma and electron beam irradiation on some food borne pathogenic bacteria contaminating meat products. *The Egyptian Journal of Pure and Applied Science*, *60*(1), 62–72.

Fan, X., & Song, Y. (2020). Advanced oxidation process as a postharvest decontamination technology to improve microbial safety of fresh produce. *Journal of Agricultural and Food Chemistry*, *68*(46), 12916–12926.

François, G. A., de Moraes Pontes, J. G., Pereira, A. K., & Fill, T. P. (2022). Exploring the citrus sour rot pathogen: Biochemical aspects, virulence factors, and strategies for disease management-A review. *Fungal Biology Reviews, 41*, 70–83

Frieman, A., Vorsatz, C., & Daum, D. (2021). Effect of nitrogen supply, plant spacing, and cultivar on hollow stem development in broccoli. In *IV International Symposium on Horticulture in Europe-SHE2021* (Vol. 1327 pp. 149–154).

Gammage, S., & Marangoni, A. G. (2025). Safety of edible coatings on fruits and vegetables. Comprehensive Reviews in Food Science and Food Safety, 24(2), e70108.

Gardas, B. B., Raut, R. D., & Narkhede, B. (2018). Evaluating critical causal factors for post-harvest losses (PHL) in the fruit and vegetables supply chain in India using the DEMATEL approach. *Journal of Cleaner Production*, *199*, 47–61.

Gimeno, D., Gonzalez-Buesa, J., Oria, R., Venturini, M. E., & Arias, E. (2021). Effect of modified atmosphere packaging (MAP) and UV-C irradiation on postharvest quality of red raspberries. *Agriculture*, *12*(1), 29.

Guillén, F., Díaz-Mula, H. M., Zapata, P. J., Valero, D., Serrano, M., Castillo, S., & Martínez-Romero, D. (2013). Aloe arborescens and aloe vera gels as coatings in delaying postharvest ripening in peach and plum fruit. *Postharvest Biology and Technology*, *83*, 54–57.

Hosseinifarahi, M., & Radi, M. (2018). Evaluation of postharvest quality and organoleptic characteristics of strawberry with application of aloe vera gel, acetic acid and UV-B irradiation. *International Journal of Horticultural Science and Technology*, *19*(1), 99–114.

Huang, H., Jing, G., Wang, H., Duan, X., Qu, H., & Jiang, Y. (2014). The combined effects of phenylurea and gibberellins on quality maintenance and shelf life extension of banana fruit during storage. *Scientia Horticulturae*, *167*, 36–42.

Ilić, S. Z., Milenković, L., Dimitrijević, A., Stanojević, L., Cvetković, D., Kevrešan, Ž., & Mastilović, J. (2017). Light modification by color nets improve quality of lettuce from summer production. *Scientia Horticulturae*, *226*, 389–397.

Ilić, Z. S., Milenković, L., Šunić, L., & Manojlović, M. (2018). Color shade nets improve vegetables quality at harvest and maintain quality during storage. *Contemporary Agriculture*, *67*(1), 9–19.

Jahan, M. S., Lia, R. S., Chowdhury, M. E. K., Hasan, M. F., Islam, M. A., Sikdar, B., & Khalekuzzaman, M. (2019). Characterization of crown rot disease of banana fruit and eco-friendly quality improvement approach during storage. *Microbiology Research Journal International, 27*(3), 1–13.

Jaleel, W., Lu, L., & He, Y. (2018). Biology, taxonomy, and IPM strategies of Bactrocera tau Walker and complex species (Diptera; Tephritidae) in Asia: A comprehensive review. *Environmental Science and Pollution Research, 25*(20), 19346–19361.

Jameson, P. E., & Song, J. (2016). Cytokinin: A key driver of seed yield. *Journal of Experimental Botany*, *67*(3), 593–606.

Jeong, M. A., & Jeong, R. D. (2018). Applications of ionizing radiation for the control of postharvest diseases in fresh produce: Recent advances. *Plant Pathology*, *67*(1), 18–29.

Jeong, R. D., Shin, E. J., Chu, E. H., & Park, H. J. (2015). Effects of ionizing radiation on postharvest fungal pathogens. *Plant Pathology Journal*, *31*(2), 176.

Jyothi, V., & Shilpa, M. E. (2020). Citrus diseases and their management. In *Diseases of fruits and vegetable crops* (pp. 41–63). Apple Academic Press.

Kawai, Y., Baba, T., Yoshida, M., Agravante, J. U., & Del Carmen, D. R. (2018). Effects of benzyladenine and light on post-harvest calamondin (x Citrofortunella microcarpa) fruit color and quality. *Journal of Horticulture*, *87*(3), 324–328.

Khaliq, G. H. U. L. A. M. (2015). Effects of edible coatings enriched with calcium chloride on physiological, biochemical and quality responses of mango (*Mangifera indica* L. cv.

Choke Anan) fruit during cold storage. Universiti Putra Malaysia, Serdang, Selangor, Malaysia.

Khaliq, G., Ali, S., Gapper, N., & Nicola, S. (2023). Recent advances and approaches in the application of elicitors to enhance resistance mechanisms in fresh produce. *Frontiers in Sustainable Food Systems, 6,* 1061079.

Khan, L., Khaliq, G., Ullah, E., Abdul-Rahaman, A., El-Mogy, M. M., Kesba, H. H., Ismail, A. M., & El-Beltagi, H. S. (2025). Guggul gum and clove essential oil synergistically reduced anthracnose disease and modulated physicochemical changes in tomato during storage. *Journal of Stored Products Research, 114,* 102713.

Khanal, M., Timilsina, S., Bhatta, B. P., Bophela, K., Coutinho, T., Cochran, K., & Malla, S. (2022). *Pseudomonas uvaldensis* sp. nov., a bacterial pathogen causing onion bulb rot. *International Journal of Systematic and Evolutionary Microbiology, 72*(4).

Koprna, R., De Diego, N., Dundálková, L., & Spíchal, L. (2016). Use of cytokinins as agrochemicals. *Bioorganic and Medicinal Chemistry, 24*(3), 484–492.

Kumar, P., Sethi, S., Sharma, R. R., Srivastav, M., & Varghese, E. (2017). Effect of chitosan coating on postharvest life and quality of plum during storage at low temperature. *Scientia Horticulturae, 226*, 104–109.

Lal, S., Khaliq, G., Mastoi, A. H., Irshad, A., Jaffar, M. A., Khan, L., & Muhammad, N. (2025). Effect of *Fagonia indica* and garlic extract on *Alternaria alternata* and post-harvest quality of tomato fruit during storage. *Journal of Crop Health, 77*(6), 1–13.

Leach, A., Reiners, S., & Nault, B. (2020). Challenges in integrated pest management: A case study of onion thrips and bacterial bulb rot in onion. *Journal of Crop Protection, 133,* 105123.

Lee, J. M., Kubota, C., Tsao, S. J., Bie, Z., Echevarria, P. H., Morra, L., & Oda, M. (2010). Current status of vegetable grafting: Diffusion, grafting techniques, automation. *Scientia Horticulturae, 127*(2), 93–105.

Legg, J., Okonya, J., Wyckhuys, K., & Coyne, D. L. (2017). Integrated pest management of root and tuber crops in the tropics. In C. Rapisarda, & G. E. M. Cocuzza (Eds.), *Integrated pest management in tropical regions* (pp. 90–112). Wallingford, UK: CABI Publishing.

Mahfoudhi, N., & Hamdi, S. (2015). Use of almond gum and gum Arabic as novel edible coating to delay postharvest ripening and to maintain sweet cherry (P runus avium) quality during storage. *Journal of Food Processing and Preservation, 39*(6), 1499–1508.

Maringgal, B., Hashim, N., Tawakkal, I. S. M. A., & Mohamed, M. T. M. (2020). Recent advance in edible coating and its effect on fresh/fresh-cut fruits quality. *Trends in Food Science and Technology, 96*, 253–267.

Maruo, T., & Johkan, M. (2020). Tipburn. In *Plant factory* (pp. 231–234). Academic Press.

Massolo, J. F., Lemoine, M. L., Chaves, A. R., Concellón, A., & Vicente, A. R. (2014). Benzyl-aminopurine (BAP) treatments delay cell wall degradation and softening, improving quality maintenance of refrigerated summer squash. *Postharvest Biology and Technology, 93*, 122–129.

Mditshwa, A., Khaliq, G., Hussein, Z., & Ejaz, S. (2023). Sustainable postharvest management practices for fresh produce. Frontiers in Sustainable Food Systems, 7, 1143759.

Munaweera, T. I. K., Jayawardana, N. U., Rajaratnam, R., & Dissanayake, N. (2022). Modern plant biotechnology as a strategy in addressing climate change and attaining food security. *Agriculture and Food Security*, *11*(1), 1–28.

Murmu, S. B., & Mishra, H. N. (2018). Post-harvest shelf-life of banana and guava: Mechanisms of common degradation problems and emerging counteracting strategies. *Innovative Food Science and Emerging Technologies*, *49*, 20–30.

Ncama, K., Magwaza, L. S., Mditshwa, A., & Tesfay, S. Z. (2018). Plant-based edible coatings for managing postharvest quality of fresh horticultural produce: A review. *Food Packaging and Shelf Life*, *16*, 157–167.

Ngolong Ngea, G. L., Qian, X., Yang, Q., Dhanasekaran, S., Ianiri, G., Ballester, A. R., & Zhang, H. (2021). Securing fruit production: Opportunities from the elucidation of the molecular mechanisms of postharvest fungal infections. *Comprehensive Reviews in Food Science and Food Safety, 20*(3), 2508–2533.

Nguyen, T. Q., Kisiala, A. B., Hai, N. N., Narine, S., & Emery, R. N. (2021). Phytohormone dynamics impact fatty acid and oil accumulation during soybean seed maturation. *Seed Science Research*, *31*(4), 278–291.

Ntsoane, M. L. (2015). Effect of photo-selective nettings on plant growth, quality at harvest and after postharvest storage in lettuce varieties.

Ombodi, A., Zoltán, P. É. K., Szuvandzsiev, P., Taskovics, Z. T., Kohazi-Kis, A., Kovacs, A., & Helyes, L. (2015). Effects of external coloured shade nets on sweet peppers cultivated in walk-in plastic tunnels. *Notulae botanicae Horti Agrobotanici Cluj-Napoca*, *43*(2), 398–403.

Parwez, S. (2013). Food supply chain management in Indian Agriculture: Issues, opportunities and further research. MPRA Paper 60441. Germany: University Library of Munich.

Pongsri, R., Aiamla-or, S., Srilaong, V., Uthairatanakij, A., & Jitareerat, P. (2021). Impact of electron-beam irradiation combined with shellac coating on the suppression of chlorophyll degradation and water loss of lime fruit during storage. *Postharvest Biology and Technology*, *172*, 111364.

Potter, D. (2012). Basic information on the stone fruit crops. In C. Kole, & A. G. Abbott (Eds.), *Genetics, genomics and breeding of stone fruits* (pp. 1–21). New York: CRC Press.

Purbey, S. K., Singh, S. K., & Pongener, A. (2019). Management of light for quality production of litchi. *International Journal of Bio-resource and Stress Management*, *10*(5), 529–538.

Rajapaksha, L., Gunathilake, D. C., Pathirana, S. M., & Fernando, T. (2021). Reducing post-harvest losses in fruits and vegetables for ensuring food security—Case of Sri Lanka. *MOJ Food Processing and Technology*, *9*(1), 7–16.

Ram, R. A., Rahim, M. A., & Alam, M. S. (2020). Diagnosis and management of nutrient constraints in mango. In *Fruit crops* (pp. 629–650). Elsevier.

Raveau, R., Fontaine, J., & Lounès-Hadj Sahraoui, A. (2020). Essential oils as potential alternative biocontrol products against plant pathogens and weeds: A review. *Foods*, *9*(3), 365.

Reitz, N. F., & Mitcham, E. J. (2021). Validation and demonstration of a pericarp disc system for studying blossom-end rot of tomatoes. *Plant Methods*, *17*(1), 1–10.

Salehi, F. (2020). Edible coating of fruits and vegetables using natural gums: A review. *International Journal of Fruit Science*, *20*(2), S570–S589.

Santana, J. Q., Balbino, M. A., Tavares, T. R., Bezerra, R. S., Farias, J. G., & Ferreira, R. C. (2012). Effect of photoselective screens in the development and productivity of red and yellow sweet pepper. In *VII International Symposium on Light in Horticultural Systems* (Vol. 956, pp. 493–500).

Schäfer, M., Brütting, C., Meza-Canales, I. D., Großkinsky, D. K., Vankova, R., Baldwin, I. T., & Meldau, S. (2015). The role of cis-zeatin-type cytokinins in plant growth regulation and mediating responses to environmental interactions. *Journal of Experimental Botany*, *66*(16), 4873–4884.

Searchinger, T., Waite, R., Hanson, C., Ranganathan, J., Dumas, P., Matthews, E., & Klirs, C. (2019). Creating a sustainable food future: A menu of solutions to feed nearly 10 billion people by 2050. Final report.

Selahle, K. M., Sivakumar, D., Jifon, J., & Soundy, P. (2015). Postharvest responses of red and yellow sweet peppers grown under photo-selective nets. *Food Chemistry*, *173*, 951–956.

Sharifi, M., Owen, J., Hajiaghaei-Kamrani, M., & Hammermeister, A. M. (2020). Broccoli yield and dry matter partitioning in response to application of sediment from post-harvest washing of mussels. *Canadian Journal of Plant Science*, *101*(2), 229–239.

Sharma, P., Roy, M., & Roy, B. (2021). A brief overview of major citrus diseases and pests and its management. *Current Topics Journal of Agricultural Science*, *1*, 95–110.

Shree, S., & Kumari, A. (2019). Postharvest handling, diseases and disorders in bulb vegetables. In *The vegetable pathosystem* (pp. 483–507). Apple Academic Press.

Sibanda, S. (2013). *Development of a wind and solar powered mobile evaporative cooling system for temporary storage and transportation of fruit and vegetables* (PhD thesis, Agricultural Engineering, School of Engineering University of KwaZulu-Natal Pietermaritzburg).

Siddique, K. H., Li, X., & Gruber, K. (2021). Rediscovering Asia's forgotten crops to fight chronic and hidden hunger. *Nature Plants*, *7*(2), 116–122.

Singh, A., Shukla, A. K., & Meghwal, P. R. (2020). Fruit cracking in pomegranate: Extent, cause, and management–A review. *International Journal of Fruit Science*, *20*, S1234–S1253.

Singh, K. P., & Aravind, T. (2021). Postharvest diseases and disorders of apple: Perspectives for integrated management. In *postharvest handling and diseases of horticultural produce* (pp. 267–280). CRC Press.

Singh, V. K. (2019). Internal breakdown in mango (Mangifera Indica L.)—An important physiological disorder: An overview. *Plant Abiotic Stresses Physiological Mechanisms Tools and Regulation*, 317.

Sisipo, T., Isaac, A. A., & Ogujiuba, K. (2021). The foremost challenges of post-harvest practices. A case of selected farmers in Nkomazi South Africa. *Natural Volatiles and Essential Oils*, 3260–3272.

Sivakumar, D., Jifon, J., & Soundy, P. (2018). Spectral quality of photo-selective shade nettings improves antioxidants and overall quality in selected fresh produce after postharvest storage. *Food Reviews International*, *34*(3), 290–307.

Smith, R. (2020). Celery. In *Carrots and related Apiaceae crops* (pp. 272–282). Wallingford UK: CABI.

Srivastava, J. N., Singh, A. K., & Sharma, R. K. (2020). Diseases of apples and their management. In *Diseases of fruits and vegetable crops* (pp. 19–39). Apple Academic Press.

Strecker, K., Bitzer, V., & Kruijssen, F. (2022). Critical stages for post-harvest losses and nutrition outcomes in the value chains of bush beans and nightshade in Uganda. *Food Security*, *14*(2), 411–426.

Thakur, R., Pristijono, P., Golding, J. B., Stathopoulos, C. E., Scarlett, C. J., Bowyer, M., ... & Vuong, Q. V. (2018). Development and application of rice starch based edible coating to improve the postharvest storage potential and quality of plum fruit (Prunus salicina). *Scientia Horticulturae*, *237*, 59–66.

Tinyane, P. P., Sivakumar, D., & Soundy, P. (2013). Influence of photo-selective netting on fruit quality parameters and bioactive compounds in selected tomato cultivars. *Scientia Horticulturae*, *161*, 340–349.

Tiwari, R. K., Kumar, R., Sharma, S., Sagar, V., Aggarwal, R., Naga, K. C., & Kumar, M. (2020). Potato dry rot disease: Current status, pathogenomics and management. *3 Biotech*, *10*(11), 1–18.

Torres, E., Alegre, S., Recasens, I., Asín, L., & Lordan, J. (2021). Integral procedure to predict bitter pit in 'Golden Smoothee'apples based on calcium content and symptom induction. *Scientia Horticulturae*, *277*, 109829.

Trivedi, D., & Dikshit, A. K. (2021). Irradiation technology for the food industry. In *Food processing* (pp. 165–180). CRC Press.

Ul Haq, Z., Khaliq, G., Irshad, A., El-Beltagi, H. S., Mahmoud Ismail, A., & El-Mogy, M. M. (2024). Effect of beeswax and lemon grass oil on Alternaria alternata and post-harvest quality of bitter gourd during storage. New Zealand Journal of Crop and Horticultural Science, 1-21.

Vázquez-Celestino, D., Ramos-Sotelo, H., Rivera-Pastrana, D. M., Vázquez-Barrios, M. E., & Mercado-Silva, E. M. (2016). Effects of waxing, microperforated polyethylene bag, 1-methylcyclopropene and nitric oxide on firmness and shrivel and weight loss of 'Manila' mango fruit during ripening. *Postharvest Biology and Technology*, *111*, 398–405.

Vijg, J., & Dong, X. (2020). Pathogenic mechanisms of somatic mutation and genome mosaicism in aging. *Cell*, *182*(1), 12–23.

Wang, Y., Guo, L., Zhao, X., Zhao, Y., Hao, Z., Luo, H., & Yuan, Z. (2021). Advances in mechanisms and omics pertaining to fruit cracking in horticultural plants. *Agronomy*, *11*(6), 1045.

Wien, H. C., & Stützel, H. (2020). Cauliflower, broccoli, cabbage, and Brussels sprouts. In *The physiology of vegetable crops* (pp. 357–388). Wallingford UK: CABI.

Xu, M., Zhang, X., Dhanasekaran, S., Godana, E. A., Yang, Q., Zhao, L., & Zhang, H. (2021). Transcriptome analysis of postharvest pear (Pyrus pyrifolia Nakai) in response to Penicillium expansum infection. *Scientia Horticulturae, 288*, 110361.

Yahia, E. M., Carrillo-López, A., & Sañudo, A. (2019). Physiological disorders and their control. In *Postharvest technology of perishable horticultural commodities* (pp. 499–527). Woodhead Publishing.

Yahia, E. M., Fonseca, J. M., & Kitinoja, L. (2019). Postharvest losses and waste. In *Postharvest technology of perishable horticultural commodities* (pp. 43–69).

Yasser, M. M., Marzouk, M. M., Kamel, T. A., & Naaffa, A. M. (2019). Effect of hot water treatment on postharvest fruit rots and quality of tomato fruits. *Plant Archives*, *19*(2), 2325–2334.

Zhang, D., Xu, X., Zhang, Z., Jiang, G., Feng, L., Duan, X., & Jiang, Y. (2018). 6-Benzylaminopurine improves the quality of harvested litchi fruit. *Postharvest Biology and Technology*, *143*, 137–142.

Zhang, W., & Jiang, W. (2019). UV treatment improved the quality of postharvest fruits and vegetables by inducing resistance. *Trends in Food Science and Technology*, *92*, 71–80.

Zhang, X. J., Zhang, M., Chitrakar, B., Devahastin, S., & Guo, Z. (2022). Novel combined use of red-white LED illumination and modified atmosphere packaging for maintaining storage quality of postharvest pakchoi. *Food and Bioprocess Technology*, *15*(3), 590–605.

Zhang, Y., Zeng, L., Yang, J., Zheng, X., & Yu, T. (2015). 6-Benzylaminopurine inhibits growth of Monilinia fructicola and induces defense-related mechanism in peach fruit. *Food Chemistry*, *187*, 210–217.

Index

1-Methylcyclopropene (1-MCP), 208
6-Benzylaminopurine (BAP), 395
γ-Aminobutyric acid (GABA), 208–209, 276

A

Aloe arborescens gel, 399
Arginine, 211

B

Benzyl adenine (BA), 395
Bitter pit (BP), 72–73
  fruit
    calcium-related physiological disorders, 226–227
    cracking, 314–315
Blackheart, 315–316
Blossom-end rot (BER), 74–75, 229–230, 315
Boron, 76
  corky core, 76
  deficiency disorders, 76
Browning
  calcium-related physiological disorders
    fruits, 226
    vegetables, 230–231
  core, 77
  enzymatic flesh, 78
  internal, 75, 226
  postharvest physiological disorders, storage-related, 80

C

Calcium ion, 209
Calcium-related physiological disorders, 69–70, 223–224
  cut flower
    chilling injury, 232
    stem bending, 232
  fruit
    bitter pit (BP), 226–227
    browning, 226
    chilling injury (CI), 225–226
    cork spot, 227
    cracking, 224–225
    lenticel breakdown (LB), 227
    superficial scald, 227–228
  mechanism, 70–72
  postharvest quality, effect on, 72
  vegetables
    blossom-end rot (BER), 229–230
    browning, 230–231
    chilling injury (CI), 231
    cracking, 230
Chilling injury (CI), 84, 177, 183–185
  alleviation, 85, 194–199
    chemical treatments, 204–209
    controlled atmosphere (CA), 202–203
    heat treatment, 199–201
    intermittent warming treatment, 202
    low-temperature conditioning treatment, 201–202
    modified atmosphere (MA), 202–203
    postharvest treatments and main effects for, 195–199
    ultraviolet (UV) treatment, 203–204
  alleviation
    mechanisms, 209–211
    postharvest treatments for, 209–211
  average critical safe temperature, 89–90
  calcium-related physiological disorders
    cut flower, 232
    fruit, 225–226
  chemical treatments, 87–88
  controlled atmospheres, 86
  genetic modification, 88–89
  growth regulators, 87
  intermittent heating, 86
  mechanism, 188–194
    alterations of cell membrane, 188–190
    energy metabolism imbalance, 190–192
    oxidative stress, 192–194

packaging, 88
physiological and biochemical aspects, 293–294
external factors, 295–296
factors affecting the symptom development, 294–295
fruits and vegetables, 296–297
internal factors, 295
symptoms, 294
postharvest physiological disorders, mechanism and management
control, 355
importance, 353
occurrence stage, 352
physiology, 354
symptoms, 353–354
salicylates and jasmonates, mechanism employed by, 299–302
antioxidant activity, 300–301
arginine pathways, 301
CBF pathway activation, 301–302
HSPS accumulation, 301
membrane integrity enhancing, 300
PAL and PPO enzymes activities alteration, 302
sensitivity of fruits to chilled conditions, 85
symptoms, 84–85, 185–188
temperature conditioning, 86
UV radiation, 88
vegetables
calcium-related physiological disorders, 231
waxing and other coatings, 88
Chilling stress, 181–182
effect on physiological changes, 182–183
Cold and/Low-temperature storage (LTS), 177, 259–260
chilling injury (CI), 183–185
alleviation, 194–199
mechanism, 188–194
symptoms, 185–188
chilling stress, 181–182
effect on physiological changes, 182–183
disorders of fresh produce, 262–263
CI symptoms and temperatures of, 263–264
freezing injury, 179–181
storage, 178–179
Controlled atmosphere storage (CAS), 139–140, 202–203, 267–268
flesh browning, 273
general leaf browning, 272
petal bluing, 271
physiological disorders reported, 269–270
related disorders of fresh produce, 268–274
scald-like disorder, 271
storage scald, 270
Core browning, 77
Cork (spot-like), 333
postharvest physiological disorders, mechanism and management, 372
control, 377–378
importance, 374–375
occurrence stage, 372–374
physiology, 376–377
symptoms, 375–376
Corky core
external, 76
internal, 76
Cracking
calcium-related physiological disorders
fruits, 224–225
vegetables, 230
C-repeat binding factor (CBF), 211
Cut flower
chilling injury, 232
stem bending, 232
Cytokinins (CKs), 394

D

Dynamic controlled atmosphere storage (DCA), 274–275
physiological disorders under, 275

E

Edible coatings/Food coatings, 249–251
applications, 250
postharvest physiological disorders, 398–400
Enzymatic flesh browning, 78
Ethylene, 208
gaseous composition, 340

Ethylene disorders, 318
Ethylene response factors (ERFs), 15

F

Flesh browning, 78, 273
Food supply chain
  mitigating in, 42–43
    chemical treatments, 50–52
    cold storage facilities, 47–48
    harvesting, 43–44
    physical treatments, 48–50
    precooling, 46–47
    temperature and relative humidity management, 45–46
    transportation and distribution, 52–54
Freezing injury, 179–181
  physiological and biochemical aspects, 297–298
    degradation, fresh fruits and vegetables' resistance to, 299
  postharvest physiological disorders, mechanism and management
    control, 357
    importance, 356
    occurrence stage, 355–356
    physiology, 357
    symptoms, 356
Fruits
  calcium-related physiological disorders
    bitter pit (BP), 226–227
    browning, 226
    chilling injury (CI), 225–226
    postharvest disorders, 228–229, 393
cracking, 224–225. *See also* Fruit cracking
Fruit coating. *See* Edible coatings
Fruit cracking, 305–306
  apple
    development, 306
    gene expression, 306–307
  bitter pit, 314–315
  blackheart, 315–316
  blossom-end rot, 315
  citrus fruit, 307
    environment, 307
    growth regulators, 308
  cork (spot-like) disorder, 313
  litchi, 310–311
    environment, 312
    fruit development, 311
    gene expression, 311
    peel morphology, 311–312
  postharvest physiological disorders, mechanism and management, 360–361
    control, 369–372
    importance, 366–367
    occurrence stage, 361–366
    physiology, 367–369
    symptoms, 367
  spongy tissue, 314
  sweet cherries
    cuticular properties, 309–310
    development, 309
    environment, 310
    firmness, 309
    genetic, 310
    shape, 309
    size, 308
    water-retaining capacity, 310
  tip burn and internal browning, 314
  tomato
    day and night temperatures, high differences between, 313
    genetic differences, 312–313
    high humidity, 313
Fruit ripening, 10
  epigenetic control, 16–17
  hormonal control, 13–15
    climacteric and nonclimacteric fruits, 14
    ethylene production, 14
  molecular regulation, 15–16
  physiology and biochemistry, 10
    acidity reduction, 11
    changes in color, 12
    flesh softening, 10–11
    starch metabolism and increase in soluble sugars, 11
    volatile compounds release, 12–13
  regulatory factors
    effect of CO_2 and O_2 levels, 18
    effect of water loss, 17
    low temperature, 19–22
    postharvest diseases, 18–19

Fruits and vegetables. *See also* Horticultural crops; *See also* Fruits; *See also* Vegetables
inspection and quantification, 165
chlorophyll fluorescence/UV light (nondestructive), 167
Cryo-scanning electron microscopy, 166
electrolyte leakage (indirect), 168
visual inspection with four to five degrees, 168–169
postharvest factors affecting
under different storage systems, 144–145
postharvest physiological disorders
causes, 155–156
high temperature, 161–163
low temperature, 156–161
mechanism, 163–165
temperature-related physiological disorders, 153–155
chilling injury, 158–161
freezing stress, 156–158
heat injury, 162–163
preventive measurements and recommendations, 169–170
success and perspectives, 170–172
Fruits/Protective food, 67

G

General leaf browning, 272
Gibberellins (GA3), 395
Glassiness, 75
Global food loss index (FLI), 38
Granulation, 73–74

H

Heat injury
postharvest physiological disorders, mechanism and management
control, 359–360
importance, 358
occurrence stage, 358
physiology, 359
symptoms, 358–359
Heat treatments/High temperature
hot water treatment (HWT), 276
postharvest, 199–201
physiological and biochemical aspects, injury, 302–303
Horticultural crops, 1–2. *See also* Fruits and vegetables
biological processes during postharvest handling, 3
fruit ripening, 10
epigenetic control, 16–17
hormonal control, 13–15
molecular regulation, 15–16
physiology and biochemistry, 10–13
regulatory factors, 17–22
fruits and vegetables, 2
highly perishable, 3
postharvest physiology, 2–3
respiration, 7–10
water loss (transpiration), 3–4
estimation of, 5
factors affecting, 5–7
Hot water rinsing and brushing (HWB) treatments, 303
Hot water treatment (HWT), 276
Hydrogen sulfide, 207
Hyperbaric atmosphere storage, 266–267
physiological disorders under, 267
Hypobaric storage, 266
physiological disorders under, 267

I

Intermittent warming treatment, 202
Internal browning (IB), 75, 226
Internal necrosis, 77

J

Jasmonates, 205–206

L

Lenticel breakdown (LB), 227
Lenticel discoloration (LD), 90
Lipid-based coatings, 251
Low-temperature conditioning treatment, 201–202

M

Melatonin, 207–208, 211
Methyl jasmonate (MeJA), 206
Mineral imbalance
  physiological and biochemical aspects, 303–304
    calcium and boron deficiencies, 304–305
    fruit cracking, 305–316
Modified atmosphere (MA), 202–203
Modified atmosphere packaging (MAP), 264–265
  related disorders of fresh produce, 265–266

N

Nitric oxide, 206–207

O

Oxalic acid, 209
Oxidative damage, 210
Oxidative stress, 192–194

P

Petal bluing, 271
Phenylalanine ammonia-lyase (PAL), 230
Phenylurea, 395
Physiological disorders control
  postharvest treatments for, 275–278
Polyamines (PAs), 204–205
Polyphenol oxidase (PPO) activity, 226
Postharvest losses (PHL), 37–38
  causes, 38
  factors contributing in fruits and vegetables, 39
    attitude, 41
    infrastructure, 42
    knowledge, 40–41
    market availability, 39–40
  food supply chain, mitigating in, 42–43
    chemical treatments, 50–52
    cold storage facilities, 47–48
    harvesting, 43–44
    physical treatments, 48–50
    precooling, 46–47
    temperature and relative humidity management, 45–46
    transportation and distribution, 52–54
  mitigation strategies, 54–57
Postharvest physiological disorders, 67, 393
  bitter pit, 72–73
  blossom-end rot, 74–75
  boron-related disorders, 76
  calcium-related disorders, 69–70
    mechanism, 70–72
    postharvest quality, effect on, 72
  causes, 327–328, 351
    chilling injury (CI), 352–355
    freezing injury, 355–357
    gaseous composition, 339–340
    heat injury, 358–360
    humidity, 341–342
    light, 342–343
    mechanism and management of, 347–350
    postharvest factors affecting, 339
    temperature, 340–341
  chilling injury (CI), 84
    alleviation, 85
    average critical safe temperature, 89–90
    chemical treatments, 87–88
    controlled atmospheres, 86
    genetic modification, 88–89
    growth regulators, 87
    intermittent heating, 86
    packaging, 88
    sensitivity of fruits to chilled conditions, 85
    symptoms, 84–85
    temperature conditioning, 86
    UV radiation, 88
    waxing and other coatings, 88
  classification, 69, 71
  colored shade nets use, 400–401
    photo-selective netting system, 401
  core browning, 77
  defined, 105
  detection of symptoms, 91
  edible coatings, 398–400
  environmental and genotype-associated factors, 329–330
    atmospheric humidity, 337–338
    climate variable influencing, 333
    genotypes/cultivars, 330–331

light, 334
rootstocks, 331–333
soil types, 335
temperature, 335–337
wind/storms/cyclones during production phase, 338–339
factors affecting, 133–135
controlled atmospheric gases, 139–140
ethylene, 140–141
fruits and vegetables under different storage systems, 144–145
relative humidity, 141
responsible for quality loss, 141–144
temperature, 135–138
freezing damage, 83
control, 84
symptoms, 83–84
freezing temperature, 68
fruits and vegetables, 393–394
future aspects, 391–393, 401–404
glassiness, 75
granulation, 73–74
heat injury, 81–82
internal browning (IB), 75
internal necrosis, 77
ionizing radiations control, 396–398
lenticel discoloration (LD), 90
low-temperature, 82
management, 80–81
metabolic processes, 68
nitrogen and phosphorus-related disorders, 78
oxygen and CO_2 injury, 81
physiological and biochemical aspects, 291–293
low-temperature disorders, 293–302
pre-and postharvest disorders, 292
plant growth regulators and, 394–395
preharvest factors affecting, 103–108
genetic composition, cultivar, and variety, 118–121
irrigation, 116–117
maturity stage, 122–124
nutrition, 112–116
pruning and thinning, 121–122
radiation, 110–112
rainfall, 117–118
rootstock scion relationship, 121
temperature, 108–110
preharvest factors influencing, 328–329
preharvest management of mineral, 68
pulp spot, 75–76
related to gases, 316
ethylene disorders, 318
high carbon dioxide, 317–318
high-oxygen, 317
low-oxygen, 316–317
soft nose, 74
storage-related
browning, 80
superficial scald, 73, 78–79
symptoms, 68
Pulp spot, 75–76

Q

Quantitative trait locus (QTL), 170

R

Respiration, 7
estimation, 7–8
factors influencing, 8
atmospheric gas composition, 9
temperature, 9
wounding, 9–10
Rootstocks, 331–333

S

Salicylates, 206
Scald-like disorder, 271
Soft nose, 74
Stem bending
cut flower, calcium-related physiological disorders, 232
Storage atmospheres, 261
simple cold storage, 261–262
Storage scald, 270
Superficial scald, 73, 78–79
calcium-related physiological disorders, 227–228
postharvest physiological disorders, storage-related, 78–79

V

Vegetables
calcium-related physiological disorders
blossom-end rot (BER), 229–230
browning, 230–231
chilling injury (CI), 231
cracking, 230
postharvest disorders, 393

W

Water loss (transpiration), 3–4
estimation of, 5
factors affecting, 5–7
product surface area, 6
relative humidity, 7
temperature, 6–7
Water loss and humidity-related physiological disorders, 239–241
fruits and vegetables
chemical treatment effects, 251–252
controlled atmospheres (CA), 246
edible coatings, 249–251
harvest technique and physical damage, 242
humidity, 244–245
lipid-based coatings, 251
modified atmospheres (MA), 246
modified humidity (MH), 246–247
postharvest factors, 242–243
pre-harvest factors, 241–242
storage duration, 245–246
temperature, 243–244
methods to reduce in fruits and vegetables
packaging treatment, 248–249
temperature and transportation, 247–248

For Product Safety Concerns and Information please contact our EU representative GPSR@taylorandfrancis.com Taylor & Francis Verlag GmbH, Kaufingerstraße 24, 80331 München, Germany